Protein–Protein Interactions

METHODS IN MOLECULAR BIOLOGY™

John M. Walker, Series Editor

286. **Transgenic Plants:** *Methods and Protocols,* edited by *Leandro Peña, 2004*
285. **Cell Cycle Control and Dysregulation Protocols:** *Cyclins, Cyclin-Dependent Kinases, and Other Factors,* edited by *Antonio Giordano and Gaetano Romano, 2004*
284. **Signal Transduction Protocols,** *Second Edition,* edited by *Robert C. Dickson and Michael D. Mendenhall, 2004*
283. **Bioconjugation Protocols,** edited by *Christof M. Niemeyer, 2004*
282. **Apoptosis Methods and Protocols,** edited by *Hugh J. M. Brady, 2004*
281. **Checkpoint Controls and Cancer, Volume 2:** *Activation and Regulation Protocols,* edited by *Axel H. Schönthal, 2004*
280. **Checkpoint Controls and Cancer, Volume 1:** *Reviews and Model Systems,* edited by *Axel H. Schönthal, 2004*
279. **Nitric Oxide Protocols,** *Second Edition,* edited by *Aviv Hassid, 2004*
278. **Protein NMR Techniques,** *Second Edition,* edited by *A. Kristina Downing, 2004*
277. **Trinucleotide Repeat Protocols,** edited by *Yoshinori Kohwi, 2004*
276. **Capillary Electrophoresis of Proteins and Peptides,** edited by *Mark A. Strege and Avinash L. Lagu, 2004*
275. **Chemoinformatics,** edited by *Jürgen Bajorath, 2004*
274. **Photosynthesis Research Protocols,** edited by *Robert Carpentier, 2004*
273. **Platelets and Megakaryocytes, Volume 2:** *Perspectives and Techniques,* edited by *Jonathan M. Gibbins and Martyn P. Mahaut-Smith, 2004*
272. **Platelets and Megakaryocytes, Volume 1:** *Functional Assays,* edited by *Jonathan M. Gibbins and Martyn P. Mahaut-Smith, 2004*
271. **B Cell Protocols,** edited by *Hua Gu and Klaus Rajewsky, 2004*
270. **Parasite Genomics Protocols,** edited by *Sara E. Melville, 2004*
269. **Vaccina Virus and Poxvirology:** *Methods and Protocols,*edited by *Stuart N. Isaacs, 2004*
268. **Public Health Microbiology:** *Methods and Protocols,* edited by *John F. T. Spencer and Alicia L. Ragout de Spencer, 2004*
267. **Recombinant Gene Expression:** *Reviews and Protocols, Second Edition,* edited by *Paulina Balbas and Argelia Johnson, 2004*
266. **Genomics, Proteomics, and Clinical Bacteriology:** *Methods and Reviews,* edited by *Neil Woodford and Alan Johnson, 2004*
265. **RNA Interference, Editing, and Modification:** *Methods and Protocols,* edited by *Jonatha M. Gott, 2004*
264. **Protein Arrays:** *Methods and Protocols,* edited by *Eric Fung, 2004*
263. **Flow Cytometry,** *Second Edition,* edited by *Teresa S. Hawley and Robert G. Hawley, 2004*
262. **Genetic Recombination Protocols,** edited by *Alan S. Waldman, 2004*
261. **Protein–Protein Interactions:** *Methods and Applications,* edited by *Haian Fu, 2004*
260. **Mobile Genetic Elements:** *Protocols and Genomic Applications,* edited by *Wolfgang J. Miller and Pierre Capy, 2004*
259. **Receptor Signal Transduction Protocols,** *Second Edition,* edited by *Gary B. Willars and R. A. John Challiss, 2004*
258. **Gene Expression Profiling:** *Methods and Protocols,* edited by *Richard A. Shimkets, 2004*
257. **mRNA Processing and Metabolism:** *Methods and Protocols,* edited by *Daniel R. Schoenberg, 2004*
256. **Bacterial Artifical Chromosomes, Volume 2:** *Functional Studies,* edited by *Shaying Zhao and Marvin Stodolsky, 2004*
255. **Bacterial Artifical Chromosomes, Volume 1:** *Library Construction, Physical Mapping, and Sequencing,* edited by *Shaying Zhao and Marvin Stodolsky, 2004*
254. **Germ Cell Protocols, Volume 2:** *Molecular Embryo Analysis, Live Imaging, Transgenesis, and Cloning,* edited by *Heide Schatten, 2004*
253. **Germ Cell Protocols, Volume 1:** *Sperm and Oocyte Analysis,* edited by *Heide Schatten, 2004*
252. **Ribozymes and siRNA Protocols,** *Second Edition,* edited by *Mouldy Sioud, 2004*
251. **HPLC of Peptides and Proteins:** *Methods and Protocols,* edited by *Marie-Isabel Aguilar, 2004*
250. **MAP Kinase Signaling Protocols,** edited by *Rony Seger, 2004*
249. **Cytokine Protocols,** edited by *Marc De Ley, 2004*

METHODS IN MOLECULAR BIOLOGY™

Protein–Protein Interactions

Methods and Applications

Edited by

Haian Fu

Department of Pharmacology

Emory University School of Medicine, Atlanta, GA

Humana Press **Totowa, New Jersey**

Softcover reprint of the hardcover 1st edition 2004
999 Riverview Drive, Suite 208
Totowa, New Jersey 07512

www.humanapress.com

This publication is printed on acid-free paper. ∞
ANSI Z39.48-1984 (American Standards Institute)

Permanence of Paper for Printed Library Materials.

Cover illustration: Foreground—Figure 3 from Chapter 1, R. C. Liddington. Background—Figure 3, Chapter 28, I. Remy and S. W. Michnick.

Cover design by Patricia F. Cleary.

For additional copies, pricing for bulk purchases, and/or information about other Humana titles, contact Humana at the above address or at any of the following numbers: Tel.: 973-256-1699; Fax: 973-256-8341; E-mail: humana@humanapr.com; or visit our website: www.humanapress.com

10 9 8 7 6 5 4 3 2 1

Library of Congress Cataloging in Publication Data
Protein-protein interactions : methods and applications / edited by Haian Fu.
p. ; cm. — (Methods in molecular biology ; 261)
Includes bibliographical references and index.
ISBN 978-1-61737-373-2 e-ISBN 978-1-59259-762-8
1. Protein-protein interactions—Laboratory manuals.
[DNLM: 1. Protein Binding—physiology—Laboratory Manuals. 2. Proteins—metabolism—Laboratory Manuals. QU 25 P96695 2004] I. Fu, Haian. II. Series: Methods in molecular biology (Clifton, N.J.) ; v. 261.
QP551.5.P76 2004
572'.6—dc22 2003020640

Preface

As the mysteries stored in our DNA have been more completely revealed, scientists have begun to face the extraordinary challenge of unraveling the intricate network of protein–protein interactions established by that DNA framework. It is increasingly clear that proteins continuously interact with one another in a highly regulated fashion to determine cell fate, such as proliferation, differentiation, or death. These protein–protein interactions enable and exert stringent control over DNA replication, RNA transcription, protein translation, macromolecular assembly and degradation, and signal transduction; essentially all cellular functions involve protein–protein interactions. Thus, protein–protein interactions are fundamental for normal physiology in all organisms. Alteration of critical protein–protein interactions is thought to be involved in the development of many diseases, such as neurodegenerative disorders, cancers, and infectious diseases. Therefore, examination of when and how protein–protein interactions occur and how they are controlled is essential for understanding diverse biological processes as well as for elucidating the molecular basis of diseases and identifying potential targets for therapeutic interventions.

Over the years, many innovative biochemical, biophysical, genetic, and computational approaches have been developed to detect and analyze protein–protein interactions. This multitude of techniques is mandated by the diversity of physical and chemical properties of proteins and the sensitivity of protein–protein interactions to cellular conditions. In order to provide scientists with practical tools to address their vital biological questions in the post-genome era, *Protein–Protein Interactions: Methods and Applications* presents a collection of frequently employed techniques for identifying protein interaction partners, qualitatively or quantitatively measuring protein–protein interactions in vitro or in vivo, monitoring protein–protein interactions as they occur in living cells, and determining interaction interfaces. It is hoped that this book will be useful to a broad spectrum of researchers who are interested in studying protein–protein interactions in various systems.

Protein–Protein Interactions: Methods and Applications consists of five sections. It begins with two concise overviews of the fundamental principles of protein–protein interactions. They illustrate the structural diversity of protein interactions and some common experimental design considerations that are important for quantification of these interactions. Part Two describes a wide range of biochemical and biophysical methods for detecting and measuring protein–protein interactions in vitro. Commonly used spectroscopic, electrophoretic, and affinity matrix–based techniques are presented, many of which allow quantitative analysis of protein–protein interactions. This section ends with tools for analyzing structural interfaces and the design of peptide inhibitors of protein interactions. Advances in genetics and molecular biology have revolutionized the way that we study protein–protein interactions. Part Three reflects these changes and covers popular methods for studying protein–protein interactions in heterologous cell systems, including various bacterial, yeast, and mammalian two-hybrid systems and co-immunoprecipitation studies. These methods provide a simple solution for analyzing protein–protein interactions in an in vivo environment. Part Four presents state of the art methodologies to monitor protein–protein interactions in living cells, including applications using the fluorescence resonance energy transfer (FRET) technology. These approaches share a common bond in that they allow the capture and visualization of protein–protein interactions as they occur. In the post-genome age, it is expected that conventional methods for studying protein–protein interactions will still play an important role. However, it is hoped that proteomics and bioinformatics-based approaches will be rapidly developed to study protein–protein interactions on a large scale. Part Five begins to address this issue, focusing on high-throughput methods and on computational approaches. This section ends with a tutorial on using Internet resources, which serves as a springboard to additional information and techniques for studying protein–protein interactions.

Both basic and clinical researchers will find this book valuable for its broad coverage from simple affinity-based pull-down assays to cutting edge technologies such as FRET and solid-phase isotope tagging–based mass spectrometry. The basic theory and practical application of these widely used, representative methods are described in detail by experienced researchers. Examples are also incorporated to illustrate each method, along with notes and explanations for sensitive procedures and potential pitfalls. These features, together with our broad coverage of the topic, are designed to empower readers in their quest to decipher the functions of proteins and complex biological regulatory systems.

It is truly a great pleasure to put together a book on this important, exciting, and timely topic! I have been privileged to work with the many leaders in the protein–protein interaction field who willingly contributed their valuable time and effort to making this book possible. I am grateful to each of the contributors for their enthusiasm and tremendous efforts and to John Walker, the series editor, and Craig Adams, James Geronimo, and all at Humana Press for their guidance. My sincere thanks go to Shane Masters, Keith Wilkinson, and Jonathan Cooper for their invaluable advice and suggestions and to Lisa Cockrell and Robert Fu for assistance in preparing this book. Finally, I express my deepest appreciation to Robert, Emily, and Guo-hua, who provide constant encouragement and support.

Haian Fu

Contents

Contributors

RUEDI AEBERSOLD • *Institute for Systems Biology, Seattle, WA*
FABRICE AGOU • *Département de Biologie Structurale et Chimie, Institut Pasteur, Paris, France*
KARIN BARNOUIN • *Department of Biochemistry and Molecular Biology, University College London and Ludwig Institute for Cancer Research, London, UK*
MATTHEW A. BENNETT • *Department of Biochemistry, Emory University School of Medicine, Atlanta, GA*
KEITH M. BERLAND • *Physics Department, Emory University, Atlanta, GA*
ROSEMARY BOYLE • *Department of Genome Sciences, University of Washington, Seattle, WA*
SHARON L. CAMPBELL • *Department of Biochemistry and Biophysics, University of North Carolina-Chapel Hill, NC*
VICTORIA E. CENTONZE • *Department of Cellular & Structural Biology, University of Texas Health Science Center, San Antonio, TX*
FRANCIS KA-MING CHAN • *Department of Pathology, University of Massachusetts Medical School, Worcester, MA*
R. JOHN COLLIER • *Department of Microbiology and Molecular Genetics, Harvard Medical School, Boston, MA*
RAINER CRAMER • *Department of Biochemistry and Molecular Biology and Ludwig Institute for Cancer Research, University College London, London, UK*
SIMON L. DOVE • *Division of Infectious Diseases, Children's Hospital, Harvard Medical School, Boston, MA*
JOHN F. ECCLESTON • *Division of Physical Biochemistry, National Institute for Medical Research, London, UK*
ERNESTO FREIRE • *Department of Biology and Biocalorimetry Center, Johns Hopkins University, Baltimore, MD*
GUANGHUA GAO • *Department of Biochemistry and Biophysics, University of North Carolina-Chapel Hill, NC*
NORMA J. GREENFIELD • *Department of Neuroscience and Cell Biology, UMDNJ-Robert Johnson Medical School, Piscataway, NJ*
KUN-LIANG GUAN • *Department of Biological Chemistry, University of Michigan, Ann Arbor, MI*

RANDY A. HALL • *Department of Pharmacology, Emory University, Atlanta, GA*
BRIAN HERMAN • *Department of Cellular & Structural Biology, University of Texas Health Science Center, San Antonio, TX*
ANN HOCHSCHILD • *Department of Microbiology and Molecular Genetics, Harvard Medical School, Boston, MA*
DALLAS E. HUGHES • *Cetek Corporation, Marlborough, MA*
RICHARD A. KAHN • *Department of Biochemistry, Emory University School of Medicine, Atlanta, GA*
ELENA KOTOVA • *Division of Medical Oncology, Fox Chase Cancer Center, Philadelphia, PA*
R. V. KRISHNAN • *Department of Cellular & Structural Biology, University of Texas Health Science Center, San Antonio, TX*
STEPHANIE A. LEAVITT • *Department of Biology and Biocalorimetry Center, Johns Hopkins University, Baltimore, MD*
JAE WOON LEE • *Department of Medicine, Division of Diabetes, Endocrinology, and Metabolism, and Department of Molecular and Cellular Biology, Baylor College of Medicine, Houston, TX*
SOO-KYUNG LEE • *Gene Expression Laboratory, The Salk Institute, San Diego, CA*
ROBERT C. LIDDINGTON • *The Burnham Institute, La Jolla, CA*
NICK LOIZOS • *Department of Protein Chemistry, ImClone Systems Inc., New York, NY*
CAROL MACKINTOSH • *Medical Research Council Protein Phosphorylation Unit, School of Life Sciences, University of Dundee, Dundee, UK*
SHANE C. MASTERS • *Medical College of Georgia, Augusta, GA*
STEPHEN W. MICHNICK • *Département de Biochimie, Université de Montréal, Montréal, Québec, Canada*
JOHN MILLER • *Department of Genome Sciences, University of Washington, Seattle, WA*
TOSHIYUKI MIYASHITA • *Department of Genetics, National Research Institute for Child Health and Development, Tokyo, Japan*
JEREMY MOGRIDGE • *Department of Laboratory Medicine and Pathobiology, University of Toronto, Toronto, Ontario, Canada*
GREG MOORHEAD • *Department of Biological Sciences, University of Calgary, Calgary, Alberta, Canada*
MICHAEL MOUREZ • *Department of Microbiology and Molecular Genetics, Harvard Medical School, Boston, MA*
ELIZABETH A. MYATT • *Biosciences Division, Argonne National Laboratory, Argonne, IL*

John Northup • *Laboratory of Cellular Biology, NIDCD, National Institutes of Health, Rockville, MD*
John C. Obenauer • *Center for Cancer Research, Massachusetts Institute of Technology, Cambridge, MA*
Sang-Hyun Park • *Department of Cellular and Molecular Pharmacology, University of California, San Francisco, CA*
William E. Pierceall • *Cetek Corporation, Marlborough, MA*
Ronald T. Raines • *Departments of Biochemistry and Chemistry, University of Wisconsin-Madison, Madison, WI*
Ingrid Remy • *Département de Biochimie, Université de Montréal, Montréal, Québec, Canada*
Katrin Rittinger • *Division of Protein Structure, National Institute for Medical Research, London, UK*
Malcolm Saxton • *Ludwig Institute for Cancer Research, Royal Free and University College London, London, UK*
Ilya G. Serebriiskii • *Division of Basic Science, Fox Chase Cancer Center, Philadelphia, PA*
Jack F. Shern • *Department of Biochemistry, Emory University School of Medicine, Atlanta, GA*
Igor Stagljar • *Institute of Veterinary Biochemistry and Molecular Biology, University of Zurich-Irchel, Zurich, Switzerland*
Fred J. Stevens • *Biosciences Division, Argonne National Laboratory, Argonne, IL*
Ian A. Taylor • *Division of Protein Structure, National Institute for Medical Research, London, UK*
Safia Thaminy • *Institut of Veterinary Biochemistry and Molecular Biology, University of Zuich-Irchel, Zuich, Switzerland*
Adrian Velazquez-Campoy • *Department of Biology and Biocalorimetry Center, Johns Hopkins University, Baltimore, MD*
Michel Veron • *Département de Biologie Structurale et Chimie, Institut Pasteur, Paris, France*
Haris G. Vikis • *Department of Biological Chemistry, University of Michigan, Ann Arbor, MI*
Keith D. Wilkinson • *Department of Biochemistry, Emory University School of Medicine, Atlanta, GA*
Jason G. Williams • *National Institute of Environmental Health Sciences, Research Triangle Park, NC*
Rosemarie Wilton • *Biosciences Division, Argonne National Laboratory, Argonne, IL*

MICHAEL B. YAFFE • *Center for Cancer Research, Massachusetts Institute of Technology, Cambridge, MA*
FEI YE • *Département de Biologie Structurale et Chimie, Institut Pasteur, Paris, France*
LIXIN ZHANG • *Cetek Corporation, Marlborough, MA*
HUILIN ZHOU • *Department of Chemistry & Biochemistry, University of California-San Diego, San Diego, CA*

I

Overview

1

Structural Basis of Protein–Protein Interactions

Robert C. Liddington

Abstract

Regulated interactions between proteins govern signaling pathways within and between cells. Although it is possible to derive some general principles of protein–protein recognition from experimentally determined structures, recent structural studies on protein complexes formed during signal transduction illustrate the remarkable diversity of interactions, both in terms of interfacial size and nature. There are two broad classes of complexes: "domain–domain," in which both components comprise prefolded structural units, and "domain–peptide," in which one component is a short motif that is unstructured in the absence of its binding partner. Signaling complexes often involve multidomain proteins whose multifaceted binding functions are regulated by intramolecular domain interactions. The structural basis of regulation, via steric and allosteric mechanisms, is discussed.

Key Words

Protein structure; protein complex; crystallography; allostery; regulation.

1. Introduction

I will begin with two observations. First, we do not currently understand the structural basis of protein–protein interactions; by which I mean that, given the structure of two proteins known to interact, we cannot confidently predict how they will do it with atomic precision. The problem is closely related to (and nearly as difficult as!) the "folding problem," of predicting the three-dimensional structure of a protein from its sequence alone. In the best case, there is a homologous protein–protein interaction system for which a crystal structure is known, and a homology model can be built with a reasonable level of confidence. Even then, there are now enough counterexamples to make experimental verification advisable.

The second observation is the astounding diversity of protein–protein interactions. Until very recently, protein–protein complexes lodged in the Pro-

From: *Methods in Molecular Biology, vol. 261: Protein–Protein Interactions: Methods and Protocols*
Edited by: H. Fu © Humana Press Inc., Totowa, NJ

tein Data Base consisted mostly of antibody–antigen and protease–inhibitor complexes ("classical" complexes). Although these structures have been useful in developing general principles of protein–protein recognition (reviewed in ref. *1*), the explosion of structures in the past 4–5 yr on complexes of proteins involved in signal transduction forces us to rethink and broaden these rules. For example, "domain–peptide" interactions, in which one component is a folded domain while the other is a short unstructured (in the absence of its partner) stretch of sequence attached to a larger folded domain, are at least as common as the classical domain–domain interaction. Moreover, conformational changes induced by one protein binding to another have become more the rule than the exception, and these changes often lie at the heart of mechanisms of signal transduction. Such conformational changes require that some of the binding energy is used to carry out work—the conformational change—typically requiring a larger or more specialized (e.g., phosphopeptide, metal-mediated) interface than a classical contact.

Crystallography is the major tool for determining protein complexes at atomic resolution. An example of what is possible in the analysis of protein–protein interactions is provided by the crystallographic analysis of the five-component E3 ubiquitin–protein ligase complex by Pavletich and colleagues *(2,3)* (**Fig. 1**). NMR has traditionally been limited to complexes below 30 kDa, but recent advances in methodology promise the study of protein–protein interactions in very large complexes *(4)*.

Any attempt to provide a comprehensive review of this field is doomed to become rapidly outdated. I will therefore only attempt to provide some classification and to illustrate the diversity of protein–protein interactions, even within closely homologous families; finally, I will discuss intramolecular domain–domain interactions within multidomain proteins, and their regulation by steric and allosteric mechanisms.

2. Domain–Domain Interactions

In domain–domain interactions, two independently folded domains form an interface in which elements in each surface typically contain residues from more than one segment of the polypeptide chain. A number of general principles have arisen from studies of "classical" domain–domain interactions (*see* ref. *1*) for a recent review), which I will summarize here.

1. In general, there is good shape and charge complementarity between apposing surfaces, such that interfaces are typically as well packed as protein interiors.
2. There are no general rules about the hydrophobicity of interfaces: some are hydrophobic, others are not; on average, interfaces are not distinct from the protein surface as whole.
3. Domain–domain interfaces have a typical layout (**Fig. 2**) with a central solvent-excluded region surrounded by a less-well-packed and often solvent-mediated

region. The central region has been called the "hot spot" *(5)*—mutations here typically have a large effect on the K_d, while those at the periphery have a negligible effect.

4. "Classical" complexes have a typical interface size of 1600 ± 400 Å^2. The lower limit reflects the need to form a solvent-excluded patch of sufficient size to provide a dissociation constant in the submicromolar range.

Given these observations, one might expect that predicting the structure of a complex would be feasible. Indeed, given the crystal structure of a complex, one can separate the components *in silico* and "predict" how they will recombine, but this has no practical value. Given the crystal structures of the isolated components, prediction is not generally possible, and at least two factors cause problems here. First, protein surfaces always move by at least an angstrom or two—both main chains and sidechains move—on complex formation. Modern methodologies have flexibility in the surfaces to allow for this movement, but this dramatically increases the computational load. A second problem arises from the existence of water molecules that play essential roles in shaping the interface and mediating interactions, because their location cannot be reliably predicted from the structures of the component domains.

Signal transduction complexes show a much larger variation in size and shape than do classical complexes, perhaps reflecting the need for reversible binding and the prevalence of conformational changes underlying their biological function. A particularly well-studied field at the structural level is the G proteins and their interactions with effectors, activators, and inhibitors (reviewed in ref. *6*). For example, effectors, defined as proteins that recognize the guanosine 5'-triphosphate (GTP)-bound, but not the guanosine 5'-diphosphate (GDP)-bound, form of the G-protein, show a great variety in their structure and mode of interaction (**Fig. 3**). Some effectors present a pre-formed interface that is complementary to the GTP-bound form (signaling by recruitment), but others show large conformational changes on binding (allosteric signaling). Although they use at least part of one or both of the "switch" regions (the regions that undergo conformational changes on hydrolysis of GTP), different effectors use different parts of the surface and occupy a large range of interface sizes that extend well beyond the classical range.

The integrin family of plasma membrane receptors illustrates a distinct type of protein–protein interaction, one that is mediated in part by a metal ion (Mg^{2+} or Ca^{2+}) that forms a bridge between integrin and ligand (**Fig. 4**). The I domain of integrins contains a metal-ion dependent adhesion site (MIDAS) *(7)*. Ligand binding is strictly metal-dependent, and mutation of any of the metal-coordinating residues abrogates ligand binding. The structure of a complex with a fragment of triple helical collagen shows that a glutamic acid from the collagen completes the coordination sphere of the

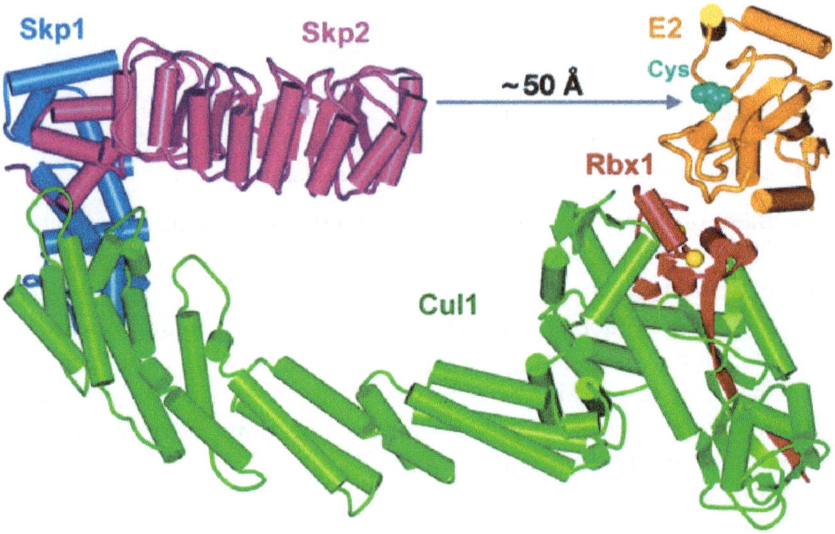
Skp1
Skp2
E2
Cys
~50 Å
Rbx1
Cul1

Fig. 1.

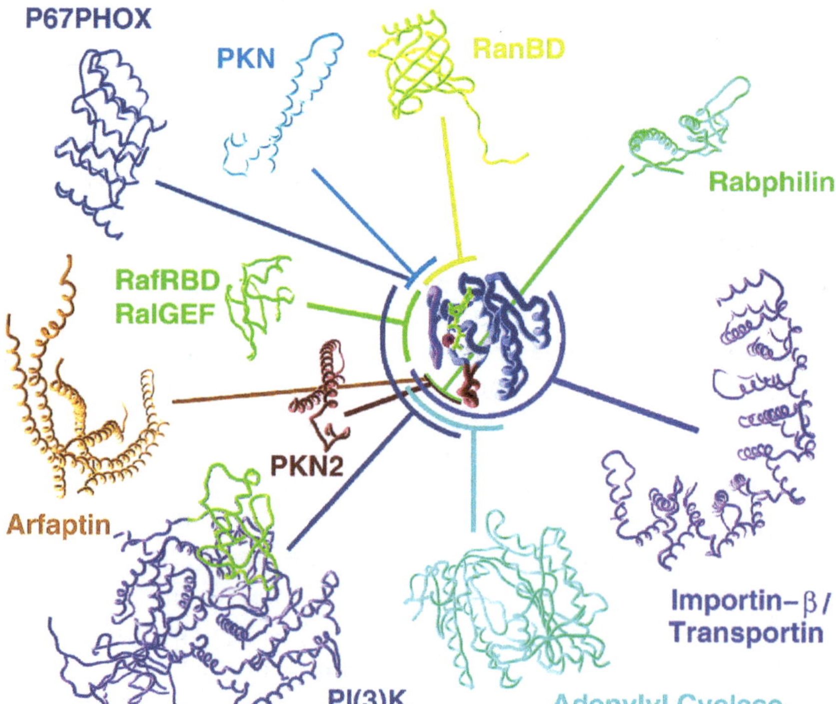
P67PHOX
PKN
RanBD
Rabphilin
RafRBD
RalGEF
PKN2
Arfaptin
PI(3)K
Adenylyl Cyclase
Importin–β /
Transportin

Fig 3.

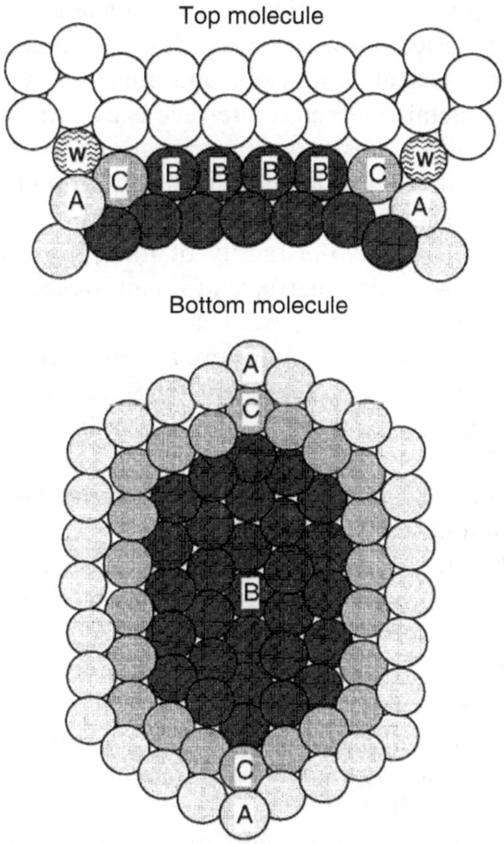

Fig. 2. Schematic (side and top views) of a classical domain–domain interface (from ref. ***38***). All the atoms of the lower molecule that lose accessibility to solvent in the presence of the upper molecule are interface atoms. Type B is fully buried, types A and C retain partial accessibility. Types B and C are contact atoms; they make van der Waals contact with the upper molecule, while type A may form water (W)-mediated contacts.

Fig. 1. *(opposite page)* Tour de force: Model of the five-component SCF^{Skp2}–E2 complex derived by crystallography ***(2,3)***. Adapted from ref. ***2***.

Fig. 3. The variety of interactions made between G proteins and their effectors, indicating the location and size of the interfaces (adapted from ref. ***6***). All effectors recognize the GTP-bound form by binding to at least one of the switch regions of the G protein (red and purple).

metal *(8)*. This bond is also crucial, because mutagenesis to aspartate abrogates binding. The buried surface area—1200 Å—is at the lower limit of interfaces found in classical complexes, and presumably reflects the strong bond formed by the metal. The small interface is even more surprising given that ligand binding causes substantial conformational changes in the integrin I domain, changes that are ultimately transduced through the plasma membrane, forming the basis of signal transduction. This mode of interaction is likely to be conserved across the family of integrins, which bind diverse ligands from the extracellular matrix and counterreceptors on other cells. Indeed, the recent structure of the αL I domain in complex with ICAM-1 supports this *(9)*. This is an example where a conserved feature (also a "hot spot") provides a general mode of ligand binding, while specificity arises from additional contacts with the surrounding (variable) residues. A similar principal, called "dual recognition," has been developed to explain the structural basis of binding and specificity between bacterial endonuclease colicins and a family of immunity proteins *(10)*.

Integrins are αβ heterodimers, and the behavior of an I-like domain in the β-subunit (which shows weak sequence homology but conserved metal-binding residues), in binding Arg-Gly-Asp- (RGD)-style ligands, was successfully predicted from the α-I domain result *(8,11)*. The integrin I domain is also homologous to the A domains found in von Willebrand factor. Although by overall sequence homology A domains are as similar to I domains as I domains are to each other, A domains lack one or more of the MIDAS residues and do not bind metal. As a consequence, their protein–protein interactions are quite different—they do not involve the same surface or large conformational changes within the A domain *(12,13)*. These observations illustrate the potential benefits and pitfalls of building models of complexes based on apparently homologous systems.

3. Domain–Peptide Interactions

As noted above, the interaction between a small unstructured (in the absence of its partner) portion of one protein and a folded domain of another is a very common theme. Early examples of this kind of interaction are the major histocompatibility complex (MHC)–antigen complexes *(14)*, and assembly interactions in certain viral capsids (e.g., ref. *15*). Recent years have witnessed the structural characterization of a huge array of such interactions among signaling molecules. The simplest generalization on the nature of these complexes is that almost anything is possible. Nevertheless, there are some popular classes that are found repeatedly in homologous and modified forms.

For example, upon binding its target domain, the peptide may adopt an extended structure that lies across a folded domain or between helices [e.g.,

14-3-3 ***(16,17)***; β-catenin/E-cadherin ***(18)***; Rb/E7 ***(19)***]; form a β-strand that augments a β-sheet [e.g., phosphotyrosine binding (PTB) ***(20,21)***, PDZ ***(22)***], sometimes followed by a β-turn (PTB); form an α-helix that augments a helical bundle [e.g., FAK-paxillin ***(23)***; calmodulin ***(24)***; β-catenin/α-catenin ***(18)***]; or even form a polyproline helix [e.g., SH3 ***(25)***; EVH1 ***(26)***]. As pointed out by Harrison ***(27)***, this kind of interaction allows for rapid evolution, because the peptide sequence is not constrained by the need to be part of a preexisting three-dimensional folded structure. For example, a peptide sequence can be inserted at the terminus of a domain, allowing for the subcellular targeting. A recent example is the γ isoform of phosphatidylinositol phosphate (PIP) kinase, which has an additional short segment at its N-terminus that binds to the PTB domain of the cytoskeletal protein, talin, thus providing the new function of focal adhesion targeting ***(28,29)***.

One superfamily of small domains illustrates the diversity of recognition modes that a single domain class can provide. Classical PTB domains, as noted above, recognize an NPxY motif that forms a reverse or helical turn, preceded by a short β-strand where elements of specificity occur, for example, through the recognition of hydrophobic side chains by pockets at different points on the domain surface. Some PTB domains (as their name would suggest!) recognize or require tyrosine phosphorylation of the NPxY motif, while others do not. The F3 subdomain of FERM domain proteins closely resembles the PTB domain. The F3 domain of the cytoskeletal protein, talin, for example, recognizes a nonphosphorylated NPxY motif in the integrin β3 tail in a variant of the canonical binding mode ***(30)***. The F3 domain of radixin recognizes the cytoplasmic tail of ICAM-2 ***(31)*** in a mode that is structurally homologous; however, the turn sequence, VLAA, bears no sequence resemblance to the NPxY motif. The EVH1 domain, found in Ena and VASP, has the same topology as the PTB domain. However, its mode of peptide recognition is quite distinct (**Fig. 5**). EVH domains recognize a polyproline sequence that binds in a hydrophobic groove formed across the face of the β-sheet ***(26)***. Finally, the pleckstrin homology (PH) domain also shares the PTB topology, but it recognizes phospholipids rather than peptides.

Phosphorylation is often required for domain–peptide complexes, where the phosphopeptide forms an energetically crucial interaction with a positively charged cluster of residues on the target domain. Nevertheless, remarkable specificity can still be achieved through additional contacts via the flanking peptide sequence. This allows for tight and rapid regulation of the complex via the actions of kinases and phosphatases. A striking example of such regulation occurs in the interaction of the cytoplasmic tail of β3 integrin with Shc and talin. Talin recognizes the unphosphorylated tail via a PTB–peptide interaction, as noted above; phosphorylation of the tail leads to a switch in recogni-

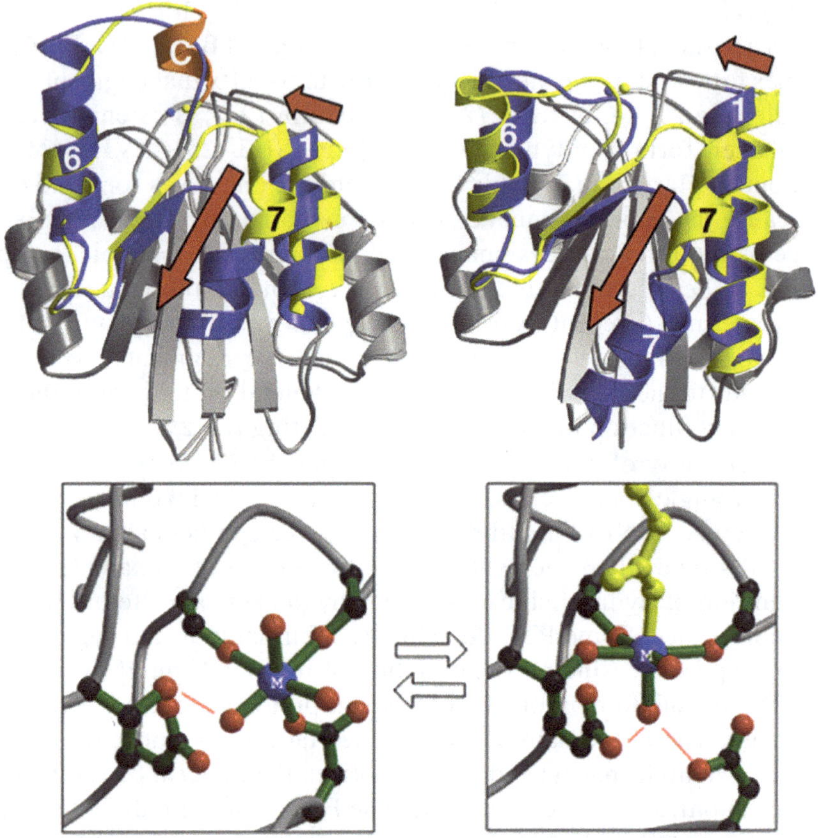
C
6
1
7
7
6
1
7
7
M
M

Fig. 4.

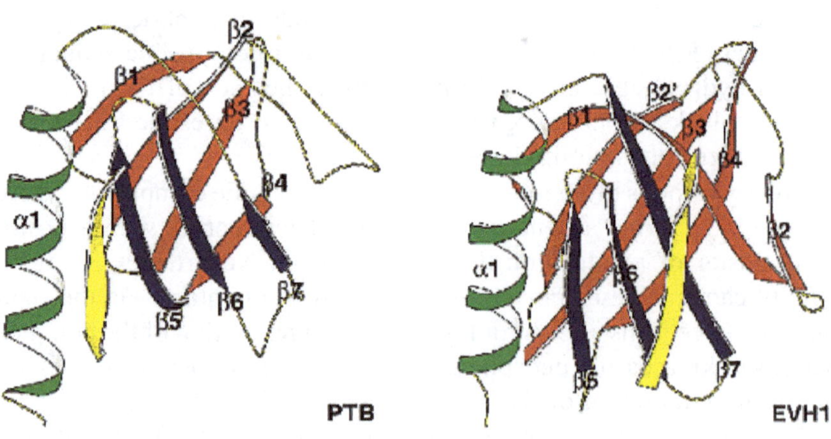
β2
β1
β3
β4
α1
β5
β6
β7
PTB
β2'
β1
β3
β4
β2
α1
β6
β5
β7
EVH1

Fig. 5.

tion to the "classical" PTB domain of Shc, resulting in a shift from cell adhesion to migration ***(32)***.

4. Intramolecular Protein–Protein Interactions and Signal Transduction

Signal transduction involves much more than the sequential association of proteins, one to the next. These associations must be regulated, of course, but their regulation typically involves more than a simple binary on–off switch such as phosphorylation–dephosphorylation. Many proteins are multidomain molecular switches that respond to multiple upstream signals and output a combinatorial signal.

A recurrent theme is the regulation of activity of multidomain proteins by intramolecular association between domains at the head and tail of the molecule that hold the molecule in a default "off" position. Two enzymes involved in signal transduction pathways illustrate two different paradigms of regulation. The tyrosine kinase, Src, consists of a tandem SH2–SH3 pair followed by a catalytic domain and a C-terminal tail. In the crystal structure ***(33)***, the N-terminal SH2 domain binds a phosphotyrosine in the C-terminal tail, while the binding site of the SH3 domain is also sequestered by a linker sequence between the SH2 and catalytic domains (i.e., both SH domains are masked by intramolecular interactions that are ligand mimetics). The SH2–3 domain pair wraps around the surface of the kinase domain distal to the active site, and inhibits the catalytic activity allosterically, by pulling on an α-helix, which disrupts the active site machinery. Dephosphorylation of the C-terminal tyrosine, or exogenous ligands to either the SH2 or SH3 domain, have some activating activity, while bivalent ligands containing both SH2- and SH3-binding motifs, are the most potent activators.

In the structure of the phosphatase, SHP-2 ***(34)***, the inhibitory mechanism is somewhat different: here, a tandem pair of SH2 domains at the N-terminus wraps around the C-terminal catalytic domain, such that the N-proximal SH2 domain masks the active site. In so doing, the SH2 domain is distorted such that its phosphotyrosine-binding pocket is still exposed but is disrupted allosterically. The most potent activator is a ligand that contains two phosphotyrosines with spacing to match the SH2–SH2 module.

Fig. 4. *(opposite page)* The MIDAS motif of integrins (lower panel). Ligand binding is dominated by a metal-mediated bridge to a critical glutamate or aspartate from the ligand, which switches the metal coordination, leading to tertiary changes in the domain (upper panel) that underlie signal transduction (adapted from ref. ***8***). Numbers refer to alpha helices and M is a metal ion.)

Fig. 5. Structures of PTB and EVH1 domains, showing their shared topology but distinct peptide binding modes (modified from ref. ***26***).

The FERM family of proteins are also regulated by head–tail associations. For example, in the structure of moesin *(35)*, the C-terminal tail of moesin adopts a series of α-helical elements that bind against one side of the FERM domain, masking a large surface. The beginning of the tail extends the β-sheet of the F3 domain in a PTB:peptide mode, and thus provides direct competition for binding of the target protein *(31)*. A phosphorylation site in the tail close to the head-tail interface activates the molecule, by disrupting head–tail association. In other FERM domains head–tail dissociation is triggered by phospholipid binding *(36)* or proteolytic cleavage *(37)*.

5. Conclusion

The complexity of these conformational switches provides perhaps the clearest imperative for further careful biochemical and structural studies at atomic resolution. That is, while modern proteomics approaches produce molecular "wiring diagrams" of signal transduction pathways, these cannot by themselves explain the molecular basis of regulation of these pathways. We must learn the nature of each switch (e.g., phosphorylation, phospholipid, or even mechanical stress), and precisely how each of the binding functions or catalytic activity of each protein is turned on by each switch.

References

1. Wodak, S. J. and Janin, J. (2003) Structural basis of macromolecular recognition. *Adv. Protein Chem.* **61,** 9–73.
2. Zheng, N., Schulman, B. A., Song, L., et al. (2002) Structure of the Cul1-Rbx1-Skp1-F boxSkp2 SCF ubiquitin ligase complex. *Nature* **416,** 703–709.
3. Schulman, B. A., Carrano, A. C., Jeffrey, P. D., et al. (2000) Insights into SCF ubiquitin ligases from the structure of the Skp1-Skp2 complex. *Nature* **408,** 381–386.
4. Fiaux, J., Bertelsen, E. B., Horwich, A. L., and Wuthrich, K. (2002) NMR analysis of a 900K GroEL GroES complex. *Nature* **418,** 207–211.
5. Bogan, A. A. and Thorn, K. S. (1998) Anatomy of hot spots in protein interfaces. *J. Mol. Biol.* **280,** 1–9.
6. Vetter, I. R. and Wittinghofer, A. (2001) The guanine nucleotide-binding switch in three dimensions. *Science* **294,** 1299–1304.
7. Lee, J.-O., Rieu, P., Arnaout, M. A., and Liddington, R. C. (1995) Crystal structure of the A-domain from the the α subunit of integrin CR3 (CD11b/CD18). *Cell* **80,** 631–635.
8. Emsley, J., Knight, C. G., Farndale, R. W., Barnes, M. J. and Liddington, R. C. (2000) Structural basis of collagen recognition by integrin α2β1. *Cell* **101,** 47–56.
9. Shimaoka, M., Tsan Xiao, T., Liu, J.-H., et al. (2003) Structures of the αL I Domain and its complex with ICAM-1 reveal a shape-shifting pathway for integrin regulation. *Cell* **112,** 99–111.

10. Kuhlmann, U. C., Pommer, A. J., Moore, G. R., James, R., and Kleanthous, C. (2000) Specificity in protein-protein interactions: the structural basis for dual recognition in endonuclease colicin-immunity protein complexes. *J. Mol. Biol.* **301,** 1163–1178.
11. Xiong, J. P., Stehle, T., Zhang, R., et al. (2002) Crystal structure of the extracellular segment of integrin αVβ3 in complex with an Arg-Gly-Asp ligand. *Science* **296,** 151–155.
12. Huizinga, E. G., Tsuji, S., Romijn, R. A., et al. (2002) Structures of glycoprotein Ibalpha and its complex with von Willebrand factor A1 domain. *Science* **297,** 1176–1179.
13. Nishida, N., Sumikawa , H., Sakakura , M., et al. (2003) Collagen-binding mode of vWF-A3 domain determined by a transferred cross-saturation experiment. *Nat. Struct. Biol.* **10,** 53–58.
14. Bjorkman, P. J., Saper, M. A., Samraoui, B., Bennett, W. S., Strominger, J. L., and Wiley, D. C. (1987) The foreign antigen binding site and T cell recognition regions of class I histocompatibility antigens. *Nature* **329,** 512–518.
15. Liddington, R. C., Yan, Y., R., S., Benjamin, T., and Harrison, S. C. (1991) Crystal structure of Simian Virus 40 at 3.8 Å resolution. *Nature* **354,** 278–284.
16. Yaffe, M. B., Rittinger, K., Volinia, S., et al. (1997) The structural basis for 14-3-3:phosphopeptide binding specificity. *Cell* **91,** 961–971.
17. Petosa, C., Masters, S. C., Bankson, L. A., et al. (1998) 14-3-3ζ binds a phosphorylated Raf peptide and an unphosphorylated peptide via its conseved amphipathic groove. *J. Biol. Chem.* **273,** 16,305–16,310.
18. Pokutta, S. and Weis, W. I. (2000) Structure of the dimerization and β-catenin-binding region of α-catenin. *Molecular Cell* **5,** 533–543.
19. Lee, J. O., Russo, A. A., and Pavletich, N. P. (1998) Structure of the retinoblastoma tumour-suppressor pocket domain bound to a peptide from HPV E7. *Nature* **391,** 859–865.
20. Eck, M. J., Shoelson, S. E., and Harrison, S. C. (1993) Recognition of a high-affinity phosphotyrosyl peptide by the Src homology-2 domain of p56lck. *Nature* **362,** 87–91.
21. Zhou, M. M., Ravichandran, K. S., Olejniczak, E. F., et al. (1995) Structure and ligand recognition of the phosphotyrosine binding domain of Shc. *Nature* **378,** 584–589.
22. Doyle, D. A., Lee, A., Lewis, J., Kim, E., Sheng, M., and MacKinnon, R. (1996) Crystal structures of a complexed and peptide-free membrane protein-binding domain: molecular basis of peptide recognition by PDZ. *Cell* **85,** 1067–1076.
23. Hayashi, I., Vuori, K., and Liddington, R. C. (2002) The focal adhesion targeting (FAT) region of focal adhesion kinase is a four-helix bundle that binds paxillin. *Nat. Struct. Biol.* **9,** 101–106.
24. Meador, W. E., Means, A. R., and Quiocho, F. A. (1993) Modulation of calmodulin plasticity in molecular recognition on the basis of x-ray structures. *Science* **262,** 1718–1721.

25. Musacchio, A., Saraste, M., and Wilmanns, M. (1994) High-resolution crystal structures of tyrosine kinase SH3 domains complexed with proline-rich peptides. *Nat. Struct. Biol.* **1,** 546–551.
26. Fedorov, A. A., Fedorov, E., Gertler, F., and Almo, S. C. (1999) Structure of EVH1, a novel proline-rich ligand-binding module involved in cytoskeletal dynamics and neural function. *Nat. Struct. Biol.* **6,** 661–665.
27. Harrison, S. C. (1996) Peptide-surface association: the case of PDZ and PTB domains. *Cell* **86,** 341–343.
28. Di Paolo, G., Pellegrini, L., Letinic, K., et al. (2002) Recruitment and regulation of phosphatidylinositol phosphate kinase type 1 by the FERM domain of talin. *Nature* **420,** 85–89.
29. Ling, K., Doughman, R. L., Firestone, A. J., Bunce, M. W., and Anderson, R. A. (2002) Type I phosphatidylinositol phosphate kinase targets and regulates focal adhesions. *Nature* 420, 89–93.
30. Garcia-Alvarez, B., de Pereda, J. M., Calderwood, D. A., et al. (2003) Structural Determinants of Integrin Recognition by Talin. *Mol. Cell* **11,** 49–58.
31. Hamada, K., Shimizu, T., Yonemura, S., Tsukita, S., Tsukita, S., and Hakoshima, T. (2003) Structural basis of adhesion-molecule recognition by ERM proteins revealed by the crystal structure of the radixin-ICAM-2 complex. *EMBO J.* **22,** 502–514.
32. Cowan, K. J., Law, D. A., and Phillips, D. R. (2000) Identification of shc as the primary protein binding to the tyrosine-phosphorylated beta 3 subunit of alpha IIbbeta 3 during outside-in integrin platelet signaling. *J. Biol. Chem.* **275,** 36,423–36,429.
33. Xu, W., Harrison, S. C., and Eck, M. J. (1997) Three-dimensional structure of the tyrosine kinase c-Src. *Nature* **385,** 595–602.
34. Hof, P., Pluskey, S., Dhe-Paganon, S., Eck, M. J., and Shoelson, S. E. (1998) Crystal structure of the tyrosine phosphatase SHP-2. *Cell* **92,** 441–450.
35. Pearson, M. A., Reczek, D., Bretscher, A., and Karplus, P. A. (2000) Structure of the ERM protein moesin reveals the FERM domain fold masked by an extended actin binding tail domain. *Cell* **101,** 259–270.
36. Martel, V., Racaud-Sultan, C., Dupe, S., et al. (2001) Conformation, localization, and integrin binding of talin depend on its interaction with phosphoinositides. *J. Biol. Chem.* **276,** 21,217–21,227.
37. Yan, B., Calderwood, D. A., Yaspan, B., and Ginsberg, M. H. (2001) Calpain cleavage promotes talin binding to the beta 3 integrin cytoplasmic domain. *J. Biol. Chem.* **276,** 28,164–28,170.
38. Lo Conte, L., Chothia, C. and Janin, J. (1999) The atomic structure of protein–protein recognition sites. *J. Mol. Biol.* **285,** 2177–2198.

2

Quantitative Analysis of Protein–Protein Interactions

Keith D. Wilkinson

Abstract

Numerous authors, including contributors to this volume, have described methods to detect protein–protein interactions. Many of these approaches are now accessible to the inexperienced investigator thanks to core facilities and/or affordable instrumentation. This chapter discusses some common design considerations that are necessary to obtain valid measurements, as well as the assumptions and analytical methods that are relevant to the quantitation of these interactions.

Key Words

Ligand binding; protein–protein interaction; fluorescence; binding equations; binding equilibria.

1. Introduction

In the post-genomic era, the importance of protein–protein interactions is becoming even more apparent *(1)*. We are coming to recognize that most, if not all, catalytic and regulatory pathways operate as networks, with frequent and extensive input from signaling pathways, feedback, and cross-talk. Replication, transcription, translation, signal transduction, protein trafficking, and protein degradation are all accomplished by protein complexes, often temporally assembled and disassembled to accomplish vectoral processes. Often these interactions are driven by interaction of recognized domains in the constituent proteins (Chapter 1). We must identify and understand these domain interactions in order to discern the patterns and logic of cellular regulation *(2)*.

2. Assumptions

There are several assumptions inherent to any analysis of a simple ligand–receptor interaction (http://www.panvera.com/tech/fpguide/FP7.pdf).

From: *Methods in Molecular Biology, vol. 261: Protein–Protein Interactions: Methods and Protocols*
Edited by: H. Fu © Humana Press Inc., Totowa, NJ

1. The interactions are assumed to be reversible. In the simplest case, the association reaction is bimolecular while the dissociation reaction in unimolecular.
2. All receptor molecules are equivalent and independent.
3. The measured response is proportional to the number of occupied receptor sites.
4. The interactions are measured at equilibrium.
5. The components do not undergo any other chemical reactions and exist only in the free or bound states.

Any or all of these assumptions may prove to be unfounded in a more complex case. In fact, it is the deviation from simple behavior that is often the first indication of a more complex binding event, and each assumption should be explored to explain deviations from simple behavior. Outlined below are treatments for simple cases. A general method to obtaining binding formulas for more complex cases has been derived from statistical thermodynamic principles *(3)*.

1.2. Binding to One Site

The receptor–ligand terminology is useful, even if artificial, in the case of protein–protein interactions. Either protein could be considered the receptor or the ligand. For the purposes of this chapter we will refer to the protein present in fixed and limiting amounts as the receptor and the component that is varied as the ligand. Thus, for one molecule of L binding to one molecule of R:

$$R_f + L_f \underset{k_2}{\overset{k_1}{\rightleftarrows}} RL \tag{1}$$

where R_f is the concentration of free receptor, L_f is the concentration of free ligand, RL is the concentration of the complex, k_1 is the association rate constant, and k_2 is the dissociation rate constant. At equilibrium,

$$\frac{[R_f]\,[L_f]}{[RL]} = \frac{k_2}{k_1} = K_\mathrm{d} \tag{2}$$

where K_d is the dissociation constant.

Rewriting eq. (2) in terms of total ligand $[L_t]$ and receptor concentrations $[R_t]$ and applying the conservation of mass assumption, $[L_f] = [L_t]–[RL]$ and $[R_f] = [R_t] – [RL]$, gives

$$\frac{([L_t]-[RL])\;([R_t]-[RL])}{[RL]} = K_\mathrm{d} \tag{3}$$

We can rearrange eq. (3) to give the fractional saturation $[RL]/[R_t]$:

$$\frac{[RL]}{[R_t]} = \frac{[L_t]-[RL]}{K_\mathrm{d}+[L_t]-[RL]} = \frac{[L_f]}{K_\mathrm{d}+[L_f]} \tag{4}$$

Thus, a plot of fraction saturation $[RL]/[R_t]$ vs $[L_f]$ will give the familiar rectangular hyperbola if only one type of binding site is present **(Fig. 1A)**. Alternatively, a plot of fractional saturation vs $\log[L_t]$ can be used. If free concentrations are actually measured (instead of calculated): we can use the Klotz plot *(4)*, a plot of fractional saturation vs $\log[L_f]$ **(Fig. 1B)**, or the Scatchard plot, a plot of ligand bound/ligand free vs ligand free **(Fig. 1C)**.

1.3. Binding to Multiple Sites

It should be noted that if more than one ligand molecule binds to R, then the behavior may be more complex. For n multiple binding sites we get:

$$[RL]-[RL_1]+[RL_2]+\cdots+[RL_n]=\frac{[R_t][L_f]}{K_{d1}+[L_f]}+\frac{[R_t][L_f]}{K_{d2}+[L_t]}+\cdots+\frac{[R_t][L_f]}{K_{dn}+[L_f]} \tag{5}$$

where n different sites can be occupied by ligand with the corresponding binding constants.

1.3.1. Identical, Noninteracting Binding Site(s)

If all binding sites are identical and noninteracting (i.e., all bind with the same K_d), then eq. (5) reduces to

$$\frac{[L_b]}{n[R_t]}=\frac{[L_f]}{K_d+[L_f]} \tag{6}$$

where $n[R_t] = [R_f] + [L_b]$.

Note that this equation is similar to eq. (4) except for the inclusion of the stoichiometry, n. A Klotz plot of fractional saturation vs $\log[L_f]$ will be sigmoidal and symmetrical about the midpoint. The curve is nearly linear from 0.1 to 10 × K_d and 99% saturation is achieved when $[L_t]$ is two orders of magnitude above K_d. A complete description of binding and accurate estimation of the plateau values requires that $[L_f]$ vary from two log units below to two log units above K_d. A steeper curve is indicative of positive cooperativity, while a flatter curve could be due to negative cooperativity or the presence of an additional binding site. The stoichiometry is calculated from the plateau value and $[R_t]$, whereas the K_d is calculated from the midpoint *(5)*, or, more accurately, using a nonlinear least squares fit to eq. (6).

If free ligand is not measured, then we must use a plot of fractional saturation vs $\log[L_t]$, and the curve will deviate from sigmoidal by the difference between $\log[L_f]$ and $\log[L_t]$. This condition is often referred to as ligand depletion *(6,7)*. It should be recognized, however, that it may not be possible to cover such a large range of concentrations with proteins. At the low end, we are often limited by the sensitivity of the technique and, at the high end, limited

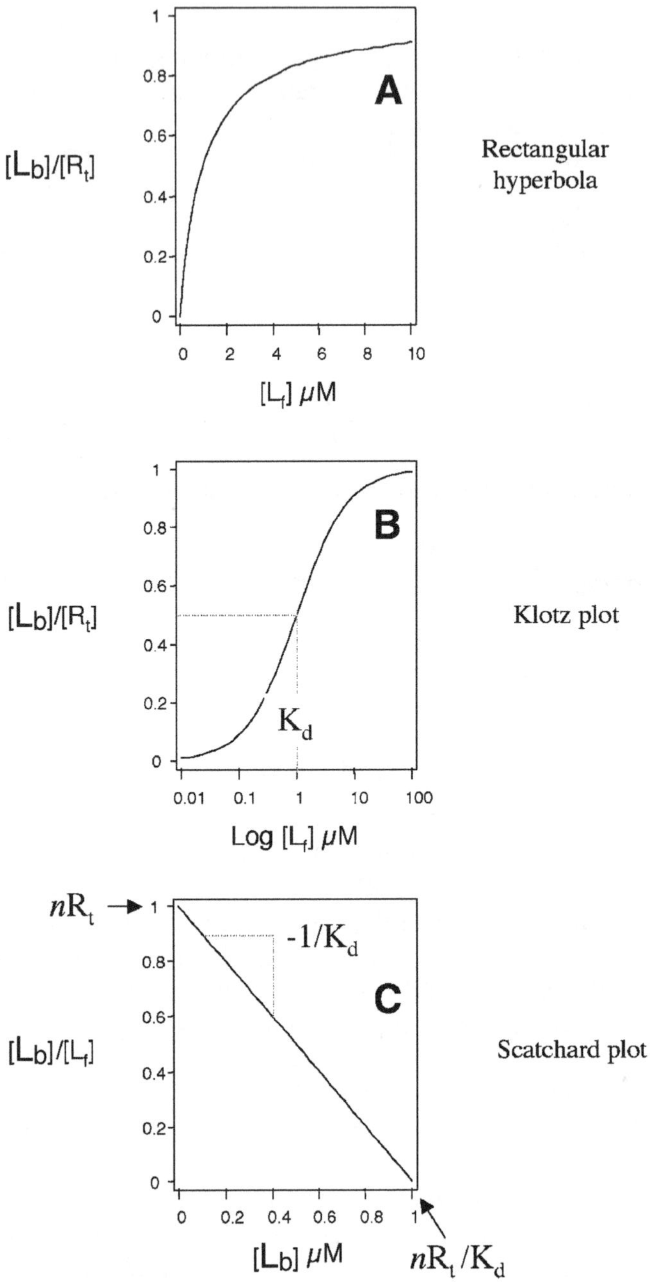

Fig. 1. Plots of simulated data for simple binding. In all cases $n = 1$ and $K_d = 1$ μ*M*. (**A**) Direct plot of fractional saturation vs free ligand; (**B**) Klotz plot of the same data, note the log scale; (**C**) Scatchard plot of the same data. The parameters nR_t and nR_t/K_d are estimated from the intercepts.

solubility or sample amounts may prevent us from attaining concentrations necessary to reach the plateau.

An alternative way to plot the data is with a Scatchard plot. For the last 30 yr this has been the traditional method for the analysis of binding data where $[L_f]$ is measured. The Scatchard plot is described by

$$\frac{[L_b]}{[L_f]} = \frac{-[L_b]}{K_d} + \frac{n[R_t]}{K_d} \tag{7}$$

In the simple model, a plot of ligand bound vs ligand bound/ligand free gives a straight line with the x-intercept = $n[R_t]$, a y-intercept of $n[R_t]/K_d$, and a slope of $-1/K_d$ **(Fig. 1C)** *(5)*.

Before the advent of computers, estimates of K_d and n were obtained by any of a number of transformations of the relevant equations to give linear plots. These transformations included the double reciprocal plot and the Scatchard plot. These linearizations are notoriously difficult to fit and generally fraught with problems . *The preferred method of obtaining K_d and n from binding data is direct fitting of the data using a nonlinear least squares fitting algorithm.* Many commercial packages for doing such fits are available today. If we do not explicitly measure the concentration of ligand free, an appropriate solution of the binding equation to obtain the dissociation constant requires that we determine and fit the fractional saturation as a function of the concentration of total L added. The solution of the equation for $[RL]/[R_t]$ as a function of $[L_t]$ is a quadratic equation with the following real solution:

$$\frac{[L_b]}{n[R_t]} = \frac{([L_t]+n[R_t]+K_d)-\sqrt{([L_t]-n[R_t]-K_d)^2-4[L_t]n[R_t]}}{2n[R_t]} \tag{8}$$

1.3.2. Nonidentical Binding Sites

Although the most common reason for observing multiple nonidentical binding sites in a protein–protein interaction is likely to be nonspecific binding (see below), it is always possible that there are two independent and noninteracting sites with different affinities. Either case will manifest itself as a deviation from the expected behavior for a simple binding model. The Scatchard plot is a useful diagnostic tool to point out such deviations **(Fig. 2)**. A Scatchard plot that is concave upward is indicative of nonspecific binding, negative cooperativity, or multiple classes of binding sites. A concave downward plot suggests either positive cooperativity or instability of the ligand. In any case, proper analysis of this behavior requires other information (for instance, stoichiometry or stability), and the data are best fitted using nonlinear least squares fitting of the data according to an appropriate model.

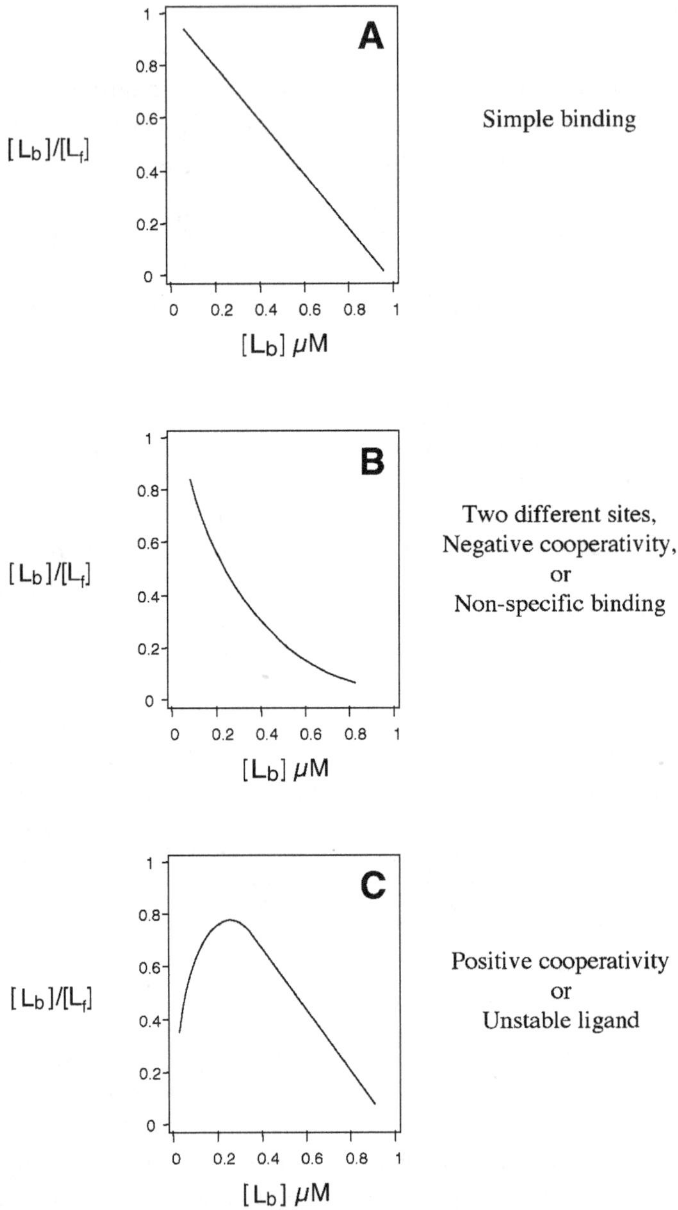

Fig. 2. Effects of complexities on the appearance of the Scatchard plot. **(A)** Represents the expected behavior in the simple case; **(B)** a concave upward deviation as shown in this panel could be caused by the presence of two different sites, the presence of negative cooperativity, or a significant nonspecific binding component; **(C)** Positive cooperativity or ligand instability would lead to the curvature shown in this panel.

Most deviations from simple binding are expected to be due to either multiple sites or nonspecific binding, which, as discussed below, may be difficult to distinguish ***(8)***. Either case can be fitted with appropriate modifications of the simple binding expressions. Note that a satisfactory analysis of such complicated binding will require measurement of $[L_f]$.

1.4. Cooperativity

Cooperativity is the term used to describe the situation where occupancy of one site changes the affinity for ligand at another site. There have been many treatments of cooperative binding interactions, including analysis by Scatchard and Hill plots, but these are beyond the scope of this discussion. In general, models explaining cooperativity invoke subunit–subunit interactions in oligomeric protein structures and may well be important in cases where multiple proteins are being assembled into a multimeric complex. The reader is referred to any of several other treatments of such binding if complications of this sort are indicated ***(9–11)***. However, it may be simpler to restrict the measurements to conditions where individual subcomplexes are assembled at saturating concentrations before measuring the binding of a subsequent protein.

2. Materials

The only materials relevant to this chapter are a computer and a program to mathematically fit the data. Many commercial and shareware packages capable of nonlinear fitting of equations are available for all platforms; i.e., Prism (GraphPad Software, Inc., San Diego, CA), SigmaPlot (SPSS Science, Chicago, IL), Mathematica (Wolfram Research Inc., Champaign, IL), and DynaFit (BioKin, Ltd., Pullman, WA). There are also published solutions using the popular spreadsheet Microsoft Excel ***(12,13)***. The choice is largely a personal preference.

3. Methods

Several chapters in this book describe techniques for determining fractional saturation and/or binding parameters. These basically fall into two categories: direct methods that measure the actual concentration of bound or free ligand and indirect methods that infer the concentrations from some measured signal. The choice of which technique to use may be limited by the strength of the interactions and the inherent sensitivity of the technique. For instance, NMR may be a poor choice to monitor binding constants tighter than micromolar since one commonly needs millimolar concentrations of protein to see a signal. Thus, $[R_t]$ may be $>>K_d$, and we would be restricted to measuring only the stoichiometry under these conditions (see below). Similarly, with an interaction of millimolar affinity it may be difficult to determine the stoichiometry, because it may not be possible to attain a concentration of $[R_t] >> K_d$. *See* below for a discussion of the relationships between K_d and $[R_t]$.

3.1. Direct Measurement of Free Ligand

Direct methods require that we accurately determine the concentrations of free and bound ligand. Examples of techniques that yield such information include gel filtration, ultracentrifugation, ultrafiltration, or equilibrium dialysis. For binding with slow dissociation rates pull-downs, band shift, or electrophoresis techniques *may* be appropriate. If the process of separating the bound and free ligand is fast compared to the rate of dissociation of the complex, such methods can yield directly the concentrations of bound and free ligand. If dissociation and separation of bound and free reactants occur on similar time scales, such methods are not appropriate for quantitation as the equilibrium will be disturbed by the separation of the reactants. For the same reasons techniques such as cross-linking may overestimate the concentration of *RL* as the removal of *RL* will disturb the equilibrium.

3.2. Indirect Measurements of Bound Ligand

More commonly, an indirect measure of saturation is used to monitor binding. These include optical methods such as fluorescence, absorbance, and resonance techniques. These methods all assume that the output signal is directly proportional to the concentration of *RL* present. For instance, if a fluorescence change is being monitored, it is assumed that there are only two states, the bound and the free, and that each has a characteristic value. If S_0 is the signal in the absence of binding, S_L the signal in the presence of total ligand concentration L, and S_∞ is the value at saturation, then

$$\text{fraction saturation} = \frac{S_L - S_0}{S_\infty - S_0} \tag{9}$$

The concentration of free ligand can be calculated by assuming a stoichiometry n and using the expression $[L_f] = [L_t] - n[R_t]$. Note that if n is incorrect, then the calculated $[L_f]$ will be incorrect also and this will be apparent in the deviation of the data from the theoretical rectangular hyperbola. This is one reason why the determination of n is an important exercise in most binding studies. Alternatively, and preferably, data are fitted using nonlinear least squares methods and n is determined directly from this analysis.

3.3. Competition Methods

Direct methods measure either bound ligand $[RL]$ or free ligand $[L_f]$ as a function of $[L_t]$, and indirect methods usually involve measuring fractional saturation $[RL]/n[R_t]$ as a function of $[L_t]$. However, one of the most useful variations of the binding experiment is the use of competitive binding assays where a single labeled indicator ligand can be bound and subsequently displaced by any of a variety of competitive inhibitors ***(14–19)***. Such experiments are particularly useful if the affinity of a series of inhibitors is to be determined. Methods such as fluorescence depolarization or fluorescence resonance energy transfer are particularly well suited for such measurements. A small amount of the labeled ligand is first bound to the receptor and subsequently displaced by titrating with unlabeled inhibitor. The K_i of the unlabeled inhibitor is then calculated. The labeled ligand does not have to be physiological or bound with a physiological affinity because we are always comparing the K_i of the unlabeled inhibitor. Thus, any adverse effects of labeling the indicator ligand will be unimportant.

The IC_{50} is the concentration of inhibitor necessary to displace half the labeled ligand. If $[R_t] << K_d$, IC_{50} is related to K_i, the affinity of the unlabeled ligand by

$$K_i = \frac{IC_{50}}{1 + L_t/K_d} \tag{10}$$

where $[L_t]$ is the concentration of labeled ligand and K_d is its dissociation constant. If only relative affinities are to be measured, then comparing IC_{50} directly is sufficient. If absolute affinities are desired, then we must also determine the concentration and affinity of the labeled ligand in the assay.

If $[R_t]$ is similar to or greater than K_d and/or K_i, it follows that the concentrations of free ligand and inhibitor are not equal to their respective total concentrations. For this reason, it is simplest to work at conditions where $[R_t] \sim 0.1 \times K_d$ so that less than 10% of the labeled ligand is bound to the receptor at the start of the experiment.

If higher concentrations of receptor are necessary or if inhibitor binds much tighter than ligand, then one has to fit with a more complex equation ***(6,15,17–19)***. The following treatment was first published by Wang in 1995 ***(19)*** and is suitable for fitting the data from competitive displacement experiments where absorbance, fluorescence, or fluorescent anisotropy are measured using commercially available fitting programs. Consider, for example, the binding of a fluorescent probe *A* to a nonfluorescent protein *P* in the presence or absence of a competitive inhibitor *B* that prevents binding of *A*.

Given: $K_a = [A_f][P_f]/[PA]$

$[A_f] + [PA] = [A_t]$

$[P_f] + [PA] + [PB] = [P_t]$

$K_b = [B_f][P_f]/[PB]$

$[B_f] + [PB] = [B_t]$

then eq. (11) describes the fractional saturation:

$$\frac{S - S_0}{S_\infty - S_0} = \frac{2\sqrt{(a^2 - 3b)}\cos(\theta/3) - a}{3K_a + \left[2\sqrt{(a^2 - 3b)}\cos(\theta/3) - a\right]} \quad (11)$$

where

$$\theta = arccos\frac{-2a^3 + 9ab - 27c}{2\sqrt{(a^2 - 3b)^3}}$$

$$a = K_a + K_b + [A_t] + [B_t] - [P_t]$$

$$b = K_b([A_t] - [P_t]) + K_a([B_t] - [P_t]) + K_a K_b$$

$$c = -K_a K_b [P_t]$$

The experiment requires the measurement of the fractional saturation at various concentrations of A_t, B_t, and P_t. Only a small range of measurements are useful: the ones where fractional saturation is >0.05 and <0.95. Fractional saturation of P with the probe A is determined by indirect measurements where it is the fluorescence or the anisotropy of AP that gives rise to the signal. The usual experiment is to measure the full binding curve, i.e., $(S - S_0)/(S_\infty - S_0)$ as a function of P_t. This experiment should then be repeated at three or more concentrations of B_t to calculate K_b. Although this may seem like its only giving you three data points, if the curve is fitted, the actual number of useful data points is equal to the total measurements made where fractional saturation is in a useful range.

3.4. Parameters of Reversible Binding

3.4.1. Stoichiometry

Quantitation of binding often requires accurate estimates of the binding stoichiometry n. Many methods are appropriate for this purpose including cross-linking, pull-downs, and electrophoretic methods (when off rates are slow). If association and dissociation rates are fast, these techniques will perturb the equilibrium and give erroneous results. In these cases stoichiometry must be determined from more conventional titrations measuring the equilibrium amounts of RL. To determine stoichiometry an excess of ligand is present and one of the components must be present at concentration well above the K_d in order to ensure saturation. Often this is the first experiment that is done as it helps greatly in fitting the data to more complete titrations.

3.4.2. Kinetics

The analysis of binding requires that we conduct the measurements after binding has reached equilibrium or that we measure individually the rate constants involved. The binding constant can then be calculated from the relationship $K_d = k_2/k_1$. From a practical standpoint, ensuring that the reaction has reached equilibrium often involves measuring a time course for binding at low ligand concentrations and making all measurements after sufficient time to allow attainment of equilibrium. Several examples of each type of analysis are given in subsequent chapters.

In any case, it is instructive to consider the magnitudes of association and dissociation rates. The association rate constants expected for protein–protein interactions are limited by diffusion. If we assume reasonable numbers for the diffusion rate of an average protein, the diffusion limit in aqueous solution is around 10^8–10^9 $M^{-1}s^{-1}$. There are also additional steric constraints, as only a fraction of the collisions occurring at this rate are oriented properly, and it is commonly assumed that the rate limiting association rate (k_1) for two proteins binding to each other is around 10^8 $M^{-1}s^{-1}$.

It can be shown that the rate of approach to equilibrium is determined by the sum of the association rate and the dissociation rate constants. Furthermore, the concentrations of reagents must be at or near the binding constant for accurate determination of both stoichiometry and affinity in the same experiment (*see* below). If the dissociation constant (K_d) for such an interaction is moderate (10^{-6} M), then the dissociation rate for such a complex will be $k_2 = k_1 \times K_d = 10^2$ s^{-1}. Thus binding will be complete in seconds and the half-life of the bound state will be tens of milliseconds. If, however, the binding constant is very tight, as may occur in antibody–antigen interactions, the overall equilibrium may take some time. Consider a binding interaction with a free energy of –16 Kcal/mole, an affinity exhibited by many antibodies and other protein–protein interactions ***(20)***. This represents a dissociation constant of 10^{-13} M. Here, binding may take as long as hours and the half-life of the bound state could be as long as 20 h. The latter fact is the reason that tight binding can be detected using techniques like immunoprecipitation and pull-down experiments, but tight binding complicates the determination of accurate binding constants.

3.5. Concentrations of Components to Use

3.5.1. Ligand Concentration

Equation (6) is the equation for the familiar rectangular hyperbola with a horizontal asymptote corresponding to 100% saturation and half-maximal saturation occurring at $L_f = K_d$. This equation points out that *the concentrations of free ligand present must be similar to the dissociation constant in order to vary the fractional saturation of receptor*, i.e., to measure the strength of binding. The most common form of the experiment, then, is to titrate a fixed amount of receptor with variable amounts of ligand and to fit the experimental data to the appropriate binding equation to determine the stoichiometry n and the binding constant K_d.

3.5.2. Receptor Concentration

If we consider the concentration of the fixed protein in this binding equation, i.e., $[R_t]$, we can define three limiting conditions; $[R_t] \ll K_d$, $[R_t] \gg K_d$, and $[R_t] \sim K_d$. **Figure 3** illustrates the interrelationships between K_d and $[R_t]$ in such experiments.

3.5.2.1. $[R_t] \ll K_d$

Under these conditions saturation is achieved by varying $[L]$ at concentrations from 0.1 to 10 times K_d. Because $[L_t]$ is always much greater that $[RL]$ under these conditions, then $[L_f] \sim [L_t]$. Thus, eq. (6) can be simplified to give

$$\frac{[L_b]}{n[R_t]} = \frac{[L_t]}{K_d + [L_t]} \tag{12}$$

If we only measure the fractional saturation (i.e., the *ratio* $[L_b]/n[R_t]$) as a function of L_t, then we cannot calculate $[L_b]$ because $[L_b] = [L_t] - [L_f]$ and we have not measured $[L_b]$. Note that even if we use direct methods and measure free ligand concentration, the calculation of bound ligand is subject to large errors because the bound is the difference between total and free and, under these conditions, they are about equal ***(6,7)***. *Thus, under these conditions, we can accurately determine K_d but not n.* Determination of accurate values for n requires that the concentration of R_t be similar to or larger than K_d.

3.5.2.2. $[R_t] \gg K_d$

If the concentration of R_t is much greater than K_d, then eq. (6) can be rearranged to give

$$\frac{[L_b]}{[L_f]} = \frac{n[R_t]}{K_d + [L_f]} \tag{13}$$

In the first part of the titration curve, when $[L_f]$ is less than K_d (and much less than $n[R_t]$ in this example), the ratio of bound/free ligand is determined solely by the ratio of $n[R_t]/[K_d]$.

If we only measure $[L_t]$, the limiting slope for a plot of saturation vs $[L_t]$ is $n[R_t]/[K_d]$. For example, if $n[R_t]/[K_d] = 100$, then only about 1% of the added

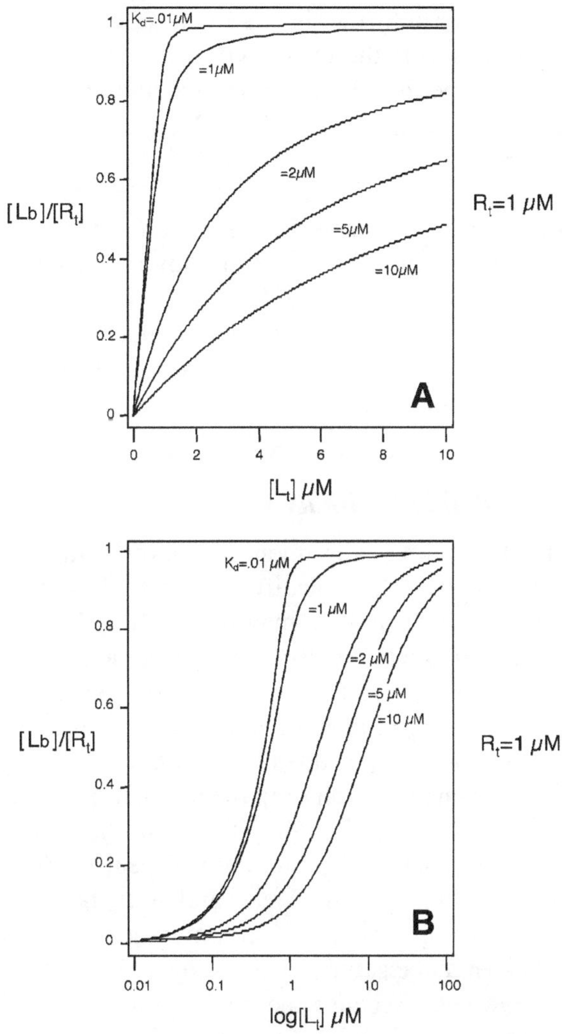

Fig. 3. Binding isotherms for a simple binding equilibrium where $n = 1$ and the total concentration of receptor is 1 μ*M*. **(A)** Direct plot of fractional staturation vs total ligand added and **(B)** the same data plotted on a log scale. Note that as the K_d approaches $[R_t]$ there is a significant deviation from the rectangular hyperbolic behavior.

ligand is free at low ligand concentrations. In order to saturate binding, $[L_t]$ must exceed $100 \times K_d$. When $[R_t] >> K_d$, the saturation curve is really an end-point determination consisting of two lines (first a slope of approx n and then 0 intersecting at $[L_t] = n[R_t]$) with little curvature **(Fig. 3)**. *Under these conditions we can accurately determine n, but not* K_d.

If direct methods to measure free ligand are used, we can, in theory, calculate K_d, but in practical terms the curve will only deviate from its biphasic nature near $L_t = R_t$ and generally there will not be enough data in this region to obtain accurate estimates of K_d.

3.5.2.3. $[R_t]$~K_d

The most useful conditions for determining both K_d and n are when $[R_t] \sim K_d$. The binding curve still resembles a rectangular hyperbola but with small deviations due to the fact that $[L_t] = [L_f] + n[RL]$. Because $[L_f]$ is similar in magnitude to $[L_b]$, each can be measured (or calculated) with good accuracy. *Under these conditions we can determine both* K_d *and n with a good degree of accuracy from the same experiment.*

4. Notes

4.1. Nonspecific Binding: Specificity vs Affinity

Almost any real-life binding experiment will show some low-affinity binding that is often attributed to "nonspecific binding." If indirect methods are used to monitor binding, one may or may not see this binding step and one must evaluate if the technique being used will reveal nonspecific binding (i.e., does the detection of binding require occupancy of a specific site such as in fluorescence resonance energy transfer techniques). Nonspecific binding usually presents as an additional slope added to the familiar rectangular hyperbola apparent at high ligand concentrations and the temptation is to simply subtract the linear phase from the observed binding to obtain the specific binding profile. The ambiguity as to whether this binding is "specific" (but just low affinity) or whether this is "nonspecific" has, and will, bedevil many studies ***(6,16,21)***. Numerous hydrophobic and ionic interactions can lead to nonspecific binding, but these may be saturable and show a defined n value when two large proteins are involved. Because the binding may well be saturable, the linear subtraction of nonspecific binding may not be appropriate. If we restrict ourselves to consider only two classes of sites, one tight site binding n_1 molecules with affinity K_{d1} and a second weaker site (either due to another specific site or nonspecific binding) binding n_2 molecules with affinity K_{d2}, then we can modify eq. (5) to give

$$[L_b] = \frac{n_1 [R_t][L_f]}{K_{d1} + [L_f]} + \frac{n_2 [R_t][L_f]}{K_{d2} + [L_f]} \qquad (14)$$

Direct fitting of the data to this expression will allow assessment of both classes of sites. If $K_{d2} >> [L_f]$, then the second term is approximately linear with $[L_f]$ and this is similar to the usual case of nonspecific binding. But if $K_{d2} \sim [L_f]$, then the second term will not be linear. Thus, it is preferable to simply fit the binding as though there are two different but independent binding sites. After the data are analyzed with no assumptions, one can question if this interaction occurs at a defined site and in a physiological range of concentrations and is therefore relevant.

4.2. Curve Fitting and Adequacy of the Models

Deviations from the simple binding expressions indicate complexity such as multiple sites or cooperativity. However, the simplest model that explains the data is to be preferred. If the data fit a model with two independent binding sites no better than that with one, the one-site model should be chosen unless there is independent evidence to suggest two sites. Methods of evaluating the goodness of fitting are beyond the scope of this chapter, but are often provided with available fitting programs and should be evaluated before proposing a more complicated expression *(7)*.

4.3. Procedures and Problems

To summarize, the determination of K_d and n for a protein–protein interaction requires that we select a technique appropriate for the binding affinity to be measured. The best concentration of receptor is near the K_d and the concentration of ligand should be varied from two orders of magnitude below to two orders of magnitude above the K_d. The concentration of bound ligand should be determined as a function of the free ligand and the data should be fit to the simplest appropriate model. Generally, n can be determined with a precision of ± 20% and K_d within a factor of 2.

Several experimental limitations and errors can limit the accuracy and correctness of the observed fits. Common problems (http://www.panvera.com/tech/fpguide/FP7.pdf) are

1. Incorrect correction for nonspecific binding or additional loose binding sites. The suggested solution is to fit to eq. (14).
2. Pooling data from experiments with different receptor concentrations. This will be a problem if the receptor concentrations are near K_d. To avoid this, collect enough data from each titration to do an independent fit and compare the fitted parameters from independent determinations.
3. Presence of a nonbinding contaminant in the receptor or labeled ligand. This may be relevant when labeling the ligand damages the protein, when recombinant proteins are used, and when there is undetected heterogeneity due to misfolded protein.
4. Use of a labeling method for the ligand that alters the binding behavior of that ligand. Use of truncated constructs or incorporation of epitope tags or fluorescent labels may be particularly troublesome. Such problems may be revealed if one compares the apparent affinity from direct experiments using titration with

labeled ligand to experiments where unlabeled ligand is used to displace labeled ligand.

5. Inadequate number of data points or range of ligand concentrations. This is avoided by collecting enough data points, especially at high ligand concentrations.

References

1. Auerbach, D., Thaminy, S., Hottiger, M. O., and Stagljar, I. (2002) The post-genomic era of interactive proteomics: facts and perspectives. *Proteomics* **2,** 611–623.
2. Pawson, T., Raina, M., and Nash, P. (2002) Interaction domains: from simple binding events to complex cellular behavior. *FEBS Lett.* **513,** 2–10.
3. Johnson, M. L. and Straume, M. (2000) Deriving complex ligand-binding formulas. *Meth. Enzymol.* **323,** 155–167.
4. Klotz, I. M. (1985) Ligand–receptor interactions: facts and fantasies. *Q. Rev. Biophys.* **18,** 227–259.
5. Munson, P. J. and Rodbard, D. (1983) Number of receptor sites from Scatchard and Klotz graphs: a constructive critique. *Science* **220,** 979–981.
6. Swillens, S. (1995) Interpretation of binding curves obtained with high receptor concentrations: practical aid for computer analysis. *Mol. Pharmacol.* **47,** 1197–1203.
7. Motulsky, H. J. and Ransnas, L. A. (1987) Fitting curves to data using nonlinear regression: a practical and nonmathematical review. *FASEB J.* **1,** 365–374.
8. Mendel, C. M. and Mendel, D. B. (1985) "Non-specific" binding. The problem, and a solution. *Biochem. J.* **228,** 269–272.
9. Tuk, B. and van Oostenbruggen, M. F. (1996) Solving inconsistencies in the analysis of receptor-ligand interactions. *Trends Pharmacol. Sci.* **17,** 403–409.
10. Koshland, D. E., Jr. (1996) The structural basis of negative cooperativity: receptors and enzymes. *Curr. Opin. Struct. Biol.* **6,** 757–761.
11. Forsen, S. and Linse, S. (1995) Cooperativity: over the Hill. *Trends Biochem. Sci.* **20,** 495–497.
12. Hedlund, P. B. and von Euler, G. (1999) EasyBound—a user-friendly approach to nonlinear regression analysis of binding data. *Comput. Methods Programs Biomed.* **58,** 245–249.
13. Brown, A. M. (2001) A step-by-step guide to non-linear regression analysis of experimental data using a Microsoft Excel spreadsheet. *Comput. Methods Programs Biomed.* **65,** 191–200.
14. Jezewska, M. J. and Bujalowski, W. (1996) A general method of analysis of ligand binding to competing macromolecules using the spectroscopic signal originating from a reference macromolecule. Application to *Escherichia coli* replicative helicase DnaB protein nucleic acid interactions. *Biochemistry* **35,** 2117–2128.
15. Schwarz, G. (2000) A universal thermodynamic approach to analyze biomolecular binding experiments. *Biophys. Chem.* **86,** 119–129.

16. van Zoelen, E. J. (1992) Analysis of receptor binding displacement curves by a nonhomologous ligand, on the basis of an equivalent competition principle. *Anal. Biochem.* **200,** 393–399.
17. van Zoelen, E. J., Kramer, R. H., van Moerkerk, H. T., and Veerkamp, J. H. (1998) The use of nonhomologous scatchard analysis in the evaluation of ligand-protein interactions. *Trends Pharmacol. Sci.* **19,** 487–490.
18. van Zoelen, E. J., Kramer, R. H., van Reen, M. M., Veerkamp, J. H., and Ross, H. A. (1993) An exact general analysis of ligand binding displacement and saturation curves. *Biochemistry* **32,** 6275–6280.
19. Wang, Z. X. (1995) An exact mathematical expression for describing competitive binding of two different ligands to a protein molecule. *FEBS Lett.* **360,** 111–1114.
20. Brooijmans, N., Sharp, K. A., and Kuntz, I. D. (2002) Stability of macromolecular complexes. *Proteins* **48,** 645–653.
21. Rovati, G. E., Rodbard, D., and Munson, P. J. (1988) DESIGN: computerized optimization of experimental design for estimating Kd and Bmax in ligand binding experiments. I. Homologous and heterologous binding to one or two classes of sites. *Anal. Biochem.* **174,** 636–649.

II

In Vitro Techniques

3

Characterization of Protein–Protein Interactions by Isothermal Titration Calorimetry

Adrian Velazquez-Campoy, Stephanie A. Leavitt, and Ernesto Freire

Abstract

Isothermal titration calorimetry (ITC) is a powerful technique to study both protein–ligand and protein–protein interactions. This methods chapter is devoted to describing protein–protein interactions, in particular, the association between two different proteins and the self-association of a protein into homodimers. ITC is the only technique that determines directly the thermodynamic parameters of a given reaction: ΔG, ΔH, ΔS, and ΔC_P. Isothermal titration calorimeters have evolved over the years and one of the latest models is the VP-ITC produced by Microcal, Inc. In this chapter we will be describing the general procedure for performing an ITC experiment as well as for the specific cases of porcine pancreatic trypsin binding to soybean trypsin inhibitor and the dissociation of bovine pancreatic α-chymotrypsin.

Key Words

Protein–protein interaction; calorimetry; thermodynamics; binding; dimerization.

1. Introduction

1.1. Isothermal Titration Calorimetry (ITC)

ITC has been used extensively to study protein–ligand binding and is being increasingly used to study protein–protein interactions ***(1–4)***. In the future the role of ITC as a basic quantitative biochemical tool for characterizing intermolecular interactions should be recognized considering the new directions in the post-genomic era for biology and medicine (proteomics, pharmacogenomics, etc). When two proteins bind, there are changes in the thermodynamic potentials (ΔG, ΔH, ΔS), which can be measured directly by highly sensitive calorimetry. Other methods for studying biomolecular reactions such as surface plasmon resonance (SPR) or analytical ultracentrifugation either require the

From: *Methods in Molecular Biology, vol. 261: Protein–Protein Interactions: Methods and Protocols*
Edited by:H. Fu © Humana Press Inc., Totowa, NJ

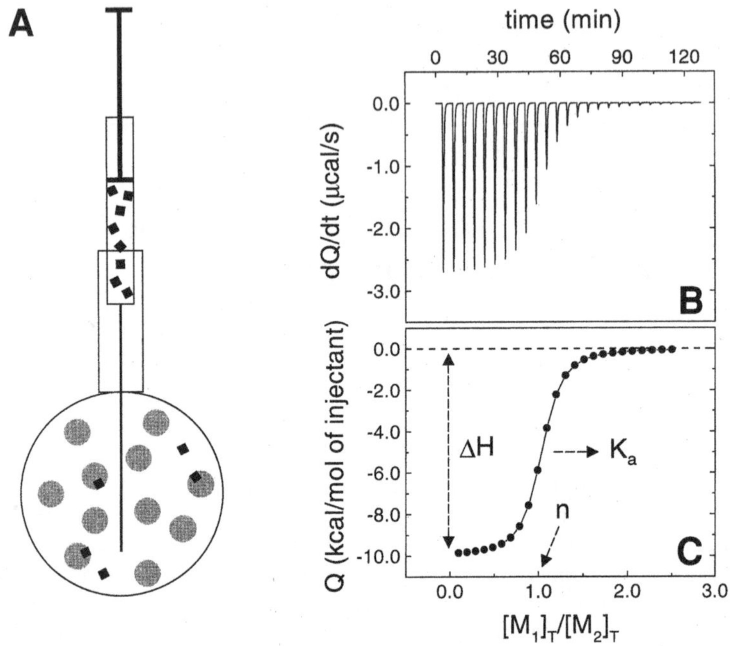

Fig. 1. **(A)** Illustration of the configuration of an ITC reaction cell. The cell volume is 1.4 mL and initially is filled with the macromolecule solution (gray). The injection syringe is filled with the ligand solution (black). At specified time intervals (400 s), a small volume (10 μL) of the ligand solution is injected into the cell triggering the binding reaction and producing the characteristic peak sequence in the recorded signal **(B)**. After saturating the macromolecule, the residual heat effects (the so-called "dilution peaks"), if any, are due to mechanical and dilution phenomena. After integration of the area under each peak (and subtraction of the dilution heat effects and normalization per mol of injected ligand) the individual heats are plotted against the molar ratio **(C)** from which, through nonlinear regression, it is possible to estimate the thermodynamic parameters n, K_a, and ΔH.

proteins to be attached to a surface, which can interfere with the binding or can take a long time to carry out the experiment. In addition, all other techniques only measure binding affinity whereas ITC also measures the enthalpic and entropic contributions to the binding affinity. ITC uses stepwise injections of one reagent into a calorimetric cell containing the second reagent to measure the heat of the reaction for both exothermic and endothermic processes (*see* **Fig. 1**). Analysis of the reaction heat as a function of concentration provides a complete thermodynamic characterization of a binding reaction. ITC is simple to run and results can be obtained in less than 2 h. Using the guidelines outlined below experiments can be designed for most systems.

1.2. Protein–Protein Interactions

Two situations are possible when characterizing intermacromolecular interactions in binding reactions: the binding partners are (a) different (heterodimeric complex) or (b) identical (homodimeric complex). Even though experimentally they require different methodologies, the underlying principles are the same in both cases, that is, they follow the same chemical scheme based on a reversible association equilibrium:

$$M_1 + M_2 \leftrightarrow M_1M_2 \tag{1}$$

where M_1 and M_2 are the interacting macromolecules. The strength of the interaction is described by the association constant K_a or the dissociation constant K_d:

$$K_a = \frac{[M_1M_2]}{[M_1][M_2]} = \frac{1}{K_d} \tag{2}$$

where $[M_1]$ and $[M_2]$ are the concentrations of the free reactants and $[M_1M_2]$ is the concentration of the complex. These constants are related to the Gibbs energy of association ΔG_a and dissociation ΔG_d and can be expressed in terms of the enthalpy, ΔH, and entropy, ΔS, change in the process:

$$\begin{aligned} \Delta G_a &= -RT \, \text{In} K_a = \Delta H_a - T\Delta S_a \\ \Delta G_d &= -RT \, \text{In} K_d = \Delta H_d - T\Delta S_d \end{aligned} \tag{3}$$

where R is the gas constant (1.9872 cal/Kmol) and T is the absolute temperature (kelvins).

Both enthalpic and entropic contributions to the Gibbs energy reflect different types of interactions underlying the overall process. Accordingly, complexes predominantly stabilized enthalpically or entropically will respond differently to environmental changes or mutations in the binding species. A number of reports have shown the importance of knowing the magnitude of the enthalpic and entropic contributions to the binding affinity ***(5–11)***.

1.2.1. Heterodimeric Interactions

Typical ITC experiments characterize the interaction between two different binding partners. A solution of one of the reactants is placed in the syringe and a solution with the other interacting macromolecule is located in the calorimetric cell. The stepwise addition of the macromolecule in the injection syringe solution triggers the binding reaction, leading the system through a sequence of equilibrium states, the composition of each one being dictated by the association constant. Given the total concentrations of reactants, $[M_1]_T$ and $[M_2]_T$, in the calorimetric cell, the association constant K_a determines the partition between the different species in each state:

$$K_a = \frac{[M_1 M_2]}{[M_1][M_2]} \tag{4}$$

ITC measures the heat q_i associated with each change of state after each injection and it is proportional to the increment in the concentration of complex in the calorimetric cell after the injection i:

$$q_i = V\Delta H_a \left([M_1 M_2]_i - [M_1 M_2]_{i-1}\right) \tag{5}$$

where ΔH_a is the enthalpy of binding and V is the calorimetric cell volume. The sequence of injections proceeds until no significant heat is detected, that is, the macromolecule in the cell is saturated and the concentration of complex reaches its maximum. Throughout the experiment the total concentrations of reactants, $[M_1]_T$ and $[M_2]_T$, are the known independent variables. Nonlinear regression analysis of q_i, the dependent variable, allows estimation of the thermodynamic parameters (K_a and ΔH_a and, therefore, ΔS_a) (see **Subheading 3.3.**).

1.2.2. Homodimeric Interactions

The self-association of a protein leading to the formation of homodimers has usually not been studied by ITC. From a practical point of view, it is impossible to isolate individual partners and perform a standard mixing assay. However, the strength of the interaction can be measured in dilution experiments ***(12)***. A solution of reactant is placed in the syringe and a buffer solution is located in the calorimetric cell. The stepwise addition of the solution in the injection syringe, with the subsequent dilution of the macromolecule, triggers the dissociation reaction, leading the system through a sequence of equilibrium states, the composition of each one being dictated by the dissociation constant K_d. Given the total macromolecule concentration $[M]_T$ in the calorimetric cell, the dissociation constant determines the partition between the different species, monomer $[M]$ and dimer $[M_2]$:

$$K_d = \frac{[M]^2}{[M_2]} \tag{6}$$

Again, the heat q_i associated with each injection is proportional to the increment in the concentration of monomer $[M]$ in the calorimetric cell after the injection i:

$$q_i = V\Delta H_d \left([M]_i - [M]_{i-1} - F_0 [M]_0 \frac{v}{V}\right) \tag{7}$$

where ΔH_d is the enthalpy of dissociation (per monomer), V is the calorimetric cell volume, v is the injection volume, $[M]_0$ is the total concentration of macromolecule (per monomer) in the syringe, and F_0 is the fraction of monomer in the concentrated solution placed in the syringe. The last term in the parenthesis is a correction that accounts for the increment of monomer concentration in the cell due to the injection of monomers from the syringe and, therefore, not contributing to the heat. As the protein concentration in the calorimetric cell progressively increases, the dissociation process is less favored and the sequence of injections proceeds until no significant heat is detected. Throughout the experiment the total concentration of reactant $[M]_T$ is the known independent variable. Nonlinear regression analysis of q_i, the dependent variable, allows the estimation of the thermodynamic parameters (K_d and ΔH_d and, therefore, ΔS_d) (*see* **Subheading 3.3.**).

1.3. Information Available by ITC and Experimental Design

1.3.1. Simultaneous Determination of the Association Constant, the Enthalpy of Binding and the Stoichiometry

Every equilibrium binding technique requires the reactant concentrations to be in an appropriate range in order to obtain reliable estimations of the association constant. A practical rule of thumb for ITC is given by the parameter $c = K_a \times [M_2]_T$, which must lie between 1 and 1000, thus, imposing a limit to the lowest and largest association constant measurable at a given macromolecule concentration ***(13)***. This phenomenon is illustrated in **Fig. 2**. As the c value increases, the transition from low to high total titrant concentration is more abrupt. In the case of very high association constants (macromolecule concentration much higher than the dissociation constant), all the titrant added in any injection will bind to the macromolecule until saturation occurs and, therefore, all the peaks, except the last ones after saturation, exhibit the same heat effect. For low association constants (macromolecule concentration far below the dissociation constant), from the very first injection only a fraction of the titrant added will bind, producing a less steep titration in which saturation is hardly reached. In order to obtain accurate estimates of the association constant, an intermediate case is desirable.

To obtain a satisfactory titration curve, the concentration of titrant should be enough to exceed the stoichiometric binding after completion of the injection sequence (i.e., for a cell volume and an injection volume of approx 1.5 mL and approx 10 μL, respectively, the reactant concentration in the syringe should be 10–20 × n times the concentration of macromolecule in the cell, where n is the number of binding sites per molecule). In case of poor solubility, the reactant with lower solubility should be placed in the cell.

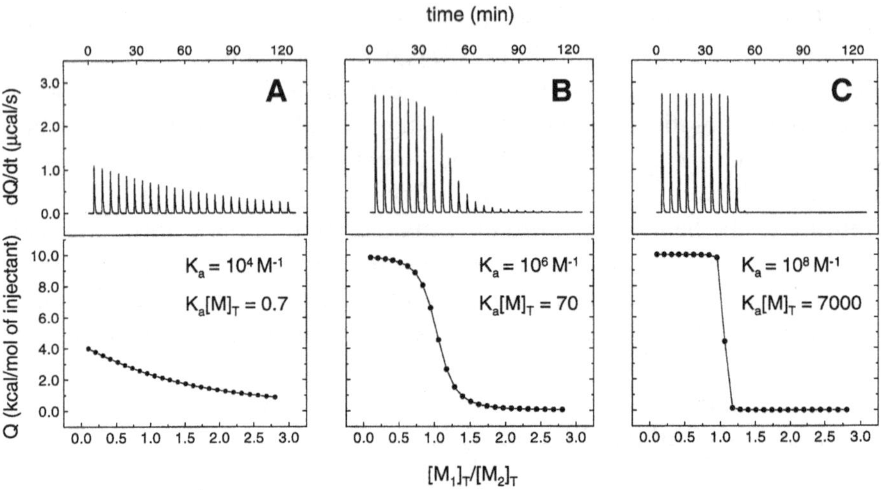

Fig. 2. Heterodimer formation. Illustration of the effect of the association constant value on the shape of a titration curve. The plots represent three titrations simulated using the same parameters (concentrations of reactants and association enthalpy), but different association constants. Low **(A)**, moderate **(B)** and high affinity **(C)** binding processes are shown. In order to obtain accurate estimates of the association constant, an intermediate case is desirable (center, $1 < c = [M_2]_T \times K_a < 1000$). When the parameter c is large enough, a good estimate of the enthalpy can be determined from the curve when the molar ratio reaches zero and the heat effect associated with dilution is used as a reference, limit given by $c/(c+1) \times \Delta H$ *(27)*.

The constraints dictated by the parameter c impose an experimental limitation, i.e., for very high association constants optimal concentrations are too small to be practical, and for very low association constants the concentrations may be prohibitively high (possibility of aggregation or economic consideration).

Sometimes it is possible to change the experimental conditions (temperature or pH, without compromising stability against aggregation or unfolding) in order to modify the association constant toward accessible experimental values ***(14–16)***. To extrapolate to the original conditions appropriate equations will need to be applied.

However, there is an extension of the ITC protocol designed to overcome such drawbacks. Without changing experimental conditions at all, displacement experiments implemented in ITC extend without limits the useful range for the association constant determination ***(7,17)***. Basically, a displacement

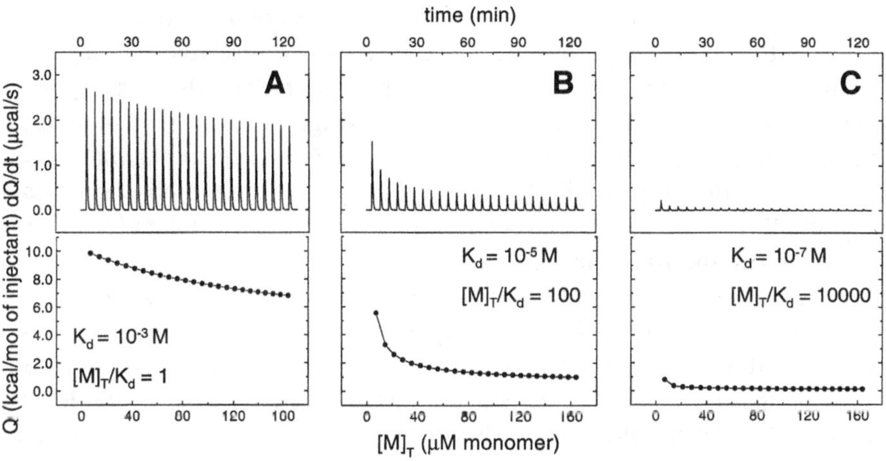

Fig. 3. Homodimer dissociation. Illustration of the effect of the dissociation constant value on the shape of a dissociation curve. The plots represent three titrations simulated using the same parameters (concentration of reactant and dissociation enthalpy), but different dissociation constants. Low **(C)**, moderate **(B)**, and high **(A)** dissociation constant processes are shown. In order to obtain accurate estimates of the association constant, an intermediate case is desirable (center, $10 < c = [M]_T/K_d < 10{,}000$).

experiment consists of a titration of the high-affinity ligand into a solution of the macromolecule prebound to a weaker ligand, therefore, decreasing the apparent affinity of the potent ligand. The thermodynamic parameters for the binding of the high-affinity ligand are calculated from the apparent binding parameters of the displacement titration and the known binding parameters for the weak ligand. This approach can be used to measure extremely high affinity binding processes as well as very low affinity binding reactions ***(18)***.

1.3.2. Simultaneous Determination of the Dissociation Constant and the Enthalpy of Dissociation

For self-associating systems that can be measured by ITC dissociation experiments, another dimensionless parameter can be used to define a practical limit for the accurate measurement of dissociation constants for a dimeric system. Now, $c = [M]_T/K_d$, which must lie between 10 and 10000, imposes a limit to the lowest and largest dissociation constant measurable at a given macromolecule concentration. This phenomenon is illustrated in **Fig. 3**.

As the c value increases, the transition from low to high total protein concentration in the cell is more abrupt. In the case of very low dissociation constants (macromolecule concentration much higher than the dissociation constant) most of the macromolecule in the syringe is forming dimers and, when diluted in the cell, the strength of the interactions determines that only a small fraction of the dimers will dissociate. For that reason the observed heats are very small. For very high dissociation constants (macromolecule concentration far below the dissociation constant), the weak interaction within the dimer makes it possible for a large fraction of the dimers to dissociate upon dilution; however, because the monomer–monomer interaction is weak, the population of dimers in the syringe will be small. In order to obtain accurate estimates of the dissociation constant, an intermediate case is desirable: the curvature of the plot should be enough for a reliable estimation of the dissociation constant and the size of the peaks should give an acceptable signal-to-noise ratio.

1.3.3. Determination of the Enthalpy of Binding

Binding enthalpy is usually determined from titrations in the optimal range of reactant concentrations. However, a greater accuracy can be achieved if enthalpy is measured by performing injections of titrant into a large excess of macromolecule (far above the dissociation constant). Therefore, although there will be no saturation, all the titrant injected will be fully associated giving a good measurement of the binding reaction heat. Blank experiments (*see* below) must be considered in order to eliminate the contribution of different phenomena (dilution of reactant, solution mixing, and other nonspecific effects) occurring simultaneously during the injections and to obtain the heat effect associated with the binding process.

1.3.4. Blank Experiment

In order to estimate the heat effect associated with dilution and mechanical phenomena (where binding does not take place), blank experiments can be performed. In this case, the experiment is similar to the one described in **Subheading 1.3.3.**: titrant is injected into a buffer solution without macromolecule. However, this may be not the ideal way to estimate such heat effects, because the degree of solvation of the titrant and the chemical composition would be different compared to the real titration with the other interacting macromolecule present. For that reason, some researchers usually consider the average effect of the last injection peaks as the reference heat effect. If during the experiment it is not possible to reach complete saturation, precluding such averaging, it is possible to include in the fitting function a term accounting for such dilution heat effects.

Blank experiments, also called "heat of dilution experiments," are required to measure binding enthalpies under excess macromolecule concentration, as mentioned previously.

1.3.5. Determination of Heat Capacity Change of Binding

The temperature derivative (at constant pressure) of the enthalpy, i.e., the heat capacity change in the process (association or dissociation), ΔC_P, is defined as

$$\Delta C_P = \left(\frac{\partial \Delta H}{\partial T} \right)_P \tag{8}$$

and can be determined performing the same experiment at several temperatures. The slope of an enthalpy versus temperature plot gives the heat capacity of the reaction.

ΔC_P has been shown to originate from surface dehydration upon binding or hydration upon dissociation and, to a less extent, from the difference in vibrational modes between the complex and the free species ***(19,20)***. It provides information about the nature of the interactions driving the binding, in addition to the temperature dependence of the thermodynamic functions (Gibbs energy, enthalpy, and entropy of binding).

1.3.6. Determination of Coupled Proton Transfer Process

Strong dependency of the thermodynamic parameters of the association or dissociation process on pH is an indication of the presence of proton exchange between the binary complex and the bulk solution. This is due to changes in the pK_a of some ionizable groups in any of the binding partners: the microenvironment of these groups is altered upon binding or dissociation, and so does their pK_a.

ITC is one of the more suitable techniques for the assessment of the existence of protonation/deprotonation processes coupled to binding. When a binding process is coupled to proton transfer between the bulk solution and the bound complex, the enthalpy of binding will depend on the ionization enthalpy of the buffer molecule, ΔH_{ion}:

$$\Delta H = \Delta H_0 + n_H \Delta H_{ion} \tag{9}$$

where ΔH_0 is the buffer-independent enthalpy of binding and n_H is the number of protons being exchanged. Therefore, by repeating the titration under the same conditions, but using several buffers with different ionization enthalpies, there will be a linear dependence of ΔH vs ΔH_{ion}, from which it is possible to estimate n_H (slope) and ΔH_0 (intercept with the y axis) ***(21,22)***. If n_H is zero, there is no net proton transfer; if n_H is positive, there is a protonation, i.e., a proton transfer from the solution to the complex; if n_H is negative, there is a deprotonation. After determination of these two parameters, n_H and ΔH_0, linkage equations can be used to couple the binding or dissociation process to the proton transfer process and allow the estimation of the thermodynamic parameters for the proton exchange event (enthalpy of ionization and pK_a for each ionizable group involved) ***(15,23,24)***. Analogous linkage equations couple the binding or dissociation reaction to the transfer of other ions [e.g., metals or salts ***(25,26)***].

2. Materials

2.1. Reagents and Supplies

1. Porcine pancreatic trypsin (Sigma cat. no. T0303).
2. Soybean trypsin inhibitor (Sigma cat. no. T9003).
3. Bovine pancreatic α-chymotrypsin (Sigma cat. no. C7762).
4. Dialysis membrane of 10,000 molecular weight cutoff (MWCO).
5. Filters (0.22 μm) with low protein binding properties.
6. Pyrex tubes for ITC syringe: 6 × 50 mm.
7. Buffer solutions.
   a. Potassium acetate/calcium chloride buffer: 25 m*M* potassium acetate, pH 4.5, 10 m*M* calcium chloride.
   b. Sodium acetate/sodium chloride buffer: 20 mM sodium phosphate, pH 3.9, 180 mM sodium chloride.

2.2. Isothermal Titration Calorimeter

The VP-ITC model calorimeter made by Microcal, Inc. (Northampton, MA) is the most advanced and sensitive isothermal titration calorimeter on the market today. The VP-ITC is simple to use, compact, and computer-controlled. The calorimeter unit consists of two cells, the reference cell and the sample cell embedded in an adiabatic chamber. The system holds the reference cell at a constant temperature. A constant power is applied to the sample cell in order to activate a feedback control mechanism whose purpose is to maintain the temperature difference between the two cells as close to zero as possible. As the reaction, initiated with the injection of titrant, occurs in the sample cell, the system adjusts the power applied to the sample cell up or down depending on whether it is an endothermic or exothermic reaction. This power change is recorded by the computer and corresponds to the signal seen in the characteristic form of a peak. The area under each peak corresponds to the heat released or absorbed during the reaction after each injection.

2.3. Biological Systems

Two different systems, but functionally and structurally closely related, have been selected to illustrate the two main types of association reactions that can be studied by ITC: formation of heterocomplexes (trypsin/trypsin inhibitor) and homocomplexes (chymotrypsin/chymotrypsin). These complexes have been characterized enzymatically, energetically, and structurally.

1. Porcine pancreatic trypsin, PPT (E.C. 3.4.21.4) is a 23.8 kDa protein consisting of a single polypeptide chain with six disulfide bonds. It hydrolyzes peptide bonds with basic side chains (Arg or Lys) on the carboxyl end of the bond.
2. Soybean trypsin inhibitor, STI, is a 20.0 kDa protein consisting of a single polypeptide chain with two disulfide bonds. Its predominant secondary structure is also β-sheet. It inhibits competitively the peptidase activity of trypsin.
3. Bovine pancreatic α-chymotrypsin, BP-α-CT (E.C. 3.4.21.1) is a 25.2 kDa protein consisting of three polypeptide chains interconnected and linked by two of the existing five disulfide bonds. It hydrolyzes peptide bonds with aromatic or large hydrophobic side chains (Tyr, Trp, Phe, Met, Leu) on the carboxyl end of the bond. This enzyme exhibits a monomer–dimer equilibrium in solution.

The experiments were performed under slightly acidic conditions to minimize the autocatalysis of both enzymes. Also, at high pH (approx 8) the interaction between trypsin and its inhibitor is so strong that the affinity is above the practical limits ($K_a \sim 10^{10}$–10^{11} M^{-1}).

3. Methods

3.1. Sample Preparation

1. PPT: Prepare PPT at 400 μ*M* in potassium acetate buffer (pH 4.5) dissolving the protein in buffer (approx 1 mL) and dialyze overnight at 4°C against 4 L of the same buffer. Filter the solution after dialysis. Measure concentration after dilution (1:20) in the same buffer using an extinction coefficient of 35,700 M^{-1} cm^{-1} at 280 nm.
2. STI: Prepare STI at 30 μ*M* in potassium acetate buffer (pH 4.5) dissolving the protein in buffer (approx 5 mL) and dialyze overnight at 4°C against 4 L of the same buffer. Filter the solution after dialysis. Measure concentration without dilution using an extinction coefficient of 18,200 M^{-1} cm^{-1} at 280 nm.
3. BP-α-CT: Prepare BP-α-CT at 200 μ*M* in potassium acetate buffer (pH 5.5) dissolving the protein in buffer (approx 5 mL) and dialyze overnight at 4°C against 4 L of the same buffer. Filter the solution after dialysis. Immediately prior to the experiment concentrate sample to about 1 m*M*. Measure concentration after dilution (1:50) in the same buffer using an extinction coefficient of 50,652 M^{-1} cm^{-1} at 280 nm.

3.2. Experimental Procedure

This protocol was designed for experiments to be performed using the VP-ITC titration calorimeter. For other calorimeters this method may be used as guidance. Other biological or instrumental systems may require different parameter values.

What follows is the step-by-step protocol for running the ITC experiment to study heterodimer formation. When studying homodimer dissociation, the only differences are in **step 5** (cell is filled with buffer solution only) and the data analysis (*see* **Subheading 3.4.**).

1. For experiments at 25°C and 30°C equilibrate the calorimeter at 24°C or 29°C. Typically the system is allowed to equilibrate one degree below the experimental temperature.
2. Prepare 2.2 mL of STI solution, 0.5 mL of PPT, glass tube (6 × 50 mm), and 10 mL of buffer solution. All solutions should be degassed for 15 min with a vacuum pump (*see* **Note 1**). A vacuum pump system is supplied with the VP-ITC.
3. Meanwhile, thoroughly clean the calorimetric reaction cell. The cell can be washed with a 5% Contrad 70™ solution or 0.1 *M* NaOH if necessary, and rinsed thoroughly with water.
4. Fill the reference cell with water. Usually, water should be replaced every week.
5. Rinse the reaction cell with buffer. Slowly load the STI solution into the reaction cell, and carefully remove bubbles. Concentration of sample should be determined again after loading because some dilution can take place due to residual buffer in the cell.
6. Fill the 250-μL injection syringe (*see* **Note 2**). Rinse the syringe tip with buffer or water and dry.
7. Carefully insert the injection syringe into the reaction cell. Avoid bending the needle or touching any surface with the needle tip.
8. Equilibrate the calorimeter at 25°C, or experimental temperature.
9. Set the running parameters for the experiment: number of injections (30), temperature (25°C), reference power (10 μcal/s), initial delay (180 s), concentration in syringe (approx 300 μ*M*), concentration in cell (approx 20 μ*M*), stirring speed (490 rpm), file name (*.itc), feedback mode (high), equilibration options (fast), comments, injection volume (first injection 3 μL, the rest 10 μL), duration (automatically set according to the injection volume), spacing between injections (400 s), filter (2 s) (*see* **Notes 3–7**).
10. Start the experiment (*see* **Note 8**). There will be thermal and mechanical equilibration stages. Initiate the injection sequence after a stable no-drift noise-free baseline (as seen in the 1 μcal/s scale) (*see* **Note 9**).
11. At the end of the experiment the system should be cleaned thoroughly with water and the syringe rinsed and dried.
12. Use the software provided by Microcal for data analysis according to the equations described previously (*see* **Note 10**). Results should be consistent with the information shown in **Fig. 4**.

3.3. Data Analysis

Equilibrium equations are very simple when using free concentration of reactants [eqs. (4) and (6)]. However, they become more complex when expressed in terms of total concentration of reactants, the known indepen-

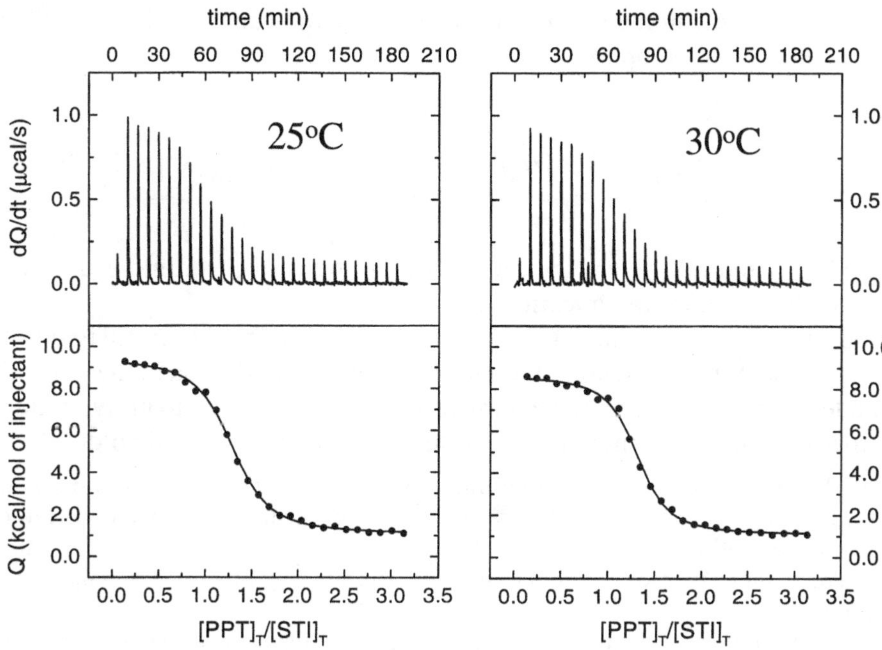

Fig. 4. Titrations of STI with PPT. The experiments were performed in 25 m*M* potassium acetate, pH 4.5, 10 m*M* calcium chloride, at 25°C (left) and 30°C (right). The concentrations of reactants are 21 μM STI (in cell) and 312 μM PPT (in syringe). The inhibitor was placed in the calorimetric cell due to its low solubility. The solid lines correspond to theoretical curves with $n = 1.26$, $K_a = 1.5\times10^6\ M^{-1}$, and $\Delta H_a = 8.4$ kcal/mol (left) and $n = 1.28$, $K_a = 2.2\times10^6\ M^{-1}$, and $\Delta H_a = 7.6$ kcal/mol (right).

dent variable in ITC. Data processing through transformation of the experimental data in order to linearize the equilibrium equations must be avoided, because it often provokes a systematic propagation of errors and uneven distribution of statistical weights over the binding curve. Nonlinear fitting procedures should be implemented to analyze directly the raw calorimetric data.

The analysis should be performed with the individual or differential heat plot, not with the cumulative or total heat plot. The analysis in terms of the individual heat is more convenient since it eliminates error propagation associated with cumulative data. *See* **Notes 11** and **12** for additional considerations.

3.3.1. Heterodimeric Interaction

The heat q_i released or absorbed during injection i is proportional to the change in concentration of binary complex after that injection and is given by eq. (5). The 1:1 stoichiometric model permits an explicit analytical solution for the concentration of complex ***(1,11)***:

$$[M_1M_2]_i = \frac{1+[M_2]_{T,i}K_a+K_a[M_1]_{T,i}-\sqrt{\left(1+[M_2]_{T,i}K_a+K_a[M_1]_{T,i}\right)^2-4[M_2]_{T,i}K_a^2[M_1]_{T,i}}}{2K_a} \tag{10}$$

where $[M_1]_{T,i}$ and $[M_2]_{T,i}$ are the total concentrations of macromolecules M_1 and M_2 in the cell after the injection i.

Even when assuming a model obeying 1:1 stoichiometry, it is useful to introduce a parameter n representing the number of binding sites, as in the general 1:n model. When estimating the parameters, n should be equal to unity, within the experimental error, otherwise the following statements would hold:

1. $n > 1$: There is an error in the determination of the reactants concentration (real titrating molecule M_1 concentration is lower and/or real titrated macromolecule M_2 concentration is higher).
   There is more than one binding site (specific binding or not) per macromolecule M_2.
2. $n < 1$: There is an error in the determination of the reactants concentration (real titrating macromolecule M_1 concentration is higher and/or real titrated macromolecule M_2 concentration is lower).
   There is less than one binding site per macromolecule M_2, that is, the sample is not chemically or conformationally homogeneous.

The 1:n stoichiometric model (n identical and independent binding sites) also permits an explicit analytical solution for the concentration of complex ***(1)***:

$$[M_1M_2]_i = \frac{1+n[M_2]_{T,i}K_a+K_a[M_1]_{T,i}-\sqrt{\left(1+n[M_2]_{T,i}K_a+K_a[M_1]_{T,i}\right)^2-4n[M_2]_{T,i}K_a^2[M_1]_{T,i}}}{2K_a} \tag{11}$$

Therefore, knowing the total concentrations $[M_1]_T$ and $[M_2]_T$ in the cell after the injections i and i–1, it is possible to evaluate the heat q_i according to eq. (5).

The dilution effect on the concentrations, due to the addition of the syringe solution, must be considered. The experiment proceeds at constant cell volume, i.e., when injecting a certain volume v, the same volume of liquid is lost from the cell. Therefore, the total concentrations of reactants after the injection i are given by:

$$\begin{aligned}[M_2]_{T,i} &= [M_2]_0\left(1-\frac{v}{V}\right)^i \\ [M_1]_{T,i} &= [M_1]_0\left[1-\left(1-\frac{v}{V}\right)^i\right]\end{aligned} \tag{12}$$

where $[M_2]_0$ is the initial concentration of macromolecule M_2 in the cell, $[M_1]_0$ is the concentration of macromolecule M_1 in the syringe, v is the injection volume, and V is the cell volume. Because the dilution of reactants lowers the effective concentration of macromolecules in the cell, it will also affect the heat signal measured in each injection. Therefore, a correction to eq. (5) is used:

$$q_i = V\Delta H_a \left| [M_1M_2]_i - [M_1M_2]_{i-1}\left(1 - \frac{v}{V}\right) \right| \quad (13)$$

Nonlinear fitting of q_i as a function of $[M_1]_{T,i}$ or $[M_1]_{T,i}/[M_2]_{T,i}$ (the so-called *molar ratio*), provides n, K_a, and ΔH_a as adjustable parameters. It is possible to avoid performing blank experiments to estimate the dilution heat effect and include in eq. (13) a constant (but floating in the fitting procedure) term, q_d, representing such contribution.

In general, models with different sets of binding sites require a numerical approach in which the model parameters (n_j, $K_{a,j}$ and ΔH_j, where j stands for each set of binding sites) are determined iteratively by numerical solution of the binding equations.

For the sample experiment of porcine pancreatic trypsin and soybean trypsin inhibitor, the data were analyzed using a model that allows the heat of dilution to be fitted simultaneously. The enthalpy was plotted against temperature and the Gibbs energy dependence on temperature is shown in **Fig. 5.** When using Origin to analyze ITC data, one should be aware that the baseline in the individual heat plot will need to be adjusted so the areas under each peak can be determined correctly.

3.3.2. Homodimeric Interaction

The heat q_i released or absorbed during injection i is proportional to the change in concentration of monomer after that injection is given by eq. (7). As already stated, the third term in the parenthesis accounts for a contribution to the increase in monomer concentration in the cell that does not contribute to the heat measured.

The monomer–dimer equilibrium model permits an explicit analytical solution for the monomer concentration ***(10)***:

$$[M]_i = \frac{K_d}{4}\left(\sqrt{1 + \frac{8[M]_{T,i}}{K_d}} - 1\right) \quad (14)$$

where $[M]_{T,i}$ is the total concentration of macromolecule M (per monomer) in the cell after the injection i. Therefore, knowing the total concentration $[M]_T$ in the cell after the injections i and i–1 it is possible to evaluate the heat q_i according to eq. (7).

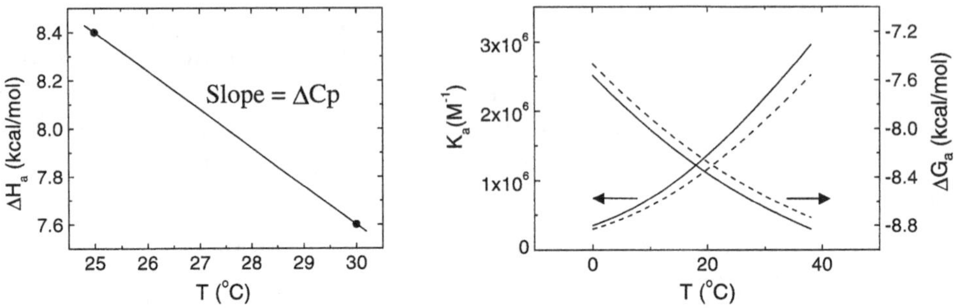

Fig. 5. (Left panel) Estimation of the change of heat capacity upon association. Once the enthalpy of association is determined at different temperatures (*see* **Fig. 4**), these values can be plotted versus temperature. The slope of this plot corresponds to the heat capacity change. In the case of the PPT/STI interaction, the heat capacity was determined to be –160 cal/Kmol. This negative value reflects the dehydration of molecular surfaces upon binding. (Right panel) Outline of the dependency of the association constant and the Gibbs energy of association on the temperature according to eq. (2) and $\Delta G_a(T) = \Delta H_a(T_0) + \Delta C_{P,a}(T - T_0) - T[\Delta S_a(T_0) + \Delta C_{P,a} \ln(T/T_0)]$, where T_0 is a given reference temperature at which the enthalpy and the affinity have been simultaneously measured. The curves were calculated using the thermodynamic parameters obtained at 25°C (dashed line) and 30°C (solid line). The estimation of the heat capacity change allows the determination of the thermodynamic parameters (affinity, Gibbs energy, enthalpy, and entropy) at any temperature.

Considering the dilution effect on the concentration, the total concentration of reactant after the injection i is given by

$$[M]_{T,i} = [M]_0 \left[1 - \left(1 - \frac{v}{V} \right)^i \right] \tag{15}$$

where $[M]_0$ is the concentration of macromolecule M in the syringe, v is the injection volume, and V is the cell volume. Again, a correction due to the dilution of reactants has to be included in eq. (7):

$$q_i = V\Delta H_d \left[[M]_1 - [M]_{i-1} - \left(1 - \frac{v}{V} \right) - F_0 [M]_0 \frac{v}{V} \right] \tag{16}$$

Nonlinear fitting of q_i as a function of $[M]_{T,i}$ provides K_d and ΔH_d as adjustable parameters. It is not possible to avoid performing blank experiments. To estimate the dilution heat effect a constant (but floating in the fitting procedure) term q_d representing such contribution is included in eq. (16).

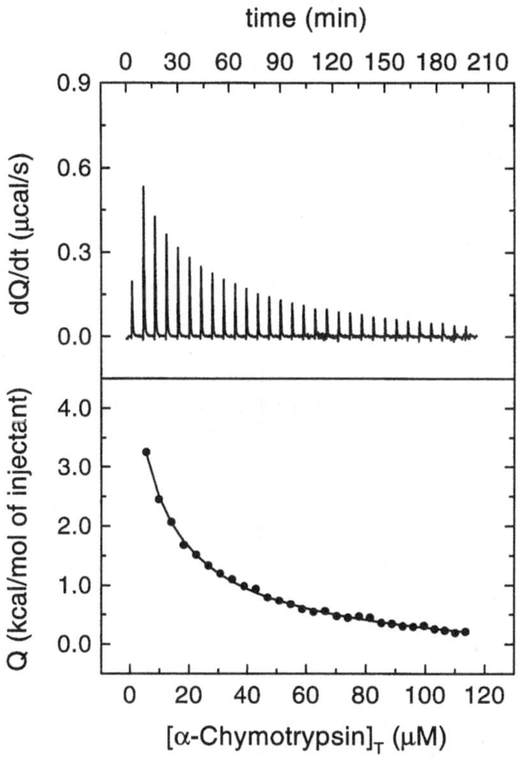

Fig. 6. Dissociation experiment of BP-α-CT. The experiment was performed in 20 m*M* sodium acetate, pH 3.9, 180 m*M* sodium chloride, at 25°C. The concentration of BP-α-CT in the syringe is 608 μM. The solid lines correspond to a theoretical curve with $K_d = 53 \times 10^{-6}\ M^{-1}$ and $\Delta H_d = 5.5$ kcal/mol of monomer. The value of the dissociation constant is in agreement with the one determined by ultracentrifugation *(28)*. However, in the ITC experiment the possibility of autocatalysis is minimized because it takes only 2 h and the experiment by ultracentrifugation lasted for 2 d. The short time required to do an experiment is one of the advantages of ITC compared to other binding techniques (e.g., dialysis or ultracentrifugation). Another experiment performed at 30°C (not shown) yielded a dissociation constant and a dissociation enthalpy of 63×10^{-6} M^{-1} and 6.2 kcal/mol. The estimated heat capacity of dissociation is 140 kcal/Kmol. A positive value reflects the solvent exposure of molecular surfaces upon dissociation.

For the sample experiment of bovine pancreatic α-chymotrypsin, the data were analyzed using a dimer dissociation model (**Fig. 6**). This fitting function is not provided by Microcal and will need to be implemented in Origin. *See* Origin manual for instruction on creating fitting functions.

4. Notes

1. When performing experiments below room temperature, samples should be kept on ice or one degree below the experimental temperature while degassing.
2. Although the syringe nominally indicates 250 μL, the available total injection volume is 309 μL. Incompletely filling the syringe would generate nonuniform injections.
3. If the reaction is expected to be so exothermic and/or the concentration of ligand is so high that the heat effect in the first injections exceeds –200 μcal, the reference power should be raised or the injection volume lowered in order to prevent the signal from going below zero.
4. In the equilibration options, compared to the other choices, fast equilibration is more convenient because it permits simultaneous thermal and mechanical equilibration and it allows the user to start the injection sequence manually whenever the baseline is good.
5. The first injection is usually erroneous because there is some mixing of the solutions inside and outside of the needle when inserting the injection syringe and/or during the time required to reach thermal and mechanical equilibration. Therefore, the first injection is usually set smaller in order to waste the minimum amount of ligand possible. In the data analysis, the first injection will be taken into consideration for the calculation of the total ligand concentration in the cell. However, its associated heat effect will not be included in the fitting procedure.
6. The feedback should be set to high feedback gain. Otherwise a performance drop will occur: given a heat effect, the peak corresponding to the injection would be smaller and broader (sensitivity loss and increase of the time for running an experiment).
7. The stirring speed should be set at 490 rpm when the molecule injected is of high molecular weight in order to ensure proper mixing. Occasionally, lower stirring speeds are required to avoid mechanical denaturation or destabilization of the macromolecule.
8. The experimental conditions and buffer solutions should carefully be determined for each system, depending on the characteristics (thermal stability, isoelectric point, solubility, propensity to interact with organic molecules, etc.) of the interacting macromolecules.
9. The time in between injections may need to be adjusted if the system is slow to reach equilibrium.
10. The data analysis software provided by Microcal (Origin) is straightforward and sufficient for simple systems (e.g., 1:1 binding reaction). However, for the analysis of more complex systems such as the dissociation reaction presented here, it is necessary to create additional fitting routines.
11. Given the values of enthalpy and association or dissociation constant it is possible to determine the optimal range of concentrations for reactants in an experiment. One is encouraged to carry out simulations in the experimental design stage.
12. The expressions for the correction due to the dilution after injection in the concentrations and the heat effect are slightly different than the ones used in the calorimeter manual.

Acknowledgments

Supported by grants from the National Institutes of Health (GM 57144) and the National Science Foundation (MCB9816661).

References

1. Freire, E., Mayorga, O. L., and Straume, M. (1990) Isothermal titration calorimetry. *Anal. Chem.* **62,** 950A–959A.
2. Doyle, M. L. (1997) Characterization of binding interactions by isothermal titration. *Curr. Opini. Biotechnol.* **8,** 31–35.
3. Jelessarov, I. and Bosshard, H. R. (1999) Isothermal titration calorimetry and differential scanning calorimetry as complementary tools to investigate the energetics of biomolecular recognition. *J. Mol. Recogn.* **12,** 3–18.
4. Leavitt, S. and Freire, E. (2001) Direct measurement of protein binding energetics by isothermal titration calorimetry. *Curr. Opin. Struct. Biol.* **11,** 560–566.
5. Velazquez-Campoy, A., Todd, M. J., and Freire, E. (2000) HIV-1 protease inhibitors: enthalpic versus entropic optimization of the binding affinity. *Biochemistry* **39,** 2201–2207.
6. Todd, M. J., Luque I., Velazquez-Campoy A., and Freire E. (2000) Thermodynamic basis of resistance to HIV-1 protease inhibition: calorimetric analysis of the V82F/I84V active site resistant mutant. *Biochemistry* **39,** 11,876–11,883.
7. Velazquez-Campoy, A., Kiso, Y., and Freire, E. (2001) The binding energetics of first- and second-generation HIV-1 protease inhibitors: Implications for drug design. *Arch. Biochem. Biophys.* **390,** 169–175.
8. Velazquez-Campoy, A. and Freire, E. (2001) Incorporating target heterogeneity in drug design. *J. Cell. Biochem.* **S37,** 82–88.
9. Ward, W. H. and Holdgate, G. A. (2001) Isothermal titration calorimetry in drug discovery. *Prog. Med. Chem.* **38,** 309–376.
10. Parker, M. H., Lunney, E. A., Ortwine, D. F., Pavlovsky, A. G., Humblet, C., and Brouillette, C. G. (1999) Analysis of the binding of hydroxamic acid and carboxylic acid inhibitors to the stromelysin-1 (matrix metalloproteinase-3) catalytic domain by isothermal titration calorimetry. *Biochemistry* **38,** 13,592–13,601.
11. Myszka, D. G, Sweet, R. W., Hensley, P., et al. (1997) Energetics of the HIV gp120-CD4 binding reaction. *Proc. Natl. Acad. Sci. USA* **97,** 9026–9031.
12. Burrows, S. D., Doyle, M. L., Murphy, K. P., et al. (1994) Determination of the monomer-dimer equilibrium of interleukin-8 reveals it is a monomer at physiological concentrations. *Biochemistry* **33,** 12,741–12,745.
13. Wiseman, T., Williston, S., Brandts, J. F., and Nin, L.N. (1989) Rapid measurement of binding constants and heats of binding using a new titration calorimeter. *Anal. Biochem.* **179,** 131–137.
14. Doyle, M. L., Louie, G. L., Dal Monte, P. R., and Sokoloski, T. D. (1995) Tight binding affinities determined from linkage to protons by titration calorimetry. *Meth. Enzymol.* **259,** 183–194.

15. Baker, B. M. and Murphy, K. P. (1996) Evaluation of linked protonation effects in protein binding using isothermal titration calorimetry. *Biophys. J.* **71,** 2049–2055.
16. Doyle, M. L. and Hensley, P. (1998) Tight ligand binding affinities determined from thermodynamic linkage to temperature by titration calorimetry. *Meth. Enzymol.* **295,** 88–99.
17. Sigurskjold, B. W. (2000) Exact analysis of competition ligand binding by displacement isothermal titration calorimetry. *Anal. Biochem.* **277,** 260–266.
18. Zhang, Y.-L. and Zhang, Z.-Y. (1998) Low-affinity binding determined by titration calorimetry using a high-affinity coupling ligand: a thermodynamic study of ligand binding to protein tyrosine phosphatase 1B. *Anal. Biochem.* **261,** 139–148.
19. Gomez, J., Hilser, V. J., and Freire, E. (1995) The heat capacity of proteins. *Proteins* **22,** 404–412.
20. Murphy, K. P. and Freire, E. (1992) Thermodynamics of structural stability and cooperative folding behavior in proteins. *Adv. Prot. Chem.* **43,** 313–361.
21. Biltonen, R. L. and Langerman, N. (1979) Microcalorimetry for biological chemistry: experimental design, data analysis and interpretation. *Meth. Enzymol.* **61,** 287–319.
22. Gomez, J. and Freire, E. (1995) Thermodynamic mapping of the inhibitor site of the aspartic protease endothiapepsin. *J. Mol. Biol.* **252,** 337–350.
23. Velazquez-Campoy, A., Luque, I., Todd, M. J., Milutinovich, M., Kiso, Y., and Freire, E. (2000) Thermodynamic dissection of the binding energetics of KNI-272, a potent HIV-1 protease inhibitor. *Prot. Sci.* **9,** 1801–1809.
24. Baker, B. M. and Murphy, K. P. (1997) Dissecting the energetics of a protein-protein interaction: the binding of ovomucoid third domain to elastase. *J. Mol. Biol.* **268,** 557–569.
25. Wyman, J. and Gill, S. J. (1990) *Binding and linkage: functional chemistry of biological macromolecules*. University Science Books, Mill Valley, CA.
26. Edgcomb, S. P., Baker, B. M., and Murphy, K. P. (2000) The energetics of phosphate binding to a protein complex. *Prot. Sci.* **9,** 927–933.
27. Indyk, L. and Fisher, H. F. (1998) Theoretical aspects of isothermal titration calorimetry. *Meth. Enzymol.* **295,** 350–364.
28. Patel, C. N., Noble, S. M., Weatherly, G. T., Tripathy, A., Winzor, D. J., and Pielak, G. J. (2002) Effects of molecular crowding by saccharides on α-chymotrypsin dimerization. *Prot. Sci.* **11,** 997–1003.

4

Circular Dichroism Analysis for Protein–Protein Interactions

Norma J. Greenfield

Abstract

Circular dichroism (CD) spectroscopy is a useful technique for studying protein–protein interactions in solution. CD in the far ultraviolet region (178–260 nm) arises from the amides of the protein backbone and is sensitive to the conformation of the protein. Thus CD can determine whether there are changes in the conformation of proteins when they interact. CD bands in the near ultraviolet (350–260 nm) and visible regions arise from aromatic and prosthetic groups. There are also changes in these regions when proteins bind to each other. Because CD is a quantitative technique, changes in CD spectra are directly proportional to the amount of the protein–protein complexes formed, and these changes can be used to estimate binding constants. Changes in the stability of the protein complexes as a function of temperature or added denaturants, compared to the isolated proteins, can also be used to determine binding constants.

Key Words

Conformation; secondary structure; binding constants; thermodynamics of folding.

1. Introduction

Circular dichroism (CD) is a valuable spectroscopic technique for studying protein–protein interactions in solution. There are many review articles in the literature on the theory and general applications of CD ***(1–8)***, so this chapter will focus on how CD can be used to follow protein–protein interactions. CD is best known as a method of determining the secondary structure of proteins in solution. It thus can be used to determine whether there are changes in protein conformation when proteins interact with each other. However, more importantly, CD can be used to estimate binding affinities for protein interactions. Because CD is a quantitative technique, changes in CD spectra are directly proportional to the amount of the protein–protein complexes formed, and these

From: *Methods in Molecular Biology, vol. 261: Protein–Protein Interactions: Methods and Protocols*
Edited by: H. Fu © Humana Press Inc., Totowa, NJ

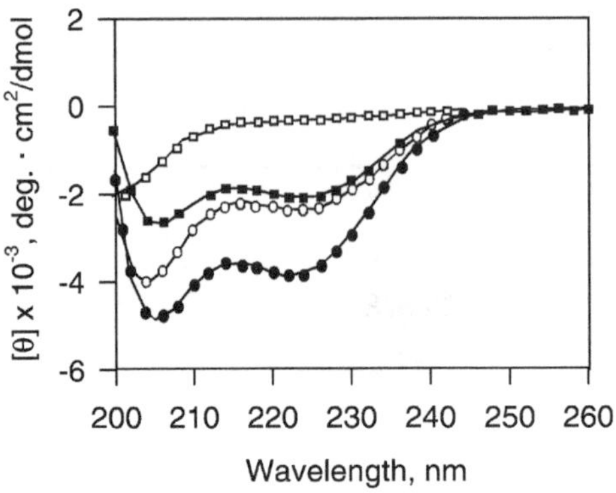

Fig. 1. Gain of signal for the Act-EF34-Zr7 complex (complex of peptides of titin, Zr7, and actinin, EF34) recorded by far-UV CD spectroscopy. The concentration of each component was 40 μ*M*, and the path length was 2 mm. The spectra of the isolated components, Zr7 (□), Act-EF34 (■), and the complex (●), are shown. The sum of the spectra of the two isolated components is also reported for comparison (○). Redrawn from data of Joseph et al. (9) with permission from American Chemical Society.

changes can be used to estimate binding constants. One can either follow changes in the "intrinsic" or "extrinsic" CD of the proteins. Changes in the thermodynamics of folding upon protein–protein interaction can also be used to determine binding constants.

1.1. Intrinsic Circular Dichroism

The intrinsic CD of a protein is the spectrum arising from the amide backbone and is sensitive to the secondary structure of the protein. The amide bond absorbs in the far ultraviolet (UV). Most commercial CD machines can collect far UV spectra between 260 and 178 nm. To a first approximation, increases in negative ellipticity at 222 and 208 nm and positive ellipticity at 293 nm usually indicate an increase in α-helical content, while increases in a single negative band near 218 nm and a positive band at 295 nm indicate an increase in β-structure. **Figure 1** illustrates the increase in α-helix that occurs when a peptide from titin, a Z-repeat, interacts with a fragment of the C-terminal domain of α-actinin ***(9)***. **Figure 2** illustrates the increase in β-structure when two cell cycle control proteins, Arf and Hdm2 interact ***(10)***. Both of these proteins are random in the absence of the interactions.

1.2. Extrinsic CD

Extrinsic CD bands are due to aromatic chromophores or to prosthetic groups such as hemes or flavins. **Figure 3** illustrates a change in the aromatic spec-

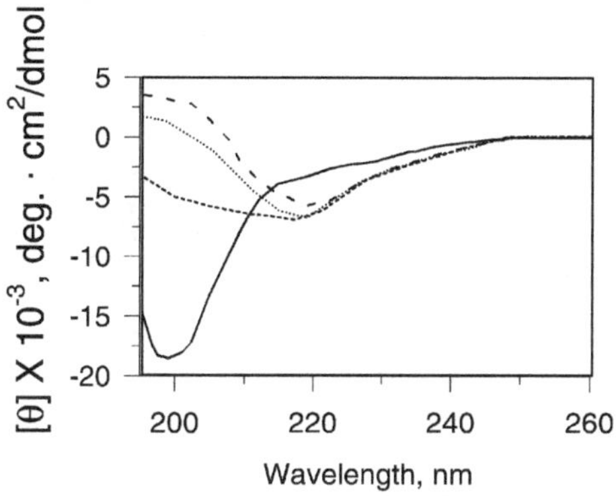

Fig. 2 The complex of two cell cycle control proteins, Arf and Hdm2, are comprised of β-strands and are thermally stable. CD spectra for $Hdm2_{210-304}$ alone (——) and mixed with 1(- - -), 1.7 (·), and 2 (– – – –) molar equivalents of mArfN37. Redrawn from data of Bothner et al. ***(10)***, © 2001 Elsevier, with permission.

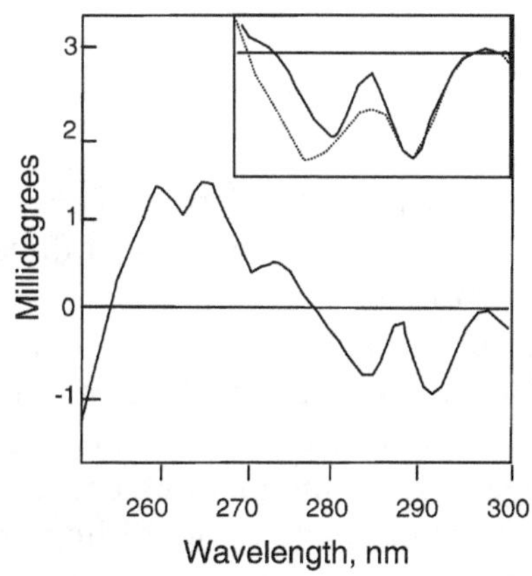

Fig. 3. Near UV CD of the catalytic subunit of adenosine cyclic 5'-monophospate–dependent protein kinase. (Inset) CD spectrum of the catalytic subunit from 275–300 nm (——) enzyme alone; (- - -) enzyme plus 250 μ*M* Kemptide, a synthetic peptide substrate (buffer plus Kemptide base line subtracted). Redrawn from data from Reed and Kinzel ***(11)*** with permission from American Chemical Society.

trum of the catalytic domain of a cAMP-dependent protein kinase when it binds to an artificial peptide substrate ***(11)***, and **Fig. 4** illustrates the change in the heme spectrum of cytochrome c when it binds to cytochrome c oxidase ***(12)***.

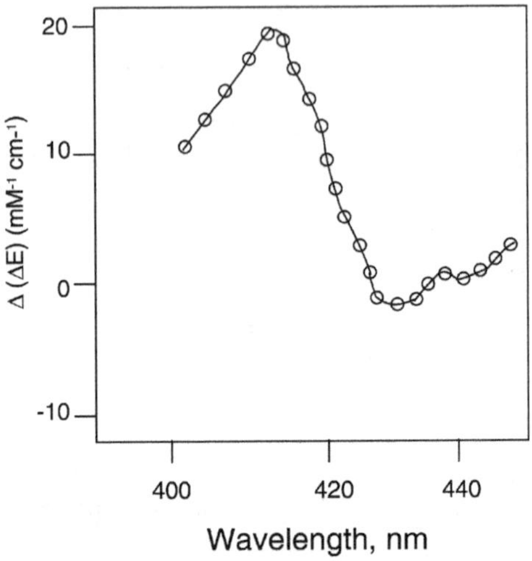

Fig. 4. CD difference spectra of cytochrome c upon binding to cytochrome c oxidase. Redrawn form data of Michel et al. ***(12)*** (note that the ellipticity in deg. $cm^2/dmol = 3300\ \Delta E$).

1.3. Binding Constants

There are two main methods that can be used to extract binding (or dissociation) constants from CD data. If formation of the complex changes the protein secondary structure, or causes changes in the environment of optically active aromatic or prosthetic groups of one or both of the interacting proteins, one protein can be titrated by the other. The change in CD upon binding is directly proportional to the amount of complex formed, and thus the binding constant can be directly determined. However, even if there are no conformational changes, if the interactions cause a change in stability, one can determine the thermodynamics of folding of the individual and complexed proteins and determine the binding constant by assuming that all of the change in free energy of folding is due to the binding. The binding constant can then be determined using the relationship $k = \exp(-\Delta\Delta G/RT)$, where ΔΔG is the difference in free energy of folding of the mixture versus the unmixed proteins, R is the gas constant, T is the temperature, and k is the association constant.

2. Materials

2.1. Proteins

CD depends on concentration of chromophores, rather than protein concentrations *per se*. For measurements of protein folding, usually performed by following changes in ellipticity at 222 nm as a function of temperature or denaturants, protein concentrations may range from approx 0.005 to 1 mg/mL. For concentrations ranging from 0.005 to 0.02 mg/mL, one needs 2–3 mL of solution, as the samples will be studied in 1 cm cuvets, with a total volume capacity of 3.5 mL. For higher concentrations, one would use 0.1 or 0.2 cm cuvets with volume capacities of 0.3 and 0.6 mL, respectively.

For measurements in the near UV, for following changes in aromatic groups of proteins, or visible regions, e.g., following the CD of prosthetic groups such as flavins or hemes, higher concentrations of 1–2 mg/mL in a 1 cm cell are generally necessary, depending on the number and asymmetry of the chromophores.

2.2. Buffers

For the most precise estimates of protein secondary structure from CD spectra, it is important to collect data to low wavelengths, 178 nm has been recommended ***(3,13,14)***. In this case one should use buffers with very low absorption, such as 10 m*M* potassium fluoride. If salt is necessary to stabilize the protein, potassium phosphate, 100 m*M* or K_2SO_4, 50 m*M*, are preferred as they are relatively transparent. NaCl should be avoided. However, for most purposes an adequate estimate of protein conformation can be obtained using data collected between 260 and 200 nm. For these measurements one can use buffers such as phosphate-buffered saline (PBS), or Tris-HCl, 20 m*M*, or Hepes, 2 m*M* containing low concentrations of EDTA or EGTA, e.g., 1 m*M*, and dithiotreitol, 1 m*M*, and up to 500 m*M* NaCl if necessary.

For folding experiments, e.g., monitored at 222 nm, almost any buffers can be utilized provided that the total absorption of the sample (protein plus buffer) does not exceed 1 at the wavelength of interest.

2.3. Cuvets

Measurements are usually performed in 0.1 or 1 cm cuvets. Both standard rectangular and cylindrical cuvets are available for CD machines, and the choice depends on the instrument utilized. If one needs to use samples with high concentrations or buffers with high absorbance, 0.01 cm cuvets are available.

2.4. Instrumentation

The following is a list of CD machines that are currently commercially available. All are adequate to perform thermal denaturations and titration experiments. They are listed alphabetically:

1. Applied Photophysics Ltd, 203/205 Kingston Road, Leatherhead, Surrey KT22 7PB United Kingdom.

2. Aviv Instruments from Proterion Corporation, One Possumtown Road, Piscataway, NJ 08854, USA.
3. Jasco Inc, 8649 Commerce Drive, Easton, MD 21601 USA.
4. Olis, Inc. (On-Line Instrument Systems, Inc.), 130 Conway Drive, Suites A & B, Bogart, GA 30622 USA.

3. Methods

3.1. Analyzing Changes in Conformation Accompanying Protein–Protein Interactions

While changes at 222 or 218 nm can give an estimate of increases in α-helical or β-structure content, more precise estimates of changes in structure accompanying protein–protein can be made utilizing computer programs. Computer programs for analyzing CD spectra are not available commercially, but many are available from their authors without charge (*see* **Note 1** for sources of software).

If changes in conformation are large, it is possible to analyze the secondary structure of the complex directly. There are several very good programs for estimating protein secondary structure if the concentration of the complex is known. Recommended programs include CDNN and SELCON (*see* below). If the changes are small, relative to the spectrum of the mixture, it is best to subtract the spectra of the unmixed components from the spectrum of the complex, and analyze the difference spectrum. In this case one needs to use a program, such as nonconstrained multiple linear regression (MLR) that does not require knowledge of the concentration of residues involved in the structural change.

The use of the various methods to determine protein conformation from CD spectra have been reviewed with a detailed comparison of the advantages and disadvantages of each ***(5)***, so only a brief description of each method is given below.

3.1.1. Constrained Multilinear Regression (LINCOMB)

Constrained linear regression deconvolutes CD spectra into component spectra characteristic of specific secondary structures ***(15)*** by fitting the data to a set of reference spectra using the methods of least squares. Standards include spectra of polypeptides with known conformations ***(15,16)***, or reference spectra deconvoluted from the spectra of proteins with known conformations by the method of least squares ***(17,18)***, or the convex constraint algorithm ***(19)***. The sum of the fractions of each component must equal 1. The LINCOMB program of Perczel et al. ***(20)*** uses constrained linear regression to determine the percent α-helix, β-sheet, and β-turns contributing to a spectrum. The method uses invariant standards, so it is useful for quantifying changes in each type of secondary structure when one protein is titrated with another.

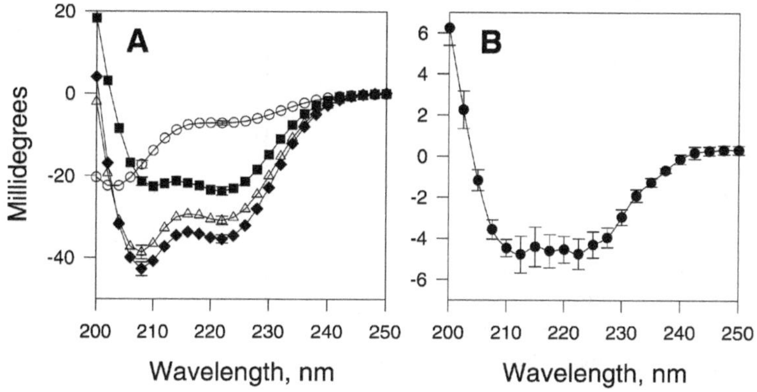

Fig. 5. **(A)** CD spectra of (○) E-Tmod1-130, (■) AcTM1bZip, (Δ) the sum of the spectra of E-Tmod1-130 and AcTM1bZip, and (◆) the spectrum of the mixture E-$Tmod_{1\text{-}130}$ and AcTM1bZip. Data are the average and standard error of four to seven measurements. **(B)** Difference between the spectrum of the mixture of the components and the sum of the unmixed components in **A**. Data are the average and standard error of four measurements. Redrawn from data of Greenfield and Fowler use with permission from Biophysical Society ***(21)***.

3.1.2. Nonconstrained Multilinear Regression (MLR)

Nonconstrained multilinear regression (MLR) ***(16)*** analyzes the CD spectrum of a protein by fitting the spectrum to those of standards using the method of least squares. The sum of the components is not constrained to equal 100%. The method is independent of the intensity of the spectrum, so it is the only method that can be used if the protein concentration isnt known precisely. The method is useful to analyze difference spectra, obtained when the sum of the spectra of the unmixed components is subtracted from the spectrum of the mixture, because one doesnt need to know the number of residues that change conformation. The method gives relatively poor estimates of β-sheets and β-turns, however, and overestimates β-sheets if data collected below 200 nm are utilized.

Figure 5 illustrates the change in CD that occurs when a fragment of E-tropomodulin, $Tmod_{1\text{-}130}$, binds to a peptide containing the N-terminus of a nonmuscle tropomyosin ***(21)***, TM1bZip. **Figure 5A** shows the raw spectra of the two unmixed peptides, the sum of the spectra of the unmixed peptides and the spectrum of the complex. **Figure 5B** shows the difference spectrum of the complex minus the sum of unmixed peptides. The difference spectrum exhibits increases in ellipticity at 208, 218, and 222 nm suggesting that complex formation causes increases in both α-helix and β-sheet. Analyzing the

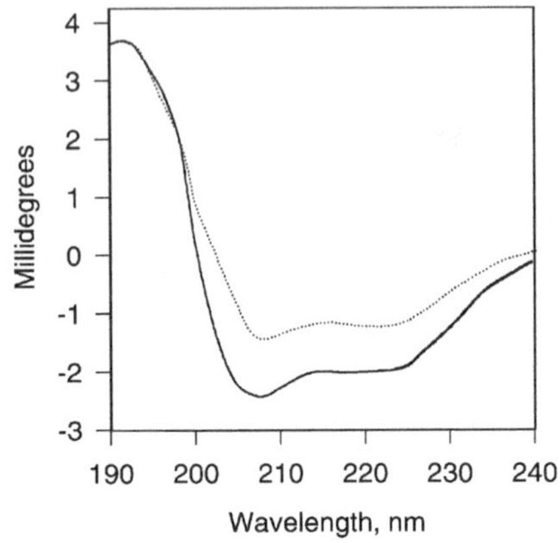

Fig. 6. Far-UV CD of the catalytic subunit of adenosine cyclic 5'-monophospate–dependent protein kinase. CD spectrum of the catalytic subunit from 190 to 240 nm (——) with enzyme alone; (- - -) enzyme plus 250 μ*M* Kemptide, a synthetic peptide substrate (buffer plus Kemptide base line subtracted). Redrawn from data of Reed and Kinzel ***(11)*** with permission from American Chemical Society.

data using the MLR program, with peptides with known conformations as references, suggests that the CD change is 58% due to an increase in α-helix formation, 41% due to an increase in β-sheet, and 1% due to an increase in turns.

3.1.3. Ridge Regression (CONTIN)

CONTIN, developed by Provencher and Glökner ***(22)***, fits the CD of unknown proteins by a linear combination of the spectra of a large database of proteins with known conformations. In this method, the contribution of each reference spectrum is kept small, unless it contributes to a good agreement between the theoretical best fit curve and the raw data. The method usually gives very good estimates of β-sheets and β-turns and gives good fits with polypeptides. **Figure 6** illustrates the use of CONTIN ***(11)*** to determine the changes in secondary structure of the catalytic subunit of adenosine cyclic 5'-monophosphate–dependent protein kinase when it bound a synthetic substrate peptide. In the absence of peptide, the protein was largely helical, but a significant proportion became β-structure when the peptide bound. CONTIN is not recommended for quantifying changes in conformation when one protein is titrated with another, because it may use a different data set of proteins for every fit.

3.1.4. Singular Value Decomposition (SVD, VARSLC, SELCON)

The application of singular value decomposition to determine the secondary structure of proteins by comparing their spectra to the spectra of a large number of proteins with known structure was first developed by Hennessey and Johnson *(14)*. This method extracts basis curves, with unique shapes, from a set of spectra of proteins with known structures. Each basis curve is then related to a mixture of secondary structures, which are then used to analyze the conformation of unknown proteins. The method is excellent for estimating the α-helical content of proteins, but is poor for sheets and turns unless the data are collected to very low wavelengths, at least 184 nm.

Several newer programs have improved the method of singular value decomposition by selecting references that have spectra that closely match the protein of interest. They include VARSLC *(13)* and SELCON *(23–27)*. These give good estimates of β-sheet and β-turns in proteins and work with data collected only to 200 nm but give poor fits to polypeptides with high β-structure content, because reference sets for such compounds are not in their database.

3.1.5. Neural Network Programs (CDNN and K2D)

A neural network is an artificial intelligence program that can detect patterns and correlations in data. Two widely used programs are CDNN *(28)* and K2D *(29)*. A neural network is first trained using a set of known proteins so that the input of the CD at each wavelength results in the output of the correct secondary structure. The trained network is then used to analyze unknown proteins. The method works very well and the fits seem to be relatively independent of the wavelength range that is analyzed.

3.1.6. Convex Constraint Algorithm (CCA)

The CCA algorithm *(19,20,30)* deconvolutes a set of spectra into basis spectra that when recombined generate the entire data set with a minimum deviation between the original data set and the reconstructed curves. It is very useful for determining whether there are intermediate states in thermal- and denaturant-induced unfolding. The method has also be used to estimate protein conformation, but is poorer than least squares, SVD, or neural net analyses.

3.1.7. Recommendations

1. For determination of globular protein conformation in solution and evaluating the conformation of protein–protein complexes: SELCON, CDNN, CONTIN, and K2D.
2. For evaluating conformational changes upon protein–protein interactions when the change in ellipticity is small compared to the sum of the ellipticities of the unmixed components: MLR.
3. For deconvoluting sets of CD spectra, e.g., to follow the effects of binding, denaturants, ligands, or changes in temperature on protein and peptide conformation: the CCA algorithm.

Table 1
Secondary Structure Analysis of Fragments and Cleaved and Uncleaved Thioredoxin

Protein	Method	α-Helix	β-Sheet	Turns	Others
N-terminal	K2D	6.0	29.0	[a]	65.0
Fragment	Contin	4.0	14.0	10.0	72.0
	MLR[b]	2.5	28.7	10.3	58.5
C-terminal	K2D	2.0	16.0	[a]	82.0
Fragment	CONTIN	3.0	11.0	1.0	86.0
	MLR[b]	0.0	15.7	0.0	84.3
Peptide complex	K2D	30.0	20.0	[a]	50.0
	CONTIN	20.0	40.0	7.0	34.0
	MLR[c]	22.8	23.7	9.2	21.8; 22.5[d]
	SELCON	24.7	21.5	16.9	22.5[c]
Thioredoxin	K2D	27.0	25.0	[a]	48.0
	CONTIN	18.0	37.0	8.0	37.0
	MLR[c]	31.8	22.9	7.2	16.9; 21.2[c]
	SELCON	28.4	24.8	24.4	21.4
	X-ray data[e]	40.0	28.25	14.0	17.8
	X-ray data[f,h]	34.3	28.1	11.3	27.3
	NMR data[g,h]	36.1	25.9	9.2	28.8

Adapted from *ref.* ***31***.
[a] Undetermined.
[b] Polypeptide reference ***(16)***.
[c] Protein reference ***(19)***.
[d] Estimate of aromatic and disulfide bond contributions.
[e] Structure calculations of Georgescu et al. ***(31)*** based on molecule *A* of the PDB file 2trx ***(40)***.
[f] Based on molecule *A* of the PDB file 2trx ***(40)***.
[g] Based on PDB file 1XOA ***(41)***.
[h] Percentage of each structure calculated using MolMol ***(42)***.

Table 1 compares the results when CONTIN, SELCON, K2D, and MLR programs were use to analyze the spectra **(Fig. 7)** of N- and C-terminal fragments of thioredoxin, the binary complex of the two thioredoxin fragments and the intact protein ***(31)***. All of the methods gave reasonable fits to the data. [Note that the amount of helix in the intact protein appeared to be underestimated by all of the programs. This occurred because approx 50% of the helix in thioredoxin is 3–10 helix rather than α-helix ***(32)***.]

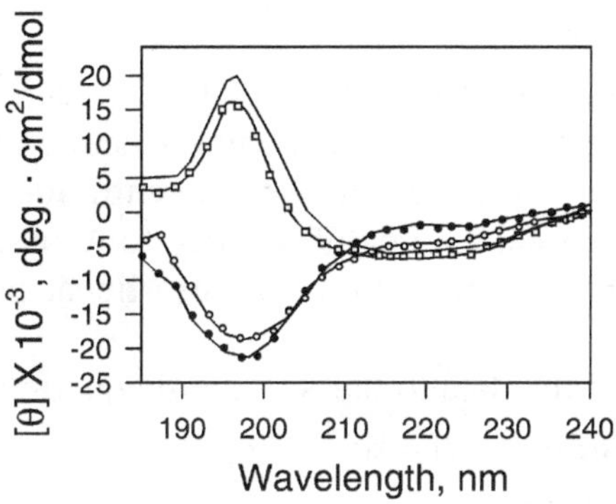

Fig. 7. Circular dichroism of an artificial heterodimer of cleaved peptides of thioredoxin from *E. coli* versus the intact protein. The protein concentration was 20 μ*M* in potassium phosphate. **(A)** Far-UV CD spectra of the isolated N- and C-fragments (○), N; (●), C; (□), cleaved and (—) uncleaved Trx. Redrawn from data of Georgescu et at. ***(31)***, used with permission from American Chemical Society.

3.2. Determining Protein–Protein Association Constants Using CD Data

3.2.1. Determination of Binding Constants by Direct Titration of One Protein by Another at Constant Temperature (Isothermal Titrations)

When two proteins interact, there are often changes in either intrinsic or extrinsic CD. CD is a quantitative technique and the change in CD is directly proportional to the amount of complex formed. The change can then be used to determine the association or dissociation constant of the complex.

If one titrates A with B to form a complex AB, the association constant is k.

$$k = [AB]/[A][B] = [AB]/([A_0] - [AB])([B_0] - [AB]) \tag{1}$$

where $[A_0]$ and $[B_0]$ are the initial concentrations of A and B and $[AB]$ is the amount of complex formed. The saturation fraction is s:

$$s = [AB]/[A_0] \tag{2}$$

If there is a change is the intrinsic or extrinsic CD accompanying binding, the change, $\Delta[\theta]$, is proportional to the amount of complex formed:

$$\Delta[\theta] = \varepsilon[AB] \tag{3}$$

where ε is the proportionality constant. When all of protein A is bound by protein B,

$$\Delta[\theta]_{max} = \varepsilon[A_0] \text{ and } s = \Delta[\theta]/\Delta[\theta]_{max} = \Delta[\theta]/\varepsilon[A_0] \tag{4}$$

3.2.1.1. Weak Binding Constants

If the binding of one protein to the other is relatively weak, under conditions where the dissociation constant, $1/k$, of the complex is more than 100-fold the concentration of A, one can assume that the concentration of unbound B is approximately equal to the total amount of B added to A. In this case it is easy to determine the binding constant and one can use linear plots to determine $\Delta[\theta]$max and K_d.

For example one can fit the data to the Scatchard equation ***(33)***, where

$$\Delta[\theta]/[A_0][B] = k[\varepsilon - (\Delta[\theta]/[A_0])] \tag{5}$$

One plots $\Delta[\theta]/[B]$ vs $\Delta[\theta]$. The y intercept $=k\varepsilon[A_0]$ and the slope $= -k$.

The Scatchard equation ***(33)*** described above works well for studying interactions where there is one or multiple equivalent binding sites of one protein for another. However, when there are multiple sites, and the binding is cooperative, it is much more difficult to estimate binding constants from spectroscopic data, and usually only apparent constants can be estimated. The Hill equation ***(34)*** may be used to fit the change in CD as a function of titrant in a phenomenological fashion to give an estimate of the apparent binding affinity. Here the first protein A is titrated with a second protein B, where $[B_0]$ is the total concentration of added protein giving an observed change in ellipticity $\Delta[\theta]$:

$$\Delta[\theta] = [\Delta[\theta]_{max}\, k^h[B_0]^h/\, (1 + k^h[B_0]^h)] + C \tag{6}$$

where k is the binding constant, $[\theta]_{max}$ is the maximal changes in ellipticity upon protein–protein interaction, h is a constant describing the apparent cooperativity of the interaction, and C is a constant correcting for baseline offsets, When $h = 1$, the binding is noncooperative.

3.2.1.2. Tight Binding Constants

If the dissociation constant is close to the concentrations of the proteins being studied, one cannot assume that the free concentration of the added protein B is the same as the total protein. One must correct the concentration of the added B protein for the amount that is bound. In this case, $[B] = [B_0] - [AB]$. From eq. (4), the fraction of protein A which is bound equals

$$[A]_{bound} = [A_0]\, (\Delta[\theta]/\Delta[\theta]_{max}) \tag{7}$$

Therefore the free concentration of B, $[B]$ at any added concentration of B_0

$$[B] = [B_0] - [A_0]\, (\Delta[\theta]/[A_0]_{max}) \tag{8}$$

Solving for the change of circular dichroism $\Delta[\theta]$ at any added concentration of protein B, $[B_0]$,

$$\begin{aligned}\Delta[\theta] = [A_0]_{max}\, \{&((1 + k[B_0] + k[A_0]) / 2k[A_0]) \\ &- (((1 + k[B_0] + k[A_0]) / 2\, k[A_0])^2 - [B_0]/[A_0])^{1/2}\}\end{aligned} \tag{9}$$

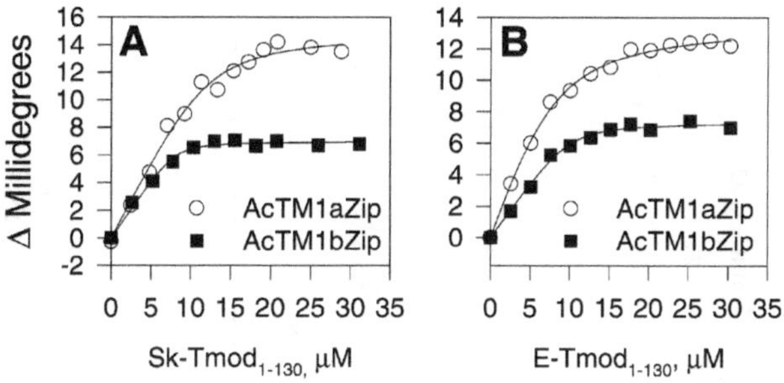

Fig. 8. Increase in ellipticity at 222 nm, 30°C, when *A*, fragments of SK- and E-tropomodulin, Sk-$Tmod_{1–130}$ and B, E-$Tmod_{1–130}$ bind to chimeric peptides containing the N-termini of long and short tropomyosins, AcTM1aZip (O—O) and AcTM1bZip (■—■) ***(21)***. 10 nmol of the TMZip peptides in 0.5 mL of 100 m*M* NaCl, 10 m*M* sodium phosphate, pH 6.5 in 2 mm path-length cells, were titrated with concentrated solutions of the Tmod constructs. The ellipticity changes were corrected for the ellipticity of the Tmod peptides alone and for dilution. The data were fit to the equation:

$$\Delta[\theta] = \Delta[\theta]_{max}\{((1+k[B_0] + k[A_0]) / 2\,k[A_0]) - (((1+k[B_0] + k[A_0])/ 2k[A_0])^2 - [B_0]/[A_0])^{1/2}\}$$

using the Levenburg–Marquardt algoritm ***(38)*** implemented in the commercial program SigmaPlot to yield dissociation constants of $SkTMod_{1–130}$ for AcTM1aZip, 0.9 ± 0.7 μ*M*, and for AcTM1bZip, 0.3 ± 0.2 μ*M*, respectively, and dissociation constants of $ETmod_{1–130}$ for AcTM1aZip, 2.8 ± 1.2 μ*M*, and for AcTM1bZip, 0.5 ± 0.3 μ*M*, respectively. Redrawn from data of Greenfield and Fowler ***(21)***, used with permission from Biophysical Society.

as described by Engel ***(35)***, who originally developed the equation for determining binding constants of enzyme–ligand complexes from fluorescence titrations.

Equation (9) can be fit by many available commercial curve fitting programs, e.g., SigmaPlot or Microcal Origin among others, that find the best fits to nonlinear equations. To use these programs one inputs starting values of the parameters to be fit, here ΔCD_{max} and *k*, and has the program find the values of A_{max} and *k* that best fit the data.

Figure 8 illustrates the binding of a fragments of E- and Sk-tropomodulin to peptides containing the N-terminus of two tropomyosin isoforms, where the binding data were fit to eq. (9).

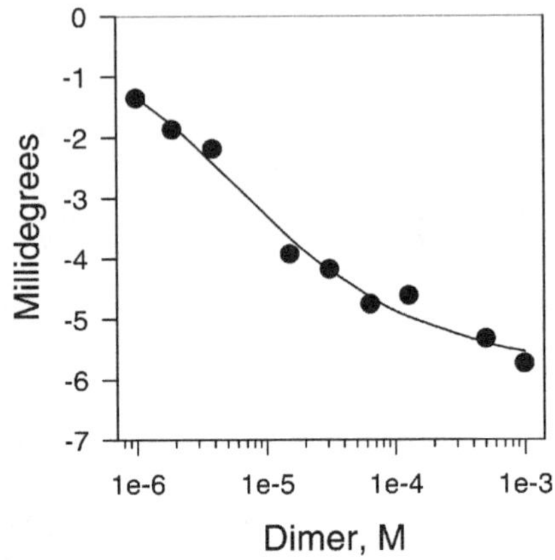

Fig. 9. Association of two complementary fragments of the immunoglobulin binding domain B1 of streptococcal protein G, PGB1(1–40) and PGB1(41–56), evaluated by molecular ellipticity at 222 nm. Equimolar concentrations of the fragments were dissolved in 50 m*M* phosphate buffer (pH 5.5) at 298 K. The closed circles and solid line show observed data and best fitted curve fitted to the equation $[\theta]_{obs} = [\theta]_F + ([\theta]_U - [\theta]_F)\,[(-k_D + (k_D^2 + 4P_t k_D)^{1/2})/(2P_t)]$ to determine the apparent dissociation constant of the two fragments (Kapp = 9×10^{-6} *M*). Redrawn from data of Honda et al. ***(36)*** used with permission from American Chemical Society.

3.2.1.3. Estimating Binding Constants When Association of the Proteins is Coupled With Folding

When two unfolded or partially unfolded proteins fold upon binding to each other, a simple method of determining the association constants is to follow the mean residue or molar ellipticity of a 1:1 mixture of the proteins as a function of concentration. In this case

$$[\theta]_{obs} = [\theta]_F + ([\theta]_U - [\theta]_F)\,[(-k_D + (k_D^2 + 4P_t k_D)^{1/2})/(2P_t)] \quad (10)$$

where k is the association constant of the two proteins, $[\theta]_{obs}$ is the ellipticity of the mixture of the proteins, at any total concentration, P_t, $[\theta]_F$ is the ellipticity of the fully folded complex, and $[\theta]_U$ is the sum of the ellipticity of the fully unfolded individual proteins ***(36)***. **Figure 9** illustrates the use of this equation to determine the dissociation constant of two fragments of the immunoglobulin binding domain B1 of streptococcal protein G ***(36)***.

3.2.2. Estimating Protein–Protein Association Constants From the Thermodynamics of Folding

One can determine the thermodynamics of folding of protein–protein complexes by measuring the unfolding of the complex either by thermal denaturation or by chemical denaturation using guanidine-HCl or urea. In some cases, the proteins are only folded when associated. In these cases, determining the ΔG of unfolding directly gives the dissociation constants of the protein–protein complexes. In other cases, the proteins are both folded, but complex formation increases their stability. In this case, the free energy of binding can be determined by subtracting the free energies of unfolding of the individual proteins from the free energy of unfolding of the complex. The difference, ΔΔG, can then be use to calculate the dissociation constant of the complex.

3.2.2.1. Chemical-Induced Unfolding

In the simplest case, two proteins *A* and *B* form a complex with 1:1 stoichiometry. The isolated proteins are unfolded, but fold to form the complex. When this complex is treated with denaturants, the complex (native form) dissociates to give the disordered proteins. At any concentration of denaturant $[D]$ the association constant is:

$$k = [AB]/[A][B] = [P_t]\alpha/([P_t]\,(1-\alpha))^2 \qquad (11)$$

where $[P_t]$ is the total concentration of complex and α is the fraction folded. The fraction folded expressed in terms of the protein concentration and association constant is

$$\alpha = \{2P_t^2k + P_t - ((-\,2P_t^2k - P_t)^2 - 4P_t^4k^2)^{1/2}\}/(\,2P_t^2k) \qquad (12)$$

If one assumes that the binding constant is equal to the folding constant, then ΔG can be calculated from the equation:

$$\Delta G = -\,RT\,\mathrm{Ink} \qquad (13)$$

It has been shown for many systems that the ΔG of folding of a protein is linearly dependent on the concentration of denaturant, see, e.g., Santoro and Bolen ***(37)***:

$$\Delta G_D = \Delta G_0 + m[D] \qquad (14)$$

where ΔG_D is the free energy of folding in the presence of detergent, ΔG_0 is the free energy of folding in the absence of the detergent, $[D]$ is the concentration of detergent, and m is the slope.

To calculate the binding constant, one then studies the denaturation of the protein complex with denaturant using several different protein concentrations by following the changes in molar or mean residue ellipticity $\Delta[\theta]_{obs}$ as a function of denaturant concentration. The fraction folded can be calculated at any concentration of detergent using the equation

$$\alpha = ([\theta]_{obs} - ([\theta]_U - m_U[D]))\,/\,(([\theta]_F - m_F[D]) - ([\theta]_U - m_U[D])) \qquad (15)$$

where $[\theta]_U$ and $[\theta]_F$ are the ellipticities of the fully unfolded and folded complexes, respectively, and m_U and m_F are any necessary linear baseline correc-

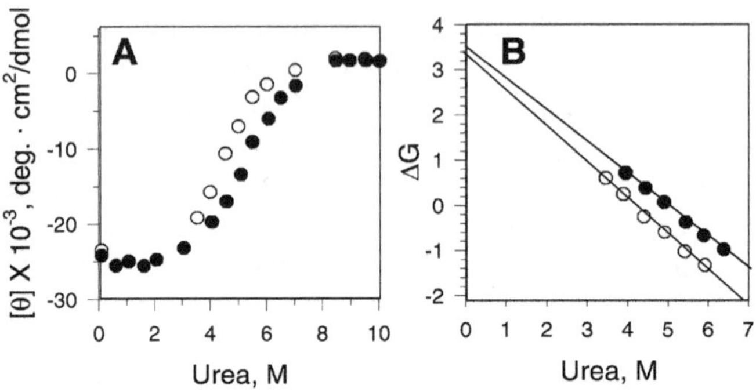

Fig. 10. Urea denaturation curves to the interactions of two designed heterodimeric coiled-coil peptides. **(A)** Acid-KeN-Base-EgC (●) and Acid-KeN-Base-EgN (○). **(B)** The data yield apparent ΔG of unfolding values of 3.4 ± 0.3 and 3.2 ± 0.3 kcal/mol, respectively. Data of McClain et al. ***(38)***, © 2001 Elsevier, used with permission.

tions for the change in ellipticity of the fully unfolded and folded peptides with denaturant observed before and after the folding transition.

The *k* at any detergent concentration can be calculated using eqs. (11) and (15) and ΔG can be calculated from eq. (13) and can be plotted as a function of total detergent concentration. By linear extrapolation ΔG can be determined at zero detergent, and the dissociation constant K_D for the protein–protein complex can then be calculated by rearranging eq. (13), $K_D = \exp(-\Delta G_0/RT)$.

Figure 10 illustrates determination of the free energy of dissociation of two designed heterodimeric coiled coils using urea denaturation ***(38)***. (Note that in this example the peptides were cross-linked by disulfide bonds and unfolded as monomers.) The raw data in **Fig. 10A** was used to determine the free energy of unfolding at each urea concentration. The linear curve of free energy as a function of urea in **Fig. 10B** was extrapolated to zero urea concentration to determine the free energy of dissociation in the absence of urea.

It is also possible to fit the unfolding data as a function of denaturant concentration directly (*see* **Note 2**).

3.2.2.2. Thermally Induced Unfolding

A change in either the intrinsic or extrinsic CD as a function of temperature can be used to determine the vant Hoff enthalpy of folding, because changes in ellipticity are directly proportional to the changes in concentration of the native and denatured forms. When a protein is fully folded, $[\theta]_{obs} = [\theta]_F$, and when it is fully unfolded, $[\theta]_{obs} = [\theta]_U$. In the simplest case of a monomeric protein, the equilibrium constant of folding, k = folded/unfolded. If we define α as the fraction folded at a given temperature T, then

$$k = \alpha/(1 - \alpha) \tag{16}$$

$$\Delta G = - RT \text{ Ink} \tag{17}$$

where ΔG is the free energy of folding and R is the gas constant. From the Gibbs–Helmholtz equation:

$$\Delta G = \Delta H + (T - T_M)\, \Delta C_p - T\, [\Delta S + \Delta C_p\, (\ln\, (T/T_M) \tag{18}$$

where ΔH is the vant Hoff enthalpy of folding and ΔS is the entropy of folding, T_M is the observed midpoint of the thermal transition, and ΔC_p is the change in heat capacity for the transition. At the T_M, $k = 1$, therefore $\Delta G = 0$ and $\Delta S = \Delta H / T_M$. Rearranging these equations one obtains:

$$\Delta G = \Delta H\, (1 - T - T_M) - \Delta C_p[(T - T_M]) + T\, (\ln\, (T/T_M))] \tag{19}$$

$$k = \exp\, (- G/RT) \tag{20}$$

$$\alpha = k/(1 + k) \tag{21}$$

$$[\theta]_{obs} = ([\theta]_F - [\theta]_U)\, \alpha + [\theta]_U \tag{22}$$

To calculate the values of ΔH and T_M that best describe the folding curve, initial values of ΔH, ΔC_p, T_M, $[\theta]_F$, and $[\theta]_U$ are estimated, and eq. (22) is fitted to the experimentally observed values of the change in ellipticity as a function of temperature, by a curve fitting routine such as the Levenberg–Marquardt algorithm ***(39)***. Because at $k = 1$, $\Delta G = 0$, the entropy of folding can therefore be calculated using eq. (18), where $\Delta S = \Delta H/T_M$. The free energy of folding at any other temperature can then be calculated using eq. (19). Similar equations can be used to estimate the thermodynamics of folding of proteins and peptides that undergo folded multimer to unfolded monomer transitions (*see* **Note 3**). The association constant k of protein–protein complexes can be obtained by subtracting the free energies of folding of the sum of the uncomplexed proteins from the free energy of folding of the complex and then $k = \exp(-\Delta\Delta G/nRT)$.

Figure 11 illustrates the increased stability when a fragment of E-tropomodulin binds to a peptide containing the N-terminus of an isoform of α-tropomyosin. The change in stability was used to estimate a dissociation constant of 0.23 ± 0.15 μM.

4. Notes

1. Sources of programs for determining secondary structure from CD data: CD software can be downloaded from the following sites:
   a. Programs adapted for DOS including SELCON, VARSLC, K2D, MLR, LINCOMB CONTIN, and CCA algorithm : http://www2.umdnj.edu/cdrwjweb.

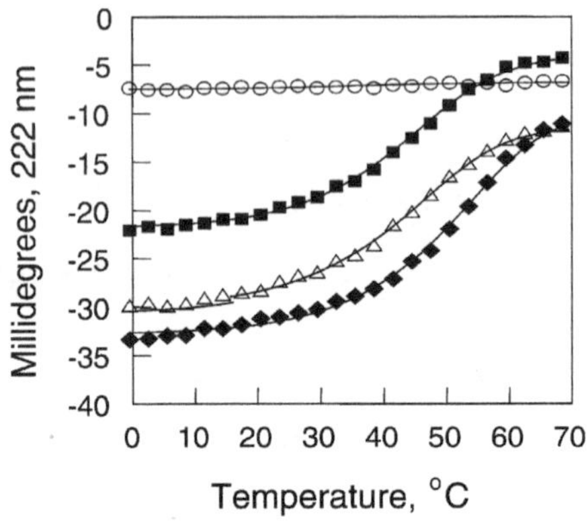

Fig. 11. Thermal denaturation curves for (○) E-Tmod$_{1-130}$, (■) AcTM1bZip, (Δ) the sum of the spectra of E-Tmod$_{1-130}$ and AcTM1bZip, and (◆) the spectrum of the mixture E-Tmod$_{1-130}$ and AcTM1bZip. The lines are the best fits of the data assuming two-state transitions between folded and unfolded conformations. The data were used to estimate dissociation constants for the complex of the peptides of 0.23 ± 0.15 μ*M*. Redrawn from data in Greenfield and Fowler ***(21)***, used with permission from Biophysical Society.

b. Programs including MLR, VARSLC, SELCON and CONTIN in a Windows environment: ftp://ftp.ibcp.fr/pub/C_DICROISM/.
c. The latest versions of the individual programs as described in the text:
CCA algorithm: http://www2.chem.elte.hu/protein/programs/cca/
CDNN: http://bioinformatik.biochemtech.uni-halle.de/cd_spec/index.html
CONTIN: http://s-provencher.com/index.shtml
K2D: http://www.embl-heidelberg.de/%7Eandrade/k2d.html
SELCON: http://lamar.colostate.edu/~sreeram/CD
VARSLC: http://oregonstate.edu/dept/biochem/faculty/johnsondownload.html

2. Protocol for fitting the unfolding of a heterodimeric protein complex by detergents from CD data. One can fit the change in ellipticity $[\theta]_{obs}$ as a function of detergent concentration $[D]$ to obtain the free energy of folding of a protein complex and the binding affinity in the absence of detergent. The variables are $[D]$ and $[\theta]_{obs}$. The known constant is the protein concentration $[P_t]$. The parameters to be fit are ΔG_0, m, $[\theta]_F$, $[\theta]_U$ and m_U and m_F, where ΔG_0 is the free energy of folding in the absence of detergent, m is the constant relating the linear change in free energy to the detergent concentration, $[\theta]_F$ and $[\theta]_U$ are the ellipticities of the fully folded and unfolded protein complexes, respectively, and m_F and m_U are the slopes describing any linear changes in ellipticity of the folded and unfolded

proteins before and after the folding transition. The equations that are used in the curve fitting procedure are:

$$\Delta G = \Delta G_0 + m[D]$$
$$k = \exp(-\Delta G/1.987T)$$
$$\alpha = \{P_t^2 k + P_t - ((-2P_t^2 k - P_t)^2 - 4P_t^4 k^2)^{1/2}\} / (2P_t^2 k)$$
$$[\theta]_{obs} = \alpha([\theta]_F - m_F[D] - [\theta]_U - m_U[D]) + [\theta]_U$$

In these equations, ΔG is the free energy of folding in the presence of detergent, k is the folding/association constant, and α is the fraction folded. The difference between the calculated change in ellipticity and the observed change in ellipticity are minimized to give the best values of the fitting parameters using a nonlinear least squares curve fitting program. Such curve fitting algorithms are usually included in most commercial graphics plotting programs. If one knows the values of $[\theta]_F$, $[\theta]_U$, m_U and m_F, they may be input as constants rather than parameters to be determined. Once ΔG_0 is determined, the binding constant is calculated using the equation $k = \exp(-\Delta G/RT)$.

3. Protocols for fitting CD data to determine the thermodynamics of folding of monomers, dimers, heterodimers, trimers, and heterotrimers from changes in CD as a function of temperature. Unless otherwise noted, in all of these equations, ΔG is the Gibbs free energy of folding, ΔH is the vant Hoff enthalpy of folding, ΔC_p is the change in heat capacity going from the folded to the unfolded form, T is the absolute temperature, kelvin, T_M is the temperature where $k = 1$, $[\theta]_{obs}$ is the observed ellipticity at any temperature T, α is the fraction folded, $[\theta]_F$ is the ellipticity of the fully folded protein and $[\theta]_U$ is the ellipticity of the unfolded protein, and p is the concentration of the folded complex. The parameters to be fit are ΔH, ΔC_p, T_M, $[\theta]_F$, and $[\theta]_U$. One inputs the protein concentration $[P_t]$ as a constant. For complicated fits, to reduce the number of parameters, ΔC_p may be set to 0 and the values of $[\theta]_F$ and $[\theta]_U$ may be set as constants rather than parameters to be minimized if they are known.

A. Monomers

$$\Delta G = \Delta H(1 - T/T_M) - \Delta C_p\{(T_M - T) + T[\ln(T/T_M)]\}$$
$$k = \exp(-\Delta G/RT)$$
$$\alpha = k/(1 + k)$$
$$[\Delta]_{obs} = ([\theta]_F - [\theta]_U)\,\alpha + [\theta]_U$$

B. Homodimers

$$\Delta G = \Delta H + [\Delta C_p(T - T_m)] - T[(\Delta H/T_m) + \{\Delta C_p[\ln(T/T_m)]\}$$
$$k = \exp[-\Delta G/(1.987T)]$$
$$a = 4kP_t^2$$
$$b = -8kP_t^2 - P_t$$
$$c = 4kP_t^2$$
$$\alpha = [-b - (b^2 - 4ac)^{1/2}]/2a$$
$$[\theta]_{obs} = \alpha([\theta]_F - [\theta]_U) + [\theta]_U$$

C. Heterodimers

Here the equations are set up in terms of unfolding and P_t is the total concentration of protein in terms of monomers and the ΔH and ΔCp values are those of unfolding rather than folding:

$$\Delta G = \Delta H(1 - T/T_m) + C_p\{T - T_m - T[\ln(T/T_m)]\} - 1.987T\ln(P_t/4)$$
$$k_D = \exp(-\Delta G/1.987T)$$
$$a = P_t$$
$$b = -2P_t - 2k$$
$$c = P_t$$
$$\alpha = [-b - (b^2 - 4ac)^{1/2}]/2a$$
$$[\theta]_{obs} = \alpha([\theta]_F - [\theta]_U) + [\theta]_U$$

This expression is set up to evaluate the apparent T_m at $\alpha = 0.5$. To evaluate it at $k = 1$, the equation to calculate the free energy of folding simplifies to:

$$\Delta G = \Delta H(1 - T/T_m) + C_p\{T - T_m - T[\ln(T/T_m)]\}$$

Note that the equations give exactly the same T_M and enthalpy of folding if the unfolding data were analyzed as if the complex was a homodimer, above.

D. Homotrimers

$$k = \exp\{(\Delta H/1.987T)[(T/T_m) - 1] - \ln[0.75P_t^2]\}$$
$$z = 3P_t^2$$
$$q = [(3kz) + 1]/(kz)$$
$$e = q - 3$$
$$d = q - 3$$
$$A = \{[(-1)(d/2)] + [(d^2/4) + (e^3/27)]^{(1/2)}\}^{(1/3)}$$
$$B = (-1)\{(d/2) + [(d^2/4) + (e^3/27)]^{(1/2)}\}^{(1/3)}$$
$$X = A + B$$
$$\alpha = X + 1$$
$$[\theta]_{obs} = \alpha([\theta]_F - [\theta]_U) + [\theta]_U$$

Note that in this treatment ΔC_p is set at 0 and T_M is the observed midpoint of the unfolding transition where $\alpha = 0.5$.

E. Heterotrimers

$$k = \exp\{[(\Delta H/1.987T)(T/T_M - 1)] - \ln(P_t^2)\}$$
$$Q = -1/(12\,P_t^2k)$$
$$R = -1/(8\,P_t^2k)$$
$$A = -\{R + [(R^2) - (Q^3)]^{1/2}\}^{(1/3)}$$
$$B = Q/A$$

$$y = A + B$$
$$\alpha = 1 - y$$
$$[\theta]_{obs} = \alpha\,([\theta]_F - [\theta]_U) + [\theta]_U$$

Note that in this treatment ΔC_p is set at 0 and T_m is the observed midpoint of the unfolding transition where $\alpha = 0.5$. The equations give the same enthalpy and T_M of folding as the equations for analyzing the unfolding of a homotrimer, above *(21)*.

References

1. Adler, A. J., Greenfield, N. J., and Fasman, G. D. (1973) Circular dichroism and optical rotatory dispersion of proteins and polypeptides. *Meth. Enzymol.* **27,** 675–735.
2. Johnson, W. C., Jr. (1988) Secondary structure of proteins through circular dichroism spectroscopy. *Annu. Rev. Biophys. Biophys. Chem.* **17,** 145–166.
3. Johnson, W. C., Jr. (1990) Protein secondary structure and circular dichroism: a practical guide. *Proteins* **7,** 205–214.
4. Woody, R. W. (1995) Circular dichroism. *Meth. Enzymol.* **246,** 34–71.
5. Greenfield, N. J. (1996) Methods to estimate the conformation of proteins and polypeptides from circular dichroism data. *Anal. Biochem.* **235,** 1–10.
6. Venyaminov, S. Y. and Yang, J. T. (1996) Determination of protein secondary structure, in *Circular Dichroism and the Conformational Analysis of Biomolecules* (Fasman, G. D., ed.), Plenum Press, New York and London.
7. Greenfield, N. J. (1999) Applications of circular dichroism in protein and peptide analysis. *Trends Anal. Chem.* **18,** 236–244.
8. Greenfield, N. J. (2000) Biomacromolecular applications of circular dichroism and ORD, in *Encylopeidea of Spectroscopy and Spectrometry* (Lindon, J. C., Tranter, G. E., and Holmes, J. L., eds.), Academic Press, London, pp 117–130.
9. Joseph, C., Stier, G., OBrien, R., et al. (2001) A structural characterization of the interactions between titin Z-repeats and the alpha-actinin C-terminal domain. *Biochemistry* **40,** 4957–4965.
10. Bothner, B., Lewis, W. S., DiGiammarino, E. L., Weber, J. D., Bothner, S. J., and Kriwacki, R. W. (2001) Defining the molecular basis of Arf and Hdm2 interactions. *J. Mol. Biol.* **314,** 263–277.
11. Reed, J. and Kinzel, V. (1984) Near- and far-ultraviolet circular dichroism of the catalytic subunit of adenosine cyclic 5'-monophosphate dependent protein kinase. *Biochemistry* **23,** 1357–1362.
12. Michel, B., Amanda E.I. Proudfoot, A.E.I., Wallace, C.J.A., and Bosshard, H. R. (1989) The cytochrome c oxidase-cytochrome complex: spectroscopic analysis of conformational changes in the protein-protein interaction domain. *Biochemistry* **28,** 456–462:
13. Manavalan, P. and Johnson, W. C., Jr. (1987) Variable selection method improves the prediction of protein secondary structure from circular dichroism spectra. *Anal. Biochem.* **167,** 76–85.

14. Hennessey, J. P., Jr. and Johnson, W. C., Jr. (1981) Information content in the circular dichroism of proteins. *Biochemistry* **20,** 1085–1094.
15. Greenfield, N. and Fasman, G. D. (1969) Computed circular dichroism spectra for the evaluation of protein conformation. *Biochemistry* **8,** 4108–4116.
16. Brahms, S. and Brahms, J. (1980) Determination of protein secondary structure in solution by vacuum ultraviolet circular dichroism. *J. Mol. Biol.* **138,** 149–178.
17. Saxena, V. P. and Wetlaufer, D. B. (1971) A new basis for interpreting the circular dichroic spectra of proteins. *Proc. Natl. Acad. Sci. USA* **68,** 969–972.
18. Chang, C. T., Wu, C.-S. C., and Yang, J. T. (1978) Circular dichroic analysis of protein conformation: inclusion of the b-turns. *Anal. Biochem.* **91,** 13–31.
19. Perczel, A., Park, K., and Fasman, G. D. (1992) Analysis of the circular dichroism spectrum of proteins using the convex constraint algorithm: a practical guide. *Anal. Biochem.* **203,** 83–93.
20. Perczel, A., Park, K., and Fasman, G. D. (1992) Deconvolution of the circular dichroism spectra of proteins: the circular dichroism spectra of the antiparallel beta-sheet in proteins. *Proteins* **13,** 57–69.
21. Greenfield, N. J. and Fowler, V. M. (2002) Tropomyosin requires an intact N-terminal coiled coil to interact with tropomodulin. *Biophys. J.* **82,** 2580–2591.
22. Provencher, S. W., and Glökner, J. (1981) Estimation of globular protein secondary structure from circular dichroism. *Biochemistry* **20,** 33–37.
23. Sreerama, N. and Woody, R. W. (2000) Estimation of protein secondary structure from circular dichroism spectra: comparison of CONTIN, SELCON, and CDSSTR methods with an expanded reference set. *Anal. Biochem.* **287,** 252–260.
24. Sreerama, N. and Woody, R. W. (1994) Poly(pro)II helices in globular proteins: identification and circular dichroic analysis. *Biochemistry* **33,** 10,022–10,025.
25. Sreerama, N. and Woody, R. W. (1994) Protein secondary structure from circular dichroism spectroscopy. Combining variable selection principle and cluster analysis with neural network, ridge regression and self-consistent methods. *J. Mol. Biol.* **242,** 497–507.
26. Sreerama, N. and Woody, R. W. (1993) A self-consistent method for the analysis of protein secondary structure from circular dichroism. *Anal. Biochem.* **209,** 32–44.
27. Sreerama, N., Venyaminov, S. Y., and Woody, R. W. (2001) Analysis of protein circular dichroism spectra based on the tertiary structure classification. *Anal. Biochem.* **299,** 271–274.
28. Bohm, G., Muhr, R., and Jaenicke, R. (1992) Quantitative analysis of protein far UV circular dichroism spectra by neural networks. *Protein Eng.* **5,** 191–195.
29. Andrade, M. A., Chacon, P., Merelo, J. J., and Moran, F. (1993) Evaluation of secondary structure of proteins from UV circular dichroism spectra using an unsupervised learning neural network. *Protein Eng.* **6,** 383–390.
30. Perczel, A., Hollosi, M., Tusnady, G., and Fasman, G. D. (1991) Convex constraint analysis: a natural deconvolution of circular dichroism curves of proteins. *Protein Eng.* **4,** 669–679.

31. Georgescu, R. E., Braswell, E. H., Zhu, D., and Tasayco, M. L. (1999) Energetics of assembling an artificial heterodimer with an alpha/beta motif: cleaved versus uncleaved Escherichia coli thioredoxin. *Biochemistry* **38,** 13,355–13,366.
32. Katti, S. K., LeMaster, D. M., and Eklund, H. (1990) Crystal structure of thioredoxin from *Escherichia coli* at 1.68 Å resolution. *J. Mol. Biol.* **212,** 167–184.
33. Scatchard, G. (1949) The attractions of proteins for small molecules and ions. *Ann. NY Acad. Sci.* **51,** 660–672.
34. Hill, A. V. (1910) The possible effects of the aggregation of the molecules of haemoglobin on its dissociation curves. *J. Physiol. (Lond.)* **40,** iv–vii.
35. Engel, G. (1974) Estimation of binding parameters of enzyme-ligand complex from fluorometric data by a curve fitting procedure: seryl-tRNA synthetase-tRNA Ser complex. *Anal. Biochem.* **61,** 184–191.
36. Honda, S., Kobayashi, N., Munekata, E., and Uedaira, H. (1999) Fragment reconstitution of a small protein: folding energetics of the reconstituted immunoglobulin binding domain B1 of streptococcal protein G. *Biochemistry* **38,** 1203–1213.
37. Santoro, M. M. and Bolen, D. W. (1988) Unfolding free energy changes determined by the linear extrapolation method. 1. Unfolding of phenylmethanesulfonyl alpha-chymotrypsin using different denaturants. *Biochemistry* **27,** 8063–8068.
38. McClain, D. L., Binfet, J. P., and Oakley, M. G. (2001) Evaluation of the energetic contribution of interhelical Coulombic interactions for coiled coil helix orientation specificity. *J. Mol. Biol.* **313,** 371–383.
39. Marquardt, D. W. (1963) An algorithm for the estimation of non-linear parameters. *J. Soc. Indust. Appl. Math.* **11,** 431–441.
40. Holmgren, A., Soderberg, B. O., Eklund, H., and Branden, C. I. (1975) Three-dimensional structure of Escherichia coli thioredoxin-S2 to 2.8 A resolution. *Proc. Natl. Acad. Sci. USA* **72,** 2305–2309.
41. Jeng, M. F., Campbell, A. P., Begley, T., et al. (1994) High-resolution solution structures of oxidized and reduced Escherichia coli thioredoxin. *Structure* **2,** 853–868.
42. Koradi, R., Billeter, M., and Wüthrich, K. (1996) MOLMOL: a program for display and analysis of macromolecular structures. *J. Mol. Graph.* **14,** 51–55, 29–32.

5

Protein–Protein Interaction Analysis by Nuclear Magnetic Resonance Spectroscopy

Guanghua Gao, Jason G. Williams, and Sharon L. Campbell

Abstract

Nuclear magnetic resonance (NMR) is a powerful technique to study protein–protein interactions in solution. Various methods have been developed and applied successfully for locating binding sites on proteins. The application of NMR chemical-shift perturbation to map the protein–protein interfaces is described in this chapter, providing a practical guideline that can be followed to carry out the experiments.

Key Words

Nuclear magnetic resonance (NMR); protein–protein interaction; chemical-shift perturbation; heteronuclear single quantum correlation (HSQC); paramagnetic relaxation.

1. Introduction

Over the last two decades there has been an increasing interest in studying protein–protein interactions in solution by nuclear magnetic resonance (NMR) *(1)*. Such noncovalent interactions are of crucial importance in many biological molecular recognition processes. NMR and X-ray crystallography are the primary techniques to determine high-resolution protein structures. Although structure determination of proteins by NMR is currently limited to proteins with molecular weights up to approx 70 kDa, NMR can provide both specific and qualitative information on much larger systems. For example, NMR can (1) give specific information on physical properties of individual functional groups such as ionization states, pK_a, and hydrogen bonds; (2) provide site-specific information on backbone and side-chain dynamic motions in solution; and, most relevant to this chapter, (3) be used to identify contacts between individual atoms of a protein and its binding partner, as well as to study the kinetics of ligand binding.

From: *Methods in Molecular Biology, vol. 261: Protein–Protein Interactions: Methods and Protocols*
Edited by: H. Fu © Humana Press Inc., Totowa, NJ

One clear advantage of NMR is the ability to conduct studies on biomolecules under conditions that are close to the natural physiological environment. Moreover, samples can be investigated under various conditions, such as different pH and/or temperatures and varied ligands. NMR is also well suited to work with weakly interacting systems, and it has become increasingly popular in the generation and optimization of lead compounds for structure-based drug-design efforts *(2)*. Furthermore, NMR is the only technique, at present, that can give useful information at the atomic level in the solution state. So, it is not a surprise that NMR has found widespread application in the field of protein–protein interactions.

One of the fundamental questions often encountered in protein–protein interaction studies is the location of the binding interface. Numerous biochemical or biophysical techniques are currently available to address this question including a collection of NMR spectroscopic methods *(1)*. Among them, chemical-shift perturbation is the most widely used NMR method to map protein interfaces because of its ease of execution and sensitivity to subtle changes in the chemical environment of proteins.

The basis of the NMR phenomenon is that magnetically active nuclei oriented by a strong magnetic field absorb electromagnetic radiation at characteristic frequencies (in the radiofrequency region) governed by their chemical environment. This environment is influenced by various factors such as chemical bonds, molecular conformations, and dynamic processes. So even the nuclei of the same element give rise to distinct spectral lines *(3)*. The relative positions of these spectral lines are therefore called chemical shifts. Chemical-shift mapping is a sensitive indicator of the protein binding sites, because the environments within and proximal to the protein interfaces are changed upon binding, consequently affecting the chemical shifts of the nuclei proximal to the binding interface, but in most cases not those remote from the interface. In other words, upon addition of a ligand, shifting peaks, line width, and/or intensity changes indicate which residues experience a change in their environment, and this in turn allows identification of the binding interface. In practice, this can be achieved simply by titrating the unlabeled ligand into an isotopically enriched (^{15}N and/or ^{13}C) protein sample and monitoring peak perturbations in spectra of the labeled protein.

The exact effects on the spectrum during the titration depend on the time scale by which the process is observed. A rule of thumb is that protein–protein interactions with $K_d < 1$ μM are generally in slow exchange (tight binding) and protein–protein interactions with $K_d > 1$ mM are in fast exchange (weak binding). In detail, assuming a nucleus has chemical shifts of ν_{bound} and ν_{free}, respectively, and if the lifetime of the complex is long compared with $(\nu_{bound} - \nu_{free})^{-1}$, it is considered to be in slow exchange (tight binding) on the NMR time scale and two signals are seen for the bound and free form, respectively. During the titration, the free form will disappear gradually. If the lifetime of the complex is short compared with $(\nu_{bound} - \nu_{free})^{-1}$, then this regime is referred to as fast chemical exchange (weak binding). One observes only one averaged signal whose chemical shift is fractionally weighted according to the populations and chemical shifts of the bound and free form. Finally, if the lifetime is around $(\nu_{bound} - \nu_{free})^{-1}$, then it is in the intermediate exchange regime. This will give rise to broadened signals from both the free and bound forms.

In this chapter, we will use an example from our own research to demonstrate the application of NMR chemical-shift perturbation technique to map the binding site of the Ras protooncogene protein onto the cysteine-rich domain of the Raf kinase (Raf-CRD, Raf-1 residues 136–187) *(4)*.

2. Materials

1. ^{15}N-labeled Raf-CRD (*see* **Notes 1** and **2**), and unlabeled Ras.
2. NMR buffer: 30 m*M* d_{11}-Tris-d_3-acetate (Isotec, Miamisburg, OH), pH 6.5, 100 m*M* NaCl, 10 μ*M* $ZnCl_2$, 1 m*M* d_{10}-DTT (Isotec, Miamisburg, OH), 10% D_2O, 0.01% NaN_3.
3. Bruker AMX 500 NMR spectrometer or Varian Inova 600 MHz NMR spectrometer.
4. 5.0 mm Wilmad standard NMR sample tube (*see* **Note 3**).
5. FELIX version 97.2 (Accelrys, San Diego, CA).

3. Methods

3.1. NMR Sample Preparation

The Raf-CRD was expressed as glutathione S-transferase (GST) fusion protein (Chapter 13) and purified by affinity chromatography. Uniformly ^{15}N-enriched Raf-CRD protein was obtained by growing *Escherichia coli* in a supplemented minimal media with 99.8% $^{15}NH_4Cl$ (Isotec, Miamisburg, OH) as the sole nitrogen source *(5)*. The purified protein was concentrated to 2 mL in 30 m*M* Tris acetate at pH 6.5, 75 m*M* Na_2SO_4, 10 μ*M* $ZnCl_2$, 1 m*M* DTT and was exchanged into NMR buffer [30 m*M* d_{11}-Tris-d_3-acetate (Isotec, Miamisburg, OH) at pH 6.5, 100 m*M* NaCl, 10 μ*M* $ZnCl_2$, 1 m*M* d_{10}-DTT, 10% D_2O, 0.01% NaN_3] using a Amersham Pharmacia Biotech PD-10 gel filtration column (*see* **Note 4**). All NMR samples had the final volume of 500 μL. The Raf-CRD NMR sample contained 0.28 m*M* uniformly ^{15}N-enriched Raf-CRD alone or mixed with wild-type unlabeled Ras at concentrations ranging from 0.30 to 0.77 m*M*.

3.2. NMR Spectroscopy

NMR experiments were recorded on a Bruker AMX 500 (500 MHz) spectrometer at 12°C or a Varian Inova 600 (600 MHz) spectrometer at 25°C (*see* **Note 5**). Two-dimensional ^1H–^{15}N heteronuclear single quantum coherence spectroscopy (HSQC) experiments were performed with pulse field gradient and water flip-back methods *(6)* (*see* **Notes 6** and **7**). Data were acquired with 1024 × 256 complex data points and a spectral width of 7042.25 Hz for the ^1H dimension and 1000 Hz for the ^{15}N dimension on the Bruker AMX 500. HSQC experiments conducted on the Varian Inova 600 spectrometer were acquired with 1024 × 128 complex data points and a spectral width of 8000 Hz for the ^1H dimension and 1700 Hz for the ^{15}N dimension. NMR data were processed and analyzed using the program FELIX (Accelrys, San Diego, CA) (*see* **Note 8**).

3.3. NMR Chemical Shift Titration

Proton and ^{15}N resonance assignments of the Raf-CRD were determined previously in this laboratory *(5)*. Two-dimensional ^1H–^{15}N HSQC data were acquired on the Raf-CRD-Ras complexes at four different molar ratios: 1:2.75, 1:1.65, 1:1.1, and 1:0. The final concentration of ^{15}N-Raf-CRD was 0.28 m*M* and unlabeled Ras ranged from 0.77 to 0.30 m*M*. Samples containing the highest molar ratio (1:2.75) of the Raf-CRD–Ras complex (0.28:0.77 m*M*) and Raf-CRD alone (0.28 m*M*) were prepared first. To prevent changes in sample volume and buffer composition, the other two complexes (molar ratio 1:1.65 and 1:1.1) were made by exchanging 200 μL of the two samples containing the highest molar ratio (1:2.75) and the sample alone (1:0) to achieve a final sample volume of 500 μL.

A fluorescence-based binding assay was used to determine the binding affinity between the Raf-CRD and Ras. Owing to solubility limitations, a quantitative measure of the K_d could not accurately be determined. However, our estimated K_d of 300 μ*M* indicated that the binding interaction between Raf-CRD and Ras was likely to be in fast exchange on the NMR time scale. During the titration, we expected to see only one peak that represents a weighted population of free and bound states *(7)*. Given an estimated K_d of 300 μ*M* for the Ras/Raf-CRD interaction, the NMR experiments were conducted under conditions ranging from approx 0 to 66% occupancy, and the full extent of chemical-shift changes (full saturation binding) will not be sampled. However, if our K_d is underestimated, our fractional occupancy will be less than 66%. Even though the chemical-shift changes will be smaller than would be anticipated for a titration that reaches saturation binding, these experiments, nevertheless, report chemical-shift changes that allow mapping of the binding interface.

3.4. Data Analysis

To elucidate the residues of the Raf-CRD important for Ras binding, we evaluated ^1H–^{15}N HSQC NMR spectral changes associated with binding of

Ras to the ^{15}N-enriched Raf-CRD. The ^{1}H–^{15}N HSQC experiment will give rise to a spectrum of proton chemical shifts in one dimension and the chemical shifts of the covalently attached ^{15}N nucleus in the second dimension, so only peaks corresponding to ^{15}N-enriched Raf-CRD will be observed. Among the various HN resonances detected are the amide protons. Therefore, the HSQC peaks provide a site-specific probe for every residue in Raf-CRD with the exception of proline. NMR spectral changes in the Raf-CRD after the addition of Ras (molar ratio 1:2.75) are shown in **Fig. 1**, as a superposition of a portion of ^{1}H–^{15}N HSQC spectra of uniformly ^{15}N-labeled Raf-CRD in the presence (black) and absence (gray) of unlabeled Ras. Raf-CRD residues that show distinguished changes in ^{1}H and ^{15}N chemical shift and/or peak intensity are labeled (*see* **Note 9**). Chemical shift changes are considered insignificant if the changes in peak position are less than the ^{1}H line width. Most of the chemical-shift differences observed between the free Raf-CRD and the complex are localized to Raf-CRD residues 144, 145, 148–150, 158–164, and 174–176. The largest ^{1}H chemical-shift changes were observed for residues 144, 145, 148–150, 158–160, 174, and 175. Residues 161, 163, 164, and 176 showed slightly smaller chemical-shift changes. The average change of the ^{1}H chemical shifts for the rest of the peaks was less than the line width, indicating that the global structure of Raf-CRD is unchanged due to the association with Ras (*see* **Note 10**). The same trend was observed in the ^{15}N chemical shift changes. The largest chemical shifts were observed for residues 145, 148, 149, 158, 163, and 175, with smaller ^{15}N chemical-shift changes found for residues 136, 150, 162, 174, and 183 in the presence and absence of a 2.75-fold excess of Ras. Uniformly reduced peak intensities (approx 55%) were also observed for a number of peaks, which likely reflects additional relaxation mechanisms due to interaction with the Ras protein. The largest variation in intensity changes for the amide peaks were found for residues 141, 145, 148–150, 158, and 170. The Raf-CRD residues experiencing significant chemical shift and/or intensity changes in the HSQC spectra upon binding of unlabeled Ras are highlighted on the structure of free Raf-CRD in **Fig. 2**. Clearly, residues 148–150 and 158–164 at the top of the molecule are identified as part of the binding interface. Consistent with these observations, a L149T/F151Q Raf-CRD mutant exhibits reduced binding affinity to Ras in vitro. Moreover, the L149T/F151Q mutation when placed in the context of full-length Raf, showed impaired Ras-mediated Raf activation in vivo *(4)*. Although spectral differences were also observed for resonances associated with Raf-CRD residues 144–145 and 174–177, mutations at positions K144 and T178 (unpublished data) do not affect Ras binding. The observed spectral changes associated with these residues may reflect changes in their chemical environment due to indirect effects of Ras binding at nearby residues (*see* **Note 11**).

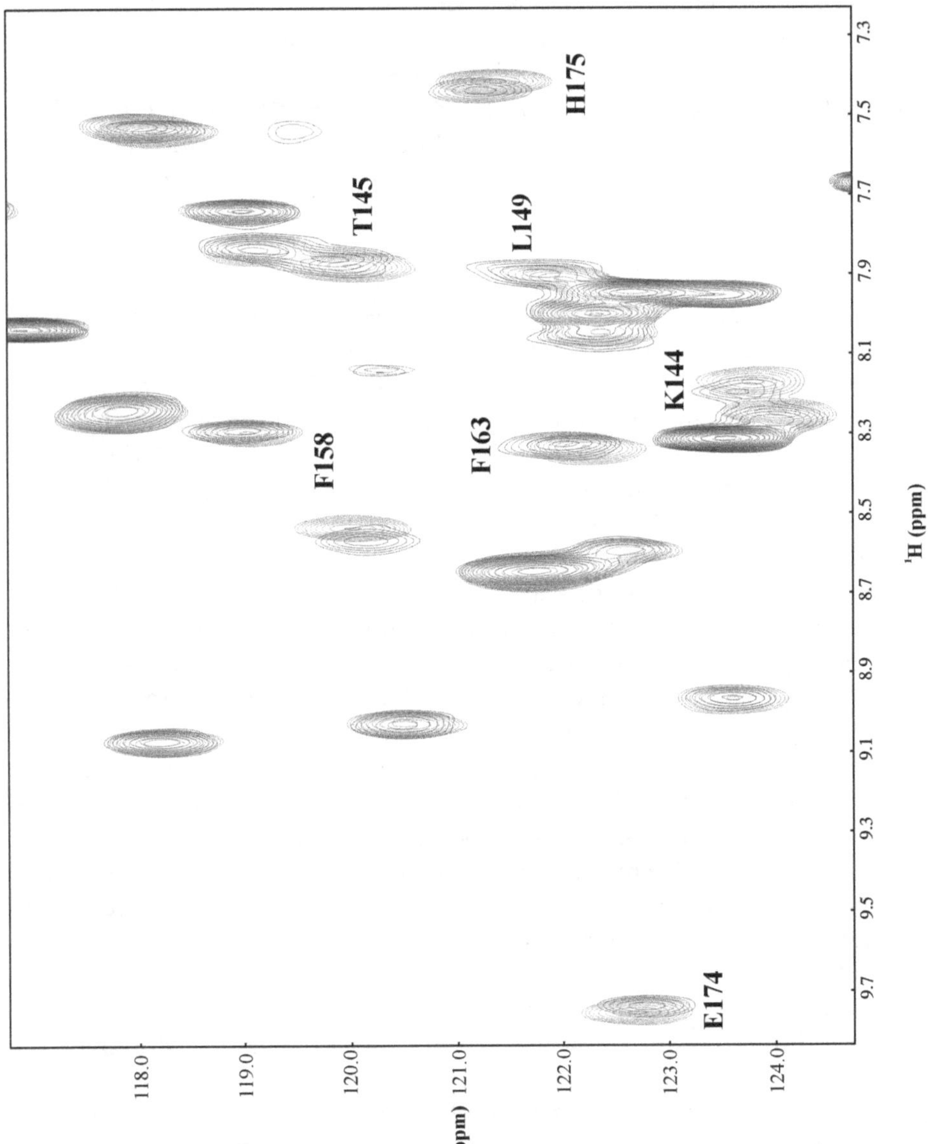

H175
T145
L149
K144
F158
F163
E174
^{1}H (ppm)
^{15}N (ppm)
7.3
7.5
7.7
7.9
8.1
8.3
8.5
8.7
8.9
9.1
9.3
9.5
9.7
118.0
119.0
120.0
121.0
122.0
123.0
124.0

4. Notes

1. NMR chemical-shift mapping applications generally use isotope-edited NMR methods. The proton is the most important magnetically active nucleus for the studies of biomolecules, because it has the highest natural abundance (99.98%) and it is the most sensitive to detection by NMR. Proton detection is also necessary for quantitative measurement of interproton distances via NOEs. However, proton-only methods become ineffective for detailed NMR analyses of proteins greater than 10–15 kDa. In addition to protons, a protein contains other magnetically active nuclei. For NMR of proteins, ^{15}N and ^{13}C are of special importance, as these NMR-active nuclei can be used as a vehicle to detect and assign resonances in larger proteins. The natural abundances of ^{15}N (0.37%) and ^{13}C (1.108%) are very low and their gyromagnetic ratio is also 10- and 4-fold lower than that of protons, respectively. With larger proteins, it is necessary to use stable isotopic labeling to enrich these low natural abundance nuclei so as to increase sensitivity of detection. Stable isotope labeling is not necessarily expensive, especially for ^{15}N enrichment, provided the protein can be produced in good yield from simple labeled precursors like $^{15}NH_4Cl$ or $(^{15}NH_4)_2SO_4$. The expense is often compensated for by the significant simplification of the spectrum and the extra information provided by the isotope-enriched protein sample. In addition, neither ^{13}C nor ^{15}N are radioactive, so there are no safety risks associated with its use.
2. Owing to the intrinsically poor sensitivity of NMR, the amount of material required to prepare a sample is often considered large by biochemical standards. For protein–protein interaction studies, protein concentrations ranging from 0.1 to 1 m*M* are generally used. However, recent technology changes, such as the advent of higher magnetic fields (900 MHz, 1 GHz) and cryoprobes, should improve sensitivity substantially and permit NMR studies of proteins at concentrations in the micromolar range. Given the sensitivity limitation of NMR, NMR signals are usually averaged many times in order to improve the signal-to-noise (S/N) ratio. The S/N ratio increases as the square root of the number of times the signal is averaged, so that a 0.5 m*M* sample will take four times as long to obtain a spectrum with the same sensitivity as for a 1 m*M* sample. For a commonly used 5-mm NMR tube, the ideal sample volume should be around 500 μL (350 μL for a special Shigemi NMR tube). Depending on the molecular weight of the protein, several milligrams to tens of milligrams of protein are required. For example, 5 mg of a 10 kDa protein is required to make a 500 μL, 1 m*M* sample, which is feasible due to the application of recombinant DNA technology. In practice, purifying a sufficient quantity of material is not the only difficulty.

Fig. 1. *(opposite page)* Expanded region of $^1H–^{15}N$ HSQC spectra of 0.28 m*M* uniformly ^{15}N-enriched Raf-CRD, free (black contours) and in the presence of 0.77 m*M* unlabeled Ras-GDP (gray contours). The peaks experiencing the largest chemical shifts and/or intensity changes are labeled.

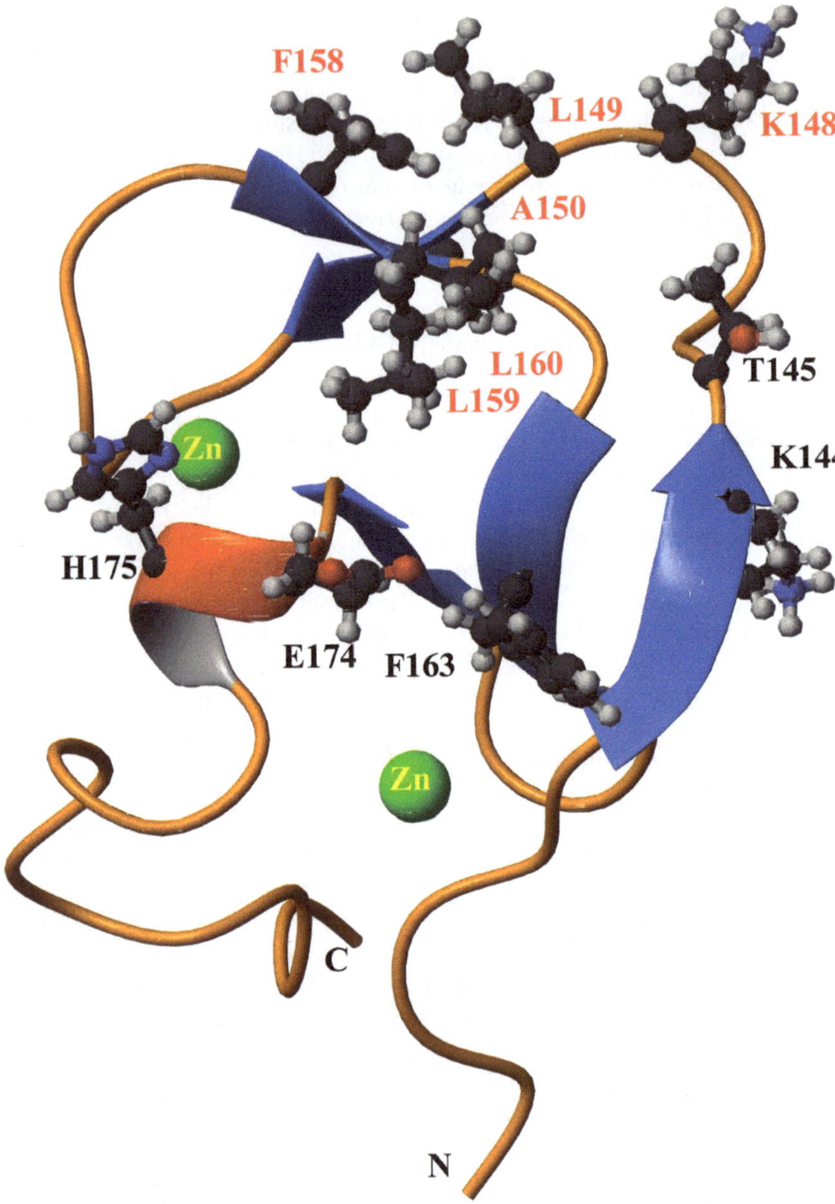

Fig. 2. Changes in amide chemical shifts or peak intensities in Raf-CRD induced by Ras binding mapped on the ribbon plot of the minimized average solution structure of the Raf-CRD (PDB accession code 1FAR), using the program MOLMOL ***(16)***. Residues (144, 145, 148–150, 158–160, 163, and 174–175) that show HN chemical shift changes greater than 0.02 ppm in ^{1}H or more than 0.2 ppm in ^{15}N are labeled with the side chains displayed in ball-and-stick representation.

At millimolar concentrations many biomolecules have unfavorable aggregation or solubility properties. It may be necessary to vary pH, protein, and/or salt concentration to obtain a stable solution amenable to NMR analysis *(8)*.

3. NMR tubes should be washed with water and/or an organic solvent such as acetone. The acetone can be evaporated off, or the sample tube can be simply rinsed thoroughly with distilled, deionized water. To avoid scratching the inside of the NMR tube, a brush or other abrasive material should not be used. NMR tubes can be dried at < 125°C for 30–40 min or by blowing with dry N_2 gas (to remove residual acetone). For tougher cleaning problems, nitric acid can be used, followed by a distilled water or acetone rinse. However, chromic acid should be avoided, as oxidative paramagnetic byproducts can be produced.
4. Chemical-shift mapping studies are generally conducted in 90% H_2O/10% D_2O in order to observe labile protons such as amide protons in proteins. The D_2O is necessary so that there is a sufficient deuterium signal for the NMR lock system to function. If possible, the NMR buffer should have a total salt concentration less than 500 m*M*, as higher salt concentration may cause detuning of the NMR probe and lead to decreased sensitivity. Any organic component of the buffer solution, for example, Tris and acetate, in excess of 500 μ*M* should be deuterated to avoid background proton signals. Many use nonprotonated buffers such as phosphate to avoid additional costs associated with isotropic enrichment of the buffer medium. For aqueous samples, an internal chemical-shift standard such as 5,5-dimethylsilapentanesulfonate (DSS) or 3,3,3-trimethylsilylpropionate (TSP) is usually used with the final concentration less than 50 μ*M*. Suppression of bacterial growth in order to prevent the NMR sample from degrading over a period of time can be achieved by adding NaN_3 (< 50 μ*M*) to the NMR buffer. Addition of 50 μ*M* EDTA to NMR samples is commonly used to chelate paramagnetic ions, as paramagnetic impurities such as Fe^{3+} and Mn^{2+} will cause large changes in chemical shifts and relaxation times if they interact with the protein. Soluble paramagnetic ions will also cause broadening of the water signal, which will severely compromise solvent suppression.
5. The macromolecule must be stable during the course of the NMR titration. It is important to make sure that the sample has not changed during the experiment by running a quick 1D ^1H or 2D ^1H–^{15}N HSQC spectrum and comparing it to the spectrum collected on a control sample at the beginning of data acquisition. An identical spectrum indicates the sample has not changed.
6. Two strategies are used for increasing the low sensitivity of heteronuclei (nonproton nucleus): isotopic enrichment of these nuclei in proteins as described above, and enhancement of the S/N by the use of so-called inverse NMR experiments in which the magnetization is transferred from protons to the heteronucleus. The most widely used inverse experiment for biomolecular NMR applications is the HSQC. The ^1H–^{15}N HSQC spectrum correlates the nitrogen atom of a ^{15}NH with the directly attached proton, thereby providing a spectral fingerprint of the protein because there is at least one HN, including only one amide resonance, for every amino acid residue with the exception of proline.

An analogous experiment (^1H–^{13}C HSQC) can be performed for ^{13}C and ^1H. The ^{13}CH is less sensitive to the variations in buffer and temperature compared with the amide NH. However, in order to interpret the spectra, not only the backbone but also the aromatic and side-chain assignments may be needed.

7. For detailed structural studies, there is a limitation on the size of a macromolecule that can be studied by NMR, even though the size limit has been increased significantly (from approx 20 kDa in 1992 to approx 70 kDa in 2002) in the past decade by the introduction of more powerful NMR spectrometers, cryoprobes, and more-complicated pulse sequences. A large molecule can increase the complexity of the spectrum simply because larger molecules have more NMR signals in a given spectral width and therefore more overlap in the spectrum. In addition, the line widths of the individual resonances will also increase because the resonance line width is proportional to the rotational correlation time τ_c of the molecule. Larger molecules have longer τ_c and therefore broader line widths. The advent of TROSY-HSQC has enabled application of NMR chemical shift mapping to larger proteins (> 60 kDa) as TROSY methods select for the longest relaxation component giving rise to HSQC spectra with narrower lines and higher sensitivity detection. In fact, the TROSY technique has been applied successfully to investigate proteins with a molecular weight even up to 900 kDa ***(9)***.
8. The NMR signal is called a free induction decay (FID). The FID is a time domain signal that contains a set of sine/cosine waves measured as a function of time and decaying at an exponential rate. The FID is digitized by sampling the intensity of the signal at discrete time intervals using a device called the ADC (analog-to-digital converter), and the intensity is stored as an integer value on a computer. Typical data processing includes: (1) Conversion of the raw FID from spectrometer format to the processing program format. (2) Apodization of the conventional FID. Apodization is a term which refers to manipulation of the FID to increase either the S/N or the resolution. It is usually achieved by applying appropriate window functions to the FIDs. A typical window function is a 90° sine-bell function. (3) Zero-filling the FID, which refers to the addition of data points that all have zero values to the end of the FID prior to Fourier transformation to yield more data points in the spectrum. (4) Fourier transformation of the time domain signal to a frequency domain spectrum. (5) Phasing of the spectrum to generate symmetrical or Lorenzian peaks. (6) Baseline correction and spectrum referencing are then performed. For 2-, 3-, or 4-dimensional data, the procedures described above need to be repeated in every dimension ***(10)***. There are many programs available now for NMR data processing and/or spectra visualization and analysis, including commercially available programs such as FELIX (Accelrys, San Diego, CA) and SYBYL TRIAD (Tripos Inc., St. Louis, MO), and free programs for academic use such as NMRPipe and NMRDraw from NIH, NMRView from Merck Research Laboratories, and Sparky from UCSF.
9. The size of spectral change in amide peak position can be quantified as the Pythagorean relation: $\Delta NH = [\Delta H^2 + (\Delta N/6)^2]^{1/2}$, where ΔH and ΔN are chemical-shift changes in parts per million (ppm) for a particular amide peak.

10. In some cases, the entire protein may experience a conformational change upon ligand binding, so additional peaks outside the binding interface also shift upon the addition of the binding partner. This makes it difficult to define the interaction surface by chemical shift mapping. Fortunately, other methods can be used to define the interaction surface such as the observation of the intermolecular NOEs ***(1)***. If remote affects are observed, without a large-scale conformational change taking place in the protein upon ligand binding, measurement of the hydrogen exchange rates, and mapping changes in backbone or side-chain dynamics may prove useful ***(1)***. We have had particular success using relaxation agents to map the binding interface ***(11–13)***.

    In this approach, soluble paramagnetic agents, or so-called relaxation reagents such as hydroxy-TEMPO and chelated gadolinium(III), are able to "bleach" the NMR resonances exposed to the solvent. These agents have unpaired electrons that can interact with the nuclei in the protein through the transient dipolar interactions and efficiently broaden the proton resonances from surface residues of the protein. Comparison of line-broadening effects produced by the paramagnetic probe in ^{1}H–^{15}N HSQC spectra of a protein, in the presence and absence of its binding partner, has been proved to be a convenient and successful tool for the delineation of the intermolecular interface.

    Recently in our lab, an attempt to identify the interaction sites of a peptide deduced from paxillin (LD2) with the focal adhesion targeting domain (FAT) of focal adhesion kinase (FAK) by chemical-shift mapping failed, because the addition of the peptide induced dramatic chemical shifts in the majority of the residues in the FAT sequence. This is presumably due to the existence of two peptide-binding sites in the FAT protein (submitted). As an alternative strategy, Gd(III)–EDTA was used as a paramagnetic probe to map the binding interface. A 0.1–0.2 m*M* ^{15}N-labeled FAT sequence sample, alone or in complex with the LD2 peptide (molar ratio 1:8), was analyzed in 20 m*M* perdeuterated Tris maleate, 50 m*M* NaCl, 0.1 m*M* EDTA, 10% D_2O in the presence of different concentrations of 1:1 Gd(III)–EDTA complex (up to 9 m*M*). 2D ^{1}H–^{15}N HSQC spectra were collected. Line broadening was measured as a ratio of the NH peak height at a given concentration of Gd(III)–EDTA to the peak height with no paramagnetic surface probe present. Normalized heights of NH peaks of the free FAT protein in the presence of 3 m*M* Gd(III)–EDTA were subtracted from normalized heights of NH peaks of FAT protein/paxillin peptide complex in the presence of 3 m*M* Gd(III)–EDTA. The highly protected amide protons were localized to the two surfaces of the FAT sequence corresponding to two hydrophobic patches. These results are in good agreement with our isothermal titration calorimetry data and provide further evidence for the two-binding-site model of the FAT sequence/ paxillin complex. However, studies have shown that the lanthanide complex prefers to a strong association with carboxylate groups whereas nitroxyl prefers hydrophobic surface residues ***(12)***, so nonspecific interactions of paramagnetic probe with protein may occur. A combination of the use of these two agents could probably provide complimentary and therefore more meaningful results.

11. Although the case we described in detail demonstrated an example of a weak binding interaction between the Ras protein and the Raf-CRD domain that is in fast exchange on the NMR time scale, NMR chemical-shift mapping approaches are commonly used to investigate higher affinity interactions. As stated earlier, for complexes with a $K_d < 1\ \mu M$, the system is in slow exchange on the NMR time scale, and two distinct sets of peaks for the free and bound state of the labeled protein will be observed. Because NMR spectra are generally collected at sample concentrations ranging from 0.1 to 1 m*M*, all of the ligand will be bound at substoichiometric concentration of the ligand. Often resonance assignments for the free but not the bound state are known prior to the titration. Analysis of the spectrum requires characterization of assigned peaks corresponding to the free state that disappears upon ligand binding. It may be difficult to interpret changes in peaks associated with the binding interface, especially if chemical shift changes between the free and bound state are difficult to track due to spectral overlap. In these cases, it may be possible to interpret only those resonances that are highly shifted to roughly map the binding interface ***(14)***, or, alternatively, it may be necessary to obtain resonances assignments of the bound state to more precisely map which resonance are perturbed upon ligand binding ***(15)***.

Acknowledgments

The Raf-CRD/Ras work was supported by NIH grants CA72644 and CA72644-10 (to G. J. Clark), CA42978, CA55008, and CA67771 (to C. J. Der), and CA70308-01 and CA64569-01 (to S. L. Campbell). The FAT/LD2 work was supported by grants RPG-96-021-04-CSM from the American Cancer Society (to M. D. Schaller) and HL45100 (to M. D. Schaller and S. L. Campbell). We thank Melissa Starovasik (Protein Engineering Department, Genentech, Inc.) for a critical reading of the manuscript.

References

1. Zuiderweg, E. R. (2002) Mapping protein-protein interactions in solution by NMR spectroscopy. *Biochemistry* **41(1),** 1–7.
2. Moore, J. M. (1999) NMR techniques for characterization of ligand binding: utility for lead generation and optimization in drug discovery. *Biopolymers* **51(3),** 221–243.
3. Roberts, G. C. K. (1993) *NMR of Macromolecules : A Practical Approach.* The Practical approach series. IRL Press, Oxford University Press, Oxford, NY, pp. xviii, 399.
4. Williams, J. G., Drugan, J. K.,, Yi, G. S., Clark, G. J., Der C. T., and Campbell, S. J. (2000) Elucidation of binding determinants and functional consequences of Ras/Raf-cysteine-rich domain interactions. *J. Biol. Chem.* **275(29),** 22,172–22,179.

5. Mott, H. R., Carpenter, J. W., Zhang, S., Ghosh, S., Bell, R. M., and Campbell, S. L. (1996) The solution structure of the Raf-1 cysteine-rich domain: a novel ras and phospholipid binding site. *Proc. Natl. Acad. Sci. USA* **93(16),** 8312–8317.
6. Grzesiek, S. and Bax, A. (1993) The importance of not saturating water in protein NMR. Application to sensitivity enhancement and NOE measurements. *J. Am. Chem. Soc. 1993.* **115(26),** 12,593–12,594.
7. Herrmann, C., Martin, G. A., and Wittinghofer, A. (1995) Quantitative analysis of the complex between p21ras and the Ras-binding domain of the human Raf-1 protein kinase. *J. Biol. Chem.* **270(7),** 2901–2905.
8. Bagby, S., Tong, K. I., and Ikura, M. (2001) Optimization of protein solubility and stability for protein nuclear magnetic resonance. *Meth. Enzymol.* **339,** 20–41.
9. Fiaux, J., Bartelsen, E. B., Horwich, A. L., and Wuthrich, K. (2002) NMR analysis of a 900K GroEL GroES complex. *Nature* **418(6894),** 207–211.
10. Hoch, J. C. and Stern, A. S. (1996) *NMR data processing.* Wiley-Liss, New York, NY, pp. xi, 196.
11. Esposito, G., Lesk, A. M., Molinari,m H., Molta, A., Nicrolar, N., and Pastore, A. (1992) Probing protein structure by solvent perturbation of nuclear magnetic resonance spectra. Nuclear magnetic resonance spectral editing and topological mapping in proteins by paramagnetic relaxation filtering. *J. Mol. Biol.* **224(3),** 659–670.
12. Petros, A. M., Mueller, L. and Kopple, K.D. (1990) NMR identification of protein surfaces using paramagnetic probes. *Biochemistry* **29(43),** 10,041–10,048.
13. Arumugam, S., Hemme, C. L., Yoshida, N., et al. (1998) TIMP-1 contact sites and perturbations of stromelysin 1 mapped by NMR and a paramagnetic surface probe. *Biochemistry* **37(27),** 9650–9657.
14. Fairbrother, W. J., Christinger, H. W., Cochran, A., G., et al. (1998) Novel peptides selected to bind vascular endothelial growth factor target the receptor-binding site. *Biochemistry* **37(51),** 17,754–17,764.
15. Pan, B., Li, B., Russell, S. J., Torn, J. Y., Cochran, A. G., and Fairbrother, W. J. (2002) Solution structure of a phage-derived peptide antagonist in complex with vascular endothelial growth factor. *J. Mol. Biol.* **316(3),** 769–787.
16. Koradi, R., Billeter, M., and Wüthrich, K. (1996) MOLMOL: a program for display and analysis of macromolecular structures. *J. Mol. Graph.* **14(1),** 51–55.

6

Measuring Rhodopsin–G-Protein Interactions by Surface Plasmon Resonance

John Northup

Abstract

G-protein–coupled receptors (GPCRs) initiate a variety of cellular responses to diverse array of extracellular stimuli. Surface plasmon resonance detection offers a powerful approach to the study of protein–protein interactions in real time. In this chapter we outline procedures for the immobilization of the prototype GPCR structure, rhodopsin or the G-protein α and βγ subunits, for analysis of the molecular interactions initiating G-protein signaling. The attachment of rhodopsin via its extracellular carbohydrate residues provides a convenient, and universally applicable, procedure for GPCR immobilization in a form that retains full biochemical activity and ability to interact with intracellular signaling components. SPR detection then allows for the analysis of the kinetic and equilibrium binding properties of the immobilized receptor with G-protein subunits and potentially other interacting molecules.

Key Words

G-proteins; receptors; rhodopsin; surface plasmon resonance; signaling.

1. Introduction

Surface plasmon resonance (SPR) spectroscopy provides a powerful approach to investigate protein–protein interactions in real time. SPR spectroscopy offers the opportunity to make quantitative comparisons of specificity, kinetics, and affinities of protein–protein interactions without some of the complexities of other methods, and SPR spectroscopy has recently been applied to the interaction of various components of G-protein–mediated signal transduction systems *(1)*, such as those between G-protein subunits, their receptors, and regulatory proteins such as regulators of G-protein–signaling (RGS) proteins.

From: *Methods in Molecular Biology, vol. 261: Protein–Protein Interactions: Methods and Protocols*
Edited by: H. Fu © Humana Press Inc., Totowa, NJ

The detection principle in SPR is based on changes of the optical properties of a surface layer containing an immobilized protein with increasing mass of protein bound from a mobile fluid phase. These changes are monitored continuously in real time so that information about both the kinetic and equilibrium properties of the protein–protein interactions is obtained. Appropriate analysis of the time-course of binding allows determination of association rate constants, and, by removing the binding partner from the mobile phase, dissociation rate constants. Also, from the binding data in steady state, binding isotherms can be generated for the thermodynamic determination of equilibrium dissociation constants. For a general introduction in the practice of biosensing (*see* ref. *2*).

In our laboratory, we use the commercial SPR instruments Biacore 1000 and Biacore 2000, and therefore most of the following description refers to the use of Biacore instruments. However, all the techniques described can be used or adapted for other surface plasmon resonance or related sensors. In the Biacore instruments, the time-course of the SPR signal is referred to as a sensorgram (*see* **Fig. 2**), and the units of a sensorgram are termed resonance units (RU) with 1 RU being equivalent to the binding of approximately 1 pg of protein per mm^2 of SPR chip surface *(3,4)*. A variety of different sensor chips provide for alternative chemistries and strategies for the immobilization of proteins; most of them are based on a carboxymethylated dextran layer that forms a flexible hydrogel of thickness 100–200 nm deposited on the surface of a thin gold foil, which is the source of the light reflection and plasmon resonance interaction. Protein is delivered in the mobile phase to the sensor surface by a microfluidic system that can generate a constant flow of the mobile phase across one or more (Biacore 2000) flow paths in series. Pneumatic valves and an HPLC-like injection loop allow switching of the mobile phase from running solution to a sample of protein solution to be analyzed for binding. Usually, one or more of the flow paths are specifically modified for the experiment, with additional unmodified flow paths serving as a reference surface to monitor nonspecific binding and the refractive index of the mobile phase. Additional sensor flow paths in the Biacore 2000 can be used for several experiments (or controls) to be performed in parallel.

1.1. Measurement of GPCR–G-protein Interactions by SPR

Ligand activation of a G-protein–coupled receptor (GPCR) is the initial event in the cascade of G-protein signaling. The GPCR functions to catalyze the dissocitation of tightly bound GDP from the G protein and the subsequent dissociation of GTP-bound Gα from Gβγ subunits. Methods for purification and reconstitution of GPCRs and G proteins were developed some time ago, and, recently, the techniques for the introduction of the GPCR into synthetic

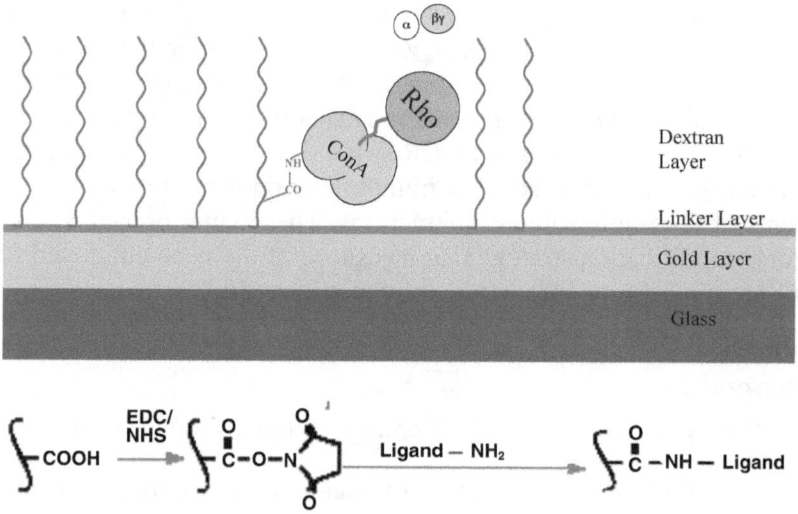

Fig. 1. Oriented immobilization of rhodopsin to SPR surfaces by ConA attachment. A diagram of the SPR surface is presented with the strategy for immobilizing rhodopsin by capture with covalently immobilized ConA, thereby orienting the rhodopsin with cytoplasmic surface available for G-protein-subunit interaction. The gold layer is responsible for the plasmon resonance interaction, which is the basis of the protein-interaction detection, and it is supported by the glass layer. Deposited on the surface of the gold is a layer of carboxymethyl dextran, which provides the hydrated environment for protein–protein interaction and the chemical sites for covalent attachment via activated carboxyl mechanisms. At the bottom of the diagram is a generic presentation of the NHS-catalyzed carbodiimide activation chemistry for primary amine coupling.

lipid vesicles or bilayers were successfully adapted to the SPR measurement of rhodopsin–retinal-G-protein interactions [for reviews *see* **5** and **6**]. These approaches have demonstrated light-regulated G protein binding to rhodopsin, as has been known since the pioneering work of Herman Kuhn *(7)*, and these SPR approaches have enabled measurement of equilibrium dissociation constants for the heterotrimer G protein binding to rhodopsin. In order to obtain kinetic information on GPCR–G-protein interaction, we have developed an alternative approach to couple functionally active GPCR to the sensor chip by covalently linking the mitogenic lectin concanavalin A (ConA) to the sensor chip and using it to adsorb detergent-dissolved rhodopsin via binding of extracellular carbohydrates (*8*; *see* **Fig. 1**). The reversible binding to ConA is a routine procedure for the purification of rhodopsin from detergent extracts of rod outer segment (ROS) discs *(9)*. Rhodopsin bound to ConA retains full biochemical activity as a catalyst for GTP exchange on retinal G-protein subunits, and the immobilized rhodopsin is stable for several days at 25°C in the Biacore instrument. These procedures are conceptually appealing in that the extracellular carbohydrate modifications for most GPCRs do not appear to be essential for ligand binding or G- protein regulation. Therefore, this approach, in theory, should be broadly applicable to GPCR structures. As an essential control and comparison, we also have adapted immobilization procedures for SPR measurements of G-protein subunit interactions. These latter procedures utilize a sulfhydryl-modification strategy that has successfully been employed for creating G-protein affinity chromatography matrices ***(10,11)***.

2. Materials

2.1. Buffers

1. 0.5 *M* MOPS, pH 7.5 (4-morpholinepropanesulfonic acid) (Calbiochem cat. no. 475898).
2. 0.5 *M* MES, pH 6.0 [2-(*N*-morpholino)ethanesulfonic acid] (Calbiochem cat. no. 475893).
3. 0.5 *M* HEPES, pH 7.5 [*N*-[2-hydroxyethyl]piperazine-*N*'-(2-ethanesulfonic acid)] (Calbiochem cat. no. 391338).
4. 0.1 *M* Na-acetate, pH 4.8.
5. 0.1 *M* potassium phosphate buffer, pH 7.0.

2.2. Salts

1. 5 *M* NaCl.
2. 0.1 *M* $MgSO_4$.
3. 0.1 *M* $MnCl_2$.
4. 0.1 *M* $CaCl_2$.

All dissolved in water.

2.3. Solutions

1. Solution A: 50 m*M* MOPS, 150 m*M* NaCl, 3 m*M* $MgSO_4$, 10 μ*M* $MnCl_2$, 10 μ*M* $CaCl_2$, pH 7.5, prepared from the following solutions:

0.5 *M* MOPS, pH 7.5	50 mL
5 *M* NaCl	15 mL
0.1 *M* $MgSO_4$	1.5 mL
0.1 *M* $MnCl_2$	50 μL
0.1 *M* $CaCl_2$	50 μL
100 m*M* CHAPS	40 mL for G-protein subunits (*see* below)
H_2O to final volume of	500 mL.

   Filter the solution through 0.22 μm pore size nitrocellulose under vacuum to remove any fine particles and to degas the solution and store at room temperature.
2. Solution B: Solution A without $MnCl_2$ or $CaCl_2$.
3. 100 m*M* CHAPS (3-[(3-cholamidopropyl)dimethylammonio]-1-propane-sulfonate; Calbiochem cat. no. 220201) dissolved in water.

4. Solution B + CHAPS: Solution B + 8 m*M* CHAPS.
5. Solution C: 50 m*M* MES, 1 m*M* $CaCl_2$, 1 m*M* $MnCl_2$, pH 6.0, made from the following solutions:

0.5 *M* MES, pH 6.0	10 mL
0.1 *M* $MnCl_2$	1 mL
0.1 *M* $CaCl_2$	1 mL

H_2O to 100 mL final volume; filtered and degassed.

6. Solution D: 10 m*M* HEPES, 150 m*M* NaCl, 3 m*M* EDTA, and 0.005% Tween20, pH 7.5, made from the following solutions:

0.5 *M* HEPES, pH 7.5	10 mL
0.5 *M* EDTA	3 mL
5 *M* NaCl	15 mL
10 % (w/v) Tween 20	0.25 mL

H_2O to 500 mL final volume; filtered and degassed with 0.22 μm filter.

7. Solution E: Solution D with 4 m*M* $MgSO_4$ and 1 m*M* DTT.

2.4. Reagents

1. 50 m*M* NHS (*n*-hydroxysuccinimide; Pierce cat. no. 24500)
2. 200 m*M* EDC [*N*-ethyl-*N*'-(dimethylaminopropyl)carbodiimide; Pierce cat. no. 22980).
3. 1 *M* Ethanolamine (ethanolamine-HCl; Sigma cat. no. E6133). Items 1 to 3 are dissolved in water and stored at –20°C as 200 μL aliquots. If well sealed, they can store for several years. Once thawed, the aliquots are discarded.
4. 2 mg/mL ConA (Concanavalin A; Sigma cat. no. C7275): Dissolved in 10 m*M* MOPS, pH 7.5, and this can be stored at 4°C for several months without loss of activity.
5. 2 m*M* GMBS (*N*-[γ-maleimidobutylryloxy]succinimide ester; Pierce cat. no. 22309): Dissolved in 100 m*M* potassium phosphate buffer, pH 7.0, immediately prior to use.
6. Sephadex G-50 fine (Amersham-Pharmacia cat. no. 17-0042-01).
7. BioRad econo column (cat. no. 735-0507).

2.5. G-proteins and Rhodopsin

The procedures for the preparation of the retinal G-proteins and rhodospin from bovine ROS have been described amply elsewhere ***(12,13)***. The isolation of recombinant Gα subunits from *Escherichia coli* ***(14)*** and Gβγ dimers using baculoviral constructs in Sf9 cells ***(15,16)*** have also been well characterized and will not be detailed here. These materials are nevertheless crucial to the SPR methods we describe and adequate care should be taken to prepare these protein reagents as described to obtain the composition and purity required for these techniques to be successful. For detailed descriptions of these procedures see the following:

1. ROS disc *(17)*.
2. Bovine retinal G-proteins and urea-washed ROS discs *(12,13,18)*.
3. Myristoylated Gα_i from *E. coli* *(14)*.
4. G$\beta\gamma$ dimers from Sf9 cells *(15,16)*.

3. Methods

3.1. Preparing G-proteins in the Running Solution

For SPR analysis of protein–protein interactions, it is highly desirable that the test proteins be prepared in the identical solution conditions as the running solution for the experiments. Otherwise, the injection of protein samples will include a signal originating from a bulk refractive index change (due to the different solution compositions) convoluted with the signal due to protein–protein interaction. To eliminate this, the G-protein samples to be used for SPR analysis are gel-filtered into Solution B (*see* **Note 1**). To maintain as high protein concentrations as possible through this procedure, it is convenient to fractionate this chromatography, rather than run batch samples. An example of this for 300 µL of G-protein is provided:

1. Hydrate about 1 g Sephadex G-50 fine by adding 10 mL H_2O and either allowing overnight at 4°C or boiling for 30 min.
2. Decant the excess water and add an equal volume of water to the swollen resin for storage.
3. Prepare a 1.0 mL bed volume column by carefully measuring 1.0 mL of liquid into the barrel of a 0.5 × 50 mm column (such as BioRad econo column) and marking the meniscus on the column barrel. Add about 2 mL of the swollen G-50 slurry to the barrel and let the resin settle. Adjust the resin amount to be 1.0 mL using the calibration of the barrel.
4. Equilibrate the 1 mL G-50 column with 5 mL of Solution B (or Solution B with CHAPS for the hydrophobic G$\beta\gamma$ dimers) and place the column in a cold cabinet or cold room. All subsequent steps should be performed at 4°C.
5. Collect samples into 12 × 75 mm test tubes (siliconized glass or polypropylene) and add 300 µL of the G-protein sample to the top of the G-50 resin. Then, changing tubes for each addition, make ten 100 µL additions of Solution B, waiting for the liquid meniscus to reach the top of the resin before each subsequent addition.
6. Assay for protein or G-protein activity in the collected fractions to identify fractions with maximal concentration and pool these. Generally, all activity will be recovered in about 0.5 mL, with about 70% of the activity in the peak 300 µL retaining nearly the initial protein concentration.

3.2. Cross Linking Concanavalin A to SPR Chips

ConA is a tetrameric protein with four identical 28 kDa subunits *(19)*. The minimal structure exhibiting high affinity binding of carbohydrates is the dimeric form. The dimer form of ConA is an active stable structure, and the tetramer–dimer transition can be promoted by low pH *(20)*. We use unmodified ConA.

Because the dissociation of dimer from tetramer leads to a continual decline in the SPR baseline, our coupling procedures first promote the tetramer–dimer dissociation prior to linkage. To accomplish this, ConA is diluted to the desired concentration (0.1–0.3 mg/mL) and stored overnight at 4°C in 100 m*M* Na-acetate, pH 4.8, prior to coupling. The low pH assists tetramer dissociation and provides for optimal concentration of the ConA dimer by ionic attraction to the unactivated carboxyl groups in the sensor chip. Linkage of the ConA is accomplished by standard NHS/EDC amine coupling ***(21)*** in the running Solution A. In the BIACore instruments, these procedures are handled internally by programming a sequence of robotic sampling and injections. Solutions and samples are prepared in vials and placed in the autosampler of the instrument, and the procedures are then performed automatically.

The stages of the coupling reaction are as follows:

1. First, the surface is activated by injection of a mixture of 0.1 *M* EDC with 25 m*M* NHS in water at a flow rate of 5 µL/min, exposure time of 5 min. Two vials of 150 µL each 0.2 *M* EDC and 50 m*M* NHS and a third empty vial are placed in the instrument. An equal volume of 100 µL is withdrawn from each of the coupling reagents and mixed in the empty vial by the autosampler. The injection of 25 µL is then made from the mixture vial.
2. This is followed by injection of ConA at 0.1–0.3 mg/mL in 100 m*M* Na-acetate, pH 4.8, with an exposure time of 5 min; 150 µL of the diluted ConA is placed in an injection vial to provide for the injection of 25 µL during the coupling reaction.
3. Activated carboxyls are then quenched by reaction with 1 *M* ethanolamine in water with an exposure time of 4 min; 150 µL of 1 *M* ethanolamine is placed in an injection vial to provide for the 20 µL injection.

An example of the coupling of ConA to a carboxymethyldextran surface is shown in **Fig. 2A**. Note that the solutions containing high concentrations of solutes, particularly the 1 *M* ethanolamine, produce very large SPR signals owing to the bulk refractive index differences between these and the running solution. These can be readily distinguished from the SPR signals originating from protein–protein interactions by a number of kinetic features. The signals arising from bulk refractive index show the time course of the mixing of the flow cells (about 3 s equilibration at 5 µL/min flow rate) and have the features of a square wave. This is clearly distinguished from the diffusion-limited binding of ConA to the carboxymethyldextran in the second step, which has the shape of an exponential approach to equilibrium. Note also that on wash-out of the coupling reagents the baseline SPR signal is increased from that measured prior to the reactions. This measures the amount of ConA covalently attached to the dextran layer (*see* **Note 2**). To stabilize the baseline we routinely wash the ConA surface for 30–60 min with running solution at a flow rate of 5 µL/

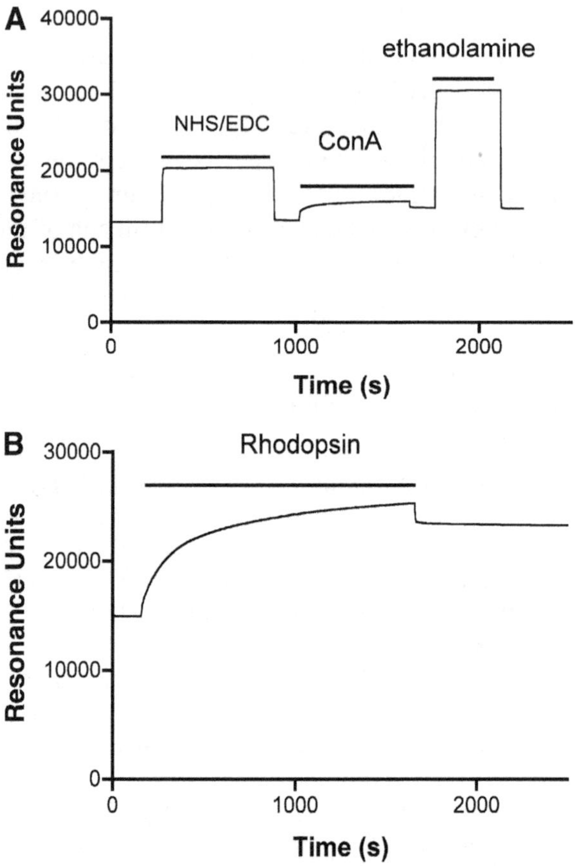

Fig. 2. Sequential coupling of ConA and rhodopsin to SPR surfaces. **(A)** The sensogram of a coupling reaction utilizing NHS/EDC chemistry for covalently attaching ConA to the SPR surface. The abscissa of this plot is the time axis and the ordinate is the SPR signal detection output in resonance units. The horizontal bars represent the sequential injection into the flow path of the NHS/EDC solution, followed by the ConA, and finally the ethanolamine quenching of activated carboxyls to prevent further covalent attachment. Note the square-wave shape of the injections of NHS/EDC and ethanolamine as compared with the exponential shape of the ConA, discriminating bulk refractive index changes from binding interactions. **(B)** The subsequent binding of detergent-dispersed rhodopsin to the covalently attached ConA. This sensogram is typical for a slowly equilibrating binding interaction. During the injection of rhodopsin, the SPR signal gradually increases as a first-order exponential approach to equilibrium. On completion of the injection, a portion of the signal due to loosely bound rhodopsin and bulk refractive index immediately decreases to a very slowly declining signal. This is the ConA-immobilized rhodopsin, which is still undergoing tetramer–dimer transition to the stable dimeric ConA–rhodopsin surface.

min to allow for complete tetramer–dimer dissociation, prior to rhodopsin immobilization.

3.3. Extraction and Immobilization of Rhodopsin on ConA-Modified SPR Chips

Urea-washed ROS discs are prepared from hypotonically washed, GTP-extracted ROS discs to remove phosphodiesterase, rhodopsin kinase, and G proteins by repeated incubation and sedimentation with 5 *M* urea at 4°C under ambient illumination as described *(18)*, resulting in the depletion of all peripheral proteins from the ROS discs and a mixture of meta-II and meta-III rhodopsin isomers in the discs. These preparations are snap frozen and stored at –80°C.

1. Prior to immobilization, rhodopsin is solubilized from a preparation of urea-washed ROS discs by addition of 25 m*M* CHAPS to a suspension of ROS discs containing 3–3.5 mg/mL protein (approx 100 μ*M* rhodopsin) in 10 m*M* TRIS pH 7.5 on ice. Typically, 25 μL of 100 m*M* CHAPS is added to 75 μL of ROS disc membrane on ice, and the sample is incubated for 30 min to allow for complete dissolution of the membrane structure.
2. After incubation, the sample is sedimented for 10 min at 14,000*g* at 4°C in an Eppendorf microfuge to remove unextracted material, and the supernatant is used for immobilization.
3. Twenty-five microliters of the solubilized rhodopsin is diluted 10-fold by addition to 225 μL of chilled Solution C. The two divalent cations are essential for the carbohydrate binding by ConA, and they must be present to retain rhodopsin bound to ConA *(22)*. The 10 μ*M* concentrations of Ca^{2+} and Mn^{2+} in the running solution are sufficient to prevent detectable dissociation of rhodopsin from ConA measured over a 10 min time period (*see* **Note 3**).
4. The diluted rhodopsin is warmed to 25°C for 1–2 min prior to injection. Because the association of rhodopsin with ConA is a slow reaction and the strategy is to saturate the ConA with rhodopsin, the flow rate is reduced to 2 μL/min to conserve the amount of receptor required for coupling. The exposure time is typically 25 min, by which time saturation of the binding equilibrium for ConA has been achieved. At this time the flow rate is returned to 5 μL/min for washing out unbound rhodopsin.

Figure 2B shows a typical capture of bovine rhodopsin by a ConA-coupled surface. Initially, the lowered pH (6.0) of the rhodopsin dilution solution assists the ionic interaction of the solubilized rhodopsin with the carboxyl groups of the carboxymethyl dextran hydrogel, concentrating it for the binding to ConA. The slow binding interaction of the rhodopsin with ConA is apparent from the rate of the exponential approach to equilibrium. Note that a large fraction of the bound rhodopsin remains associated with the immobilized ConA over the time period of this experiment (*see* **Note 4**).

3.4. Experimental Applications of Immobilized Rhodopsin

3.4.1. Synergistic Binding of G-Protein Subunits to Immobilized Rhodopsin

ConA-immobilized rhodopsin surfaces provide a basis for examining the interactions of G-protein subunits with rhodopsin independent of guanine nucleotide exchange. In a typical SPR experiment G-protein subunits are injected over the ConA–rhodopsin surface at flow rates of 5 μL/min. This flow rate balances the requirements for rapid mixing of the flow paths (60 nL volume equilibrates in about 3 s) for kinetic studies with the expense of adequate volumes of G proteins at the concentrations required for binding analysis. This trade-off is most dramatically illustrated in the experiments measuring the kinetics of G-protein subunits presented in **Figs. 3** and **4** below. For the slower binding reactions of the hydrophobic Gβγ dimmers, 5 μL/min is clearly adequate for kinetic analysis (*see* **Fig. 4**). **Figure 3** presents the binding interactions of bovine retinal-G-protein subunits independently and in concert both to the immobilized rhodopsin and to the ConA–dextran used to immobilize the rhodopsin. What is revealed is that rhodopsin displays a measurable binding of Gβγ independent of Gα and that there is a profound synergy of the binding reaction when both Gα and Gβγ subunits are present (*see* **Notes 5** and **6**).

3.4.2. Equilibrium Dissociation Constants for G-protein Subunits and Rhodopsin

The data shown in **Fig. 3** demonstrate the rapidity of the equilibration of binding of retinal-G-protein subunits to rhodopsin. A number of experiments suggest that both rates of Gα and Gβγ association with and dissociation from rhodopsin are too rapid for precise measurement under these conditions. Indeed, for Gα it appears that equilibration is completed within the mixing time of the flow path at a 5 μL/min flow rate. Therefore, the kinetics of retinal-G-protein subunit interactions have not been pursued. However, equilibrium binding is readily obtained, and these values can be determined in experiments varying the concentrations of free G-protein subunits in the injection. These experiments utilize the synergistic binding of the combination of Gα and Gβγ to examine the binding constants for each. The concentration of one of the subunits is held constant, while that for its interacting partner is varied to assess the saturation of binding to rhodopsin.

The basis for this type of experiment is provided in **Fig. 3**, demonstrating the synergistic interaction of retinal-G-protein subunits binding with rhodopsin. If the concentration of retinal Gα is held constant and varying concentrations of the Gβγ are mixed together with Gα prior to injection, a family of binding association curves are generated similar to that in **Fig. 3B** that enable

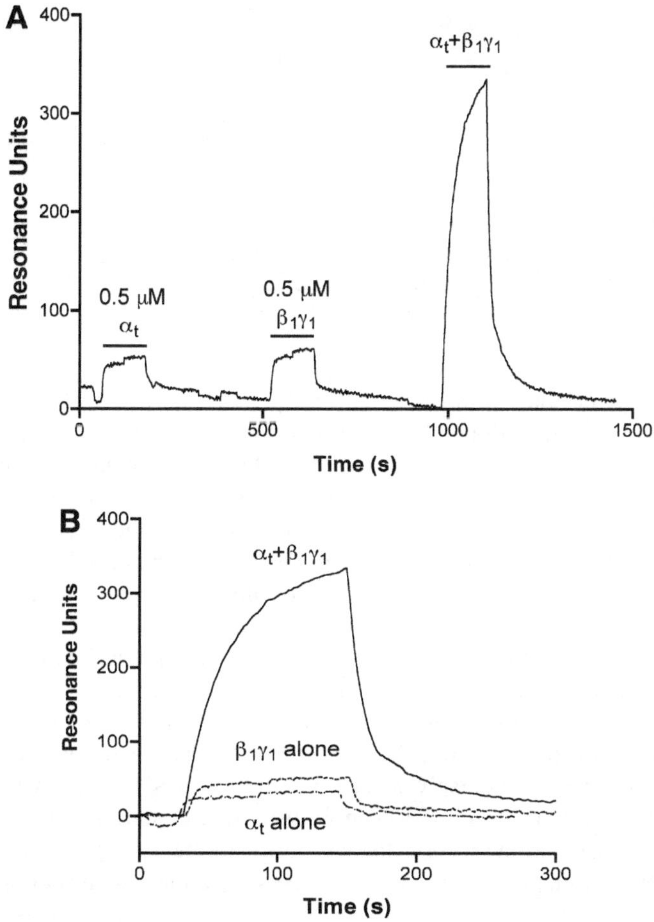

Fig. 3. Synergistic binding of retinal-G-protein subunits with immobilized rhodopsin. **(A)** The sensogram from three sequential injections over a ConA–rhodopsin surface prepared as presented in Fig. 2. The horizontal bars indicate the times during which protein samples are injected into the flow cell monitored for the SPR signal. The three injections are as indicated: 0.5 μ*M* α_t, 0.5 μ*M* βγ, and the combination of 0.5 μ*M* α_t with 0.5 μ*M* βγ. **(B)** The data from the sensogram of **A** baseline-subtracted and superimposed to compare the three binding interactions.

the calculation of the K_d for Gβγ by plotting the plateau values for SPR for each injection as a function of the concentration of Gβγ added. These data conform reasonably well to a simple bimolecular binding model, which is used to compute the binding constant. Our data find an apparent K_d value of about 300 n*M* *(8)* which is in good agreement with data for retinal Gβγ saturation of the rhodopsin-catalyzed binding of GTPγS to Gα *(13)* (*see* **Notes 7** and **8**).

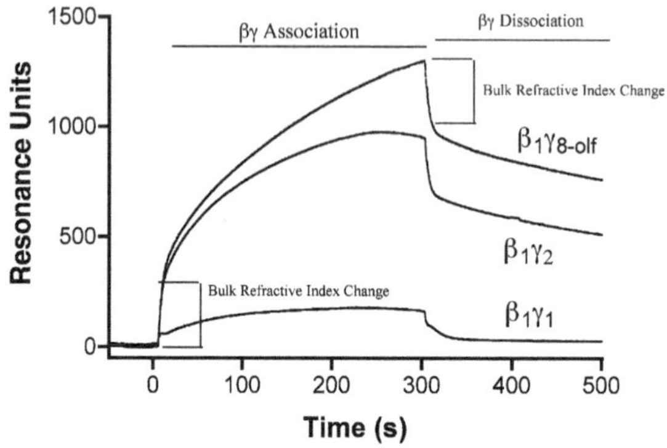

Fig. 4. Independent binding interaction of Gβγ with rhodopsin. This presents data from three independent injections of Gβγ on a surface of ConA-immobilized rhodopsin. The SPR signals have been baseline-subtracted and overlayed for comparison of the binding interactions. The significant features of the hydrophobic Gβγ dimer binding for $G\beta_1\gamma_2$ and $G\beta_1\gamma_{8-olf}$ are as follows: the initial phase of the SPR signal includes a rapid increase in the first second, which is due to the refractive index change of the sample solution (containing CHAPS) versus the running solution. This same rapid change is seen in the decrease of the SPR signal in the first second after the injections finish. These two time periods must be excluded from any numerical analysis of the data, as they originate from bulk-solution refractive index changes rather than protein–protein binding interactions. Note also that after the refractive index change, both association, and, most dramatically, the dissociation phase for these two dimers are slow. These easily present themselves for numerical analysis and, when applied to multiple concentrations of the dimers, can be used to calculate the association rate constants and the dissociation rate constants for measurement of the equilibrium binding constants for the independent binding of these Gβγ dimers with rhodopsin.

3.4.3. Independent Binding of Gβγ Subunits to Immobilized Rhodopsin

The SPR approach allows the examination of the molecular interaction of G-protein-signaling components that do not lead to functional changes. Although all of the protein interactions discussed thus far can be assessed indirectly through functional assays, the binding of Gβγ to receptor in the absence of Gα has no identified functional sequelae. **Figure 3** provides an indication of the independent binding of retinal Gβγ. As there are potentially a wide variety of Gβγ dimers of distinct Gβ and Gγ composition, SPR provides an opportunity to examine the influence of this diversity on receptor interaction. To test the specificity for rhodopsin, Gβγ dimers of defined compositions are expressed by co-infection in Sf9 cells ***(15,16)***. **Figure 4** illustrates the examination of Gβγ selectivity for rhodopsin binding determined by SPR. As seen here, the binding interactions of $G\beta_1\gamma_1$, $G\beta_1\gamma_2$, and $G\beta_1\gamma_{8-olf}$ all at 200 n*M* give distinct kinetic and equilibrium values. Most significantly, although the $G\beta_1\gamma_1$ completely dissociates from rhodopsin within seconds of terminating the injection, both $G\beta_1\gamma_2$ and $G\beta_1\gamma_{8-olf}$ dissociate slowly. Calculated half-times for their dissociations are on the order of 1000 s (*see* **Notes 9** and **10**).

3.5. Covalent Coupling of G-Protein Subunits to SPR Chips

Procedures for the immobilization of $G_0\alpha$ and bovine brain $G\beta\gamma$ have been described for affinity chromatography to purify G-protein subunits *(10,11)*. This procedure involves the cross-linking of G-protein subunits to dextran matrices via thioester coupling with the heterobifunctional reagent GMBS (*N*-[γ-maleimidobutylryloxy]succinimide ester) that contains a short spacer group. The procedure for affinity chromatography can be adapted with essentially identical chemistries for SPR. Because the flow coupling in the SPR instrument led to inconsistent results (usually poor coupling), we perform the following operations by pipetting volumes of 50 µL directly to the exposed sensor surface. This seems to have the advantage of allowing for more uniform immobilization and longer contact times.

1. First, a primary amine-linkable group is created by reacting the carboxymethlydextran hydrogel with 1 *M* diamino ethane, 200 m*M* EDC, and 50 m*M* NHS in 100 m*M* potassium phosphate buffer, pH 7.
2. After 20 min at room temperature, unreacted reagents are removed by 10 successive rinsing of the surface with 50 µL of distilled water.
3. 1 *M* ethanolamine dissolved in water is then applied and allowed to react for 10 min to passivate all remaining activated carboxyl residues, followed by 10 rinses with water.
4. The surface is then reacted with 2 m*M* GMBS in 100 m*M* potassium phosphate buffer, pH 7, for 30 min at room temperature, followed by 10 rinses of the surface with distilled water.
5. Finally, the G-protein subunits are coupled by addition to the surface in 100 m*M* potassium phosphate buffer, pH 7, with 0.2% (w/v) Lubrol 12A9 for $G\beta_1\gamma_2$ or other hydrophobic G-protein subunits (detergent is omitted for retinal $G\alpha_t$, recombinant myr-$G\alpha_0$, myr-$G\alpha_{i1}$, and $G\beta_1\gamma_1$) for 30 min at room temperature.
6. Uncoupled G protein is removed by 10 successive rinses of the surface with the 100 m*M* potassium phosphate buffer, pH 7, and 0.2 % Lubrol (for hydrophobic G-protein subunits).
7. The G-protein–coupled sensor chip is then inserted into the instrument for measurement of subunit interactions as described below. Typically this procedure yields 600–1000 RU of coupled G-protein subunit with a binding capacity of about 400–700 RU for the cognate subunit. This coupling procedure produces less heterogeneity in the orientation of immobilized protein than amine coupling because there are fewer surface-accessible cysteines than primary amines. Furthermore, it does not require the exposure of the protein to low pH in which G proteins are unstable.

3.5.1. Determination of Equilibrium Constants for Gα–βγ Binding

The major purpose for developing the immobilization of G-protein subunits is to examine the kinetics and thermodynamics of their interactions as controls for the constants determined for the interactions of the Gαβγ complex and independent subunits with rhodopsin (or other receptors). These experiments are routinely performed in the running Solutions D. G-protein subunits are prepared for injection by gel filtration in Solution E, and their concentrations are adjusted prior to injection by dilution with Solution E. We have examined the interactions of Gβγ dimers with immobilized $G\alpha_0$, $G\alpha_{i1}$, and $G\alpha_t$. Conversely, we have immobilized $G\beta_1\gamma_1$ and $G\beta_1\gamma_2$ and examined the binding of the Gα subunits. Theoretically, by analyzing a family of binding curves at differing G-protein subunit concentrations to obtain the second-order rate constant for binding association, and by determining the dissociation rate constant of the bound subunits, the equilibrium binding constant is the ratio of the two constants. In practice, because the immobilization does not produce a homogeneous orientation of the G-protein subunits, the direct interactions of individual species of G-protein subunits do not produce single, consistent equilibrium association or dissociation rate constants. We therefore use a strategy of solution competition to determine the binding constant. This has the virtue of determining the binding interaction for both subunit partners in unmodified form, and it overcomes the difficulties outlined above. The theory of the method is detailed elsewhere ***(23)***. The experiment using this approach becomes quite straightforward. The binding of a constant concentration G-protein subunit (say Gα) to immobilized Gβγ is monitored alone and in the presence of varying concentrations of Gβγ. The presence of the competing Gβγ diminishes the free concentration of Gα, and both the rate and equilibrium binding obtained are diminished based on the binding constants of the species of Gα and Gβγ. The data are then readily modeled for the equilibrium constant.

3.5.2. Regenerating the Surface by Dissociating Gαβγ

Most experimental strategies for biosensor experiments require multiple cycles of binding and dissociation of Gα–βγ. Before each cycle, the bound subunit must be completely dissociated from the immobilized subunit with little or no deterioration in its ability to rebind in subsequent cycles. Because of the relatively slow dissociation of Gβγ from Gα, regeneration of the surface by changing the chemical environment is necessary. High concentrations of Mg^{2+} lower the affinity of Gα for Gβγ, and, as a consequence, Gαβγ can be rapidly dissociated to remove free subunit from the surface by injecting 20 μL of Solution D containing 120 m*M* $MgSO_4$, 1 m*M* DTT, and 1.0% Lubrol PX at a flow rate of 10 μL/min. Including Lubrol PX facilitates the Mg^{2+}-induced dissociation of Gαβγ. Stripping the SPR chip of bound subunits by this procedure has little effect on the ability of the immobilized subunit to rebind in subsequent experimental cycles for as many as 40 cycles of stripping and rebinding (*see* **Notes 11–13**).

4. Notes

1. The running solutions used for these experiments do not contain detergent, and detergents are added only with G-protein subunits at minimal concentrations sufficient for protein solubility. The running solution utilized for these experiments is based on reagents used for measuring GTP binding exchange reactions catalyzed by rhodopsin ***(14)*** in order to provide for a direct comparison of protein interaction affinity constants determined directly by SPR and indirectly by saturation of G-protein activation measurements.
2. The coupling procedures described above lead to the immobilization predominantly of dimeric ConA. The immobilized ConA is quite stable, so that rhodopsin coupling does not have to proceed immediately. However, the coupling reactions are sufficiently reproducible and rapid that we find no advantage to prior preparation of ConA-modified surfaces. For experiments defining the independent binding of Gα or Gβγ described below, surfaces modified with 7000–12000 RU of ConA have been utilized. We routinely wash the ConA surface for 30–60 min with running solution at a flow rate of 5 μL/min to allow for complete tetramer–dimer dissociation, prior to rhodopsin immobilization.
3. Mn^{2+} can inactivate Gα by attaching to the sulfur atoms of cysteine residues. The 10 μ*M* concentration of this ion does not add appreciably to the refractive index of the running solution. Therefore, we prepare and store the G-protein subunits in the absence of the additional divalent cations to enhance their stability.
4. We have designed these procedures to be limited by the amount of ConA immobilized to address issues of nonspecific interactions of G proteins with ConA discussed below. The immobilized rhodopsin has a surprisingly long half-life at 25°C of approx 40–60 h based on the binding capacity for Gβγ. However, several factors may decrease the useful lifetime of a rhodopsin surface, most critically, repeated detergent exposures and exposure to denaturing concentrations of detergent are detrimental to the rhodopsin. All of our research has utilized either Na-cholate or CHAPS as the detergent for solubilizing G-protein subunits, and we routinely limit the concentrations of these to less than 800 μ*M* in the sample injections with exposure times of 5 min or less. Because of the variability in expected rhodopsin surface life times, we generally use a surface for only a single day of experimental tests, synthesizing a new surface for each experiment.
5. The Running solution used for these studies is based on the composition of the reagent used for the analysis of G-protein subunit saturation of rhodopsin interaction by GTPγ[^{35}S] binding ***(14)***. The binding reagent differs from the SPR Running solution in the inclusion of BSA in the binding reagent (which has been eliminated from the SPR Running solution), the addition of $MnCl_2$ and $CaCl_2$ to the SPR Running solution for retention to ConA, and the increase of NaCl from 100 m*M* in the binding reagent to 150 m*M* in the SPR Running

solution to eliminate ionic adsorption to the carboxymethyldextran. None of these changes influences the rates of GTPγS binding catalyzed by rhodopsin, except that the recovery of bound GTPγS is increased by the presence of BSA at low concentrations of Gα.

6. It is important to note that the results for the synergistic binding of Gα and Gβγ are dramatically influenced by the temperature of the samples when mixed. If the subunits are mixed and held on ice prior to injection, the synergism is profoundly reduced. Therefore, the samples are diluted into the Running solution, which has been equilibrated to 25°C about 120 s before injection, and the mixed samples are then maintained at 25°C prior to injection.
7. Ideally, a family of equilibrium binding curves will be generated co-varying both Gα and Gβγ concentrations, from which the individual equilibrium constants may be obtained. Such an evaluation is inherently model dependent, as the impact of variation of each subunit on the binding interactions of its partner are governed by the molecular mechanism of rhodopsin catalysis of G protein activation. SPR analysis could provide important clues as to the precise mechanism.
8. The limitations on this approach are the binding constants for the G-protein subunits. While the empirical determination of saturation by retinal Gβγ was obtainable in these experiments, that for the retinal Gα was not. It appears to have too low an affinity for affordable measurement by these methods (we estimate $K_d >$ 1 μ*M*). As for the determination of the rate constants for the binding interactions of Gα with rhodopsin, the repeated injection of 20–50 μL of G-protein subunits at several micromolar concentration is impractical for tissue-derived proteins.
9. These rates are not facilitated by the addition of detergent to the Running solution, nor metal ions, nor modest pH changes. Indeed, to date the only conditions that have been identified to enhance the rate of this dissociation reaction lead to denaturation of the rhodopsin. Hence, the examination of the hydrophobic Gβγ dimers is limited in that there are no convenient regeneration procedures, and sequential injections require extensive washing times to allow for complete dissociation of the Gβγ.
10. Except for the naturally occurring $G\beta_1\gamma_1$ and $G\beta_3\gamma_{8\text{-cone}}$ dimers, all of the Sf9-produced products we have tested behave as integral membrane proteins and require detergent for maintaining solubility. These samples are prepared and stored in 8 m*M* CHAPS solutions, and the running solution for these dimers is supplemented with 8 m*M* CHAPS as described above. The introduction of CHAPS with the Gβγ presents a technical difficulty in the SPR signal due to the differences in the refractive indices between running solution without and with CHAPS. This is illustrated in **Fig. 4** by the steep increases and decreases in the SPR signal at the onset and termination of injections of $G\beta_1\gamma_2$ and $G\beta_1\gamma_{8\text{-olf}}$ samples. As discussed above, the stability of the rhodopsin-binding activity for G proteins is drastically reduced in the presence of detergent, so that this refractive index change cannot be avoided by the inclusion of a constant concentration of detergent to the running solution. The contribution of CHAPS refractive index to the SPR signal for Gβγ injections can be addressed by the following strategy. For the Biacore 2000 and later instruments, it is possible to specify that all four flow paths be coupled

with ConA, while only one, two, or three paths are then further coupled to rhodopsin. Because the identical solutions are applied to the flow paths, the coupling of ConA is nearly identical in all four flow cells. Thus, the flow paths that have not been further exposed to rhodopsin serve as controls to allow for the subtraction of all contributions to the SPR signal that are not due to the protein–protein interactions with rhodopsin. For single flow path monitoring, a somewhat more cumbersome approach can be adopted using the same strategy. For these experiments, reference injections of the G-protein samples are made using the immobilized ConA surface, prior to binding the rhodopsin. These injections result in nearly square-wave SPR signals from the refractive index changes. The stored reference signals can then be subtracted from the signals obtained for the same G-protein samples with the rhodopsin-coupled surface.

11. In solution, high concentrations of Mg^{2+} normally cause the guanine nucleotide to dissociate from Gα producing an unstable state of the protein that is susceptible to denaturation ***(24)***. Denaturation is accompanied by a loss in the ability of Gα to bind Gβγ ***(25)***, and this can be prevented by GDP, GTP, or its nonhydrolyzable analogs, or by AlF_4^- at concentrations sufficient to ensure occupancy of the guanine nucleotide binding site. Consequently, 100 μ*M* GDP is routinely included in the regeneration buffer that is used to dissociate Gβγ from the immobilized Gα. However, it should be noted that this practice may not be necessary. We have observed that exposure of immobilized Gα to 120 m*M* Mg^{2+} in the absence of guanine nucleotides for a period of time that would denature the soluble protein does not prevent the immobilized protein from subsequently binding Gβγ. These data suggest that the immobilized Gα subunit is able to retain bound guanine nucleotide under conditions promoting nucleotide dissociation from the protein in solution, thus rendering the immobilized Gα surprisingly stable.
12. If NaCl is omitted from Solution E, nonspecific binding of G-protein subunits to the SPR chip surface occurs. The concentration of salt in Solution E (150 m*M*) is sufficient to prevent nonspecific binding. It is possible to substitute $NaSO_4$ for NaCl in Solution E. Magnesium has a marked effect on G-protein subunit interaction. Magnesium is routinely included in the SPR mobile phase to ensure tight binding of guanine nucleotide to Gα, but the concentration could safely be reduced as micromolar concentrations are all that is required for guanine nucleotide binding to Gα.
13. It is not necessary to use detergents when investigating all G-protein-subunit interactions by SPR spectroscopy. However, the maximum binding for a given concentration of $G\beta_1\gamma_2$ to the immobilized $G\alpha_{i1}$ is reduced several fold in the absence of detergent when compared with the binding in Solution E that contains 0.005% Tween20. It appears that either the concentration of Gβγ reaching the surface of the SPR chip is lower, or the affinity of Gβγ for the immobilized $G\alpha_{i1}$ is lower in the absence of detergent. Although the reason for the reduced binding is unclear, a reduction of Gβγ concentration due to precipitation in the absence of detergent seems unlikely because this should also result in nonspecific and irreversible binding to the surface of the SPR chip, which we have not observed.

References

1. Slepak, V. Z. (2000) Application of surface plasmon resonance for analysis of protein-protein interactions in the G protein-mediated signal transduction pathway. *J. Mol. Recog.* **13,** 20–26
2. Schuck, P., Boyd, L. F., and Andersen, P. S. (1999) Measuring protein interactions by optical biosensors in *Current Protocols in Protein Science* (Coligan, J. E., Dunn, B. M., Ploegh, H. L., Speicher, D. W., and Wingfield, P. T., eds.), John Wiley, New York, NY, Vol. 2, pp. 20.2.1–20.2.21.
3. Karlsson, R., Roos, H., Fġerstam, L., and Persson, B. (1994) Kinetic and concentration analysis using BIA technology, *Methods: A companion to Methods in Enzymology* **6,** 99–110
4. Stenberg, E., Persson, B., Roos, H., and Urbaniczky, C. (1991) Quantitative determination of surface concentration of protein with surface plasmon resonance using radiolabeled proteins. *J. Coll. Interface Sci.* **143,** 513–526
5. Bieri, C., Ernst, O. P., Heyse, S., Hofmann, K. P., and Vogel, H. (1999) Micropatterned immobilization of a G protein-coupled receptor and direct detection of G protein activation. *Nature Biotech.* **17,** 1105–1108.
6. Salamon, Z., Brown, M. F., and Tollin, G. (1999) Plasmon resonance spectroscopy: probing molecular interactions within membranes. *Trends in Biochem.* **24,** 213–219
7. Kuhn, H. (1980) Light- and GTP-regulated interaction of GTPase and other proteins with bovine photoreceptor membranes. *Nature* **283,** 587–589.
8. Mumby, S. M. and Linder, M. E. (1994) Myristoylation of G-protein alpha subunits. *Methods Enzymol.* **237,** 254–268.
9. Clack, J. W. and Stein, P. J. (1988) Opsin Exhibits cGMP-Activated Single-Channel Activity. *Proc. Natl. Acad. Sci. USA* **85,** 9806–9810.
10. Pang, I.-H. and Sternweis, P. C. (1989) Isolation of the α subunits of GTP-binding regulatory proteins by affinity chromatography with immobilized βγ subunits. *Proc. Natl. Acad. Sci. USA* **86,** 7814–7818.
11. Pang, I. H., Smrcka, A. V., and Sternweis, P. C. (1994) Synthesis and applications of affinity matrix containing immobilized beta-gamma subunits of G proteins. *Methods Enzym.* **237,** 164–174.
12. Yamazaki, A. and Bitensky, M. W. (1988) Purification of rod outer segment GTP-binding protein subunits and cGMP phosphodiesterase by single-step column chromatography. *Methods Enzymol.* **159,** 702–710.
13. Fawzi, A. B., Fay, D. S., Murphy, E. A., Tamir, H., Erdos, J. J., and Northup, J. K. (1991) Rhodopsin and the retinal G-protein distinguish among G-protein βγ subunit forms. *J. Biol. Chem.* **266,** 12,194–12,200.
14. Clark, W. C., Jian, X., Chen, L., and Northup, J. K. (2001) Independent and synergistic binding of retinal G protein subunits with bovine rhodopsin measured by surface plasmon resonance. *Biochem. J.* **358,** 389–397.
15. Kozasa, T. and Gilman, A. G. (1995) Purification of recombinant G proteins from Sf9 cells by hexahistidine tagging of associated subunits. *J. Biol. Chem.* **270,** 1734–1741.

16. Wildman D. E., Tamir, H., Leberer, E., Northup, J. K., and Dennis, M. (1993) Prenyl modification of guanine nucleotide regulatory protein γ_2 subunits is not required for interaction with transducin α subunit or rhodopsin. *Proc. Natl. Acad. Sci. USA* **90,** 794–798
17. Papermaster, D. S. and Dreyer, W. J. (1974) Rhodopsin content in outer segment membranes of bovine and frog retinal rods. *Biochemistry* **13,** 2438–2444.
18. Fawzi, A. B. and Northup, J. K. (1990) Guanine nucleotide binding characteristics of transducin: essential role of rhodopsin for rapid exchange of guanine nucleotides. *Biochemistry* **29,** 3804–3812.
19. Kalb, A. J. and Lustig, A. (1968) The molecular weight of concanavalin A. *Biochim. Biophys. Acta* **168,** 366–367.
20. Gunther, G.R., Wang, J. L., Yahara, I., Cunningham, B. A., and Edelman, G. M. (1973) Concanavalin a derivatives with altered biological activities. *Proc. Natl. Acad. Sci. USA* **70,** 1012–1012.
21. Johnsson, B., Lfå, S., and Lindquist, G. (1991) Immobilzation of proteins to a carboxymethyldextran-modified gold surface for biospecific interaction analysis in surface plasmon resonance sensors. *Anal. Biochem.* **198,** 268–277
22. Reeke, G. N., Becher, J. W., Cunningham, B. A., Gunther, G. R., Wang, J. L., and Edelman, G. M. (1974) Relationship between the structure and activities of conanavalin A. *Ann. NY Acad. Sci.* **234,** 369–382.
23. Rebois, V., Schuck, P., and Northup, J. K. (2002) Elucidating kinetic and thermodynamic constants for interaction of G protein subunits and receptors by surface plasmon resonance spectroscopy, in *Methods of Enzymology*, Vol. 344 (Iyengar, R. and Hildebrandt, J., eds.) Academic Press, San Diego, CA, pp. 15–42.
24. Katada, T., Oinuma, M., and Ui, M. (1986) Two guanine nucleotide-binding proteins in rat brain serving as the specific substrate of islet-activating protein, pertussis toxin. Interaction of the α-subunits with the βγ-subunits in development of their biological activities. *J. Biol. Chem.* **261,** 8182–8191
25. Toyoshige, M. Okuya, S., and Rebois, R. V. (1994) Choleragen catalyzes ADP-ribosylation of the stimulatory g protein heterotrimer but not its free α-subunit. *Biochemistry* **33,** 4865–4871.

7

Using Light Scattering to Determine the Stoichiometry of Protein Complexes

Jeremy Mogridge

Abstract

The stoichiometry of a protein complex can be calculated from an accurate measurement of the complex's molecular weight. Multiangle laser light scattering in combination with size-exclusion chromatography and interferometric refractometry provides a powerful means for determining the molecular weights of proteins and protein complexes. In contrast to conventional size-exclusion chromatography and analytical centrifugation, measurements do not rely on the use of molecular weight standards and are not affected by the shape of the proteins. The technique is based on the direct relationship between the amount of light scattered by a protein in solution and the product of its concentration and molecular weight. A typical experimental configuration includes a size-exclusion column to fractionate the sample, a light-scattering detector to measure scattered light, and an interferometric refractometer to measure protein concentration. The determination of the molecular weight of an anthrax toxin complex will be examined to illustrate how multiangle laser light scattering can be used to determine the stoichiometry of protein complexes.

Key Words

Light scattering; stoichiometry; molecular weight; size-exclusion chromatography; interferometric refractometer; protein complex.

1. Introduction

Elucidation of the stoichiometry of a protein complex can provide insights into its structure and molecular mechanisms of action. Stoichiometry can be calculated from the measurement of the molecular weight of a complex, but this can be challenging if there are several copies of more than one protein species. Techniques such as conventional size-exclusion (Chapter 9) chromatography and analytical centrifugation (Chapter 8) have been used to estimate molecular weights, but these techniques are limited by their reliance on pro-

From: *Methods in Molecular Biology, vol. 261: Protein–Protein Interactions: Methods and Protocols*
Edited by: H. Fu © Humana Press Inc., Totowa, NJ

tein standards and by the influence of protein shape on the measurement. In contrast, the combination of size-exclusion chromatography, laser light scattering, and interferometric refractometery is an absolute method for the determination of molecular weight ***(1)***. It is based on the theory that the amount of light scattered by a protein in solution is directly proportional to the product of the protein's concentration and molecular mass. Scattered light is measured by a light-scattering (LS) detector, protein concentration is measured by an interferometric refractometer (IR), and the molecular weight is calculated by computer software. The accuracy of this technique allows for the precise molecular weight measurements that are required to calculate the stoichiometry of a multisubunit protein complex. The determination of the stoichiometry of the *Bacillus anthracis* edema toxin complex will be used as an example to describe how multiangle laser light scattering can be used to probe protein structure ***(2)***. Edema toxin consists of two proteins, edema factor (EF) and protective antigen (PA), that are secreted from *B. anthracis* and assemble on the surface of mammalian cells ***(3)***. After PA is cleaved on the cell surface into two fragments, the remaining cell-associated fragment, PA_{63}, oligomerizes into heptamers that can bind edema factor to form a toxic complex. The molecular weight of the saturated complex measured in solution corresponded to a stoichiometry of three molecules of EF per PA_{63} heptamer.

2. Materials

1. HPLC.
2. DAWN EOS 18-angle light scattering detector (Wyatt Technology).
3. OPTILAB DSP interferometric refractometer (Wyatt Technology).
4. ASTRA software (Wyatt Technology).
5. Superdex 200 HR 10/30 column (Amersham Biosciences).
6. Column buffer: 20 m*M* Tris-HCl, pH 8.2, 200 m*M* NaCl.
7. 0.2 micron filter units (Nalgene).
8. 0.02 micron Anotop 10 filters (Whatman).

3. Methods

3.1. System Setup

Before measurements are taken, the LS detector must be calibrated with a pure solvent that has a known Rayleigh ratio, such as toluene. Calibration relates the voltage detected by the 90° detector to the measured intensity of scattered light. If the LS detector was calibrated by the manufacturer, it needs to be recalibrated only if any changes to the detector that affect the 90° signal, such as realignment of the laser, have been made.

The HPLC is connected in series to the LS detector and the IR. The IR should be placed after the LS detector to avoid subjecting the IR to high back-pressure. Connections should be made with tubing of small inner diameter (0.01 in.) to minimize the dead volume between the instruments.

A solution of 4 mg/mL bovine serum albumin (an isotropic scatterer) in column buffer can be injected onto the size-exclusion column and used to normalize the LS detector, a process that relates the measured voltages at each detector to the 90° detector in order to compensate for slight differences in electronic gain among the detectors. Normalization must be performed each time a different solvent is used.

A *dn/dc* (refractive index increment) value of 0.185 mL/g is used for nonglycosylated proteins in the molecular weight calculations made by ASTRA software.

3.2. Preparation

1. Turn on the IR approx 12 h before measurements are taken to warm the instrument. Set the temperature to approximately 10°C above room temperature.
2. Turn on the LS detector 2 h before measurements are taken.
3. Connect the size-exclusion column to the HPLC (*see* **Note 1**) and run buffer through the column while the light scattering detector is warming (*see* **Note 2**).
4. Once the column has preequilibrated with buffer (approx two column volumes), connect the waste line from the HPLC to the LS detector and the output from the LS detector to the IR (*see* **Note 3**).
5. Equilibrate the LS detector and IR with column buffer (*see* **Note 4**).
6. Flow buffer simultaneously through the reference and sample cells of the IR and set the baseline to zero. After the reference has been set, program the IR so that the column buffer bypasses the reference cell.

3.3. Injecting a Sample

Do not stop the HPLC pumps after the equilibration step or between measurements because starting and stopping the pumps may cause particles from the column to dislodge and interfere with the measurements. Start data collection by ASTRA software and inject each component of the protein complex (EF and PA_{63}; *see* **Note 5**) individually and then a mixture of the proteins (*see* **Note 6**). The results of a chromatography run in which a molar excess of EF was mixed with PA_{63} are shown in **Fig. 1**.

3.4. Analyzing the Results

The results can be analyzed using a computer program, such as ASTRA.

1. Set the baseline voltages of the signals detected by the LS detector and IR.
2. When selecting values for the light scattering and refractive index signals, use data relating to the middle of the protein peak where the signal-to-noise ratio is the highest (signal-to-noise ratio should be at least 2).
3. Use ASTRA software to calculate the molecular weights (**Table 1**; *see* **Note 7**).

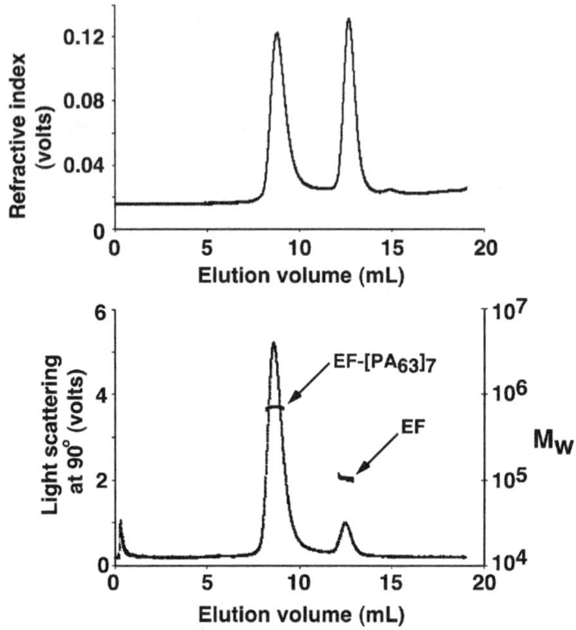

Fig. 1. Multiangle laser light scattering of EF:$[PA_{63}]_7$. A mixture (120 μL) of EF (260 μg) and $[PA_{63}]_7$ (93 μg; approximately twofold molar excess of EF over PA_{63} monomer) was chromatagraphed over a Superdex 200 size-exclusion column, which was connected to a light-scattering detector (lower panel) and an interferometric refractometer (upper panel). The values of molecular mass determined in volume increments across each peak are shown (arrows). Reprinted with permission from *Biochemistry* **41,** 1079–1082. © 2002 American Chemical Society.

4. Notes

1. The HPLC system should have a pulse dampener because small fluctuations in flow rate can affect the measurements. The Superdex 200 size-exclusion column was chosen for this study because EF and the EF:$[PA_{63}]_7$ complex are separated by this column.
2. It is important to make fresh buffer with high-quality reagents. Filter the buffer through a 0.2 μm filter and degas it thoroughly before use. Preequilibrate the column at the flow rate (e.g., 0.5 mL/min) that will be used to take measurements to avoid changes in flow rate that might cause particles to dislodge from the column.
3. New columns should be equilibrated for approx 48 h before measurements are taken. The LS detector and IR are stored in methanol, which should be replaced with filtered water before introducing buffer (solvents that are injected directly into the LS detector or IR should be filtered through a 0.02 μm filter).

Table 1
Molecular Mass Determinations Using Multiangle Laser Light Scattering

	Measured molecular mass (kDa)	Theoretical molecular mass (kDa)
EF	93 ± 0.3	91
$[PA_{63}]_7$	460 ± 3	444
EF:$[PA_{63}]_7$	720 ± 20	717[a]

[a]Theoretical value for saturated complex is based on a stoichiometry of three molecules of EF per PA_{63} heptamer.

4. Equilibrate the LS detector and IR until the signals have stabilized. Baseline noise for the LS detector should be in the 10–20 mV range. The lower-angle LS detectors (the Wyatt EOS LS detector has 18 detectors) will have a higher level of noise, but these detectors can be turned off without adversely affecting the measurements. Detectors 8–18 were used in this study.
5. EF and $[PA_{63}]_7$ were purified as described previously ***(4,5)***. Small amounts of aggregated protein in a sample can interfere with the light-scattering measurements because aggregates are very efficient at scattering light. An aggregate will produce a high signal from the LS detector and a very low signal from the IR. Recently prepared protein samples that have not been frozen may contain less aggregrate than a sample that has been frozen and thawed.
6. The maximum injected volume is determined by the column specifications (250 μL for the Superdex 200 column), but smaller volumes can be used. The amount of protein required to make an accurate measurement depends on the molecular weight of the protein—more sample is required for low-molecular-weight proteins (approx 50 kDa) than for high-molecular-weight proteins (> 100 kDa) because high-molecular-weight proteins scatter more light. Accurate measurements of the amino terminal domain of *B. anthracis* lethal factor (30 kDa) were made with 200 μg of protein ***(2)***. Accurate measurements of transferrin (75 kDa) and ovalbumin (43 kDa) were made with less than 10 μg of protein ***(6)***. An additional consideration when measuring protein complexes is that the proteins should be at a high enough concentration to ensure that they fully associate. In general, the protein concentration should exceed the dissociation constant of the interaction by approx 10-fold (taking into account that the protein may be diluted by several-fold on the size-exclusion column). The ratios of the proteins in the samples should be varied to ensure that saturation has occurred. The amount of one protein is held constant and increasing amounts of the other protein are added. Saturation has been achieved when the addition of protein does not change the molecular weight of the complex peak and a free protein peak is observed.

7. ASTRA software calculates molecular weights at multiple points across the selected section of an elution peak and uses these data to calculate the average molecular weight of the protein. These points can be displayed in a graph of molecular weight vs elution volume to aid in the analysis of the experiment (**Fig. 1**, lower panel). Calculated molecular weights across a homogeneous peak will be similar at each point, but these points may form a frowning pattern, which is indicative of peak broadening between the LS detector and IR (*see* EF:$[PA_{63}]_7$ complex peak in **Fig. 1**). Slight peak broadening will not adversely affect the molecular weight measurements. If the protein peak is not homogeneous, one might observe a linear pattern with higher molecular weights at lower elution volumes and lower molecular weights at higher elution volumes. If this is the case, a size-exclusion column that better separates the species can be used.

References

1. Wyatt, P. J. (1993) Light scattering and the absolute characterization of macromolecules. *Anal. Chim. Acta* **272,** 1–40.
2. Mogridge, J., Cunningham, K., and Collier, R. J. (2002) Stoichiometry of anthrax toxin complexes. *Biochemistry* **41,** 1079–1082.
3. Duesbery, N. S. and Woude, G. F. V. (1999) Anthrax toxins. *Cell. Mol. Life Sci.* **55,** 1599–1609.
4. Zhao, J., Milne, J. C., and Collier, R. J. (1995) Effect of anthrax toxin's lethal factor on ion channels formed by the protective antigen. *J. Biol. Chem.* **270,** 18,626–18,630.
5. Miller, C. J., Elliott, J. L. ,and Collier, R. J. (1999) Anthrax protective antigen: prepore-to-pore conversion. *Biochemistry* **38,** 10,432–10,441.
6. Folta-Stogniew, E. and Williams, K. R. (1999) Determination of molecular masses of proteins in solution: implementation of an HPLC size exclusion chromatography and laser light scattering service in a core laboratory. *J. Biomolecular. Tech.* **10,** 51–63.

8

Sedimentation Equilibrium Studies

Ian A. Taylor, John F. Eccleston, and Katrin Rittinger

Abstract

The reversible formation of protein–protein interactions plays a crucial role in many biological processes. In order to carry out a thorough quantitative characterization of these interactions, it is essential to establish the oligomerization state of the individual components first. The sedimentation equilibrium method is ideally suited to perform these studies because it allows a reliable, accurate, and absolute value of the solution molecular weight of a macromolecule to be obtained. This technique is independent of the shape of the macromolecule under investigation and allows the determination of equilibrium constants for a monomer–multimer self-associating system.

Key Words

Analytical ultracentrifugation; sedimentation equilibrium; solution interaction; quaternary protein structure; protein–protein interaction.

1. Introduction

Analysis of macromolecular interactions, in particular, protein–protein interactions, is an increasingly common goal in modern biological research. Methods vary widely, from qualitative techniques aimed at the detection of interactions in vivo to quantitative in vitro analyses that attempt to produce a detailed thermodynamic description of an interacting system. A necessary part of any quantitative in vitro method is that an initial characterization of the oligomerization state of the systems components be undertaken. This characterization is essential in order to correctly interpret binding data produced from subsequent titration experiments monitored, for example, by optical and magnetic resonance spectroscopy (see related chapters in this volume) or isothermal titration calorimetry (Chapter 3). Sedimentation equilibrium studies are unique in that they can be used at all stages of these quantitative studies. The technique allows a reliable, accurate, and absolute value of the solution

From: *Methods in Molecular Biology, vol. 261: Protein–Protein Interactions: Methods and Protocols*
Edited by: H. Fu © Humana Press Inc., Totowa, NJ

molecular weight of a macromolecule to be obtained, making it invaluable in an initial characterization of a system. Furthermore, the technique can be extended to look at self-associating systems and heterologous equilibria, making it complementary to spectroscopic and calorimetric methods (some good reviews about this technique include refs. ***1–8***).

1.1. Determination of Solution Molecular Weight

Popular methods for the determination of molecular weights are size-exclusion chromatography and dynamic laser light scattering. In these methods, estimates of the molecular weight of unknowns are obtained by interpolation, using a curve generated by plotting an experimentally determinable parameter against the molecular weight of a set of standards. In size-exclusion chromatography, K_{av}, the partition coefficient between a porous matrix and free solution, is often used. In dynamic laser light scattering, a translational diffusion coefficient D_T, derived from the measured autocorrelation function, is often the parameter of choice. Along with the fact that using these methods molecular weights have to be obtained by interpolation, the main drawback is that no account is taken of molecular shape. Consequently, there are often large errors in the values obtained. The effects of molecular shape are not accounted for because the experimentally measured parameter reported by both methods is a function of D_T rather than molecular weight. Although D_T is correlated with molecular weight, it is directly related, through eq. (1), to an important molecular parameter, the frictional coefficient, f ***(9)***.

$$D_T = \frac{RT}{N_0 f} \qquad (1)$$

The value of f can be regarded as what determines inherent capacity of a molecule to undergo translational motion, and both molecular size and shape contribute to its value. Taking this into account it is easy to see how measurements of molecular weights are so shape dependant, because it is the frictional coefficient that directly determines both the D_T measured in dynamic light scattering and the partition coefficient observed in size-exclusion chromatography. The complications due to molecular shape are serious enough in single-component systems. When dealing with protein complexes and interacting systems, it should be apparent that these problems are further exacerbated because of the need to account for the effects of equilibrium constants.

1.2. Sedimentation Equilibrium Theory

To overcome problems associated with solution molecular weight determinations, it is best to examine a phenomenon where the shape of the molecule does not contribute to the measurement. This occurs when a concentration gradient is established by a macromolecular species sedimenting in a gravitational field, referred to as "sedimentation equilibrium." This method enables accurate, shape-independent, and absolute measurements of molecular weights to be obtained. Furthermore, sedimentation equilibrium studies can be used to analyze interacting systems, allowing the determination of equilibrium constants and stoichiometries alongside molecular weights.

In order to understand how this works, it is necessary to introduce the concept of flux. Equation (2) is a general expression for flow, and can be likened to Ohm's law. Where the flux (J_i), or the flow of material, is related to a term for a generalized conductivity (L_i) and ($\partial U_i/\partial r$) a generalized gradient of potential.

$$J_i = L_i\left(\frac{\partial U_i}{\partial r}\right) \tag{2}$$

In analytical ultracentrifugation, we are concerned with the flow of mass. The total potential in this case is comprised from a component due to the applied field, the centrifugal potential energy, together with a component from the chemical potential gradient generated from the solute. The conductivity term (L_i) in this case is the manifestation of the frictional coefficient and contributes to both flow due to sedimentation and flow due to diffusion:

$$J = \left|\frac{M(1-\nu\rho)}{N_0 f}\right|\omega^2 rC - \left[\frac{RT}{N_0 f}\right]\left(\frac{\partial C}{\partial r}\right) \tag{3}$$

where M, molecular weight; v, partial specific volume; ρ, solute density; N_0, Avogadro's number; ω angular velocity; r, radial distance from center of rotation; R, Gas constant; T, absolute temperature.

Equation (3) provides a full description of the transport process occurring in the ultracentrifuge cell. The terms $M(1\text{-}v\rho)/N_0 f$ and $RT/N_0 f$ correspond to the sedimentation coefficient (s) and translational diffusion coefficient (D_T), respectively, and chemical potential U_i has been replaced by concentration C. It should be noted that the term for the strength of the applied gravitational field ($\omega^2 r$) will dominate at high rotor speeds, diffusion then only manifests itself as the boundary spreading observed in sedimentation velocity experiments. At lower rotor speeds, back diffusion due to the establishment of the chemical potential gradient counteracts transport from the applied gravitational field. Under these conditions equilibrium is established where the net transport in the system vanishes to zero at all points. Equation (3) can then be written as

$$\left|\frac{M(1-\nu\rho)}{N_0 f}\right|\omega^2 rC = \left[\frac{RT}{N_0 f}\right]\left(\frac{dC}{dr}\right) \tag{4}$$

An important result of this is that the shape term ($N_0 f$) cancels meaning, that while the concentration gradient established by the macromolecular species is dependant on the molecular weight, it is completely independent of shape. The useful form of the expression involves rearrangement to give eq. (5) and then integration with respect to C and r, between the limits C_0 and C_x and r_0 and r_x gives eqs. (6a) and (6b):

$$\int_{C_0}^{C_x} \left(\frac{1}{C}\right) dC = \frac{M\omega^2 (1 - v\rho)}{RT} \int_{r_0}^{r_x} r dr \tag{5}$$

$$\ln\left(\frac{C_x}{C_0}\right) = \frac{M\omega^2 (1 - v\rho)}{2RT} \left(r_x^2 - r_0^2\right) \tag{6a}$$

or

$$C_x = C_0 \exp\left[\frac{M\omega^2 (1 - v\rho)}{2RT} \left(r_x^2 - r_0^2\right)\right] \tag{6b}$$

1.3. Analysis of Molecular Weight Data

The absorbance optical system of the Beckman XL-A is able to measure the optical density, at a chosen wavelength, at many points in the cell between r_0 and r_x ***(10)***. Providing data are collected within the usable linear range of the optical system, absorbance can replace concentration in eq. (6b) simply using the Beer–Lambert law:

$$A_x^\lambda = \varepsilon^\lambda C_x l \tag{7}$$

Molecular weights are then simply extracted by direct nonlinear least squares fitting of the data to the absorbance form of eq. (6b) between r_0 and r_x and solving for M and A_0 (*see* **Note 1**). An offset is also included in the fitting procedure for contributions to the profile from nonexact matching of the cells, but should be treated with care (*see* **Note 2**). In the simplest case, a single sedimenting species, the data should fit to a single exponential curve in terms of r^2. Deviations in the data from a single exponential are indicative of sample heterogeneity, nonideality, or an associative system. Often this is revealed by inspection of a plot of the fit residuals. A more complex model to account these effects can then be applied.

1.4. Chapter Outline

The instrumentation necessary to carry out analytical ultracentrifugation studies is described in the **Subheading 2.**, followed by a description of sample requirements and potential practical limitations of this technique. **Subheading 3.**, describes general considerations concerning experimental conditions for equilibrium runs and contains a detailed protocol for data collection and analysis (*see* **Subheading 3.3.** and **3.4.**).

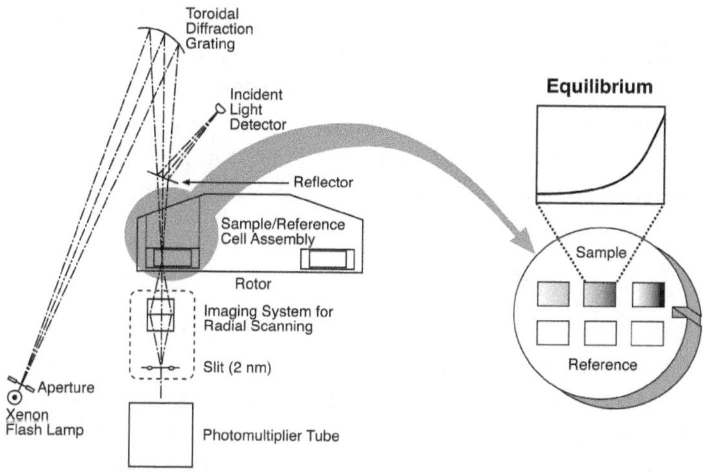

Fig. 1. A schematic diagram of the optical system of the Beckman Optima XL-A analytical ultracentrifuge. Figure courtesy of Beckman-Coulter.

2. Materials

2.1. Reagents

1. Tris (hydroxymethyl) methylamine and sodium chloride ("AnalaR," Merck).
2. Tris(2-carboxyethyl)phosphine (TCEP).
3. Dialysis cassettes or Mini dialysis units ("Slide-A-Lyser," Pierce).
4. Purified proteins (approx 1–2 mg).
5. Fluorocarbon oil FC-43 (Beckman).

2.2. Equipment

1. Optima XL-A analytical ultracentrifuge (Beckman-Coulter).
2. Six channel centerpieces for analytical ultracentrifuge cells.
3. Scanning (UV/visible) spectrophotometer.

2.3. Description of the Instrument

There are currently two models of analytical ultracentrifuges available, the Optima XL-A and Optima XL-I, both manufactured by Beckman-Coulter. The XL-A is equipped with a UV/visible absorbance optical system, while the XL-I has additional Rayleigh interference optics. We will concentrate on the XL-A in this chapter, whose basic feature is that the absorbance of the cell can be measured as a function of radial distance while the solution is being centrifuged. This is achieved by the use of cells in which the solution is contained within quartz windows. A monochromator placed above the cell allows it to be illuminated with monochromatic light between 190 and 800 nm while a slit mechanism below the cell scans radially at 0.001 cm or greater step sizes (**Fig. 1**).

Two types of centrifuge rotors that hold the cells are available—the An-50 Ti rotor (maximum speed 50,000 rpm) and the An-60 Ti rotor (maximum speed 60,000 rpm). The An-50 rotor has eight rotor holes, seven for sample cells and one for a counterbalance, whereas the An-60 has four rotor holes, three for sample cells and one for a counterbalance. The counterbalance is required in all centrifuge runs in order to coordinate the flash of the lamp with the cell position when it is above the detector.

2.4. Software and Data Analysis

The Beckman Optima XL-A/XL-I is supplied with analysis software based on the Origin® software package (MicroCal). This software has been used to fit all the data described in this chapter. The calculations of buffer density (ρ) and the partial specific volumes (v) have been carried out using SEDNTERP (*see* **Note 1**). This and several other useful programs for the analysis of equilibrium data have been developed by experts in the field and are available by download over the Internet, in most cases as Freeware. The Reversible Associations in Structural and Molecular Biology group (RASMB) maintains a web site that provides links to most of these programs, some of which are available for different platforms (http://www.bbri.org/RASMB/rasmb.html). Simulations to determine optimal run conditions can be carried out using the Beckman XL-A software, whereas the relative amounts of monomer and *n*-mers for self-associating systems can be calculated using the Ultrascan software (http://www.ultrascan.uthscsa.edu/).

2.5. Sample Preparation

When preparing a sample for sedimentation equilibrium analysis, various points should be taken into account concerning buffer composition and protein concentration.

2.5.1. Sample Buffer

The sample buffer should not absorb more than 0.3 OD referenced against water at the wavelength chosen for the experiment, as this may otherwise interfere with the range and linearity of absorbance measurements (*see* **Note 3**). Charge repulsion between macromolecules will lead to problems with nonideality, which is best addressed by using a buffer with ionic strength > 0.1 *M* and by carrying out the experiments close to the p*I* of the protein in order to minimize the charge on the surface of the molecule. Furthermore, it should be borne in mind that samples might be in the centrifuge for long periods of time (>60 h); therefore, buffer components that have significant time-dependent changes in their absorbance spectrum, for instance DTT, should be avoided.

2.5.2 Protein Concentration

The sample concentration should not exceed 0.7 OD when measured against the sample buffer, otherwise the linear range of the detector (approx 1.5 OD) will be exceeded when the absorbance rises toward the bottom of the cell during the run (*see* **Fig. 1**). If no information about the association state of the protein is available, 0.5 OD is a good starting point. If self-association is present, the loading concentration should be chosen such that there are detectable amounts of monomers and multimers (*see* **Subheading 2.4.**). The protein solution and sample buffer have to be in equilibrium, which is best achieved by gel filtration (using for example NAP-5 columns, Amersham Biosciences) or exhaustive dialysis (two or three buffer changes). We routinely use Slide-A-Lyzers (Pierce); these are available for different volumes starting at 10 μL.

Before every ultracentrifugation run, the sample should be checked for aggregation. This might be done by a separate sedimentation velocity experiment or by dynamic light scattering analysis if an instrument is available. Otherwise, a simple test is to spin the sample in a microcentrifuge for 15–20 min at maximum speed and to recheck the absorbance (*see* **Note 4**). If the sample OD has significantly decreased, the sample might not be suitable for equilibrium analysis at this time.

2.6. Limitations of Measurement for Protein–Protein Complexes

The range of molecular weights tractable by sedimentation equilibrium is extremely large ($5 \times 10^2 - 10^7$), governed only by the rotor speed attainable. However, the analysis of an interacting system is limited by the absorbance range where reliable concentration measurements can be made within the confines of the equilibrium constant under investigation. **Figure 2** shows the fraction of protein monomers expected for a set of monomer–dimer equilibria ($K_a = 10^3 - 10^9$ M^{-1}) over a concentration range of 1 n*M* to 1 m*M*. Superimposed on this figure, in grey, is the concentration range over which reliable absorbance measurements can be made.

It should be apparent that for interactions with $K_a > 10^8\ M^{-1}$, the sedimentation experiment will give an accurate value for the molecular weight of the complex and therefore the stoichiometry. For interactions much weaker than $10^4\ M^{-1}$ or where a heterogeneous mixture is present, sedimentation equilibrium will provide the weight-averaged molecular weight of the mixture showing no concentration dependency. For interactions between these two limits, $K_a > 10^4\ M^{-1}$ and $< 10^8\ M^{-1}$, sedimentation experiments will also provide a weight-averaged molecular weight, but in this case with a concentration dependence allowing the equilibrium constant for the system to be evaluated. **Figure 2** is representative of a 50 kDa protein with a molar extinction coefficient of 50,000 M^{-1} cm^{-1} at 280 nm. Using a cell with a 1.2 cm optical path

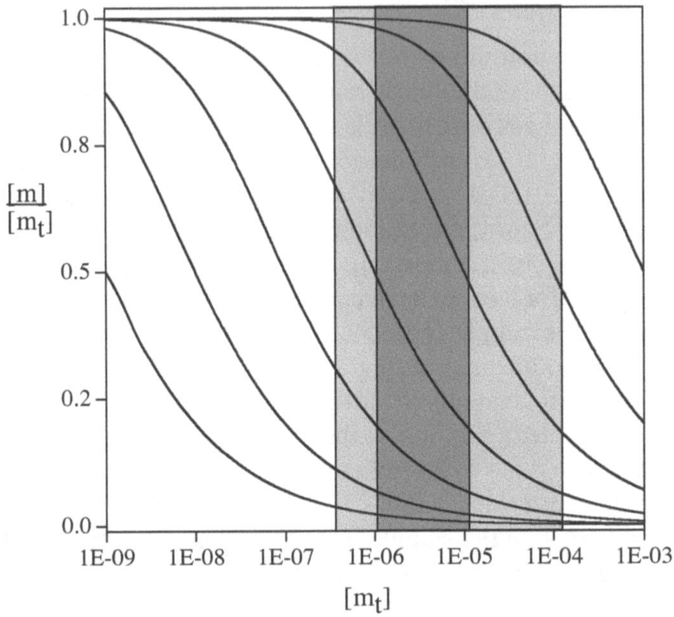

Fig. 2. Equilibrium binding isotherms for a homodimeric interaction. The fraction of monomer $[m]/[m_t]$ is plotted as a function of the total monomer concentration $[m_t]$. Curves from left to right are log order decreases in the association constant ranging from $K_a = 10^9\ M^{-1}$ to $K_a = 10^3\ M^{-1}$. The grey shaded area represents the concentration range over which reliable measurements can be made. The dark grey represents measurements made at 280 nm using standard 12 mm cells. The lighter region is where data have to be collected off peak, at shorter wavelength, or using short path length cells.

length, a concentration of 13.3 μ*M* gives a starting absorbance of 0.8 and a concentration of 1.66 μ*M* has an absorbance of 0.1. These represent the limits in absorbance for which reliable data can be collected. However, there are a number of ways to extend this range. First by collecting at a shorter wavelength where peptide bond absorbance will contribute to the extinction coefficient. The XL-A can measure at 230 nm, and this will usually increase the extinction coefficient by about three or four times, extending the absorbance range down to 0.5 μ*M* (*see* **Note 3**). To extend the range upward, shorter path length cells (3 mm) are available, extending the range up to 50 μ*M*. Finally, data can be collected away from the peak of the absorbance in the 250 nm trough or at 295 nm to give around a two- or threefold decrease in sensitivity, expanding the range up to 150 μ*M*.

3. Methods

3.1. Cell Assembly

The cells themselves are assembled for each centrifugation run. There are several different types of cell, but all consist of the following main components: a cylindrical cell housing that fits into the rotor and two window assemblies consisting of a window holder, a quartz window, a white Vinylite window gasket, and a Bakelite window liner. These two window assemblies form a sandwich around a centerpiece that contains the samples. This sandwich is assembled in the cell housing, and a screw ring and gasket is then screwed into one or both ends of the cell housing and tightened with a torque wrench (*see* **Note 5**). Full details of the cell assembly are provided in the manufacturer's instructions and should be followed very carefully. Furthermore, cell assembly videos for the An-50Ti and An-60Ti rotors can be downloaded from http://www.beckman.com/resourcecenter/labresources/sia/cellassy_video.asp.

There are many types of centerpieces constructed of different materials and dimensions. The simplest centerpieces are the double sector centerpieces, which allow for a sample solution and a buffer blank. These have a path length of 1.2 cm and are made in aluminum, aluminum-filled Epon, or charcoal-filled Epon. The aluminum double sector centerpiece has the advantage of having a maximum speed of 60,000 rpm, while the others are limited to 42,000 rpm. The charcoal-filled Epon centerpiece should be used if there is any possibility of aluminum ions interacting with the sample, otherwise all of the centerpieces are compatible with aqueous solutions, but softening of the Epon can occur if strong acids or organic solvents are used (*see* **Note 5**). A 3 mm path length charcoal-filled Epon centerpiece is also available and requires the use of two spacers. Charcoal-filled Epon centerpieces that contain six channels (three samples and three buffer blanks) have a maximum speed of 48,000 rpm, whereas the eight channel cells (four samples and four buffer blanks) can be operated up to 50,000 rpm. The double sector centerpieces that hold 450 μL per sector are generally used for sedimentation velocity measurements because of their longer column length. However, they can also be used for equilibrium runs if the volume is reduced to 110 μL. This is particularly useful if the 3 mm path length centerpiece is required, because this is not available in six channel centerpieces. In this case, the volume is reduced to 33 μL. The volumes required to fill a six sector cell are 110 μL of protein and 125 μL of sample buffer.

3.2. Speed and Duration of the Experiment

Before starting a sedimentation equilibrium experiment, two important parameters need to be determined. First, the optimal speed of centrifugation needs to be known. Beckman provides a chart relating optimal speed to expected molecular weight (http://www.beckman.com/resourcecenter/labresources/sia/ds820.asp).

Alternatively, eq. (6b) can be solved for rpm [rpm = (30/π) ω] for a given value of C_x/C_0. A good guide is that, at equilibrium, around a fivefold increase in concentration over the column length will produce an exponential distribution suitable to get a good fit to the data ($C_x/C_0 = 5$). Regardless of this calculation, it is advisable to centrifuge at speeds lower and higher than this optimal speed (allowing equilibrium to be reached first at the lower speed, then optimal speed and finally at higher speed). The lower and higher speeds can be calculated from eq. (6b), say for $C_x/C_0 = 3$ and 7. Another way of finding optimal experimental conditions in terms of protein concentration or rotor speed is to simulate equilibrium data for a particular set of parameters (*see* **Note 6**). This can be done with the Origin XL-A analysis software under "Utilities/Data Simulator."

The other important parameter that needs to be determined is the time taken for equilibrium to be reached T_{eqm}. This can be calculated from eq. (8), where, T_{eqm} (s) is directly proportional to the square of the solution height, h (cm) and inversely proportional to D_T:

$$T_{\text{eqm}} = \frac{0.7(h)^2}{D_{\text{T}}} \tag{8}$$

From eq. (8) it is apparent that smaller molecules reach equilibrium faster than larger ones. A 20 kDa protein in about 16 h and a 100 kDa protein in around 27 h. A value for D_T can be determined experimentally, but for these purposes it is reasonable to estimate a value by assuming that the macromolecule is spherical and applying eq. (9).

$$D_{\text{T}} = \frac{3 \times 10^{-5}}{\sqrt[3]{M}} \tag{9}$$

Whichever method is used, it is still necessary to show experimentally that the system has reached equilibrium. The best way to do this is to overlay or subtract successive scans, taken at 2 h intervals. When equilibrium is established, no further change in the absorbance profile should occur, the only differences being due to noise in the data.

3.3. Sedimentation Equilibrium Studies of a Monomeric Species

1. Prepare around 2 L of a suitable buffer for the sedimentation study to be carried out in, for instance, 10 m*M* Tris-HCl, 100 m*M* NaCl, 1 m*M* TCEP, pH 7.5 (*see* **Note 3**).
2. Dialyze a stock protein solution against at least three changes of 500 mL of this baseline buffer. If necessary, concentrate the sample after dialysis. Ensure that

there is enough of the stock protein to be able to prepare around 700 µL of a solution with an optical density of 0.7 at 280 nm, in a 1 cm path length cell (*see* **Note 7**).

3. Using this working solution prepare a dilution series of protein solutions, 120 µL in each, ranging from the highest optical density down to about 0.1 in a 1 cm cell. Because of the set up of the cells, it is convenient to use either three or nine samples. If sample is limiting, then three is adequate; however nine allows duplicates and coverage of a larger concentration range.
4. Place 10 µL of Fluorocarbon oil FC-43 in each channel of the preassembled bottom sections of the six channel centrifuge cells and add 110 µL of each protein solution into the sample channels and 125 µL of baseline buffer into the buffer channels (*see* **Note 8**). Finally, insert the top window assembly and tighten down the top section of each cell.
5. Balance the cells and place into the analytical rotor as described in the user manual. Carefully install the monochromator, set the run temperature, usually 20°C, and apply the vacuum (*see* **Note 9**).
6. Set the rotor speed to an initial speed of 3000 rpm. When the centrifuge has reached the required speed, temperature, and vacuum, collect a single radial scan at $\lambda = 280$ nm of each cell using a step size of 0.001 cm and two averages per scan. Refer to **Note 10** for expected appearance of this prescan.
7. Set the centrifuge speed to the lowest of the three speeds to be collected during the run, for instance, about 10,000 rpm for a 90 kDa protein. If association equilibria are expected or suspected, set the rotor speed to optimize the gradient in favor of the higher-molecular-weight species (the slowest).
8. After 16 h collect radial scans every 2 h using a step size of 0.001 cm and five averages per scan. Assess whether equilibrium has been reached by overlaying or subtracting successive scans. Do this until there is no further change in the absorbance profile. The final scan in this data set should be suitable for molecular weight analysis.
9. Set the centrifuge to the next speed in the set of three, wait for about 8 h before again collecting radial scans every 2 h using a step size of 0.001 cm and five averages per scan. Again, assess when equilibrium has been reached and collect a scan to use in the molecular weight analysis.
10. Finally, set the centrifuge to the highest speed where data will be collected. After 8 h, collect the radial scan data, assess for equilibrium as before, and collect a final scan suitable for use in data analysis.
11. When data have been collected at all the necessary speeds, set the centrifuge to 42,000 rpm. After approx 16 h, collect a radial scan using a 0.001 cm step size with two replicates. This "overspeed" scan gives a depleted boundary which should be used to provide a reasonable estimate of the cell offset during the data fitting procedure.

3.4. Analyzing the Data

All data analysis described below has been carried out using the Origin® software package. This program allows the direct fitting of the equilibrium concentration gradients to mathematical models for single species or associating systems using nonlinear least squares algorithms *(11,12)*. Furthermore, it allows global fitting of multiple data sets covering a range of concentrations and rotor speeds. This is particularly important for self-associating systems.

1. Once a whole data set is collected, select the files that will be used in the molecular weight analysis. Load the relevant "RA" files into Origin, then cut out the data corresponding to the sample absorbance profiles using the select subset command (*see* **Note 11**).
2. First, use the Origin software to individually fit each "cut" file assuming an ideal single species model. In this fitting procedure, include the offset value obtained from the "overspeed" scan along with the calculated values for ρ and v.
3. Plot the molecular weights obtained against the initial protein concentration. Is there any correlation between initial concentration and apparent molecular weight? If so, this may be indicative of an interacting system.
4. Plot ln(C) against r^2 [under Utilities/Plot] (*see* **Note 12**).
5. If the fit of the individual data files provided any indication of self-association, carry out a simultaneous (global) fit, still assuming a single species. Under these conditions there should be clear systematic deviations in the residuals plot indicating that the protein is not an ideal monomeric species.
6. Carry out global fits to monomer–dimer, monomer–trimer, and monomer–tetramer equilibria and carefully inspect for deviations between the fitted curve and the experimental data as well as for systematic deviations in the residuals plot. At this stage it is of great importance to have some previous knowledge of the monomer molecular weight of the protein under investigation as this allows the monomer molecular weight to be fixed (*see* **Note 13**).
7. Convert the association constant expressed in absorbance units into the commonly used association constant K_a (*see* **Note 14**).

3.5. Examples of Fits to Experimental Data

1. Swi6 is a 90 kDa protein. Analysis of the molecular weight using size-exclusion chromatography and dynamic light scattering provide estimates of the molecular weights of 250 and 280 kDa, respectively, suggesting that the protein is either di- or trimeric. Analysis of the solution molecular weight of Swi6 using sedimentation equilibrium, **Fig. 3**, reveals that the protein is in fact monomeric. The data fit well to the monomer molecular mass of 90 kDa *(13)*. Analysis of ln$C$ vs r^2 plots and the application of associative models provide no indication of oligomerization.
2. Sedimentation equilibrium runs of bovine dynamin I, a 100 kDa protein that oligomerizes in vivo and in vitro *(14,15)*. The runs shown in **Fig. 4** have been carried out at 4°C at a speed of 7,000 rpm and protein concentration of 0.33 mg/mL. **Figures 4A**, **B**, and **C** show fits of the data, along with residuals, assuming a monomeric, dimeric, and tetrameric species, respectively. The fit to a single species gave a molecular weight of 285 kDa, significantly larger than that of a monomeric protein. Furthermore, the nonrandom distributions of the residuals of all three fits clearly indicate that none of these models represents the correct asso-

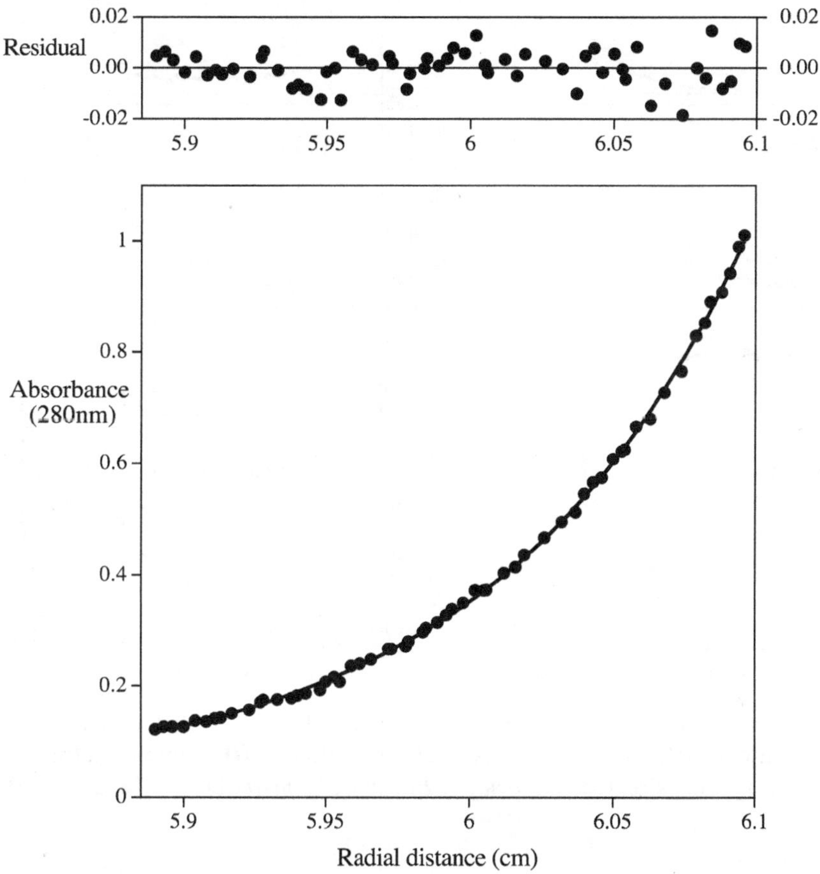

Fig. 3. Analysis of the solution molecular weight of Swi6 by sedimentation equilibrium. Lower panel is the absorbance profile produced at equilibrium by Swi6 rotor speed 12,000 rpm, $T = 293$ K, $v = 0.727$, $r = 1.003$. Upper panel shows the residuals to the plot.

ciation behavior. Only a monomer–tetramer model gave a satisfactory fit with a random distribution of residuals as shown in **Fig. 4D**, allowing the calculation of an association constant of $K_{1,4} = 1.67 \times 10^{17}\ M^{-3}$. Assuming a monomer–dimer–tetramer model resulted in the same quality of fit with no improvement over the monomer–tetramer model. The association constant $K_{1,2}$ for the monomer–dimer equilibrium calculated from this fit is so small relative to $K_{1,4}$, that there would be hardly any dimer present under the experimental conditions, thereby justifying the assumption of a monomer–tetramer model.

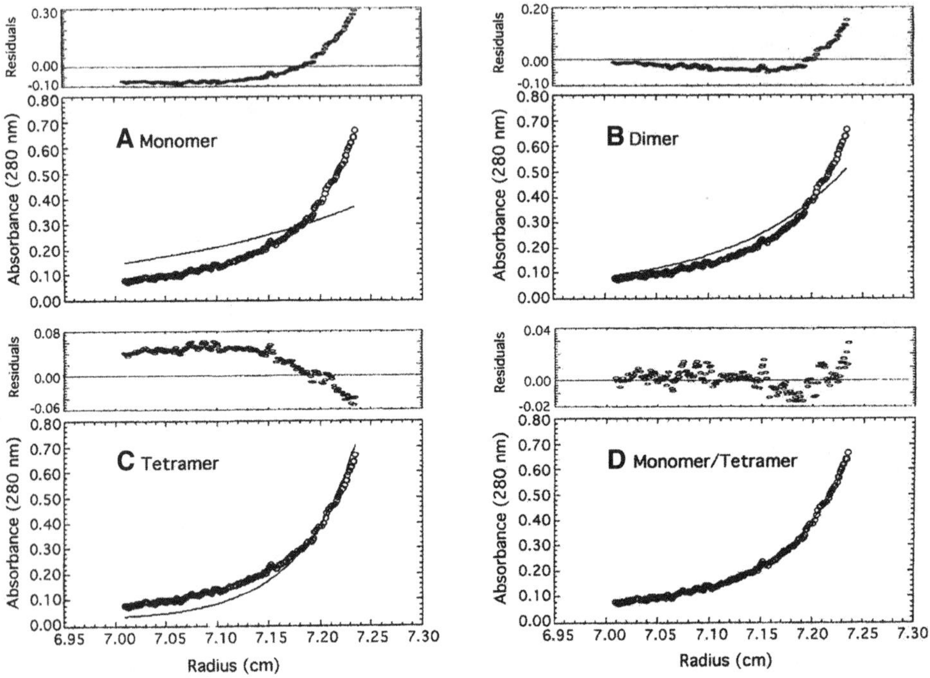

Fig. 4. Sedimentation equilibrium data of dynamin I fitted to different models: **(A)** monomer alone; **(B)** dimer alone; **(C)** tetramer alone; **(D)** monomer/tetramer. Reproduced with permission from *Journal of Protein Chemistry* (1999) **18,** 277–290.

3.6. Heterologous Systems

Sedimentation equilibrium methods could be especially useful for studying heterologous interactions, because these play a role in almost every biological process. However, the analysis of these data is complicated by the fact that the absorbance along the cell has contributions from both the free proteins and the complex(es). If the two proteins are of similar molecular weight and extinction coefficient, they can be treated as a homologous associating system *(16)*. However, this is usually not the case and so more complex strategies must be used. These are beyond the scope of this chapter and interested readers are referred to a review article by Minton and colleagues *(17)* which discusses different methods for the quantitaive characterisation of heterologous interactions and to Philo *(18)* who has discussed how the problems can be alleviated through the use of numerical constraints during data analysis.

4. Notes

1. In fact, it is the buoyant molecular weight M^* that is directly obtainable from the experimental data. The absolute molecular weight has to be calculated from eq. (10) using the solvent density ρ and the partial specific volume v of the particle.

$$M = \frac{M^*}{1 - v\rho} \tag{10}$$

Good values for the solvent density ρ can be obtained from tabulated values. An approximate value for the partial specific volume of a protein v is 0.73 mL/g. In some cases, it may be determinable experimentally using densitometry. A reasonably accurate alternative is to calculate partial specific volumes based on the amino acid composition of the protein. The SEDNTERP program does this using eq. (11), where n_i, M_i and v_i are the number, molecular mass, and partial specific volume of the ith residue:

$$v = \frac{\sum n_i M_i v_i}{\sum n_i M_i} \tag{11}$$

The need to obtain accurate values for these parameters stems from the $1 - v\rho$ relationship and, because of this, a 3% error in v results in close to a 10% error in molecular weight.

2. The offset value should be used with caution and only if it can be justified in some experimental way. It is highly correlated with A_0 and "adjustment" of its value will dramatically alter the molecular weight values obtained. The use of the overspeed at the end of a run is a reasonable estimation of the value, but may not be correct. The best ways to avoid offset problems are to avoid buffers with absorbing components and, taking care when preparing the cells for a run to avoid scratches and fingerprints.
3. The optical system of the XL-A is capable of measuring data at wavelengths between 800 and 190 nm, although below 230 nm there is very little lamp intensity and so this region is not generally useful. In all measurements it is important to try to keep the buffer absorbance to a minimum because low transmittance of the buffer blank will decrease the range, linearity, and sensitivity of any measurements. If measurements are carried out at a wavelength in the ultraviolet region, a nonabsorbing buffer has to be selected. At short wavelengths NaCl should be substituted by NaF to reduce buffer absorbance; 232 nm is a good wavelength because there is a peak in the lamp spectrum. However, the intensity of the light source of the instrument decreases with time, in particular, in the UV region, due to deposits of oil on the lamp, and a scan of the light intensity against wavelength should be done regularly to determine if the lamp needs cleaning.
4. A UV spectrum taken from 230 to 350 nm can provide a good indication of sample quality. A sloping baseline between 310 and 350 nm is indicative of protein aggregation.

5. Before assembling the cells, make sure everything is clean because salt deposits can lead to leakage or cracking of the windows. Furthermore, it is necessary to check if the buffer components are compatible with the cell components. Chemical compatibility tables are available at http://www.beckman.com/resourcecenter/labresources/centrifuges/pdf/chemres.pdf.
6. The "Simulator Setup" window allows the user to define all necessary run parameters. Note that the loading concentration is given in mg/mL and will be converted to absorbance based onto the concentration conversion factor. A detailed description of the program can be found at http://www.beckman.com/resourcecenter/literature/BioLit/BioLitList.asp.
7. The optical path length of the ultracentrifuge cells is 12 mm, rather than 10 mm used in most standard spectrophotometric cuvets. Bear this in mind when making up the sample dilution series.
8. Load the reference column slightly higher than the sample so the reference meniscus does not interfere with the sample. Fluorocarbon oil FC-43 has the same refractive index as water but is denser and forms a boundary at the bottom of the cell and thereby reduces artifacts caused by light at the bottom of the cell.
9. Do not forget to equilibrate the rotor at the experimental temperature or best at a slightly lower temperature. Also, the experimental parameters ρ and v are temperature dependent and usually hydrodynamic data are referenced to 20°C. If the experiment is not carried out at 20°C, this needs to be accounted for.
10. The 3000 rpm prescan should have a constant absorbance along the whole cell. It is worth noting these values down for any subsequent analysis of concentration dependency. The presence of a boundary or the appearance of upward curvature in the absorbance profile toward the bottom of the cell is indicative of high molecular weight or aggregated material in the sample. This probably means the sample is not suitable for further analysis. The lack of any absorbance indicates cell leakage and requires the run to be repeated.
11. If the equilibrium run has been carried out using six or eight channel centerpieces, all the data for the sample channels will be in a single file and have to be separated into individual files before any fitting procedures can be carried out. This is done via the "Select Subset" option in the Utilities menu. This feature allows the selection of a subset of data points from a whole profile to be saved as an individual file.
12. A simple model-independent method to check for heterogeneity is a plot of $\ln(C)$ vs r^2. A straight line for this plot suggests the presence of a single species, with the slope equaling M^*, the buoyant molecular weight of the protein (*see* eq. (9) and **Note 1**). An upward curving line is indicative of an associating system. However, this method can be rather insensitive to heterogeneity and should only be taken as a guide.
13. The molecular weight of the protein can easily be calculated based on the amino acid composition if the amino acid sequence of the protein is known; *see*, for example, http://www.expasy.ch/tools/protparam.html. Otherwise, the molecular weight has to be experimentally determined, for example, by mass spectrometry.

Alternatively, sedimentation equilibrium runs can be carried out under denaturing conditions.

14. Equilibrium constants determined in the Beckman Origin software have units expressed in absorbance units. Equations (12a–c) are used to convert them into concentration units for monomer–dimer, monomer–trimer, and monomer–tetramer equilibria, respectively:

$$K_a = \frac{K_a^{\text{abs}}\left(\varepsilon_m l\right)}{2} \tag{12a}$$

$$K_a = \frac{K_a^{\text{abs}}\left(\varepsilon_m l\right)^2}{3} \tag{12b}$$

$$K_a = \frac{K_a^{\text{abs}}\left(\varepsilon_m l\right)^3}{4} \tag{12c}$$

References

1. Minton, A. P. (1990) Quantitative characterization of reversible molecular associations via analytical centrifugation. *Anal. Biochem.* **190,** 1–6.
2. McRorie, D. K. and Voelker, P. (1993) *Self-associating systems in the analytical ultracentrifuge*, Beckman, Fullerton, CA.
3. Rivas, G. and Minton, A. P. (1993) New developments in the study of biomolecular associations via sedimentation equilibrium. *Trends Biochem. Sci.* **18,** 284–287.
4. Hansen, J. C., Lebowitz, J., and Demeler, B. (1994) Analytical ultracentrifugation of complex macromolecular systems. *Biochemistry* **33,** 13,155–13,163.
5. Schuster, T. M. and Toedt, J. M. (1996) New revolutions in the evolution of analytical ultracentrifugation. *Curr. Opin. Struct. Biol.* **6,** 650–658.
6. Cole, J. L. and Hansen, J. C. (1999) Ananlytical ultracentrifugation as a contemporary biomolecular research tool. *J. Biomol. Techn.* **10,** 163–176 (URL: http://www.beckman.com/resourcecenter/literature/BioLit/BioLitList.asp).
7. Laue, T. M. and Stafford, W. F., 3rd (1999) Modern applications of analytical ultracentrifugation. *Annu. Rev. Biophys. Biomol. Struct.* **28,** 75–100.
8. Liu, J. and Shire, S. J. (1999) Analytical ultracentrifugation in the pharmaceutical industry. *J. Pharm. Sci.* **88,** 1237–1241.
9. Cantor, C. R. and Schimmel, P. R. (1980) *Biophysical Chemistry, Part II*, Freeman, San Francisco.
10. Laue, T. (1996) Choosing which optical system of the Optima XL-I analytical centrifuge to use, *Beckman-Coulter Technical Application Information Bulletin* A-1821-A (URL: http://www.beckman.com/resourcecenter/literature/BioLit/BioLitList.asp).
11. Johnson, M. L., Correia, J. J., Yphantis, D. A., and Halvorson, H. R. (1981) Analysis of data from the analytical ultracentrifuge by nonlinear least-squares techniques. *Biophys. J.* **36,** 575–588.

12. Johnson, M. L. and Faunt, L. M. (1992) Parameter estimation by least-squares methods. *Methods Enzymol.* **210,** 1–37.
13. Sedgwick, S. G., Taylor, I. A., Adam, A. C., et al. (1998) Structural and functional architecture of the yeast cell-cycle transcription factor swi6. *J. Mol. Biol.* **281,** 763–775.
14. Binns, D. D., Barylko, B., Grichine, N., et al. (1999) Correlation between self-association modes and GTPase activation of dynamin. *J. Protein Chem.* **18,** 277–290.
15. Binns, D. D., Helms, M. K., Barylko, B., et al. (2000) The mechanism of GTP hydrolysis by dynamin II: a transient kinetic study. *Biochemistry* **39,** 7188–7196.
16. Silkowski, H., Davis, S. J., Barclay, A. N., Rowe, A. J., Harding, S. E., and Byron, O. (1997) Characterisation of the low affinity interaction between rat cell adhesion molecules CD2 and CD48 by analytical ultracentrifugation. *Eur. Biophys. J.* **25,** 455–462.
17. Rivas, G., Stafford, W., and Minton, A. P. (1999) Characterization of heterologous protein-protein interactions using analytical ultracentrifugation. *Methods* **19,** 194–212.
18. Philo, J. S. (2000) Sedimentation equilibrium analysis of mixed associations using numerical constraints to impose mass or signal conservation. *Methods Enzymol.* **321,** 100–120.

9

Analysis of Protein–Protein Interactions by Simulation of Small-Zone Gel Filtration Chromatography

Rosemarie Wilton, Elizabeth A. Myatt, and Fred J. Stevens

Abstract

Small-zone gel filtration chromatography, combined with analytical-scale columns and fast run times, provides a useful system for the study of protein–protein interactions. A computer simulation (SCIMMS, or Simulated Chromatography of Interactive MacroMolecular Systems) that replicates the small-zone behavior of interacting proteins has been developed. The simulation involves an iterative sequence of transport, equilibration, and diffusion steps. This chapter illustrates the use of the simulation to study the homodimerization of rapidly equilibrating immunoglobulin light chain proteins and for determination of association constants. The simulation can also be used to study heterogeneous interactions, kinetically controlled interactions, and higher-order oligomerization, and it can replicate large-zone and Hummel–Dreyer conditions.

Key Words

Protein–protein interaction; Gel filtration chromatography; small-zone; simulation; association constant; affinity.

1. Introduction

Gel filtration (or size-exclusion) chromatography is a method for separating macromolecules on the basis of differences in Stokes' radii *(1)*. The chromatographic resin consists of porous particles made up of cross-linked polymers. Many different polymers and particle pore sizes are available to accommodate proteins of different sizes and properties. Separation of proteins is achieved by partitioning between the mobile phase, the solvent that is exterior to the chromatographic particles, and the stationary phase, consisting of the solution within the porous particles themselves. The volume at which a protein elutes from the column is a direct reflection of the amount of time it spends within the

From: *Methods in Molecular Biology, vol. 261: Protein–Protein Interactions: Methods and Protocols*
Edited by: H. Fu © Humana Press Inc., Totowa, NJ

porous particle. Thus, larger proteins, which are less likely to penetrate the particle, elute earlier, and smaller proteins, which effectively spend more time in the stationary phase, elute later.

Gel filtration chromatography has been used extensively for protein purification and for estimation of protein molecular weight. Because the interaction between two or more proteins results in the formation of a complex having a larger Stokes' radius than its constituents, gel filtration chromatography can also be applied to the detection and characterization of protein–protein interactions. A number of analytical applications have been described that evaluate the affinities of protein–protein interactions, notably, the Hummel–Dreyer method *(2)*, which was originally developed to measure protein–ligand interactions, but has also been applied to protein–protein interactions *(3,4)* and large-zone equilibrium gel filtration *(5)*. Because these are equilibrium techniques, mathematically rigorous solutions for the behavior of interacting proteins are attainable. However, both of these techniques can consume significant quantities of sample, a potential disadvantage when studying expensive solutes or unique materials of limited supply.

Nonequilibrium, or small-zone, gel filtration chromatography uses sample volumes that are less than 1% of the column volume. Combined with analytical scale columns and fast run times, this mode of chromatography is attractive for the study of protein–protein interactions. During a small-zone chromatographic run there is a continual decrease in the concentration of the sample components due to dilution, dispersion, and diffusion; a mathematical solution for this behavior is not possible. To overcome this problem, we developed a computer simulation (SCIMMS, or Simulated Chromatography of Interactive MacroMolecular Systems) that replicates the elution profiles of interacting proteins (*see* **Note 1**; F.J. Stevens, unpublished). Computer simulations have been used to study a variety of transport or mass migration processes *(6)* including ultracentrifugation *(7)*, electrophoresis *(8)*, and gel filtration chromatography *(9,10)*. The computer-simulation technique developed in this laboratory is conceptionally similar to these examples; it involves an iterative sequence of transport, equilibration, and diffusion steps *(11)*. We have used the simulation to study the homodimerization of rapidly equilibrating immunoglobulin light chain proteins *(11–14)* and heterogeneous (antibody:antigen) interactions *(15–17)*. The simulation can also be used to study kinetically controlled interactions *(18)* and higher order oligomerization, and it can replicate large-zone and Hummel–Dreyer conditions (F.J. Stevens, unpublished results).

1.1. General Principles

The simulation SCIMMS consists of an iterative sequence of transport, equilibration, and diffusion steps. The simulated column, depicted in **Fig. 1**,

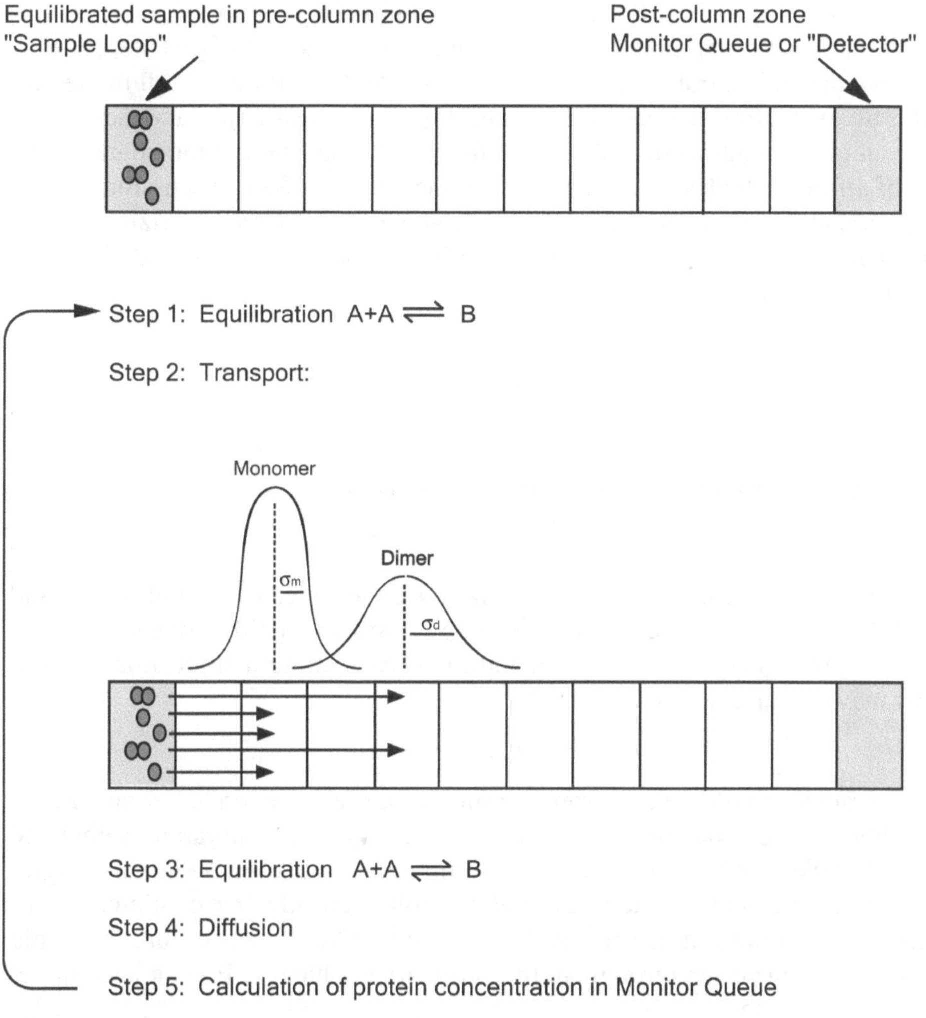

Fig. 1. Series of steps used by SCIMMS to simulate gel filtration behavior of interacting proteins. The diagram illustrates the behavior of a dimerizing protein and a simulated "column" consisting of 10 cells. The protein is shown in the pre-column zone, akin to the sample loop of the experimental system. Following the column cells is the post-column zone, or monitor queue, which represents the detector in the experimental system. In step 1, the equilibrium concentrations of all species are calculated for each cell in the column. In step 2, each species migrates according to a Gaussian distribution, the mean of which is given by the velocity (in cells/cycle) of each component and the standard deviation of which is given by the dispersion parameter assigned in the simulation for each species. In step 3, the equilibrium concentrations are again calculated. In step 4, a diffusion cycle is executed. In step 5, the concentration of protein in the monitor queue is calculated.

consists of a linear array of cells onto which the protein(s) of interest is "loaded." A five-step iteration cycle then commences. In the first step, equilibration, the concentration of each species is calculated for each cell in the column by evaluating the equilibrium and kinetic expressions that describe the protein–protein interaction. Although the simulation can accommodate a variety of different modes of interactions, for the purpose of this discussion and the experimental example in **Subheading 3.**, we will consider a dimerizing system under rapid equilibrium. Therefore, the formation of the dimer, *B*, from the monomers, *A*, is shown by

$$A + A \underset{k_r}{\overset{k_f}{\rightleftharpoons}} B \tag{1}$$

and the rate of formation of the dimer, *B*, is given by

$$\frac{db}{dt} = k_f a^2 - k_r b \tag{2}$$

where a is the concentration of monomer, b is the concentration of dimer, and k_f and k_r are the forward and reverse rate constants. For slowly equilibrating systems, the change in dimer concentration (Δb) during a short time element (Δt) may be numerically estimated by

$$\Delta b = \left(k_f a^2 - k_r b\right) \Delta t \tag{3}$$

For rapidly equilibrating systems, a simplifying assumption of "instantaneous equilibration" can be made. This means that complete equilibration is obtained at each cycle in the simulation. This assumption is valid even though the migration of the interacting proteins through the column results in a continual departure from equilibrium; the relatively slow column flow rates ensure that rapid reequilibration occurs throughout the chromatographic run. Equation (3) therefore simplifies to

$$K_a = \frac{a^2}{b} \tag{4}$$

where K_a is the equilibrium association constant. The equilibration step is followed by a transport step during which monomers and dimers are advanced through the column according to a Gaussian distribution (step 2). The movement of each species is characterized by a velocity (in cells/cycle) that defines its mean displacement, and a dispersion factor (in cells) that represents the standard deviation, σ, of the displacement about the mean position. The Gaussian transport strategy creates a device to represent the axial dispersion or band spreading that occurs during chromatography due to incomplete chromatographic partitioning, flow irregularities, edge effects, and other physical effects ***(15)***. This is followed by another equilibration step (step 3). Next, monomers and dimers are individually subjected to a diffusion cycle (step 4) according to the equation ***(11)***

$$c_i^* = c_i - f_d(c_i - c_{i-1}) + f_d(c_{i+1} - c_i) \tag{5}$$

where c_i^* represents the concentration of monomer or dimer in cell i following the diffusion step, c_i represents the initial monomer or dimer concentration of protein in cell i, and f_d is the fractional diffusion coefficient assigned to the monomeric or dimeric species. For $f_d = 0.5$, diffusion equilibrium is reached and c_i is the average of c_{i-1} and c_{i+1}. In step 5, the protein concentration in a post-column "monitor-queue," analogous to the detector in the experimental system, is recorded, generating one data point on the simulated chromatogram. The cycle is then repeated. SCIMMS has been semiautomated for determining the association constants of dimers that display rapid equilibration. The simulation will iteratively adjust the velocities and association constants of the migrating components so that the peak positions of the simulated chromatograms match the experiment. After a simulation sequence has been completed, the experimental and simulated chromatograms are overlaid and inspected by eye, parameter adjustments are made, and the process is repeated until the peak shapes and positions match. The association constant is then given in the SCIMMS output.

2. Materials

1. Chromatography system consisting of:
   a. Pump that can produce slow, stable flow rates.
   b. Injector, preferably with fixed volume sample loop (e.g., we use a 5 μL sample loop); an autosampler is helpful.
   c. Detector that can measure absorbances minimally at 214 and 280 nm.
   d. Analytical-scale gel filtration column. We typically use 0.3 × 25 cm columns containing Superose or Sephadex (Amersham Biosciences, Piscataway, NJ) gel filtration resins (*see* **Note 2**).
   e. Computer, for data collection.
2. Protein standards for gel filtration chromatography (Amersham Biosciences, Piscataway, NJ).
3. Purified proteins of interest, filtered through 0.2 μm filters or centrifuged to remove any insoluble material. Approximately 5 mg of protein is required for the complete concentration series described below.
4. Buffers, such as phosphate-buffered saline, that are compatible with the protein under study; the buffer should contain at least 50–100 m*M* salt to minimize ionic interactions of the protein with the gel filtration media (check with the manufacturer of the specific gel filtration media used). The buffers should be filtered through 0.45 μm filters.

3. Methods

The methods described below outline (1) setup and optimization of the chromatography system, (2) sample setup and determination of experimental chromatograms, (3) normalization of experimental chromatograms, and (4) analysis of protein–protein interactions by SCIMMS. Although SCIMMS is capable of simulating many different modes of protein–protein interactions, the example given below illustrates determination of the dimer association constant of the immunoglobulin light chain variable domain (V_L). The interactions between V_L monomers display rapid association and dissociation. This can best be seen in the concentration dependence of the elution profile and the notably asymmetric peak shape, especially at intermediate concentrations (**Fig. 2A**). As the sample concentration varies from low to high, the peak position shifts from that of the monomer toward that of the dimer.

3.1. Chromatography System

A standard chromatography system is acceptable for most applications. The two most important features include a pump that can produce slow, stable flow rates (*see* **Note 3**) and an injection valve that gives reproducible injection volumes. To check accuracy, the flow rate should be measured manually by collecting the column effluent in a graduated cylinder over a 10–12 h period. To produce a consistent sample injection volume, an autosampler or manual injector with a fixed-volume sample loop should be used. The detector should be capable of measuring absorbance at 214 nm, for samples of low concentration, and, minimally, at 280 nm, for higher concentration samples. Other wavelengths may be useful if the protein of interest contains a chromaphore. Analytical-scale columns can be purchased from commercial sources or can be prepared in the laboratory (*see* **Note 2**). The performance of the column and chromatography system should be tested with gel filtration calibration standards. Symmetrical peaks with reproducible elution positions and areas should be observed with the standard proteins.

Fig. 2. *(opposite page)* (**A**) Overlaid experimental elution profiles obtained with different concentrations of V_L. The elution profiles were normalized with the program KRUNCH. The concentration of protein used to generate each elution profile is (from left to right) 16.4, 13.7, 10.9, 9.6, 8.2, 6.8, 5.5, 4.1, 2.7, 1.4, 0.55, 0.41, 0.27, 0.14, 0.055, and 0.027 mg/mL. Chromatography was performed on a 0.3 × 25 cm Superdex 75 column operating at a flow rate of 0.06 mL/min. The buffer contained 20 m*M* potassium phosphate, 100 m*M* NaCl, pH 7.0. (**B**) Overlaid simulated elution profiles.

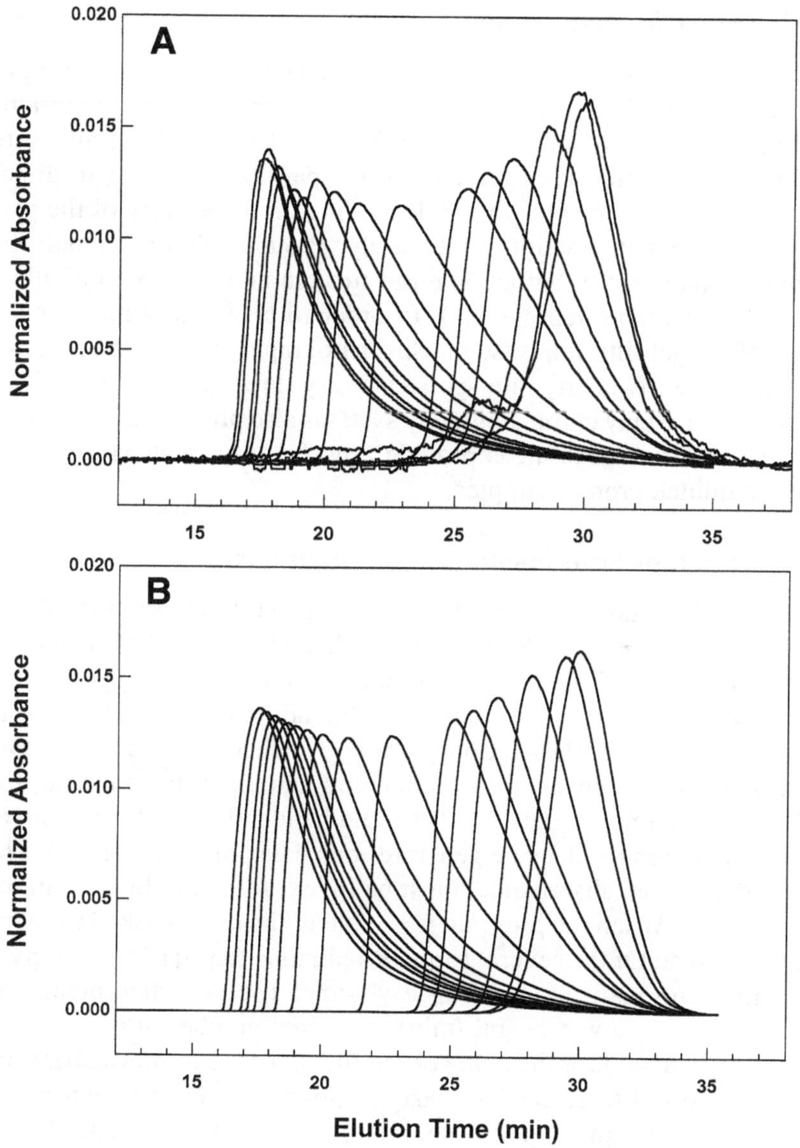

Fig. 2. *(continued)* The simulated elution profiles were normalized with the program NORM. The protein concentrations in the simulated elution profiles are identical to the experimental elution profiles. The simulation was performed in the automated mode, using the instantaneous equilibration assumption. The best fit was obtained using monomer and dimer dispersion parameters of 0.500 and 3.500 cells, respectively, and monomer and dimer diffusion parameters of 0.500 and 0.250, respectively. The association constant was determined to be 1.0×10^5 M^{-1}. Used with permission from ref. ***(14)***.

3.2. Experimental Chromatograms

Proteins under study should be of the highest purity possible to ensure unambiguous detection of interactions. For the experimental chromatograms shown in **Fig. 2A**, dilutions of V_L ranging from 16.4 to 0.027 mg/mL were prepared. These concentrations were chosen to span the monomer to dimer transition. Concentrations can be varied depending on the affinity of the particular protein–protein interaction under study. Protein dilutions should be made in the chromatography buffer. The run time should be chosen to allow complete elution of the protein peak(s). In order to continually monitor the performance of the gel filtration system, chromatographic runs of the proteins of interest should be frequently interspersed (every three or four runs) with a protein standard run. Many chromatography systems also provide tools for peak integration. The relative peak areas provide a useful way to check injection volumes of the diluted protein samples.

3.3. Normalization of Experimental Chromatograms

Once the experimental elution profiles are obtained, they are normalized with program KRUNCH (F. J. Stevens, unpublished). KRUNCH scales the data so that the integrated area under the elution profile is equal to 1. This enhances visualization of the concentration-dependent changes in the peak shape. Additionally, the KRUNCH program will reduce the number of data points. For later direct comparison of simulated and experimental data, the final number of experimental data points should be identical to the number of cycles in the simulation (each cycle generates one data point, *see* below). The experimental data (time, absorbance) must be in text format to be used in the KRUNCH program. When the program is executed, the user is asked to select the output file format. Of the several formats available, the "qT" format gives the output data as normalized absorbance vs time. The user then inputs the column volume (mL), flow rate (mL/min), and final number of data points. The user is then prompted for the name of the first data file for normalization. The user is then asked to select two baseline segments (each is entered as min1,min2; fractional minutes are acceptable). Ideally, these should lie on either side of the peak in regions where the baseline shows minimal noise. The user then selects the normalization segment (entered as min1,min2), which should as closely as possible describe the exact start (min1) and end (min2) of the peak. The normalized data file, which has the prefix "qT," is then generated, and the user is asked to input a second data file. The process proceeds until all of the data files have been normalized. The normalized data files can then be exported to any plotting program. The experimental chromatograms in **Fig. 2A** were normalized with KRUNCH.

```
708          !column length (cells)
0            !column volume (ml)
05           !monitor queu length (cells)
2            !sample size (cells)
708          !run length (cycles)
100          !pre-run length (cycles)
0.0          !time increment (seconds)   ( 0 ==> inst. equil.)
0.01         !max change in concentration (fraction)
3.0          !time per cycle (seconds)
1            !number of diffusion iterations per cycle
6            !gaussian spread
0            !flow rate (ml/min)
2            !saturate with B2   (1 = yes; 2 = no)
```

Fig. 3. Example of a column.npt file. The various parameters are discussed in the text. Information following exclamation points consists of comments, which are not used by the simulation.

3.4. Analysis of Protein–Protein Interactions by SCIMMS

The program SCIMMS is organized as two blocks of subroutine calls. Although the details of these subroutines are beyond the scope of this chapter, several of the subroutines execute user-defined parameters via several input files. These files must be edited by the user. A description of these parameter files are given below and should be sufficient for the beginning user to run the program. The program was written in FORTRAN and has been compiled to run on a UNIX platform.

3.4.1. Column Data

A sample column input file (column.npt) is shown in **Fig. 3**. User-defined parameters include the number of cells in the column and cell volume (not explicitly stated), which are adjusted to emulate the experimental column volume. For example, our 1.77 mL experimental column is simulated with 708 cells with a volume of 2.5 μL/cell. A smaller cell volume improves the resolution of the simulated chromatogram, but this is balanced by increased computational requirements. The sample size of two cells represents the 5.0 μL sample loop of the experimental system. The user also defines the number of iteration cycles (run length) and the time per cycle. In the example shown, a run length of 708 cycles and 3 s/cycle was used to specify a run time of approx 35 min. A prerun length parameter allows control of sample equilibration prior to loading on the column. For simulations using the instantaneous equilibration assumption, the time increment parameter is set to zero, and the maximum change in concentration parameter is not used. The length of the monitor queue is specified in cells. The Gaussian spread provides the number of cells to each side of

```
012600.0 001.320 001.000 000.500 000.500        !mw vel extcf dif disp (comp A)
010000.  000.000 000.000 000.000 000.000        !  (comp B1)
010000.  000.000 000.000 000.000 000.000        !  (comp B2 or B1:B1)
2.464    0.250   3.50                           !vel(0)   dif(0)   disp(0)(dimer)
00.00   00.00   00.00   00.00   00.00           !velocities(4 -->  8)
00.00   00.00   00.00   00.00   00.00           !velocities(9 --> 13)
0.00   0.00   0.00   0.00   0.00                !dispersion(4 -->  8)
0.00   0.00   0.00   0.00   0.00                !dispersion(9 --> 13)
0.00   0.00   0.00   0.00   0.00                !diff_coefs(4 -->  8)
0.00   0.00   0.00   0.00   0.00                !diff_coefs(9 --> 13)
0.                                              !buff conc B1
0.                                              !buff conc B2

! concentration in mg/ml
! mw in daltons
! migration velocity in cells per cycle
! extinction coefficient as effective absorbance units per mg/ml
! diffusion coefficient as percent equilibration per cycle
! dispersion coefficient as gaussian spread per translation
```

Fig. 4. Example of a protein.dat file that defines parameters related to the various protein species. For the purposes of the homodimerization example, only lines 1 and 4 are relevant. In practice, the diffusion and dispersion parameters for the monomer (line 1) and dimer (line 4) are varied after each round of the simulation to adjust the peak shapes.

the mean cell for which possible transport of material is calculated. This value should be at least three times the dispersion factor to account for greater than 99% of the material. The number of diffusion iterations per cycle can also be specified; more than one diffusion cycle can be used if increased band spreading is required. The saturate parameter can be used to simulate Hummel–Dreyer conditions. Note that the column volume and flow rate parameters are optional. If both are zero, then the default output of the simulation gives the elution profile as a function of time. If the flow rate term is finite while the column volume term remains zero, then the output provides the elution profile as a function of volume. If both flow rate and column volume are finite, the elution volume is generated as a function of V_e/V_t, the elution volume as a fraction of the total column volume.

3.4.2. Protein Data

An example of a protein parameter file (protein.dat) is shown in **Fig. 4**. For a simple dimerization, the relevant input parameters are found in lines 1 and 4. Line 1 gives the protein parameters for the monomer, component *A*. From left to right are the concentration (mg/mL), the molecular weight (Daltons), the velocity (cells/cycle; note that when the semiautomated version of the simulation is used, as described below, the velocity is calculated by the program; nonetheless, a finite velocity value must still be entered into this field for the program to operate properly), the extinction coefficient (for 1 mg/mL solution), the diffusion parameter (the fractional diffusion equilibrium achieved in each cycle), and the dispersion coefficient (the standard deviation of the Gaussian distribution). The fourth line contains the velocity, diffusion, and dispersion parameters that characterize the dimer of component *A*. For the purposes of the semiautomated simulation, the diffusion and dispersion parameters are most important. These fields are adjusted by the user to fit the shape of the eluting peak. A diffusion value of 0.5 represents complete diffusion equilibrium at each cycle; therefore, larger values are meaningless. As explained previously, the dispersion parameter defines the standard deviation, in cells, of the Gaussian distribution. Diffusion and dispersion parameters are oppositely related to the molecular weight of the molecule; thus, monomers diffuse rapidly but have small dispersion parameters; dimers diffuse slower but have larger dispersive properties. **Figure 5** illustrates the effects of varying dispersion and diffusion parameters on the simulated elution profiles.

3.4.3. Experimental Data

The experimental chromatogram files are linked to the simulation through an input file called chrom_dat.lst. This parameter file is illustrated in **Fig. 6**.

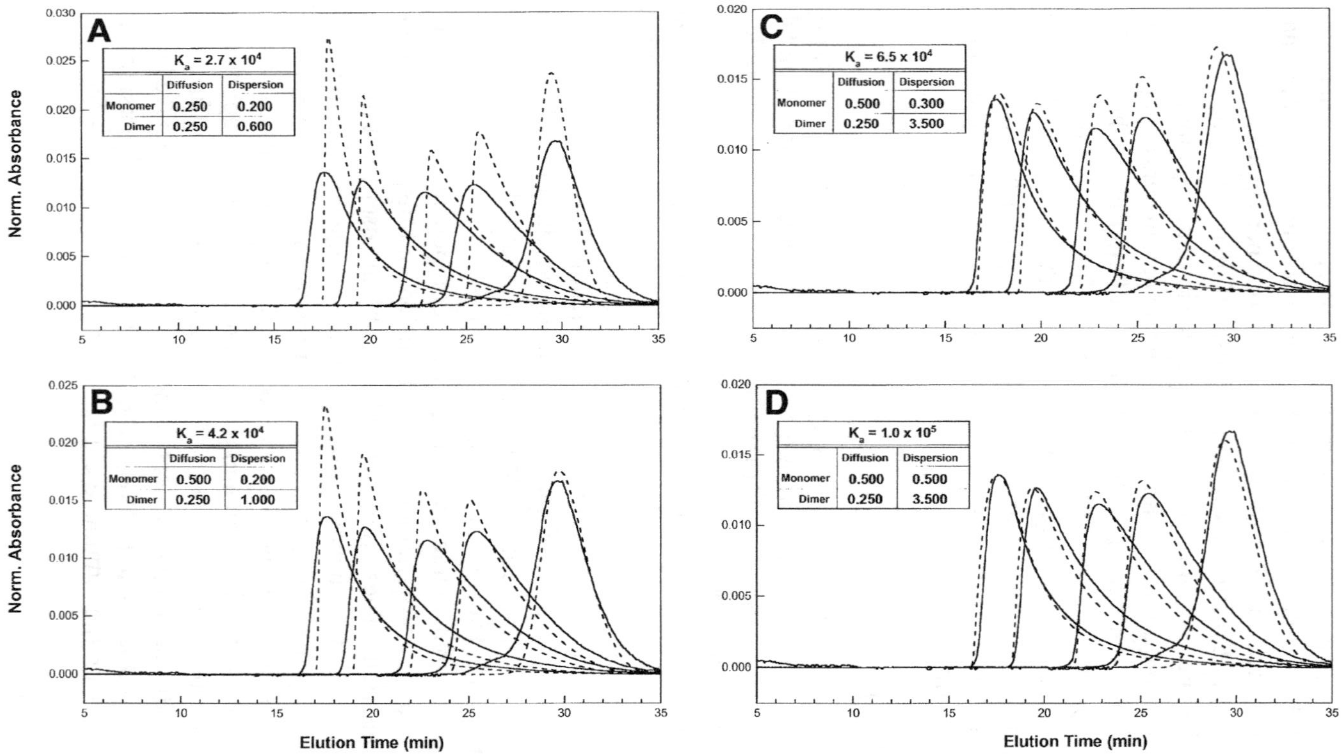

Fig. 5. Sampling of simulation results obtained during the fitting of experimental data shown in Fig. 2B. For clarity, only five of the experimental and simulated elution profiles are displayed in each panel. Experimental chromatograms are shown as solid lines and simulated results are shown as dashed lines. The table in each panel displays the monomer and dimer diffusion and dispersion parameters used in the simulation and the corresponding association constant. Panel **D** shows the best fit obtained for the experimental elution profiles. Used with permission from ref. ***(19)***.

```
16.4     0.0  0.0
DATA/qT3627c.dat
13.7     0.0  0.0
DATA/qT3627d.dat
10.9     0.0  0.0
DATA/qT3627e.dat
9.6      0.0  0.0
DATA/qT3627f.dat
8.2      0.0  0.0
DATA/qT3627g.dat
6.8      0.0  0.0
DATA/qT3627h.dat
5.5      0.0  0.0
DATA/qT3628b.dat
1.4      0.0  0.0
DATA/qT3628e.dat
2.7      0.0  0.0
DATA/qT3628d.dat
4.1      0.0  0.0
DATA/qT3628c.dat
0.55     0.0  0.0
DATA/qT3628f.dat
0.41     0.0  0.0
DATA/qT3628g.dat
0.27     0.0  0.0
DATA/qT3628h.dat
0.14     0.0  0.0
DATA/qT3629c.dat
0.055    0.0  0.0
DATA/qT3629d.dat
0.027    0.0  0.0
DATA/qT3629e.dat
```

Fig. 6. Example of a chrom_dat.lst file showing protein concentrations and experimental data files used to generate the simulated data shown in **Fig. 2B**. The simulation extracts the peak positions from the experimental chromatograms at the top and bottom of the list (which generally should correspond to the highest and lowest protein concentrations, respectively) and from the chromatogram at the center of the list. In this case, with an even number of files (16), the eighth experimental data set is used by the simulation. Note that the 1.4 mg/mL data set is placed at this central position. From examination of the experimental data in **Fig. 2A**, it can be seen that this correspond to a concentration of protein that generates a broad, asymmetric elution profile.

The first line describes the concentration, in mg/mL, of each (up to three) component in the sample. In the simple dimerization example, only the first component contains a finite value. Directly beneath this line is the path to the directory that contains the experimental data file and the name of the data file.

```
VL (20mM KPi/100mM NaCl/pH7) Superdex75    !header documentation
chrom_dat.lst               !name of file containing experimental data
protein.dat                 !name of file continaing miscellaneous protein fit values
.true.                      !auto >> true = automated simulation
0.05                        !fractional range for fit of target profile positions
0.0                         !t_inc_auto  (0.0 => instantaneous equilibration)
1.0                         !kr_0_auto 0.0
.true.                      !auto_mode_1    (simple dimerization)
   .false.                  !fit_dimer_vel   fit dimer velocity
   2.844                    !value to be used for V_dimer if above is .false.
   .false.                  !fit_mono_vel    fit monomer velocity
   1.400                    !value to be used if above is .false.
   .true.                   !fit_kf          fit dimerization constant
   0.0                      !value for t_inc_auto if above is .true.
   3.7e5                    !value for kf if above is .false.
   .true.                   !refit monomer and dimer velocities.
.false.                     !auto_mode_7    (heterologous dimerization)
   1.2e04  1.41  2.10 0.75 3.5    !kf_d1  vl_m1 vl_d1  dis_m1 dis_d1
   3.2e03  1.37  2.61 0.75 3.5    !kf_d2  vl_m2 vl_d2  dis_m2 dis_d2
```

Fig. 7. Example of a scimz.inp file which contains parameters for running the simulation in automated fashion.

In this example we have used the normalized "qT" files. During the execution of the automated simulation, the program will read the peak positions of three of the experimental data files: the first in the list (typically the highest concentration), the last in the list (the lowest concentration), and the middle data set. These data sets need not be in descending order of concentration; particularly, the middle data set was chosen to represent a protein concentration that produced a broad, asymmetric peak in the experimental elution profile. The simulation will then iteratively adjust the monomer and dimer velocities and the association constant to match the peak positions of these three experimental data sets. After a satisfactory fit has been found, SCIMMS will calculate simulated elution profiles for all concentrations given in chrom_dat.lst.

3.4.4. Simulation Data

The simulation for dimerizing systems using the instantaneous equilibration assumption has partially been automated. It determines the monomer and dimer velocities and the association constant, K_a, by fitting the peak positions of the simulated chromatograms with those of the experimental chromatograms. However, the monomer and dimer dispersion and diffusion parameters are defined by the user in the protein.dat file, as illustrated in **Fig. 4**. These parameters affect the shape of the simulated peaks; they are adjusted by trial and error until the simulated chromatogram matches the experimental chromatogram. The file scimz.inp contains the parameters relevant to the automation process (**Fig. 7**). Line 1 can be used to describe the experiment. It will appear as the header on the simulation output file, scimz####.arc, described below. Lines 2 and 3 define the files that contain the experimental data and the protein data. Line 4 defines whether the simulation should be run in the automated mode by entering "true" or "false." Line 5 defines how closely the simulated peak positions must match the experimental peak positions. It is generally advantageous (computationally) to keep this number higher at earlier rounds of the simulation. One might start out with a fit of 0.02 (or 2%) Then, as the various parameters are refined through several rounds of the simulation, one can gradually adjust the peak fit criteria to 0.005. In line 6, the t_inc_auto parameter should be set to 0.0 to invoke the instantaneous equilibration assumption, and the following line, kr_0_auto 0.0, should contain the value 1.0. Line 8 indicates that the mode of protein–protein interaction is a simple dimerization. The next four lines refer to fitting of the monomer and dimer velocities. From a computational standpoint, it is generally best to begin a series of simulations with an estimated monomer and dimer velocity (in cells/cycle) in lines 10 and 12 and enter "false" for fit_dim_vel and fit_mono_vel in lines 9 and 11. Then, in line 16, enter "true" to *refit* the monomer and dimer velocity. After a simulation has been completed, a new monomer and dimer velocity is given in the scimz####.arc file (below). These values should then be used in lines 10 and 12 for the next execution of simulation (*see* **Note 4**). Generally, one will want to enter "true" at line 13, to fit the rate constant, k_f. The last three lines pertain to heterologous dimerization ***(14)*** and are not relevant to this discussion; enter "false" at line 17.

3.4.5. Miscellaneous Input Files

Other required input files include the kinetics_in.dat file, which contains rate constant terms, but need not be edited for the automated simulation example described herein, and the ksym_rec.dat file, which is used by the program for numbering the output files.

3.4.6. Comparing the Simulated and Experimental Elution Profiles

Several files are generated as the simulation is executed. The scimz####.arc file (where #### represents a number generated with the ksym_rec.dat file cited above) contains the monomer and dimer velocities, as well as the association constant, determined by the simulation to fit the experimental peaks. This file also documents the input parameters and data files used by the simulation. The other relevant output files are the ksm_####.out files. These contain the time vs absorbance data for the simulated elution profiles. The simulation actually generates many ksm_####.out files as it fits the experimental peaks. Once a satisfactory peak fit is achieved, the simulation will generate a ksm_####.out file for each protein concentration given in the chrom_dat.lst file. For most purposes, only this final set of ksm_####.out files will be used, and the others can be deleted. As with the experimental data, it is generally convenient to normalize the simulated data. This is done by executing the program "NORM," which simply asks for the input file name, and writes the normalized file as Nksm_####.out. This file can be exported to any plotting program, and overlaid on the experimental elution profiles for comparison. The final set of simulated data, normalized for comparison with the experimental data, is shown in **Fig. 2B**.

4. Notes

1. The programs described in this chapter will be furnished upon request.
2. Gel filtration matrices with a variety of fractionation ranges are commercially available from a number of suppliers; for example, the Superdex and Superose resins from Amersham Biosciences (Piscataway, NJ) are routinely used in our laboratory. The fractionation range of the chosen resin should span the molecular weights of the proteins and complexes of interest. Suitable analytical-scale columns can easily be packed in the laboratory. Briefly, a slurry of the appropriate gel filtration matrix is placed in a donor column; the volume of the donor column should be approx 5- to 10-fold larger than the analytical column (the "recipient"

column) to allow formation of a uniform slurry. One end of the donor is connected via a short piece of narrow bore tubing to the recipient column (such as a silanized glass column); frits between the two columns must be removed. The other end of the donor column is connected to the HPLC pump. The gel filtration matrix is then pumped into the recipient column at a low flow rate, which is adjusted during the filling procedure according to the back pressure. After filling is complete, the donor column is removed and the end frit of the recipient column is replaced. The newly packed column should be allowed to equilibrate overnight at flow rates/pressures appropriate for the gel filtration media used. Any void space that forms at the top of the column after the overnight equilibration period is filled manually with a small amount of matrix. Column performance should be tested with gel filtration calibration standards, and only columns that produce symmetrical peaks are used. Aside from the occasional replacement of the top few millimeters of gel filtration matrix, these columns can be used for many months with no noticeable change in performance.

3. We typically use a flow rate of 0.06 mL/min. This slow flow rate was chosen because it provides convenient run lengths using our 1.77 mL analytical column.
4. In general, it is preferable that the simulated velocities be greater than one for any component. This will avoid significant backflow of material that can result from the Gaussian formulation for transport. If necessary, this can be achieved by increasing the total number of cells or by increasing the absolute time per cycle; both of these parameters are in the column.npt file.

Acknowledgments

This work was supported in part by the U.S. Department of Energy, Office of Biological and Environmental Research, under contract W-31-109-ENG, and by U.S. Public Health Service Grant DK43757.

References

1. Ackers, G. K., (1975) in *Molecular Sieve Methods of Analysis, Vol. 1. The Proteins.* (Neurath, H., and Hill, R. L., eds.) Academic Press, New York, NY.
2. Hummel, J. P. and Dreyer, W. J. (1962) Measurement of protein-binding phenomena by gel filtration. *Biochim. Biophys. Acta* **63,** 530–532.
3. Yong, H., Thomas, G. A., and Peticolas, W. L. (1993) Metabolite-modulated complex formation between alpha-glycerophosphate dehydrogenase and lactate dehydrogenase. *Biochemistry* **32,** 11,124–11,131.
4. Gegner, J. A. and Dahlquist, F. W. (1991) Signal transduction in bacteria: CheW forms a reversible complex with the protein kinase CheA. *Proc. Natl. Acad. Sci. USA* **88,** 750–754.
5. Winzor, D. J. and Scheraga, H. A. (1963) Studies of chemically reacting systems on sephadex. I. Chromatographic demonstration of the Gilbert theory. *Biochemistry* **2,** 1263–1267.
6. Winzor, D. J. (1981) Mass migration methods, in *Protein-Protein Interactions* (Frieden, C., and Nichol, L. W., eds.), Wiley, New York, NY, pp. 129–172.

7. Cox, D. J. and Dale, R. S. (1981) Simulation of transport experiments for interacting systems, in *Protein–Protein Interactions* (Frieden, C., and Nichol, L. W., eds.), Wiley, New York, pp. 173–211.
8. Cann, J. R. (1996) Theory and practice of gel electrophoresis of interacting macromolecules. *Anal. Biochem.* **237,** 1–16.
9. Cann, J. R., York, E. J., Stewart, J. M., Vera, J. C., and Maccioni, R. B. (1988) Small zone gel chromatography of interacting systems: theoretical and experimental evaluation of elution profiles for kinetically controlled macromolecule-ligand reactions. *Anal. Biochem.* **175,** 462–473.
10. Patapoff, T. W., Mrsny, R. J., and Lee, W. A. (1993) The application of size exclusion chromatography and computer simulation to study the thermodynamic and kinetic parameters for short-lived dissociable protein aggregates. *Anal. Biochem.* **212,** 71–78.
11. Stevens, F. J. and Schiffer, M. (1981) Computer simulation of protein self-association during small-zone gel filtration. Estimation of equilibrium constants. *Biochem. J.* **195,** 213–219.
12. Stevens, F. J., Westholm, F. A., Solomon, A., and Schiffer, M. (1980) Self-association of human immunoglobulin kappa I light chains: role of the third hypervariable region. *Proc. Natl. Acad. Sci. USA* **77,** 1144–1148.
13. Kolmar, H., Frisch, C., Kleemann, G., Gotze, K., Stevens, F. J., and Fritz, H. J. (1994) Dimerization of Bence Jones proteins: linking the rate of transcription from an *Escherichia coli* promoter to the association constant of REIV. *Biol. Chem. Hoppe Seyler* **375,** 61–70.
14. Raffen, R., Stevens, P. W., Boogaard, C., Schiffer, M., and Stevens, F. J. (1998) Reengineering immunoglobulin domain interactions by introduction of charged residues. *Protein Eng.* **11,** 303–309.
15. Stevens, F. J. (1986) Analysis of protein-protein interaction by simulation of small-zone size-exclusion chromatography: application to an antibody-antigen association. *Biochemistry* **25,** 981–993.
16. Stevens, F. J., Carperos, W. E., Monafo, W. J., and Greenspan, N. S. (1988) Size-exclusion HPLC analysis of epitopes. *J. Immunol. Methods* **108,** 271–278.
17. Myatt, E. A., Stevens, F. J., and Benjamin, C. (1994) Solution-phase binding of monoclonal antibodies to bee venom phospholipase A2. *J. Immunol. Methods* **177,** 35–42.
18. Stevens, F. J. (1989) Analysis of protein-protein interaction by simulation of small-zone size exclusion chromatography. Stochastic formulation of kinetic rate contributions to observed high-performance liquid chromatography elution characteristics. *Biophys. J.* **55,** 1155–1167.
19. Raffen, R. and Stevens, F. J. (1999) Small zone, high-speed gel filtration chromatography to detect protein aggregation associated with light chain pathologies. *Methods Enzymol.* **309,** 318–332.

10

Fluorescence Gel Retardation Assay to Detect Protein–Protein Interactions

Sang-Hyun Park and Ronald T. Raines

Abstract

A gel mobility retardation assay can be used to detect a protein–protein interaction. The assay is based on the electrophoretic mobility of a protein–protein complex being less than that of either protein alone. Electrophoretic mobility is detected by the fluorescence of a green fluorescent protein variant that is fused to one of the protein partners. The assay is demonstrated by using the interaction of the S-protein and S-peptide fragments of ribonuclease A as a case study.

Key Words

Electrophoresis; fusion protein; gel mobility shift; gel retardation; green fluorescent protein; protein–protein interaction.

1. Introduction

Gel mobility retardation is a useful tool for both qualitative and quantitative analyses of protein–nucleic acid interactions *(1)*. Here, a gel mobility retardation assay is described that can be used to detect and identify specific protein–protein interactions *(2,3)*. The assay is based on the electrophoretic mobility of a protein–protein complex being less than that of either protein alone. Electrophoretic mobility is detected by the fluorescence of a green fluorescent protein (GFP) variant that is fused to one of the protein partners.

GFP from the jellyfish *Aequorea victoria* has exceptional physical and chemical properties besides spontaneous fluorescence. These properties include high thermal stability and resistance to detergents, organic solvents, and proteases *(4–6)*. These properties endow GFP with enormous potential for biotechnical applications *(7,8)*. Since the cDNA of GFP was cloned *(9)*, a variety of GFP

From: *Methods in Molecular Biology, vol. 261: Protein–Protein Interactions: Methods and Protocols*
Edited by: H. Fu © Humana Press Inc., Totowa, NJ

variants have been generated that broaden the spectrum of its application ***(10–15)***. Among those variants, S65T GFP is unique in having increased fluorescence intensity, faster fluorophore formation, and altered excitation and emission spectra than that of the wild-type protein ***(13,16)***.

The fluorescence gel retardation assay described below uses S65T GFP to probe protein–protein interactions in vitro ***(2)***. This method requires fusing S65T GFP to a target protein (X) to create a GFP chimera (GFP–X). The interaction of this fusion protein with another protein (Y) is then analyzed by native gel electrophoresis ***(17)*** followed by the detection of the fluorescence of free (GFP–X) and bound chimera (GFP – X·Y). The fluorescence gel retardation assay is a rapid method to demonstrate the existence of a protein–protein interaction and to estimate the equilibrium dissociation constant (K_d) of the resulting complex.

2. Materials

1. 10 m*M* Tris-HCl (pH 7.5) containing glycerol (5% v/v).
2. Solution of aqueous glycerol (10% v/v).
3. Solution of aqueous acrylamide (30% w/v) and bisacrylamide (8% w/v), filtered through a 0.45-µm filter and stored at 4 °C in the dark.
4. 0.5 *M* Tris-HCl (pH 8.8), filtered through a 0.45-µm filter and stored at 4°C.
5. Electrophoresis buffer (5X stock solution): Tris base (125 m*M*) and glycine (0.96 *M*), which is diluted to a 1X working solution as needed.
6. Solution of aqueous ammonium persulfate (10% w/v).
7. *N',N',N',N'*-tetramethylethylenediamine (TEMED).
8. Native acrylamide (6% w/v) gels (8 × 8 cm, 0.75 mm thick) prepared free of detergents or reducing agents (*see* **Note 1**).
9. Mini-gel electrophoresis apparatus and power supply.
10. Purified protein GFP–X and purified protein Y.
11. Fluorimager system (Molecular Dynamics; Sunnyvale, CA).
12. UV/Vis spectrophotometer.

3. Methods

3.1. Preparation of Polyacrylamide Gel

1. In a 50 mL plastic tube, mix 1 mL of acrylamide/bisacrylamide stock solution, 1.25 mL of 0.5 *M* Tris-HCl (pH 8.8), and 2.75 mL of deionized water.
2. Add 20 µL of ammonium persulfate (10% w/v) and 5 µL of TEMED. Vortex the solution briefly to mix.
3. Using a 5-mL pipet, slowly apply the solution to a mini-gel cast (8 cm × 8 cm, 0.75 mm thick). Insert a 0.75-mm comb and allow the gel to polymerize for 30 min at room temperature.
4. Carefully remove the comb without disrupting the edges of the polymerized wells.

3.2. Gel Electrophoresis

1. Estimate the concentration of protein GFP–X by using the extinction coefficient [$\varepsilon = 39.2\ mM^{-1}cm^{-1}$ at 490 nm *(13)*] of S65T GFP.
2. To begin the gel retardation assay, mix protein GFP-X (1.0 μM) with varying amounts of protein Y in 10 μL of 10 m*M* Tris-HCl buffer (pH 7.5) containing glycerol (5% v/v) (*see* **Note 2**).
3. Incubate the mixtures at 20°C for 20 min. Using gel-loading pipet tips, apply the mixtures onto the polyacrylamide gel. Perform the electrophoresis at 4°C at 10 V/cm for 30 min using 1X electrophoresis buffer (*see* **Note 3**).
4. Immediately after electrophoresis, scan the gel with a Fluorimager SI System at 700 V using the built-in filter set (490 nm for excitation; ≥515 nm for emission) (*see* **Notes 4** and **5**).

3.3. Data Analysis

1. Quantify the fluorescence intensities of bound and free GFP–X in the gel scan by using the program ImageQuaNT (Molecular Dynamics; Sunnyvale, CA).
2. Determine the value of **R** (= fluorescence intensity of bound GFP–X/total fluorescence intensity) for each gel lane from the measured fluorescence intensities.
3. Calculate values of K_d for each lane with the equation:

$$K_d = \frac{1-\mathbf{R}}{\mathbf{R}} \times \left([\mathrm{Y}]_{\mathrm{total}} - \mathbf{R} \times [\mathrm{GFP{-}X}]_{\mathrm{total}}\right) \tag{1}$$

3.4. Case Study

The well-characterized interaction of the S-15 and S-protein fragments of ribonuclease A has been used to demonstrate the potential of the fluorescence gel retardation assay to detect protein–protein interactions *(2)*. The assay has also been used to detect the interaction of CREB and importins *(18)*, and that of cyclophilin and the capsid protein p24 of HIV-1 *(19)*. In the first example, a GFP chimera, S15-GFP(S65T)-His6, is produced by standard recombinant DNA techniques and affinity-purified from bacteria. A fixed quantity of S15–GFP(S65T)–His_6 is incubated with a varying quantity of S-protein prior to electrophoresis in a native polyacrylamide gel. After electrophoresis, the gel is scanned with a fluorimager and the fluorescence intensities of bound and free S15–GFP(S65T)–His_6 are quantified (**Fig. 1**). The value of K_d for the complex formed in the presence of different S-protein concentrations is calculated from the values of **R** and the total concentrations of S-protein and S15–GFP(S65T)–His_6 by using eq. (1). The average (± SE) value of K_d is $(6 \pm 3) \times 10^{-8}$ *M*.

4. Notes

1. Polymerized, unused gels can be store at 4°C for up to 1 mo. To prevent the gels from getting dry, they can be wrapped in wet paper towels and sealed in a plastic bag.

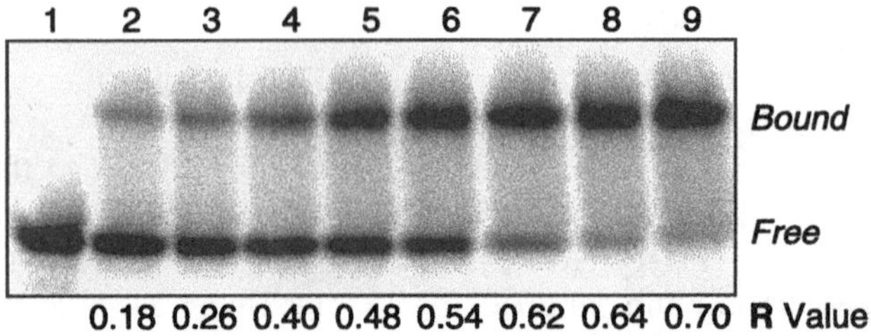

Fig. 1. Gel retardation assay of a protein–protein interaction. Gel retardation assay of the interaction of S15–GFP(S65T)–His_6 with varying amounts of S-protein. Lanes 1–9, 1 μ*M* S15–GFP(S65T)–His_6 and 0, 0.2, 0.3, 0.4, 0.5, 0.6, 0.7, 0.8, and 0.9 μ*M* S-protein, respectively. The relative mobilities of free and bound S15–GFP(S65T)–His_6 in an acrylamide (6% w/v) gel are 0.72 and 0.47, respectively. The value of **R** is obtained for each lane, and values of K_d are calculated by using eq. (1), with the average being $K_d = (6 \pm 3) \times 10^{-8}$ *M*.

2. Pilot experiments may need to be conducted to determine the proper concentration range of GFP–X and Y. A good starting point is to adjust the concentration of GFP–X to the K_d value of complex, if known, and to prepare serial dilutions of Y so that its concentrations spans from $K_d \times 10$ to $K_d/10$. For example, if K_d = 1.0 μ*M*, mix 1.0 μ*M* GFP–X with 0.1, 0.2, 0.5, 1.0, 2.0, 5.0, and 10 μ*M* Y.
3. The gel, electrophoresis apparatus, and electrophoresis buffer should be pre-equilibrated at 4°C before electrophoresis.
4. Although it is desirable to scan the gel immediately after electrophoresis, gels can be stored at 4°C for a few hours before scanning without compromising the resolution and fluorescence sensitivity.
5. It is not necessary to remove the gel from the glass plates before scanning. Most glassware used for casting gels are not fluorescent. The surface of the glass to be scanned should be rinsed with deionized water to remove any residual buffer and acrylamide, and wiped dry.

References

1. Carey, J. (1991) Gel retardation. *Methods Enzymol.* **208,** 103–117.
2. Park, S.-H. and Raines, R. T. (1997) Green fluorescent protein as a signal for protein–protein interactions. *Protein Sci.* **6,** 2344–2349.
3. Park, S. H. and Raines, R. T. (2000) Green fluorescent protein chimeras to probe protein–protein interactions. *Methods Enzymol.* **328,** 251–261.
4. Bokman, S. H. and Ward, W. W. (1981) Renaturation of *Aequorea* green fluorescent protein. *Biochem. Biophys. Res. Commun.* **101,** 1372–1380.
5. Ward, W. W. (1981) in *Bioluminescence and Chemiluminescence* (DeLuca, M., and McElroy, W., eds), Academic Press, New York, NY, pp. 235–242.

6. Ward, W. W. and Bokman, S. H. (1982) Reversible denaturation of *Aequorea* green fluorescent protein: physical separation and characterization of the renatured protein. *Biochemistry* **21,** 4535–4540.
7. Hirschberg, K., Phair, R. D., and Lippincott-Schwartz, J. (2000) Kinetic analysis of intracellular trafficking in single living cells with vesicular stomatitis virus protein G-green fluorescent protein hybrids. *Methods Enzymol.* **327,** 69–89.
8. Meyer, T. and Oancea, E. (2000) Studies of signal transduction events using chimeras to green fluorescent protein. *Methods Enzymol.* **327,** 500–513.
9. Prasher, D. C., Eckenrode, V. K., Ward, W. W., Prendergast, F. G., and Cormier, M. J. (1992) Primary structure of the *Aeqourea victoria* green-fluorescent protein. *Gene* **111,** 229–233.
10. Cubitt, A. B., Heim, R., Adams, S. R., Boyd, A. E., Gross, L. A., and Tsien, R. Y. (1995) Understanding, improving and using green fluorescent proteins. *Trends Biochem. Sci.* **20,** 448–455.
11. Delagrave, S., Hawtin, R. E., Silva, C. M., Yang, M. M., and Youvan, D. C. (1995) Red-shifted excitation mutants of the green fluorescent protein. *BioTechnology* **13,** 151–154.
12. Ehrig, T., O'Kane, D. J., and Prendergast, F. G. (1995) Green-fluorescent protein with altered fluorescence excitation spectra. *FEBS Lett.* **367,** 163–166.
13. Heim, R., Cubitt, A. B., and Tsien, R. Y. (1995) Improved green fluorescence. *Nature* **373,** 663–664.
14. Crameri, A., Whitehorn, E. A., Tate, E., and Stemmer, W. P. C. (1996) Improved green fluorescent protein by molecular evolution using DNA shuffling. *Nature Biotechnol.* **14,** 315–319.
15. Ward, W. W. (1997) *Green Fluorescent Protein: Properties, Applications and Protocols* (Chalfie, M., and Kain, S., eds.), Wiley, New York, NY.
16. OrmϸM., Cubitt, A. B., Kallio, K., Gross, L. A., Tsien, R. Y., and Remington, S. J. (1996) Crystal structure of the *Aeqourea victoria* green fluorescent protein. *Science* **237,** 1392–1395.
17. Laemmli, U. K. (1970) Cleavage of structural proteins during the assembly of the head of bacteriophage T4. *Nature* **227,** 680–685.
18. Forwood, J. K., Lam, M. H., and Jans, D. A. (2001) Nuclear import of Creb and AP-1 transcription factors requires importin-β1 and Ran but is independent of importin-α. *Biochemistry* **40,** 5208–5217.
19. Kiessig, S., Reissmann, J., Rascher, C., Kullertz, G., Fischer, A., and Thunecke, F. (2001) Application of a green fluorescent fusion protein to study protein–protein interactions by electrophoretic methods. *Electrophoresis* **22,** 1428–1435.

11

Fluorescence Polarization Assay to Quantify Protein–Protein Interactions

Sang-Hyun Park and Ronald T. Raines

Abstract

A fluorescence polarization assay can be used to evaluate the strength of a protein–protein interaction. A green fluorescent protein variant is fused to one of the protein partners. The formation of a complex is then deduced from an increase in fluorescence polarization, and the equilibrium dissociation constant of the complex is determined in a homogeneous aqueous environment. The assay is demonstrated by using the interaction of the S-protein and S-peptide fragments of ribonuclease A as a case study.

Key Words

Fluorescence anisotropy; fluorescence polarization; fusion protein; green fluorescent protein, protein–protein interaction.

1. Introduction

Fluorescence polarization can be used to analyze macromolecular interactions in which one of the reactants is labeled with a fluorophore (*see* **Note 1**). In this assay, the formation of a complex is deduced from an increase in fluorescence polarization, and the equilibrium dissociation constant (K_d) of the complex is determined in a homogeneous aqueous environment (*see* **Note 2**). Most fluorescence polarization assays have used a small molecule such as fluorescein as a fluorophore ***(1–4)***.

Here, a variant (S65T) of green fluorescent protein (GFP) is used as the fluorophore in a polarization assay ***(5,6)***. The advantages of using S65T GFP as the fluorophore is the ease with which a protein can be fused to GFP by using recombinant DNA techniques, the high integrity of the resulting chimera, and the broad chemical and physical stability of GFP compared to small-molecule fluorophores. To quantify complex formation of two proteins (X and Y),

From: *Methods in Molecular Biology, vol. 261: Protein–Protein Interactions: Methods and Protocols*
Edited by: H. Fu © Humana Press Inc., Totowa, NJ

a GFP fusion protein (GFP–X) is produced by using recombinant DNA technology. Protein GFP–X is then titrated with protein Y, and the equilibrium dissociation constant is obtained from the increase in fluorescence polarization that accompanies complex formation. Like a free fluorescein-labeled ligand, free GFP–X is likely to rotate more rapidly and therefore to have a lower rotational correlation time than does the GFP–X·Y complex. An increase in rotational correlation time upon binding results in an increase in fluorescence polarization, which can be used to assess complex formation *(7)*.

In a fluorescence polarization assay, the interaction between the two proteins is quantified in a homogeneous solution. The fluorescence polarization assay thereby allows for the determination of accurate values of K_d in a wide range of solution conditions. GFP is particularly well suited to this application because its fluorophore is held rigidly within the protein, as revealed by the three-dimensional structure of wild-type GFP and the S65T variant *(8,9)*. Such a rigid fluorophore minimizes local rotational motion, thereby ensuring that changes in polarization report on changes to the *global* rotational motion of GFP, as effected by a protein–protein interaction.

2. Materials

1. 20 m*M* Tris-HCl buffer (varying pH).
2. Solution of aqueous NaCl (varying concentration).
3. Purified GFP–X and purified protein Y.
4. Beacon Fluorescence Polarization System (PanVera, Madison, WI).
5. Graphics software capable of nonlinear regression (e.g., DeltaGraph or SigmaPlot).

3. Methods

3.1. Fluorescence Polarization Assay

1. Mix protein GFP–X (0.50–1.0 n*M*) with various concentrations of protein Y in 1.0 mL of 20 m*M* Tris-HCl buffer, pH 8.0, with or without NaCl at 20°C (*see* **Note 3**). Conditions such as buffer, pH, temperature, and salt can be varied as desired.
2. After mixing, make five to seven polarization measurements at each concentration of protein Y using a Beacon Fluorescence Polarization System (*see* **Notes 4–6**). For a blank measurement, use a mixture that contains the same components except for protein GFP–X.

3.2. Data analysis

1. Fluorescence polarization (P) is defined as

$$P = \frac{I_{\parallel} - I_{\perp}}{I_{\parallel} + I_{\perp}} \tag{1}$$

where $I_{\parallel}$ is the intensity of the emission light parallel to the excitation light plane and $I_{\perp}$ is the intensity of the emission light perpendicular to the excitation light plane. P, the ratio of light intensities, is a dimensionless number with a maximum value of 0.5. Calculate values of K_d by fitting the data to the equation:

$$P = \frac{\Delta P \cdot F}{K_d + F} + P_{min} \tag{2}$$

In eq. (2), P is the measured polarization, ΔP (= $P_{max} - P_{min}$) is the total change in polarization and F is the concentration of free protein Y (*see* **Note 7**).

2. Calculate the fraction of bound protein (f_B) by using the equation

$$f_B = \frac{P - P_{min}}{\Delta P} = \frac{F}{K_d + F} \tag{3}$$

Plot f_B vs F to show the binding isotherms.

3.3. Case Study

Fluorescence polarization is used to determine the effect of salt concentration on the formation of a complex between S15 and S-protein fragments of ribonuclease A ***(10)***. A GFP chimera of S-peptide [S15–GFP(S65T)–His$_6$] was produced from bacteria and titrated with free S-protein ***(5)***. The value of K_d increases by fourfold when NaCl is added to a final concentration of 0.10 *M* (**Fig. 1**). A similar salt dependence for the dissociation of RNase S had been observed previously ***(11)***. The added salt is likely to disturb the water molecules hydrating the hydrophobic patch in the complex between S15 and S-protein, resulting in a decrease in the binding affinity ***(12)***. Finally, the value of $K_d = 4.2 \times 10^{-8}$ *M* observed in 20 m*M* Tris-HCl buffer (pH 8.0) containing NaCl (0.10 *M*) is similar (i.e., threefold lower) than that obtained by titration calorimetry in 50 m*M* sodium acetate buffer (pH 6.0) containing NaCl (0.10 m*M*) ***(13)***.

4. Notes

1. We used the term "polarization" instead of "anisotropy" herein. Fluorescence polarization (P) and fluorescence anisotropy (A) are related [$A = 2P/(3 - P)$] and contain equivalent physical information with respect to monitoring macromolecular complex formation. Many instruments report on both polarization and anisotropy and either parameter can be used to evaluate K_d.
2. Polarization is proportional to the rotational correlation time (τ), which is defined as

$$P \propto \tau = \frac{3\eta V}{RT} \tag{4}$$

In eq. (4), rotational correlation time (τ) is the time taken for a molecule to rotate 68.5° and is related to the solution viscosity (η), molecular volume (V),

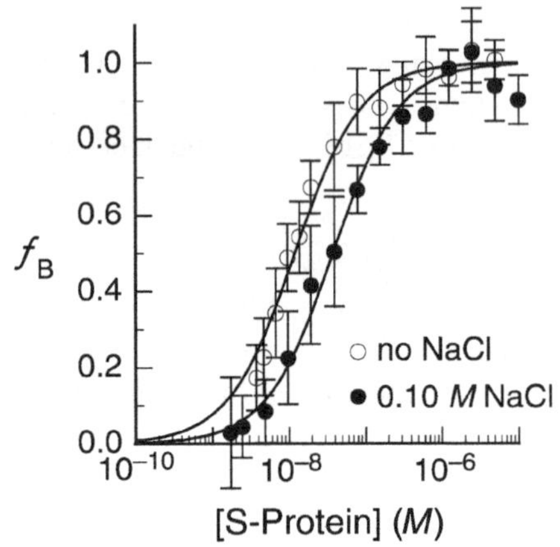

Fig. 1. Fluorescence polarization assay of a protein–protein interaction. S15–GFP(S65T)–His_6 with S-protein. S-protein is added to 20 m*M* Tris-HCl buffer (pH 8.0) in a volume of 1.0 mL. Each data point is an average of five to seven measurements. Curves are obtained by fitting the data to eq. (3). The values of K_d in the presence of 0 and 0.10 *M* NaCl are 1.1×10^{-8} and 4.2×10^{-8} *M*, respectively.

gas constant (*R*), and absolute temperature (*T*). Thus, under conditions of constant viscosity and temperature, polarization is directly proportional to the molecular volume, which increases upon complex formation.

3. In the assay solution, [GFP–X] should be significantly lower than the value of K_d ([GFP–X] << K_d) but still be high enough to generate detectable fluorescence in the spectrometer. In the case study, [GFP–X] = 1 n*M* and K_d > 10 n*M*.
4. Data collection must be done at equilibrium. To estimate the time to reach equilibrium, a pilot experiment can be performed in which Y is added at [Y] = K_d, and the polarization is monitored until it reaches a stationary value.
5. At each [Y], the sample should be blanked with an identical mixture that lacks GFP–X.
6. Although it is desirable to use a dedicated fluorescence polarization system, a conventional fluorometer equipped with polarization measurement capability can also be used to obtain values of *P* or *A*.
7. The change in polarization (ΔP) upon complex formation must be detectable. For example, if the value of τ for GFP–X does not change significantly upon formation of the GFP–X·Y complex, then the value of ΔP is small and the data analysis is difficult.

References

1. LeTilly, V. and Royer, C. A. (1993) Fluorescence anisotropy assays implicate protein-protein interactions in regulating *trp* repressor DNA binding. *Biochemistry* **32,** 7753–7758.

2. Heyduk, T., Ma, Y., Tang, H., and Ebright, R. H. (1996) Fluorescence anisotropy: Rapid, quantitative assay for protein–DNA and protein–protein interaction. *Methods Enzymol.* **274,** 492–503.
3. Malpeli, G., Folli, C., and Berni, R. (1996) Retinoid binding to retinol-binding protein and the interference with the interaction with transthyretin. *Biochim. Biophys. Acta* **1294,** 48–54.
4. Fisher, B. M., Ha, J.-H., and Raines, R. T. (1998) Coulombic forces in protein-RNA interactions: binding and cleavage by ribonuclease A and variants at Lys7, Arg10 and Lys66. *Biochemistry* **37,** 12,121–12,132.
5. Park, S.-H. and Raines, R. T. (1997) Green fluorescent protein as a signal for protein–protein interactions. *Protein Sci.* **6,** 2344–2349.
6. Park, S. H. and Raines, R. T. (2000) Green fluorescent protein chimeras to probe protein–protein interactions. *Methods Enzymol.* **328,** 251–261.
7. Jameson, D. M. and Sawyer, W. H. (1995) Fluorescence anisotropy applied to biomolecular interactions. *Methods Enzymol.* **246,** 283–300.
8. OrmöM., Cubitt, A. B., Kallio, K., Gross, L. A., Tsien, R. Y., and Remington, S. J. (1996) Crystal structure of the *Aequorea victoria* green fluorescent protein. *Science* **237,** 1392–1395.
9. Yang, F., Moss, L. G., and Phillips, G. N., Jr. (1996) The molecular structure of green fluorescent protein. *Nature Biotechnol.* **14,** 1246–1251.
10. Richards, F. M. and Vithayathil, P. J. (1959) The preparation of subtilisin modified ribonuclease and separation of the peptide and protein components. *J. Biol. Chem.* **234,** 1459–1465.
11. Schreier, A. A. and Baldwin, R. L. (1977) Mechanism of dissociation of S-peptide from ribonuclease S. *Biochemistry* **16,** 4203–4209.
12. Baldwin, R. L. (1996) How Hofmeister ion interactions affect protein stability. *Biophys. J.* **71,** 2056–2063.
13. Connelly, P. R., Varadarajan, R., Sturtevant, J. M., and Richards, F. M. (1990) Thermodynamics of protein–peptide interactions in the ribonuclease S system studied by titration calorimetry. *Biochemistry* **29,** 6108–6114.

12

Studying Protein–Protein Interactions via Blot Overlay or Far Western Blot

Randy A. Hall

Abstract

Blot overlay is a useful method for studying protein–protein interactions. This technique involves fractionating proteins on SDS-PAGE, blotting to nitrocellulose or PVDF membrane, and then incubating with a probe of interest. The probe is typically a protein that is radiolabeled, biotinylated, or simply visualized with a specific antibody. When the probe is visualized via antibody detection, this technique is often referred to as "Far Western blot." Many different kinds of protein–protein interactions can be studied via blot overlay, and the method is applicable to screens for unknown protein–protein interactions as well as to the detailed characterization of known interactions.

Key Words

Protein–protein interactions; blot overlay; Far Western blot; protein; receptor; association; nitrocellulose; SDS-PAGE; binding.

1. Introduction

During preparation for SDS-PAGE, proteins are typically reduced and denatured via treatment with Laemmli sample buffer *(1)*. Because many protein–protein interactions rely on aspects of secondary and tertiary protein structure that are disrupted under reducing and denaturing conditions, it might seem likely that few if any protein–protein interactions could survive treatment with SDS-PAGE sample buffer. Nonetheless, it is well known that many types of protein–protein interaction do in fact still occur even after one of the partners has been reduced, denatured, run on SDS-PAGE, and Western blotted. Blot overlays are a standard and very useful method for studying interactions between proteins.

From: *Methods in Molecular Biology, vol. 261: Protein–Protein Interactions: Methods and Protocols*
Edited by: H. Fu © Humana Press Inc., Totowa, NJ

In principle, a blot overlay is similar to a Western blot. For both procedures, samples are run on SDS-PAGE gels, transferred to nitrocellulose or PVDF, and then overlaid with a soluble protein that may bind to one or more immobilized proteins on the blot. In the case of a Western blot, the overlaid protein is antibody. In the case of a blot overlay, the overlaid protein is a probe of interest, often a fusion protein that is easy to detect. The overlaid probe can be detected either via incubation with an antibody (this method is often referred to as a "Far Western blot"), via incubation with streptavidin (if the probe is biotinylated), or via autoradiography if the overlaid probe is radiolabeled with ^{32}P (also see Chapter 31). The specific method that will be described here is a Far Western blot overlay that was used to detect the binding of blotted hexahistidine-tagged PDZ domain fusion proteins to soluble GST fusion proteins corresponding to adrenergic receptor carboxyl-termini *(2)*. However, this method may be adapted to a wide variety of applications.

2. Materials

1. SDS-PAGE mini-gel apparatus (Invitrogen).
2. SDS-PAGE 4–20% mini gels (Invitrogen).
3. Western blot transfer apparatus (Invitrogen).
4. Power supply (BioRad).
5. Nitrocellulose (Invitrogen).
6. SDS-PAGE prestained molecular weight markers (BioRad).
7. SDS-PAGE sample buffer: 20 m*M* Tris-HCl, pH 7.4, 2% SDS, 2% β-mercaptoethanol, 5% glycerol, 1 mg/mL bromophenol blue.
8. SDS-PAGE running buffer: 25 m*M* Tris-HCl, pH 7.4, 200 m*M* glycine, 0.1% SDS.
9. SDS-PAGE transfer buffer: 10 m*M* Tris-HCl, pH 7.4, 100 m*M* glycine, 20% methanol.
10. Purified hexahistidine-tagged fusion proteins.
11. Purified GST-tagged fusion proteins.
12. Anti-GST monoclonal antibody (Santa Cruz Biotechnology, cat. no. sc-138).
13. Goat anti-mouse HRP-coupled secondary antibody (Amersham Pharmacia Biotech).
14. Blocking buffer: 2% nonfat powdered milk, 0.1% Tween-20 in phosphate-buffered saline, pH 7.4.
15. Enhanced chemiluminescence kit (Amersham Pharmacia Biotech).
16. Blot trays.
17. Rocking platform.
18. Autoradiography cassette.
19. Clear plastic sheet.
20. Film.

3. Methods

3.1. SDS-PAGE and Blotting

The purpose of this step is to immobilize the samples of interest on nitrocellulose or an equivalent matrix, such as PVDF. It is very important to keep the blot clean during the handling steps involved in the transfer procedure, because contaminants can contribute to increased background problems later on during detection of the overlaid probe.

1. Place gel in SDS-PAGE apparatus and fill chamber with running buffer.
2. Mix purified hexahistidine-tagged fusion proteins with SDS-PAGE sample buffer to a final concentration of approx 0.1 μg/μL of fusion protein (*see* **Note 1**).
3. Load 20 μL of fusion protein (2 μg total) in each lane of the gel. If there are more lanes than samples, load 20 μL of sample buffer in the extra lanes (*see* **Note 2**).
4. In at least one lane of the gel, load 20 μL of SDS-PAGE molecular weight markers.
5. Run gel for approx 80 min at 150 V using the power supply.
6. Stop gel, turn off the power supply, remove the gel from its protective casing, and place in transfer buffer.
7. Place precut nitrocellulose in transfer buffer to wet it.
8. Put nitrocellulose and gel together in transfer apparatus, and transfer proteins from gel to nitrocellulose using power supply for 80 min at 25 V.

3.2. Overlay

During the overlay step, the probe is incubated with the blot and unbound probe is then washed away. The potential success of the overlay depends heavily on the purity of the overlaid probe. GST and hexahistidine-tagged fusion proteins should be purified as extensively as possible. If the probe has many contaminants, this may contribute to increasing the background during the detection step, making visualization of the specifically bound probe more difficult.

1. Block blot in blocking buffer for at least 30 min (*see* **Note 3**).
2. Add GST fusion proteins to a concentration of 25 n*M* in 10 mL blocking buffer.
3. Incubate GST fusion proteins with blot for 1 h at room temperature while rocking slowly.
4. Discard GST fusion protein solution and wash blot three times for 5 min each with 10 mL of blocking buffer while rocking the blot slowly.
5. Add anti-GST antibody at 1:1000 dilution (approx 200 ng/mL final) to 10 mL blocking buffer.
6. Incubate anti-GST antibody with blot for 1 h while rocking the blot slowly.
7. Discard anti-GST antibody solution and wash blot three times for 5 min each with 10 mL of blocking buffer while rocking the blot slowly.
8. Add goat anti-mouse HRP-coupled secondary antibody at 1:2000 dilution to 10 mL blocking buffer.
9. Incubate secondary antibody with blot for 1 h while rocking the blot slowly.
10. Discard secondary antibody solution and wash blot three times for 5 min each with 10 mL of blocking buffer while rocking the blot slowly (*see* **Note 4**).
11. Wash blot one time for 5 min with phosphate buffered saline, pH 7.4.

3.3. Detection of Overlaid Proteins

The final step of the overlay is to detect the probe that is bound specifically to proteins immobilized on the blot. In viewing different exposures of the visualized probe, an effort should be made to obtain the best possible signal-to-noise ratio. Nonspecific background binding will increase linearly with time of exposure. Thus, shorter exposures may have more favorable signal-to-noise ratios.

1. Incubate blot with enhanced chemiluminescence solution for 60 s (*see* **Note 5**).
2. Remove excess ECL solution from blot and place blot in clear plastic sheet.
3. Tape sheet into autoradiography cassette.
4. Move to darkroom and place one sheet of film into autoradiography cassette with blot.
5. Expose film for 5–2000 s, depending on intensity of signal.
6. Develop film in standard film developer.

4. Notes

1. The protocol described here is intended for the in-depth study of a protein–protein interaction that is already known. However, blot overlays can also be utilized in preliminary screening studies to detect novel protein–protein interactions. For this application, tissue lysates would typically be loaded onto the SDS-PAGE gel instead of purified fusion protein samples. The blotted tissue lysates would then be overlaid with the probe of interest. The advantages of this technique are (i) many tissue samples can be screened in a single blot and (ii) the molecular weight and tissue distribution of probe-interacting proteins can be immediately determined. The disadvantages of this method are (i) due to the multiple washing steps involved in the procedure, a fairly high affinity interaction is required for the interaction to be detected, (ii) detection of probe-interacting proteins is dependent on their level of expression in native tissues, and (iii) interactions requiring native conformations of both proteins will not be detected. Tissue lysate overlays have been utilized as screening tools to detect not only the interaction of the β_1-adrenergic receptor with MAGI-2 described here *(2)* (**Fig. 1**), but also the interaction of the β_2-adrenergic receptor with NHERF *(3)* and the interactions of a number of different proteins with actin *(4–6)*, calmodulin *(7,8)*, and the cyclic AMP-dependent protein kinase RII regulatory subunit *(9–12)*.
2. Because some probes can exhibit extensive nonspecific binding to blotted proteins, it is important in overlay assays to have negative controls for probe binding. When the blotted proteins are GST fusion proteins, GST by itself is a good negative control (as illustrated in **Fig. 2**). When the blotted proteins are His-tagged fusion proteins, as illustrated in **Fig. 1**, it is helpful to have one or more His-tagged fusion proteins on the same blot that will not bind to the probe. In this way, it is possible to demonstrate the specificity of binding and to rule out the possibility that the observed interaction is due to the tag.

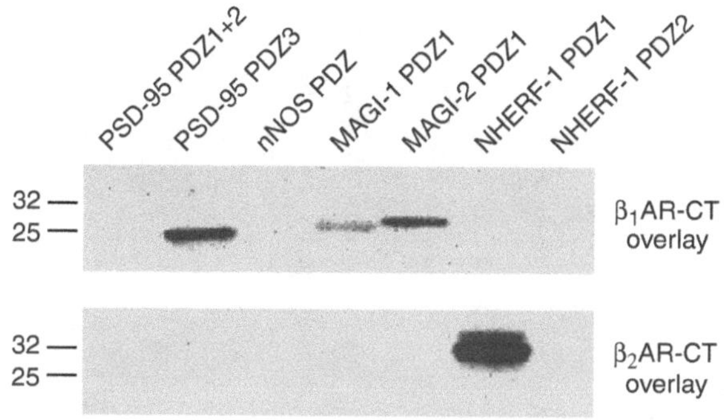

Fig. 1. Overlay of GST-tagged adrenergic receptor carboxyl-termini onto hexahistidine-tagged PDZ domains. Equal amounts (2 μg) of purified His-tagged fusion proteins corresponding to PDZ domains from PSD-95, nNOS, MAGI-1, MAGI-2, and NHERF-1 were immobilized on nitrocellulose. Overlays with the carboxyl-terminus of the β_1-adrenergic receptor expressed as a GST fusion protein (β_1AR-CT-GST) (25 n*M*) revealed strong binding to PSD-95 PDZ3 and MAGI-2 PDZ1, moderate binding to MAGI-1 PDZ1, and no detectable binding to the first two PDZ domains of PSD-95 or to the PDZ domains of nNOS or NHERF-1. In contrast, overlays with the β_2-adrenergic receptor expressed as a GST fusion protein (β_2AR-CT-GST) (25 n*M*) revealed strong binding to NHERF-1 PDZ1 but no detectable binding to any of the other PDZ domains examined. These data demonstrate that selective and specific binding can be obtained in overlay assays.

3. The blocking of the blot is a very important step in every overlay assay. The idea is to block potential nonspecific sites of protein attachment to the blot, so that nonspecific binding of the probe will be minimized. When a high amount of nonspecific background binding is observed, it is often helpful to block for a longer time or with a higher concentration of milk. Some investigators favor bovine serum albumin or other proteins in place of milk for blocking blots prior to overlay.
4. The washing of the blot is of critical importance. If the washes are not rigorous enough, the nonspecific background binding of the probe will be undesirably high. Conversely, if the washes are too rigorous, specific binding of the probe may be lost and the protein–protein interaction of interest may be difficult to detect. Thus, if a large amount of nonspecific background binding is observed, one should consider increasing the rigor of the washes, whereas, conversely, if the background is low but little or no specific binding is observed, one should consider decreasing the rigor of the washes. The rigor of the washes is dependent on (i) time, (ii) volume, (iii) speed, and (iv) detergent concentration. To make washes more rigorous, one should wash for a longer time, wash in a larger volume, increase the rate at which the gels are rocked during the washes, and/or increase the detergent concentration in the buffer used for washing.

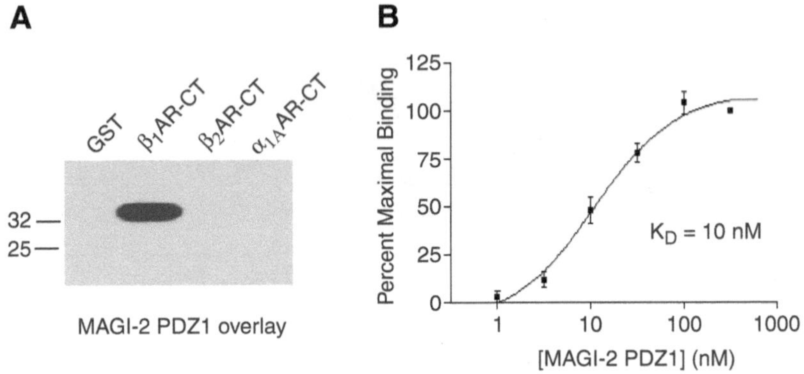

Fig. 2. Overlay of hexahistidine-tagged MAGI-2 PDZ1 onto GST-tagged adrenergic receptor carboxyl-termini. **(A)** In the reverse of the overlay experiments illustrated in Fig. 1, equal amounts (2 μg) of purified GST fusion proteins corresponding to the carboxyl-termini of various adrenergic receptor subtypes were immobilized on nitrocellulose. Overlay with MAGI-2 PDZ1 His- and S-tagged fusion protein (20 n*M*) revealed strong binding to β_1AR-CT-GST but no detectable binding to control GST, β_2AR-CT-GST or α_1AR-CT-GST. These data demonstrate that the interaction between the β_1AR-CT and MAGI-2 PDZ1 can be visualized via overlay in either direction. **(B)** Estimate of the affinity of the interaction between β_1AR-CT and MAGI-2 PDZ1. Nitrocellulose strips containing 2 μg β_1AR-CT-GST (equivalent to lane 2 in the preceding panel) were incubated with His/S-tagged MAGI-2 PDZ1 at six concentrations between 1 and 300 n*M*. Specific binding of MAGI-2 PDZ1 did not increase between 100 and 300 n*M*, and thus the binding observed at 300 n*M* was defined as "maximal" binding. The binding observed at the other concentrations was expressed as a percentage of maximal binding within each experiment. The bars and error bars shown on this graph indicate mean ± SEM ($n = 3$). The K_d for MAGI-2 PDZ1 binding to β_1AR-CT was estimated at 10 n*M* (*see* **Note 6**).

5. There are a number of ways to visualize bound probe in an overlay assay. The method described here depends on detection of the probe with an antibody, which is often referred to as a "Far Western blot." One alternative approach is to biotinylate the probe and then detect it with a streptavidin/enzyme conjugate ***(5–7)***. The appeal of this approach is that it can be quite sensitive, because the streptavidin–biotin interaction is one of the highest affinity interactions known. The main drawback of this approach is that biotinylation of the probe may alter its properties, such that it may lose the ability to interact with partners it normally binds to. An additional approach to probe detection is phosphorylation of the probe using ^{32}P-ATP, to make the probe radiolabeled ***(10–12)***. A primary advantage of this method is that once the probe is overlaid onto the blot, no

further detection steps are necessary (i.e., no incubations with antibody or streptavidin are required). This cuts down on the number of washing steps and may aid in the detection of protein–protein interactions that are of somewhat lower affinity. The main disadvantages of the phosphorylation approach are (i) radioactive samples require special handling and (ii) as with biotinylation, phosphorylation of the probe may alter its properties, such that certain protein–protein interactions may be disrupted.

6. As is illustrated in **Figs. 1** and **2**, detection of the interactions between adrenergic receptor carboxyl-termini and their PDZ domain-containing binding partners are completely reversible. Either partner can be immobilized on the blot and overlaid with the other. Many other protein–protein interactions can similarly be detected in a reversible manner, but some interactions can only be detected in one direction due to a requirement for the native conformation of one of the partners. As is also illustrated in **Fig. 2**, the affinity of a given protein–protein interaction may be estimated via blot overlay saturation binding curves. This method involves increasing the concentration of overlaid probe until a maximal amount of specific binding is obtained. An estimate for the affinity constant (K_d) of the interaction can then be determined from the slope of the binding curve, much as one would determine K_d values from ligand binding curves using a program such as GraphPad Prism. Estimates such as these must be evaluated with the caveat that they are derived under artificial conditions involving many hours of incubation time, washing, and detection. Nonetheless, affinity constant estimates derived via this method are useful in comparing affinities between proteins examined under the same conditions and overlaid with the same probe.

Acknowledgments

R.A.H is supported by grants from the National Institutes of Health (GM60982, HL64713) and a Faculty Development award from the Pharmaceutical and Manufacturers of America Foundation.

References

1. Laemmli, U. K. (1970) Cleavage of structural proteins during the assembly of the head of bacteriophage T4. *Nature* **227,** 680–685.
2. Xu, J., Paquet, M., Lau, A. G., Wood, J. D., Ross, C. A., and Hall, R. A. (2001) β_1-adrenergic receptor association with the synaptic scaffolding protein membrane-associated guanylate kinase inverted-2 (MAGI-2): differential regulation of receptor internalization by MAGI-2 and PSD-95. *J. Biol. Chem.* **276,** 41,310–41,317.
3. Hall, R. A., Premont, R. T., Chow, C. W., et al. (1998) The β_2-adrenergic receptor interacts with the Na^+/H^+-exchanger regulatory factor to control Na^+/H^+ exchange. *Nature* **392,** 626–630.
4. Luna, E. J. (1998) F-actin blot overlays. *Methods Enzymol.* **298,** 32–42.
5. Li, Y., Hua, F., Carraway, K. L., and Carraway, C. A. (1999) The p185(neu)-containing glycoprotein complex of a microfilament-associated signal transduc-

tion particle. Purification, reconstitution, and molecular associations with p58(gag) and actin. *J. Biol. Chem.* **274,** 25,651–25,658.

6. Holliday, L. S., Lu, M., Lee, B. S., et al. (2000) The amino-terminal domain of the B subunit of vacuolar H^+-ATPase contains a filamentous actin binding site. *J. Biol. Chem.* **275,** 32,331–32,337.
7. Pennypacker, K. R., Kyritsis, A., Chader, G. J., and Billingsley, M. L. (1988) Calmodulin-binding proteins in human Y-79 retinoblastoma and HTB-14 glioma cell lines. *J. Neurochem.* **50,** 1648–1654.
8. Murray, G., Marshall, M. J., Trumble, W., and Magnuson, B. A. (2001) Calmodulin-binding protein detection using a non-radiolabeled calmodulin fusion protein. *Biotechniques* **30,** 1036–1042.
9. Lohmann, S. M., DeCamilli, P., Einig, I., and Walter, U. (1984) High-affinity binding of the regulatory subunit (RII) of cAMP-dependent protein kinase to microtubule-associated and other cellular proteins. *Proc. Natl. Acad. Sci. USA* **81,** 6723–6727.
10. Bregman, D. B., Bhattacharyya, N., and Rubin, C. S. (1989) High affinity binding protein for the regulatory subunit of cAMP-dependent protein kinase II-B. Cloning, characterization, and expression of cDNAs for rat brain P150. *J. Biol. Chem.* **264,** 4648–4656.
11. Carr, D. W., Hausken, Z. E., Fraser, I. D., Stofko-Hahn, R. E., and Scott, J. D. (1992) Association of the type II cAMP-dependent protein kinase with a human thyroid RII-anchoring protein. Cloning and characterization of the RII-binding domain. *J. Biol. Chem.* **267,** 13,376–13,382.
12. Hausken, Z. E., Coghlan, V. M., and Scott, J. D. (1998) Overlay, ligand blotting, and band-shift techniques to study kinase anchoring. *Methods Mol. Biol.* **88,** 47–64.

13

Glutathione-*S*-Transferase–Fusion Based Assays for Studying Protein–Protein Interactions

Haris G. Vikis and Kun-Liang Guan

Abstract

Glutathione-*S*-transferase (GST)–fusion proteins have become an effective reagent to use in the study of protein–protein interactions. They can be produced in bacterial and mammalian cells in large quantities and purified rapidly. Given that GST can be coupled to a glutathione matrix permits its use as an effective affinity column to study interactions in vitro or to purify protein complexes in cells expressing the GST–fusion. Here we provide a technical description on the utilization of GST–fusion proteins as both a tool to studying protein–protein interactions and also as a means to purify interacting proteins.

Key Words

GST; affinity chromatography; Ras; Raf.

1. Introduction

A convenient method for the analysis of protein–protein interactions is through the use of chimeric proteins comprised of glutathione-*S*-transferase (GST) linked in frame to a protein of interest. Commonly referred to as GST–fusion proteins, these chimeras can typically be expressed in *Escherichia coli* and rapidly affinity-purified under non-denaturing conditions. Fusion of GST to a particular protein commonly enhances the protein's solubility, and, because GST has virtually no toxicity in cells, these proteins can generally be expressed in large amounts. Numerous convenient expression vectors now exist for production of GST–fusions in both bacterial and mammalian cells.

GST–fusion proteins are commonly used to study protein–protein interactions mainly because GST has a strong affinity for glutathione and thus can be coupled at high concentration to an immobilized glutathione matrix. Furthermore, the interaction of GST with glutathione is particularly robust and resis-

From: *Methods in Molecular Biology, vol. 261: Protein–Protein Interactions: Methods and Protocols*
Edited by: H. Fu © Humana Press Inc., Totowa, NJ

tant to stringent buffer conditions. The GST moiety of the fusion protein generally does not interfere with the accessibility of the fused protein, because a long flexible linker region exists that permits each half of the fusion to behave as a separate domain. All of these attributes make GST–fusion proteins bound to a glutathione matrix simple and effective affinity columns which many investigators have used as a tool to isolate novel interacting proteins (**Fig. 1**). In this chapter we will use GST–Ras expressed in *E.coli* and GST–B-Raf expressed in mammalian cells to illustrate how to express and purify GST–fusion proteins and use them in the analysis of protein–protein interactions.

2. Materials

1. pGEX-KG expression vector.
2. *E. coli* strains: DH5α, BL21(DE3)-RIL (Stratagene).
3. LB media (agar, liquid).
4. Antibiotics (Sigma).
5. 37°C, 30°C, and room temperature incubators.
6. Isopropyl β-D-1-thiogalactopyranoside (IPTG; Boehringer).
7. Spectrophotometer.
8. French Press or sonicator.
9. Complete EDTA-free protease inhibitor (Roche Molecular Biochemicals).
10. Refrigerated centrifuge.
11. SDS-PAGE equipment.
12. Glutathione (GSH)–agarose (Sigma).
13. Reduced glutathione (Sigma).
14. Liquid N_2, –80°C storage.
15. Rocking/rotating platform.
16. Lipofectamine transfection reagent (Invitrogen).
17. Antibodies: α-C-Raf (Transduction Laboratories), α-HA (Babco), α-GST (Zymed).
18. Lysis buffer A: 50 m*M* Tris-HCl, pH 7.5, 100 m*M* NaCl, 1 m*M* DTT, 0.2 m*M* PMSF, and Complete EDTA-free protease inhibitor tablet (Roche).
19. Elution buffer: 50 m*M* Tris-HCl, pH 8, 100 m*M* NaCl, 10 m*M* reduced glutathione, 1 m*M* DTT.
20. Lysis buffer B: 20 m*M* Tris-HCl, pH7.5, 100 m*M* NaCl, 1% Nonidet P-40, 5 m*M* $MgCl_2$, 1 m*M* DTT, 0.2 m*M* PMSF, and Complete EDTA-free protease inhibitor tablet.

3. Methods

Here we outline the methods involved in the general cloning, expression, and purification of GST–fusion proteins and then describe the specific use of bacterially expressed GST–Ras mutants to analyze their interaction with C-Raf. We also describe the expression of the mammalian GST–fusion protein, GST–B-Raf, and analyze its interaction with AKT.

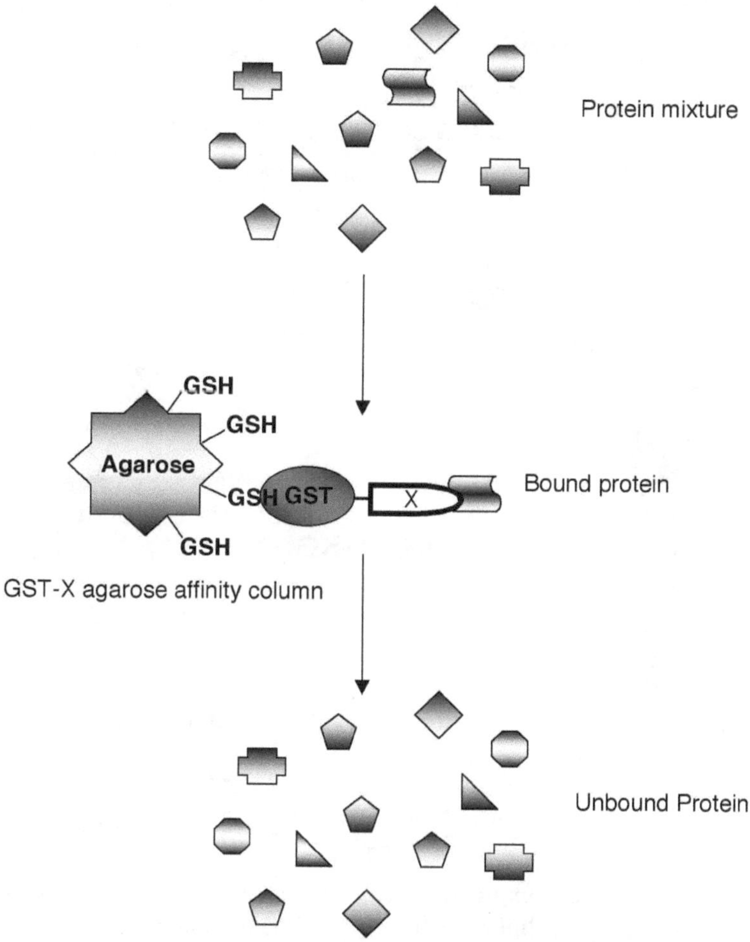

Fig. 1. A GST–fusion protein affinity column. Cell lysates comprised of numerous proteins are incubated with GST–X fusion protein bound to a glutathione (GSH)–agarose matrix. Only proteins that interact with the X portion of the protein can be retrieved from the mixture.

3.1. Expression Plasmid

3.1.1. pGEX-KG E.coli Expression Vector

We commonly use the pGEX-KG expression vector for production of GST–fusion proteins in *E. coli (1)* (**Fig. 2**). This vector is based on a set of vectors initially constructed by Pharmacia Biotech (www.apbiotech.com/product/product_index.html). It is typical of many prokaryotic expression vectors in

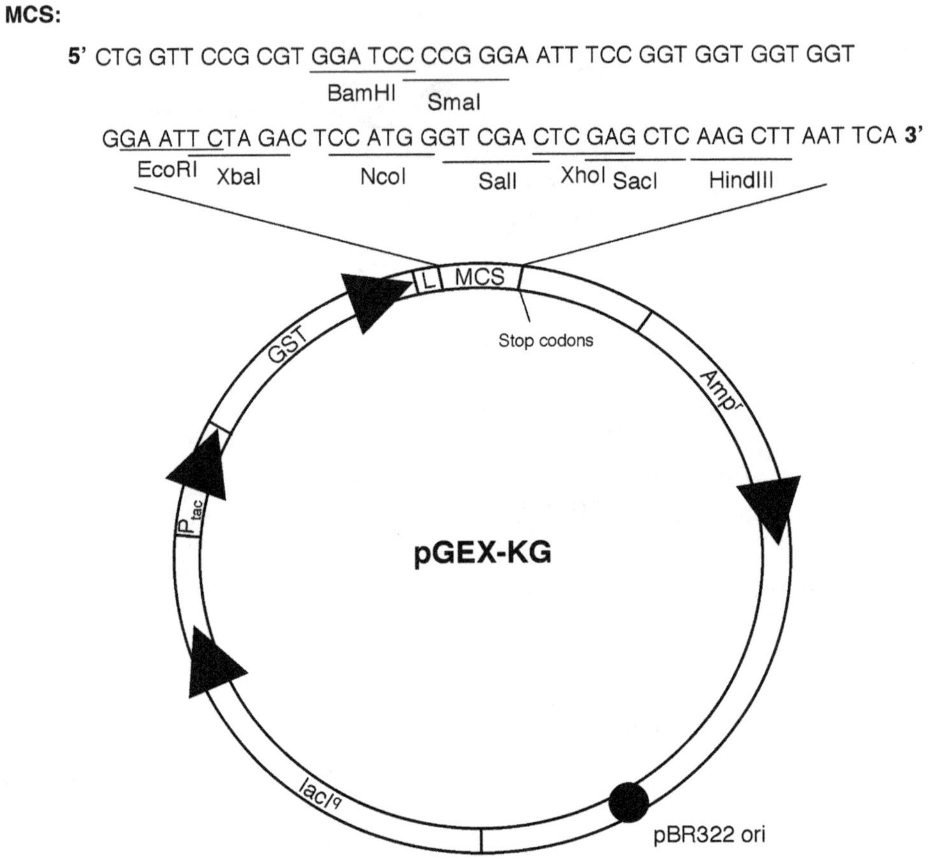

Fig. 2. pGEX-KG *E. coli* expression vector. pGEX-KG vector shown with essential components: *GST* gene, L (linker region), MCS (multiple cloning site), *Amp*^{r} (ampicillin resistance gene), pBR322 ori (origin of replication), *lacI*^{q} (lac repressor), and P_{tac} (promoter). The MCS is shown with restriction sites.

that it contains an origin of replication (pBR322 ori), an ampicillin resistance gene (*Amp*^{r}), and a strong promoter (P_{tac}), which is kept in the "off" state by the lac repressor (*lacI*^{q}). Addition of the non-hydrolyzable lactose analog, IPTG, inactivates the lac repressor and permits activation of the promoter. The *GST* gene lies downstream of the promoter and is followed by a linker region, which is a stretch of amino acids predicted to be very flexible and with no apparent structure. This is then followed by the multiple cloning site (MCS) which contains a number of commonly used restriction sites for insertion of the cDNA of interest.

3.1.2. Cloning Strategy

Insertion of the cDNA of interest into the pGEX-KG expression vector MCS is performed by conventional molecular biology manipulations ***(2,3)***. It is worth noting that the cDNA need not contain an initiator methionine codon; however, it must be inserted in the same frame as the *GST* gene. DNA sequencing may be a necessary diagnostic method to confirm the construct.

3.2. Protein Expression

Following construction and verification of the expression construct, the construct must then be transformed into the appropriate bacterial host strain. We typically use two *E. coli* strains for protein expression: (1) DH5α and (2) BL21(DE3)-RIL (*see* **Note 1**).

3.2.1. E.coli Transformation and Induction of Protein Expression

1. Transform plasmid expression vector into *E. coli* strain using standard methods ***(2,3)*** and plate on an LB/Amp agar plate overnight at 37°C.
2. Pick an individual colony the next day (after 12–18 h of growth) and innoculate 10 mL of LB/Amp (50–100 μg/mL) liquid medium and grow overnight in a 37°C shaker (*see* **Note 2**).
3. The next morning pour the 10 mL culture of *E. coli* into 1 L of LB/Amp and grow for 2–5 h until an OD_{600} of 0.5–0.8 has been reached (log phase growth).
4. Add IPTG to 0.1 m*M* and shake cells for 3–4 h at 37°C or overnight at room temperature (*see* **Note 3**).

3.2.2. Protein Extraction

1. Centrifuge cells at 5000*g* for 10 min in polypropylene bottles.
2. Decant supernatant. If desired, cells can be stored as a pellet at –20°C.
3. Resuspend cells in 15–30 mL cold lysis buffer A (*see* **Note 4**). The use of an aluminum beaker to perform the following lysis step will ensure the cells stay cold.
4. Cell lysis: There are two methods commonly used: (a) pass twice through French Press or (b) sonicate (4 × 20–30 s bursts).

   The French Press is more cumbersome than the sonicator, but it is usually preferred when purifying a protein whose activity can be destroyed by heat generated by the metal tip of the sonicator. Therefore, we recommend keeping the sample as cold as possible while sonicating (e.g., keep on ice with constant mixing) and performing four 20–30 s sonication bursts allowing 1–2 min intervals between bursts.
5. Add Triton X-100 (TX100) detergent to 1%. This is added after the lysis step to help solubilize proteins and to avoid any frothing that may occur during sonication.
6. Centrifuge lysate at 30,000*g* for 20 min at 4°C.

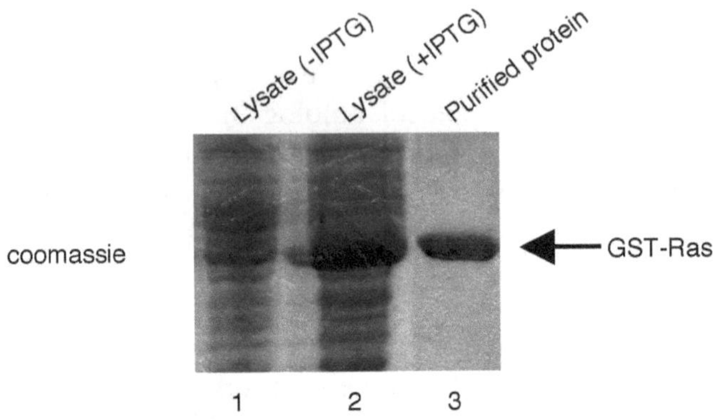

Fig. 3. IPTG induction and purity of GST–Ras protein. Samples are run on a 12.5% SDS-PAGE gel and stained with Coomassie blue. Lane 1, *E. coli* lysate prior to IPTG induction; lane 2, *E. coli* lysate after 3 h of IPTG induction; lane 3, purified GST–Ras protein. The arrow indicates the location of the GST–Ras protein in the induced lysate and after purification by glutathione–agarose.

7. At this point NaCl can be added up to 1 *M* to prevent co-purifying nonspecific proteins. Addition of NaCl will not affect the affinity of GST for the glutathione matrix.
8. Add 1 mL of GSH–agarose (50% slurry, prewashed in lysis buffer) to the *E. coli* supernatant and mix for at least 1 h at 4°C. The binding capacity of glutathione–agarose is approx 10 mg of GST protein per mL of 50% slurry.
9. After incubation, the glutathione–agarose beads must be washed (*see* **Note 5**). Wash steps: (a) twice in lysis buffer + 1 *M* NaCl; (b) twice in lysis buffer; (c) three times in lysis buffer (–TX100).

   At this point the protein can be stored bound to the glutathione beads (*see* **Subheading 3.3.**) or be eluted off the beads.

3.2.3. Elution and Dialysis of GST–Fusion Protein

1. Elute the protein twice with 0.5 mL elution buffer (10 min each).
2. To remove the glutathione from the buffer, the eluted protein can be dialyzed against a buffer of choice at 4°C. We typically dialyze 1 mL of eluted protein against 1 L of elution buffer (without glutathione) and replace 1 m*M* DTT with 0.1% β-mercaptolethanol. However, we recommend choosing whichever dialysis buffer system you deem most appropriate for your protein.
3. Quantify the amount of protein by running an SDS-PAGE gel against known protein quantities (e.g., BSA). Electrophoresis/Coomassie staining is preferable to using the Bradford assay, because of the ability to assess protein purity as well (**Fig. 3**). Performing a Western blot with α-GST antibody will also allow you to asses the purity and breakdown products.

3.3. Storage of GST–Fusion Proteins

Various methods of storage are used and each should be tested to see whether it affects protein stability, activity, etc. GST–fusions can be stored on beads or in solution. Common storage methods include:

1. Addition of glycerol to 50% and storage at –20°C (sample does not freeze).
2. Addition of glycerol to 5–10% and stored at –80°C.
3. Addition of glycerol to 5–10% and snap cooled in liquid N2, prior to storage at –80°C.

After thawing and prior to use in an experiment, the beads should be washed to remove glycerol.

3.4 Using GST–Ras to Analyze the Interaction with Raf

To illustrate the use of a GST–fusion protein in a binding reaction, we will analyze the interaction of the H-Ras GTPase with the Raf kinase. We have constructed pGEX-KG-RasV12 and -RasN17 expression vectors that express two mutants of the human H-Ras protein. The mutation G12V locks Ras in the GTP nucleotide bound conformation, while the T17N mutation locks the protein in the GDP nucleotide bound conformation *(4)*. GTP-bound Ras is in the "on" conformation and is able to interact with numerous effector molecules such as the protein kinase, Raf, through the Raf N-terminal domain. GDP bound Ras is in the "off" conformation and is unable to interact with Raf.

We have purified GST–RasV12 and GST–RasN17 using the protocol described in **Subheading 3.2.** It is important to note that these small GTPases are required to be purified in the presence of 5 m*M* $MgCl_2$ to prevent the loss of bound nucleotide. The following procedure will describe the analysis of the interaction between *E. coli*-expressed GST–Ras mutants and mammalian-expressed N-terminal domain (amino acids 1–269) of Raf. Raf 1–269 expression will be directed from the plasmid pCDNA3-cRaf-1-269, which will be transfected in mammalian HEK293 cells according to established protocols *(3)*.

1. A 10 cm dish of mammalian HEK293 cells is transfected (lipofectamine method as per manufacturer's recommendation) with 10 μg of the pCDNA3-cRaf (1-269) plasmid and grown for a further 48 h *(3)*.
2. Wash cells once with 10 mL phosphate-buffered saline (PBS) and lysed with 1 mL of lysis buffer B.
3. Cells are placed on ice for 10 min, scraped, and collected into an Eppendorf tube.
4. Lysate is then centrifuged at 15,000*g* for 15 min at 4°C and the supernatant is collected.
5. Add 10 μg GST, GST–RasV12, and GST–RasN17 into three separate tubes and to these tubes add the HEK293 cell supernatant collected in **step 4**. Mix by incubating at 4°C for 1–2 h on a rocking/rotating platform.

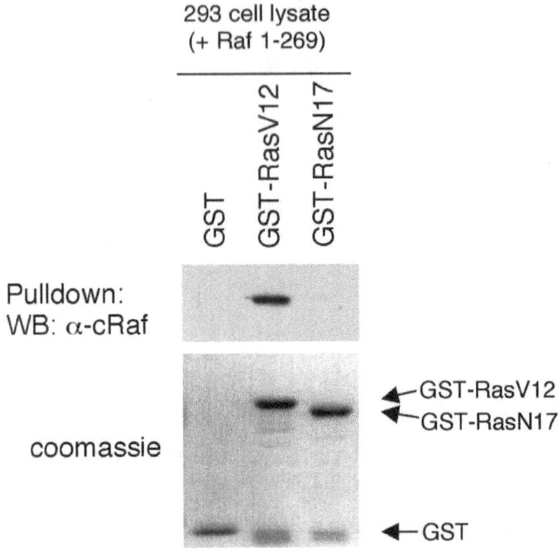

Fig. 4. Interaction of Raf and Ras. Top panel, Raf (1–269) binds GST-RasV12 as detected by α-Raf Western blot. Bottom panel, Coomassie blue stained gel of the GST–fusions used in the binding.

6. Add 10–15 μL GSH–agarose to each tube and rock for a further 0.5–1 h (*see* **Note 6**). Wash the GSH–agarose beads four times with 1 mL of the lysis buffer B.
7. Bound GST–Ras can be eluted with the elution buffer as explained in **Subheading 3.2.3.**, or with 1X SDS sample buffer (*see* **Note 7**).
8. Samples are then run on an SDS-PAGE gel and subject to a Western blot *(2,3)* using α-c-Raf antibody (**Fig. 4**).

3.5. Production of GST–Fusion Proteins in Mammalian Cell Culture

There are a limited number of commercially available GST–fusion expression vectors for use in mammalian cells. One benefit of expression in mammalian cells is that many eukaryotic proteins undergo modifications that do not occur in *E. coli*. For example, the Ras protein is C-terminal prenylated when expressed in mammalian cells, but does not undergo this modification in *E. coli* because the appropriate modification enzymes are not present in bacteria. Furthermore, mammalian proteins that typically cannot be expressed in *E. coli* or are easily degraded in *E. coli* can often be expressed in mammalian cells where conditions (tRNA, folding machinery, etc.) are more suitable. However, the fact that expression in mammalian cells results in less protein compared to *E. coli* makes it less economical.

pEBG-3X-HV multiple cloning site:

```
                                            BamHI    NdeI     EcoRV    NheI
5' ATC GAA GGT CGT GGG ATC GGA TCC CAT ATG GAT ATC GCT AGC

     XmaI     SalI     SpeI     ClaI        NotI
   CCC GGG GTC GAC ACT AGT ATC GAT GCG GCC GCT GAA TAG 3'
```

Fig. 5. pGEX-3X-HV vector multiple cloning site.

The vector pEBG-3X drives the expression of GST–fusion proteins from a very strong E1Fα promoter. We have constructed a modified version of pEBG-3X vector where more restriction sites have been added to the MCS and have named it pEBG-3X-HV (**Fig. 5**). Transfection of a 10 cm plate of HEK293 cells with pEBG-3X-HV can produce up to 5–10 μg of GST protein, which is roughly an order of magnitude more protein than what most cytomegalovirus (CMV) promoter–based vectors can direct.

Using glutathione–agarose to purify GST–fusion proteins from mammalian cells bypasses the need for using antibodies and, thus, immunoprecipitation methods. This is particularly useful because immunopurified proteins cannot easily be eluted from the antibody and must be eluted by boiling, which releases the antibody into the mixture. Because GST–fusions can easily be eluted with glutathione, there is no antibody present. This makes identification of co-purified proteins by protein sequencing or mass spectrometry much easier because of the lack of large amounts of contaminating antibody present.

Note that the purification protocol for mammalian GST–fusion proteins is the same as in **Subheading 3.2.2.** except during the lysis step detergents such as 1% TX-100 or 1% NP-40 are used to lyse the cells instead of a French Press or sonicator.

3.6. Detecting the Interaction Between B-Raf and AKT Using Mammalian GST–Fusion Proteins

1. Transfect HEK 293 cells (in a six-well plate) with the following combination of plasmids:
   i. pEBG-3X-HV (0.5 μg) + pCDNA3-HA-B-Raf (0.5 μg).
   ii. pEBG-3X-HV-AKT (0.5 μg) + pCDNA3-HA-B-Raf (0.5 μg).
   iii. pEBG-3X-HV (0.5 μg) + pDNA3-HA-AKT (0.5 μg).
   iv. pEBG-3X-HV-B-Raf (0.5 μg) + pDNA3-HA-AKT (0.5 μg).
2. Forty-eight hours post-transfection, lyse the cells in 250 μL of lysis buffer B without $MgCl_2$.

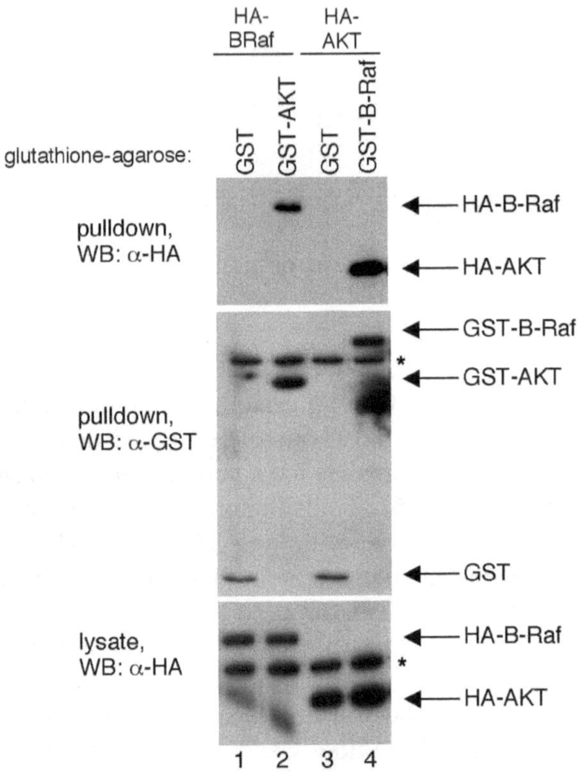

Fig. 6. Interaction of B-Raf with AKT. HEK293 cells were transfected with the plasmid combinations indicated and grown for 48 h. Cells were lysed and GST–fusion proteins were purified by addition of glutathione–agarose. To detect the presence of co-purified proteins, SDS-PAGE was performed followed by an α-HA Western blot. *, nonspecific band.

3. Centrifuge lysate at 15,000g for 10 min at 4°C.
4. Collect the supernatant and add 15 μL of GSH–agarose (50% slurry).
5. Incubate for 1–2 h on a rocker/rotating platform at 4°C.
6. Wash GSH–agarose beads four times with the above lysis buffer.
7. Elute using glutathione elution buffer (**Subheading 3.2.3.**) or boil in 1X SDS sample buffer.
8. SDS-PAGE followed by Western blot with α-HA (**Fig. 6**).

3.7. Passing Radiolabeled Lysates Over GST–Fusion Columns

A common method to identify proteins that interact with your GST–fusion protein is to metabolically label cells with ^{35}S-Met/Cys *(2,3)* and pass these cell lysates over a GST–fusion affinity column. This provides a very sensitive method of identifying novel interacting proteins, although there are several procedural changes to be aware of to make this an effective method:

1. When lysing radiolabeled cells in mammalian culture, layer the lysis buffer onto the radiolabeled cells so as to not detach the cells from the cell culture plate. Furthermore, do not scrape the cells after the lysis. Simply tilt the dish and remove as much of the soluble lysate as possible. This will help reduce any insoluble fragments that might not be completely removed in the centrifugation step.
2. Often nonspecific proteins bind the glutathione–agarose matrix during the binding step. One way to reduce this is to block the beads in lysis buffer with the addition of 1% BSA. In addition, preclearing the lysate with glutathione–agarose prior to incubating with glutathione–agarose/GST–fusion protein will help remove matrix interacting proteins. This step typically requires about 100 μL of glutathione agarose per 300 μL of lysate.

3.8. Different Lysis Buffers to Use

Reagents in the binding buffer may influence protein–protein interactions. It is often useful to try different buffer systems when analyzing binding. Here are three commonly used buffers systems in order of increasing stringency:

1. NP-40 buffer: 20 m*M* Tris-HCl, pH 7.5, 100 m*M* NaCl, 1% Nonidet P-40.
2. Triton buffer: 20 m*M* Tris-HCl, pH 7.5, 100 m*M* NaCl, 1% TX-100.
3. RIPA buffer: 0.1% SDS, 1% TX-100, 0.5% deoxycholate, 50 m*M* Tris-HCl pH 7.5, 150 m*M* NaCl.

4. Notes

1. BL21(DE3)-RIL cells contain some human tRNAs that are underrepresented in *E. coli* and hence enhance expression of certain human proteins in *E. coli*.
2. We recommend starting from a fresh colony for maximum expression. This is particularly important in the case of the RIL cells because often we see no protein expression if using colonies more than 1 day old.
3. We find 0.1 m*M* IPTG is adequate and higher concentrations do not seem to increase expression. To optimize expression, one might want to try different conditions such as length of induction and temperature of induction on small-scale cultures first. For example, if solubility is a problem, this can often be improved if cells are induced overnight at 30°C.
4. We have found that the extent of protein degradation in *E. coli* depends more on the expression conditions and less on the concentration of protease inhibitors during the purification. Hence, it is important to vary the conditions as explained in **Note 3** to minimize protein degradation.
5. For washing of 1 mL of glutathione–agarose matrix, we typically spin down beads for 20 s at 5000*g* and wash in Eppendorf tubes using multiple 1 mL washes. The washes need not contain protease inhibitors.

Because proteins expressed in *E. coli* are produced at a rapid rate, they sometimes do not fold efficiently. When this happens, the *E. coli* chaperone Hsp70 binds these misfolded proteins. Therefore, it is common to find Hsp70 (at 70 kDa) associated with purified GST–fusion proteins. One method to remove Hsp70 from the GST–fusion is to wash the beads (prior to elution) twice with 1 mL of 500 m*M* triethanolamine-HCl (pH 7.5), 20 m*M* $MgCl_2$, 50 m*M* KCl, 5 m*M* ATP, 2 m*M* DTT for 10 min at room temperature.

6. Another method is to use GST–Ras that is pre-coupled to GSH–agarose. Either way is acceptable as we haven't noted any differences between the two.
7. Eluting the protein is more specific, but will release less GST–fusion protein than boiling for 3 min in SDS sample buffer.

Acknowledgments

The authors would like to thank Huira Chong and Jennifer Aurandt for a critical review of the manuscript. This work was supported by grants from National Institutes of Health and Walther Cancer Institute (KLG). KLG is a MacArthur Fellow. HV is supported by a Rackham Predoctoral Fellowship.

References

1. Guan, K. L. and Dixon, J. E. (1991) Eukaryotic proteins expressed in *Escherichia coli*: an improved thrombin cleavage and purification procedure of fusion proteins with glutathione S-transferase. *Anal. Biochem.* **192(2),** 262–267.
2. Sambrook, J., Fritsch, E. F., and Maniatis, T. (1989) *Molecular Cloning, A Laboratory Manual*, 2nd Ed. Cold Spring Harbor Laboratory Press, Cold Spring Harbor, NY.
3. Ausubel, F. M. (1987) *Current Protocols in Molecular Biology.* Wiley, New York, NY.
4. Katz, M. E. and McCormick, F. (1997) Signal transduction from multiple Ras effectors. *Curr. Opin. Genet. Dev.* **7(1),** 75–79.

14

Affinity Capillary Electrophoresis Analyses of Protein–Protein Interactions in Target-Directed Drug Discovery

William E. Pierceall, Lixin Zhang, and Dallas E. Hughes

Abstract

Protein–protein interactions are instrumental in virtually all biological processes and their understanding will shed light on designing novel and effective drugs for therapeutic interventions targeting the pathways in which they function. Protein–protein interactions have been studied using many genetic and biochemical methods, most recently, affinity capillary electrophoresis (ACE). We used ACE as a high-throughput screening assay to establish and define binding interactions between a therapeutic target protein and chemical entities from natural product or synthetic chemical libraries. Furthermore, ACE has demonstrated its value in the measurement of binding constants, the estimation of kinetic rate constants, and the determination of the stoichiometry of protein–protein interactions. Herein, we will describe qualitatively several assay formats using ACE for detecting protein–protein interactions, and discuss their advantages and limitations.

Key Words

Protein–protein interactions; hit-to-lead discovery; capillary electrophoresis (CE), target(s), natural products; library screening.

1. Introduction

Protein–protein interactions are instrumental in virtually all biological processes and their understanding will shed light on designing novel and effective drugs for therapeutic interventions targeting the pathways in which they function ***(1–3)***. Protein–protein interactions have been studied using many genetic and biochemical methods, most recently, affinity capillary electrophoresis (ACE) ***(4–10)***. ACE is used as a high-throughput screening (HTS) assay to establish and define binding interactions between a therapeutic target protein

From: *Methods in Molecular Biology, vol. 261: Protein–Protein Interactions: Methods and Protocols*
Edited by: H. Fu © Humana Press Inc., Totowa, NJ

and chemical entities from natural product ***(11)*** or synthetic chemical libraries. Furthermore, ACE has demonstrated its value in the measurement of binding constants, the estimation of kinetic rate constants, and determining the stoichiometry of protein–protein interactions (***12,21***: *see* **Note 1**).

ACE is based on the electrophoretic separation of analytes, usually under soluble conditions in which the analyte carries a net charge. With proteins, CE conditions are established using buffers at a pH different from the protein's isoelectric point in order to impart a net charge and to prevent protein precipitation. The rate of migration through the electric field is target specific, but is generally dependent on the ratio of charge to mass. The result is that different proteins or isoforms of proteins migrate with a particular velocity toward the electrode of opposite charge. Detection of the target occurs at some point along the capillary when it passes a photomultiplier tube (PMT) and a target "CE profile" is generated (*see* **Fig. 1A**). The high resolution of ACE is due primarily to the fact that it is run at high field strengths with short migration times and small injection volumes which minimize diffusion ***(13,14)***. The use of laser-induced fluorescence detection allows the measurement of subnanomolar concentrations of dye-conjugated protein ***(15)***. If a binding event occurs between the target and an entity being screened, then the target's mobility is altered, which results in a change in the "CE profile."

ACE is a powerful technology that offers several strengths as a general screening technology. (1) ACE offers short analysis times, requires minute amounts of protein samples, involves no radiolabeled compounds, and is carried out in solution. (2) ACE is automatible for high-throughput screening. The system that we have designed and used is shown in **Fig. 1B**. (3) The conditions for ACE can be tuned such that the relative strength of binding can be established between a chemical entity and a target of interest (*see* **Note 2**). This is of special interest in the screening of crude natural extracts and prioritizing which extracts to fractionate. (4) ACE is able to separate the detected target from background fluorescence, which is a problem with many crude natural extracts and some synthetic compound libraries. As such, homogeneous assays may have difficulty in the interpretation of some data derived from fluorescent test samples.

A significant strength of ACE is the ability to assess targets in the context of protein–protein interactions ***(16–18)***. In this article, we will show data from ACE analysis of protein–protein interactions of the well-known therapeutic target vascular endothelial growth factor ***(19–23)*** and its corresponding receptor (VEGF/VEGFr1). We will then describe qualitatively several assay formats for using ACE to detect protein–protein interactions, and discuss their advantages and limitations.

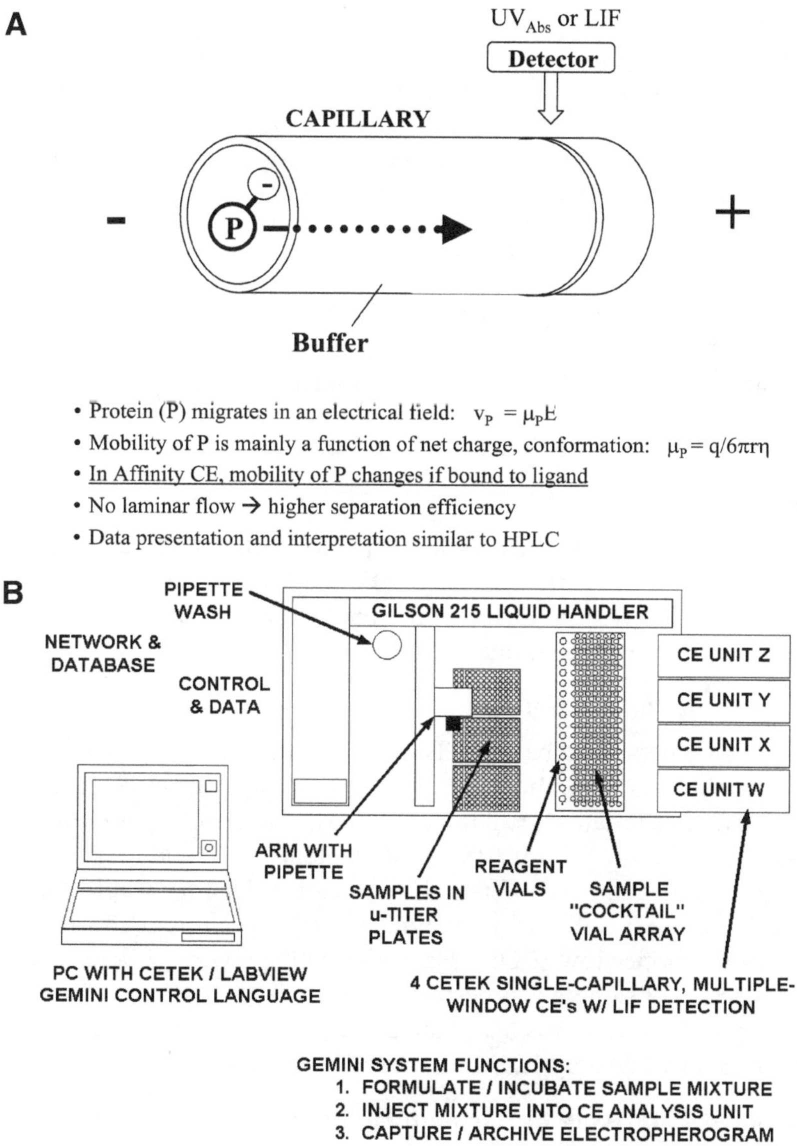

Fig. 1. Principle of ACE and automated instrumentation for HTS screening. (**A**) Schematic representation of ACE is shown where a protein carrying a charge is injected at one end of a capillary and migrates through a field in which an electric current is applied. Detection occurs at some point along the capillary via a photomultiplier tube (PMT). The migration of the protein through the field occurs as a function of the charge to mass ratio and is influenced by conformation of the target. (**B**) Shown is the instrumentation that we have designed and implemented in our HTS screening of targets versus chemical or natural product libraries.

2. Materials

1. MDQ Capillary Electrophoresis Unit (Beckman Coulter, Foster City, CA).
2. Borosillicate capillary: 50 μm, 75 μm, 100 μm diameters (Polymicro Technologies, Phoenix, AZ).
3. Surfactants: Brij-35, Triton X-100, and CHAPS (Sigma, St. Louis, MO).
4. Fluoroscein isothiocyanate (FITC).
5. Alexa protein labeling reagents (Molecular Probes).
6. Vascular endothelial growth factor (VEGF-165) (Calbiochem).
7. Vascular endothelial growth factor receptor (VEGF-r1) (Calbiochem).
8. Stock buffer: 20 m*M* Tris-HCl, pH 8.0, 5 m*M* DTT, 0.1% CHAPS, 150 m*M* NaCl.
9. Sample buffer: 20 m*M* Tris-HCl, pH 8.3, 1 mg/mL BSA.
10. Running buffer: 50 m*M* Tris-HCl, pH 8.3, 1 mg/mL BSA.
11. Micro-spin columns.
12. Bradford Assay reagent (Bio-Rad).

3. Methods

One of the strengths of ACE is the capacity to use different formats to address separate questions that are being asked or to test different hypotheses. Described below are several different methodologies that may be used in the assessment of protein–protein interactions.

3.1. General Instrument Set Up

For the studies described below, CE was performed on a P/ACE MDQ Capillary Electrophoresis instrument equipped with a diode array detector or LIF detection (488 nm excitation, 520 nm emission). The capillary is composed of fused silica with 50 μm, 75 μm, 100 μm diameters and a total length of 20 cm (the length from the sample inlet to the detector is 12 cm). Capillaries were coated on the internal walls with a polymer to prevent protein absorption and reduce electroosmotic flow (EOF). Operation voltage was 15–30 kV. Hydrodynamic injection of sample was performed at a pressure of 0.5 psi for 10–20 s. The capillary temperature was maintained at 20°C. Data retrieval and analyses were performed by P/ACE MDQ Software (Beckman Coulter).

3.2. Example of an Assay for Growth Factor/Receptor Interaction

Growth factor–growth factor receptor interactions are the initial step in cascades of signal transduction events that direct cell growth, apoptosis, and differentiation and development processes. Finding new chemical entities that may modulate these processes through the disruption of growth factor and growth factor receptor binding is an exciting application of ACE. One such example of ACE application to the study of growth factor–growth factor receptor interactions comes from our studies of vascular endothelial growth factor (VEGF) and its receptor [VEGF-r1 *(23)*]. VEGF-r1 is a favorite target for many screening programs due to its well-documented importance in inflammation and the role it plays in angiogenesis. Stopping cancer cell growth through shutting down blood vessel development within the tumor that supplies the lesion with nutrients and growth factors has garnered this target special interest over the past several years ***(24)***. In our laboratory, we have developed a protein–protein disruption assay for finding chemical entities that impede VEGF/VEGF-r1 binding.

3.2.1. Sample Preparation

In this assay, VEGF-165 was labeled with FITC in 30 m*M* HEPES buffer using a dye to protein molar ratio of 10:1. Labeled target was separated from free dye using microsep spin columns. The concentration of the purified material was obtained by using the Bradford Assay with BSA as a standard. Approximately equal molar amounts of VEGF and VEGFr1 (50 n*M* of each in stock buffer) were used in these experiments by dilution of each into Tris-based sample buffer. The sample was kept at 4°C prior to injection onto the capillary. Crude natural extracts in 100% DMSO were added to the running buffer to the level of 1%; clean DMSO was used as a control.

3.2.2. Data from the CE Assay

Within the capillary, labeled VEGF migrates within a large segment of BSA that is provided with the target protein by the vendor to assist in solubilizing the cytokine. When VEGF is labeled in a mixture containing excess BSA, then BSA will also be labeled. Thus, the unbound labeled VEGF is not visible in the CE electropherogram due to excess labeled BSA (data not shown). Upon the addition of unlabeled VEGF-r1, the VEGF/VEGF-r1 complex shifts away from the labeled BSA and is visible as a separate peak (reproducibility of consecutive injections depicted in **Fig. 2A**). This slower migration is expected as the receptor is basic and its binding to VEGF would provide for a complex of slower mobility than the growth factor alone under these conditions.

3.2.3. Data Analyses from the CE Assay

In theory, the fluorescently labeled (FL)-VEGF/VEGF-r1 complex CE profile would be sensitive to hits interfering with complex formation and one might hypothesize that molecules that interfered with binding would manifest as a loss in signal of the complex peak. To test this hypothesis, we titrated the FL-VEGF/VEGF-r1 complex against increasing amounts of unlabeled VEGF (**Fig. 2B**). As the concentration of unlabeled VEGF is increased, it competes with the labeled VEGF for binding to the VEGF-r1, resulting in increasing amounts of unlabeled-VEGF/VEGF-r1 complex formation (not visible under LIF detection) and a decrease in the amount of FL-VEGF/VEGF-r1. This

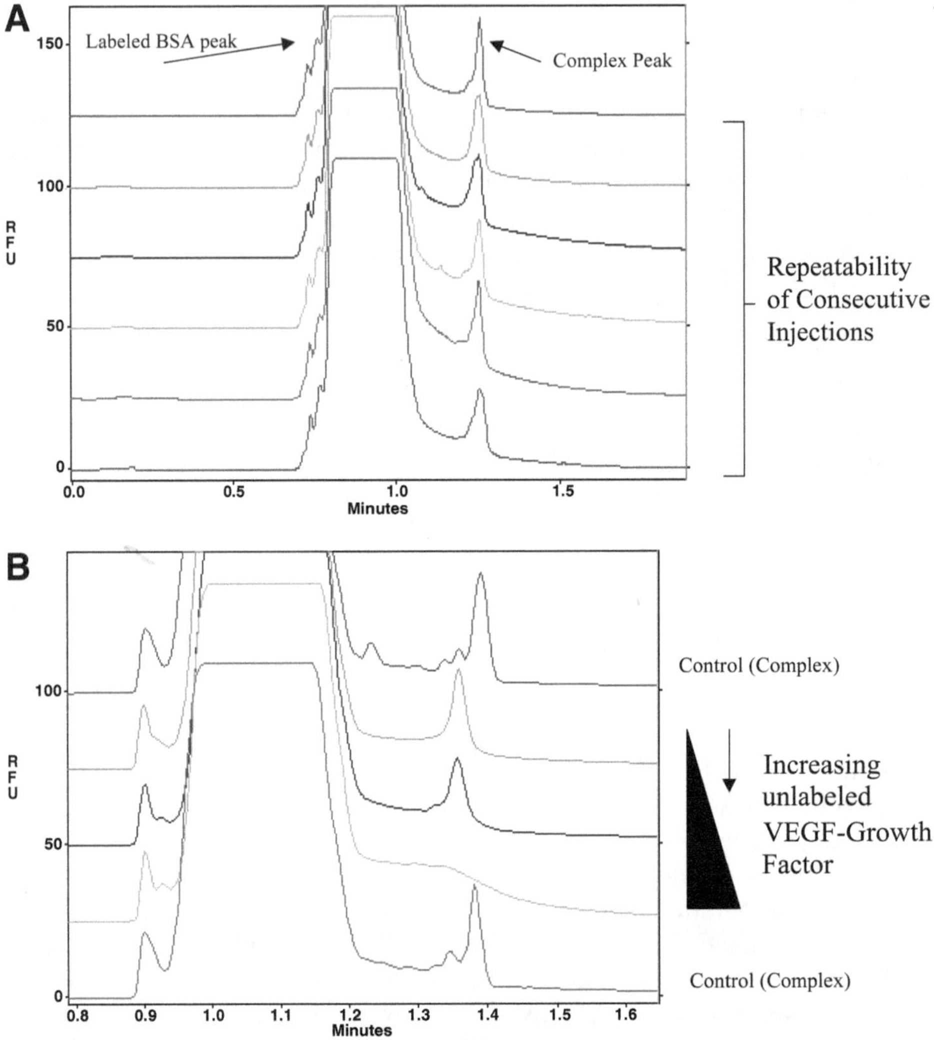

Fig. 2. Studies with VEGF/VEGFr1 utilizing ACE. (**A**) Labeled VEGF/unlabeled VEGFr1 complex. Unlabeled VEGFr1 and labeled VEGF are incubated for 10 min at room temperature and then injected onto the capillary. The figure depicts that the CE profile of the complex is stable with repeated injections. (**B**) FL-VEGF/unlabeled VEGFr1 complex disruption by addition of unlabeled VEGF. The FL-VEGF/VEGFr1 complex re-equilibrates with the addition of unlabeled VEGF. With increased unlabeled VEGF, more VEGF(unlabeled)/VEGFr1 forms and less FL-VEGF/VEGFr1 complex is formed. (**C**) *(opposite page)* Screening of crude extracts. The FL-VEGF/VEGFr1 complex was incubated with crude natural extract. In this experiment, the presence of any chemical entity inhibiting the binding of VEGF to VEGFr1 would result in a loss of complex formation and a reduction in CE profile signal. Here, extract 4 incubation with VEGF and VEGFr1 results in less complex signal and would be a candidate for further study.

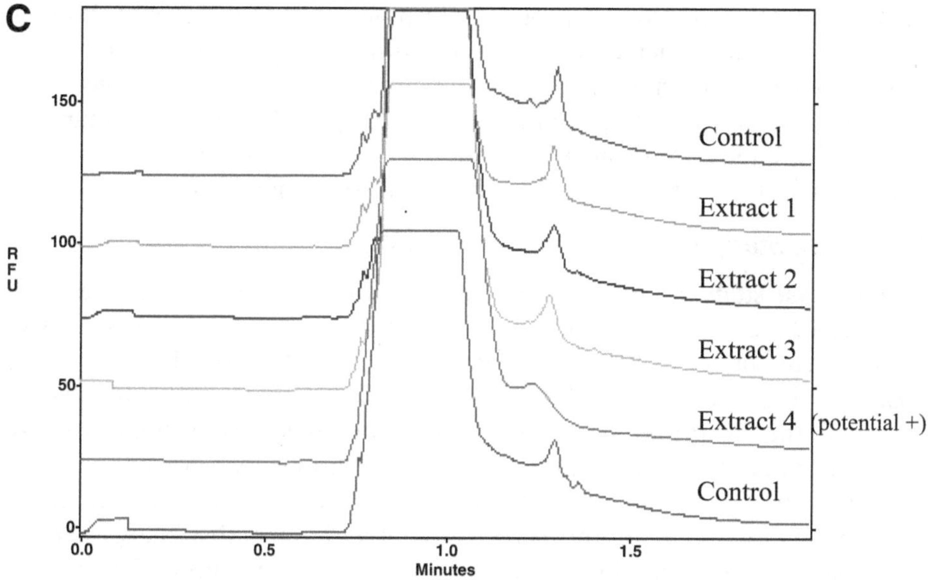

results in an overall loss of labeled complex signal. Some data from screening representative natural extracts for their ability to disrupt FL-VEGF/VEGF-r1 interactions are depicted in **Fig. 2C**. These data may be interpreted to indicate that sample 4, which caused an overall loss of labeled complex signal, would be selected for further study (fractionation, testing of fractions for activity, and isolation of chemical entity.) In screening a large number of crude extracts, we found the primary hit rate using this format to be approx 1%.

3.3. Other Protein–Protein Interactions Analyzable by ACE

Homophilic and heterophilic higher-order interactions: Many enzymes do not functionally exist as individual components. Rather, they may form higher-order structures that comprise their functional complex. The ability to assess intact complexes in ACE allows for the assay of enzymes in a state that most closely approximates how they work functionally within the cell. For example, while p53 has been well documented as a tumor suppressor *(25)*, functionally it may exert its DNA-binding effects to redirect protein synthesis as a homophilic tetramer. Thus, in searching for binders and potential inhibitors of protein function, any screen of this target would benefit from the ability to assess interactions not just with the monomer, but rather with the higher-order structure created by the homophilic complex. Thus, in screening for modulators of p53 protein, one could monitor changes in mobilities of either the p53 monomer or of the higher-order complex and a CE profile change created by disruption of the complex. Conversely, other targets may function to re-direct protein synthesis by forming higher-order structures of heterophilic interactions, such as in the interaction of jun/fos heterodimer, which binds to AP-1 transcriptional regulatory sequences ***(26)***.

3.3.1. Case Study A: Homophilic Interaction

In one case study, a CE assay with UV detection was developed for a target with DNA-binding capacity that was validated in the field of anti-infective research. This target (full length) is known to form higher-order homophilic hexameric aggregates. When the complex migrates through the capillary as an aggregate, the CE profile is stable and it retains its ability to bind to DNA. However, when the target is assayed under conditions that prevent hexamer formation, the CE profile changes dramatically and the functional ability to bind DNA is lost. Thus, in this screen any target binders that change the CE profile may be either binding to the intact complex or disrupting the complex formation. Under these conditions, ACE assay will identify binders that can bind to active sites directly or exert their inhibition through binding to other sites and manifest target inhibition through allosteric interaction.

3.3.2. Case Study B: Heterophilic Interaction

In a second case study, a CE assay with LIF detection was developed for a target validated in the study of cancer biology. This target was comprised of two different full-length proteins that migrated through the capillary as a heterophilic complex. In this study the complex was labeled with FITC and assayed for retention of functional activity by binding to an unlabeled control peptide. The labeled heterodimer peak was collected from the end of the capillary and the presence of both subunits was confirmed by SDS-PAGE. While the target was screened as a heterodimer in this particular assay, one could easily see that the heterodimer bound to the control peptide could be used to characterize binders competing for the peptide-binding site. This is clearly distinguished as a new peak is formed by the complexation of the peptide with the heterodimer.

Alternatively, a separate approach may be used to characterize compounds as competitive for a specific site using a peptide that is labeled, instead of the target. Under this format, one may bind the labeled peptide to the unlabeled target. The target/FL-peptide complex is then monitored under control conditions by CE-LIF. When a competitive compound is present, the FL-peptide is displaced and the overall amplitude of the target/FL-peptide is reduced (schematically depicted in **Fig. 3**). If the CE mobility of the target/FL-peptide com-

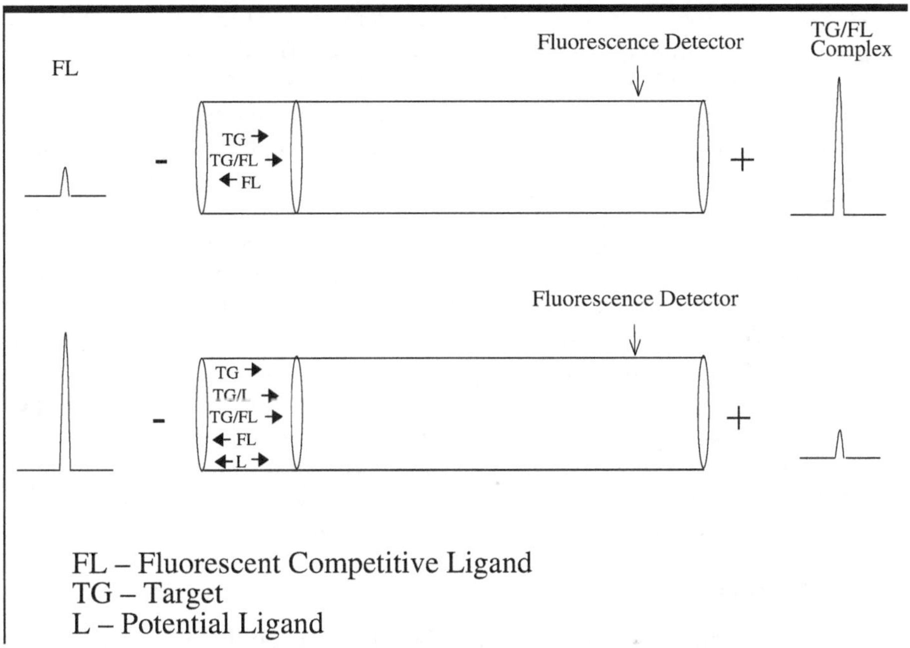

Fig. 3. Competitive CE format using a competitive synthetic ligand. ACE can be used to define binders to a specific site if a competitive ligand is available. Depicted here, an unlabeled target is incubated with a fluorescent peptide and a peak representing the complex is the net result. When an unlabeled test ligand is included, the FL-ligand is displaced resulting in an overall decrease in signal due to the labeled complex.

plex changes without a change in the overall fluorescence magnitude, then one may infer that the compound is binding to the target at a site alternative to the site of peptide binding.

4. Notes

1. CE is an emerging technique for the analysis of proteins and ACE has been widely used to study protein–protein interactions. It is a fully automated tool for rapid, highly sensitive, quantitative, and real-time analysis of minute amounts of complex samples. In binding systems where proteins associate tightly with their ligands (i.e., slow dissociation rates), binding constants can be estimated by direct integration of the peak areas of the free and bound protein (or the ligand), provided that the free and bound species have different electrophoretic mobilities. Quantitative binding constants can be determined by employing the following formula at equilibrium:

$$K_d = [\text{Ligand free}]\ [\text{Target free}]/[\text{Complex bound}]$$

In more common practice, however, data from the ACE are not usually interpreted to be absolutely quantitative owing to the difficulty of establishing that the binding has truly reached equilibrium. Thus, it is more common in estimating the general affinities of targets and corresponding ligands utilizing ACE in more semiquantitative terms of determining the IC_{50} of a ligand for a target (the concentration of ligand at which 50% effect of a saturatable binding and shiftability of the target is observed).

2. One of the great advantages of ACE is its ability to detect weak binding interactions in addition to strong binders. In our laboratory we have found the limit of binding strength that can be detected by ACE to be near 100 μ*M*. Additionally, conditions can be tuned to make the assay more stringent. After a primary assay has been run under a format designed to detect weak, moderate, and strong binders, the hits may be run under these increasingly stringent conditions and a general ranking of the strongest to weakest binders emerges. Some of the conditions that can be varied to make the assay more stringent include altering capillary temperature, addition of the extract or compound in the sample buffer versus addition in the running buffer, and capillary length.

Acknowledgments

The authors would like to thank Drs. Hans Fliri, Yuriy Dunayevskiy, Linda Engle, and Andrew Weiskopf for critical reading and comments on the manuscript. Additional gratitude is extended to Robyn Kangas, Megan Huber, Damian Houde, and Karen Jacka for their excellent technical assistance. The administrative assistance of Ms. Brenda Byron is greatly appreciated.

References

1. Perrier, V., Wallace, A. C., Kaneko, K., Safar, J., Prusiner, S. B., and Cohen, F. E. (2000) Mimicking dominant negative inhibition of prion replication through structure-based drug design. *Proc. Natl. Acad. Sci.* **97(11),** 6073–6078.
2. Hulme, E. C. (ed.) (1992) *Receptor–Ligand Interaction: A Practical Approach,* Oxford University Press, New York, NY.
3. Lakey, J. H. and Raggett, E. M. (1998) Measuring protein-protein interactions. *Curr. Opin. Struct. Biol.* **8,** 119–123.
4. Colton, I. J., Carbeck, J. D., Rao, J., and Whitesides, G. M. (1998) Affinity capillary electrophoresis: a physical-organic tool for studying interactions in biomolecular recognition. *Electrophoresis* **19,** 367–382.
5. Shimura, K. and Kasai, K.-I. (1996) Affinophoresis: selective electrophoretic separation of proteins using specific carriers. *Methods Enzymol.* **271,** 203–218
6. Chu, Y.-H. and Cheng, C. C. (1998) Affinity capillary electrophoresis in biomolecular recognition. *Cell. Mol. Life Sci.* **54,** 663–683.
7. Tanaka, Y. and Terabe, S. (2002) Estimation of binding constants by capillary electrophoresis. *J. Chromatogr. B Analyt. Technol. Biomed. Life Sci.* **768(1),** 81–92.

8. Heegaard, N. H., Nissen, M. H., and Chen, D. D. (2002) Applications of on-line weak affinity interactions in free solution capillary electrophoresis. *Electrophoresis* **23(6),** 815–822.
9. Kasicka, V. (2001) Recent advances in capillary electrophoresis of peptides. *Electrophoresis* **22(19),** 4139–4162.
10. Brocke, A., Nicholson, G., and Bayer, E. (2001) Recent advances in capillary electrophoresis/electrospray-mass spectrometry. *Electrophoresis* **22(7),** 1251–1266.
11. Knight, V., Sanglier, J. J., DiTullio, D., et al. (2003) Diversifying microbial natural products for drug discovery. *Appl. Microbiol. Biotechnol.* **62,** 446–458.
12. Tseng, W. L., Chang, H. T., Hsu, S. M., Chen, R. J., and Lin, S. (2002) Immunoaffinity capillary electrophoresis: determination of binding constant and stoichiometry for antibody-antigen interaction. *Electrophoresis* **23(6),** 836–846.
13. Karger, B. L., Chu, Y.-H., and Foret, F. (1995) Capillary electrophoresis of proteins and nucleic acids. *Annu. Rev. Biophys. Biomol. Struct.* **24,** 579–610.
14. Karger, B. L., Foret, F., and Berka, J. (1996) Capillary electrophoresis with polymer matrices: DNA and protein separation and analysis. *Methods Enzymol.* **271,** 293–319.
15. Le, X. C., Wan, Q. H., and Lam, M. T. (2002) Fluorescence polarization detection for affinity capillary electrophoresis. *Electrophoresis* **23(6),** 903–908.
16. Kiessig, S., Reissmann, J., Rascher, C., Kullertz, G., Fischer, A., and Thunecke, F. (2001) Application of a green fluorescent fusion protein to study protein-protein interactions by electrophoretic methods. *Electrophoresis* **22(7),** 1428–1435.
17. Vergnon, A. L. and Chu, Y. H. (1999) Electrophoretic methods for studying protein-protein interactions. *Methods* **19(2),** 270–277.
18. Wan, Q. H. and Le, X. C. (1999) Fluorescence polarization studies of affinity interactions in capillary electrophoresis. *Anal. Chem.* **71(19),** 4183–4189.
19. Little, J. N., Hughes, D. E., and Karger, B. L. (1999) A powerful screening technology utilizing capillary electrophoresis. *American Biotechnology Laboratory* **17,** 36.
20. Hughes, D. E. and Karger, B. L. (1998) Screening natural samples for new therapeutic compounds using capillary electrophoresis. *U.S. Patent No. 5,783,397.*
21. Greve, K. F., Hughes, D. E., Richberg, P., Kats, M., and Karger, B. L. (1996) Liquid chromatographic and capillary electrophoretic examination of intact and degraded fusion protein CTLA4Ig and kinetics of conformational transition. *J. Chromatogr. A.* **723(2),** 273–284.
22. Dunayevskiy, Y. M., Waters, J. L., and Hughes, D. E. (2001) Capillary electrophoretic methods to detect new biologically active compounds in complex biological material. *U.S. Patent No. 6,299,747.*
23. Polverini, P. J. (2002) Angiogenesis in health and disease: insights into basic mechanisms and therapeutic opportunities. *J. Dent. Educ.* **66(8),** 962–975.
24. Folkman, J. (2002) Looking for a good endothelial address. *Cancer Cell.* **1(2),** 113–115.
25. Kiyohara, C., Otsu, A., Shirakawa, T., Fukuda, S., and Hopkin, J. (2002) Genetic polymorphisms and lung cancer susceptibility. *Lung Cancer* **37(3),** 241.
26. van Dam, H. and Castellazzi, M. (2001) Distinct roles of Jun:Fos and Jun:ATF dimers in oncogenesis. *Oncogene* **20(19),** 2453–2464.

15

Mapping Protein–Ligand Interactions by Hydroxyl-Radical Protein Footprinting

Nick Loizos

Abstract

Hydroxyl-radical protein footprinting is a direct method to map protein sites involved in macromolecular interactions. The first step is to radioactively end-label the protein. Using hydroxyl radicals as a peptide backbone cleavage reagent, the protein is then cleaved in the absence and presence of ligand. Cleavage products are separated by high-resolution gel electrophoresis. The digital image of the footprinting gel can be subjected to quantitative analysis to identify changes in the sensitivity of the protein to hydroxyl-radical cleavage. Molecular weight markers are electrophoresed on the same gel and hydroxyl-radical cleavage sites assigned by interpolation between the known cleavage sites of the markers. The results are presented in the form of a difference plot that show regions of the protein that change their susceptibility to cleavage while bound to a ligand.

Key Words

Hydroxyl-radical protein footprinting; macromolecular interactions; protein end-labeling; Fe-EDTA.

1. Introduction

Hydroxyl-radical protein footprinting is a direct method to map areas of a protein that interact with a ligand. Areas of interaction are those that change their accessibility to the peptide cleavage reagent (i.e., diffusible hydroxyl radicals) when in a molecular complex. Accessibility could alter as a result of either direct binding or indirectly through conformational changes induced by the molecular partner. The hydroxyl-radical protein footprinting procedure presented here was developed by Heyduk and Heyduk ***(1,2)*** and has been the subject of recent reviews ***(3,4)***. Using this procedure, data have been obtained for numerous protein–ligand complexes. These include subunits of

From: *Methods in Molecular Biology, vol. 261: Protein–Protein Interactions: Methods and Protocols*
Edited by: H. Fu © Humana Press Inc., Totowa, NJ

Escherichia coli RNA polymerase, the AsiA protein with the σ subunit of RNA polymerase, elongation factor GreB with RNA polymerase, and the Arf1 GTPase with the guanine nucleotide exchange factor Sec7 homology domain of human ARNO ***(2,5–10)***. The footprinting of transcription elongation factor GreB on *E. coli* RNA polymerase will be used to illustrate the methods utilized for the hydroxyl-radical footprinting of a protein–ligand interaction.

2. Materials

1. Modified vectors (available on request) of the pET expression system (Novagen).
2. Heart muscle kinase (HMK) (Sigma; cat. no. P2645) reconstituted in 40 m*M* dithiothreitol.
3. [γ–^{32}P]ATP (6000 Ci/mmol), [γ–^{33}P]ATP (2000 Ci/mmol) (Perkin Elmer).
4. Buffer A: 10 m*M* MOPS/NaOH, pH 7.2, 500 m*M* NaCl, 10 m*M* $MgCl_2$.
5. Buffer B: Buffer A + 250 m*M* imidazole (1 *M* imidazole stock solution pH to 7.5 with concentrated HCl).
6. Microcon 10 microconcentrator (Amicon).
7. RNA polymerase (also called Core).
8. Footprinting reaction buffer: 10 m*M* MOPS/NaOH, pH 7.2, 200 m*M* NaCl, 10 m*M* $MgCl_2$.
9. Cleavage reagent stocks:
   a. 10 m*M* hydrogen peroxide.
   b. 40 m*M* ammonium iron (II) sulfate, hexahydride (Aldrich), in H_2O.
   c. 80 m*M* EDTA, in footprinting reaction buffer.
   d. 0.2 *M* sodium ascorbate, in footprinting reaction buffer.
10. 3X loading buffer: 150 m*M* Tris-HCl, pH 7.9, 36% glycerol, 12% SDS, 6% 2-mercaptoethanol, 0.01% bromophenol blue.
11. Buffer C: 3 *M* Tris-HCl, pH 8.45 + 0.3% SDS.
12. Acrylamide, Ultra pure (ICN).
13. Anode buffer: 24.22 g Tris (base) in 100 mL of H_2O, adjust to pH 8.9 with HCl and then dilute to 1 L with H_2O.
14. Cathode buffer: 12.11 g Tris (base), 17.92 g tricine, and 1 g SDS in 1 L H_2O.
15. CNBr (Aldrich).
16. Endoproteinases LysC, Glu-C, and AspN (Sigma).
17. Program ALIGN (written in Basic by G. Lubecki and T.Heyduk, available on request to T. Heyduk).
18. PhosphorImager (Molecular Dynamics model Storm).
19. Excel (Microsoft).

3. Methods

The major steps involved in the hydroxyl-radical protein footprinting protocol are outlined in **Fig. 1.**

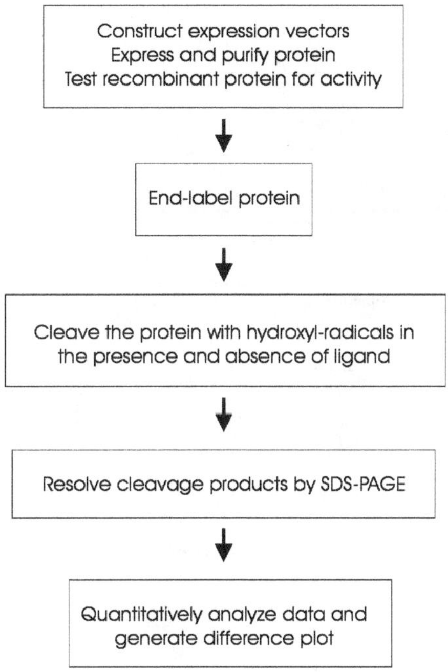

Fig. 1. Flow-chart showing the steps of a hydroxyl-radical protein footprinting experiment.

3.1. Generate Active Recombinant Protein

The production of recombinant forms of GreB that can be radioactively end-labeled is described in **Subheadings 3.1.1.–3.1.2.** This includes the (a) description of expression vectors and (b) verification of activity for the recombinant forms of GreB.

3.1.1. Vector Constructions

End-labeling GreB is achieved by introducing the recognition sequence (LRRASV) for a heart muscle kinase (HMK) to the N- or C- terminus followed by phosphorylating it with radioactive phosphate ***(11)*** (*see* **Note 1**). The pET vectors 16b and 29a (Novagen) were modified to code for HMK sites at the N- and C- termini, respectively, of expressed open reading frames (ORFs) (**Fig. 2**). The GreB ORF was inserted between the *Nde*I and *Bam*HI restriction sites in PK-pET16b to generate a vector (PK-pET16Gb) that directs the production of GreB with an N-terminal HMK site and His_6 tag (HMK-His_6-GreB). The GreB ORF was also inserted between the *Nde*I and *Sac*I restriction sites in pET1529-PK to generate a vector that directs the production of GreB with a C-terminal HMK site and His_6 tag (GreB-His_6-HMK).

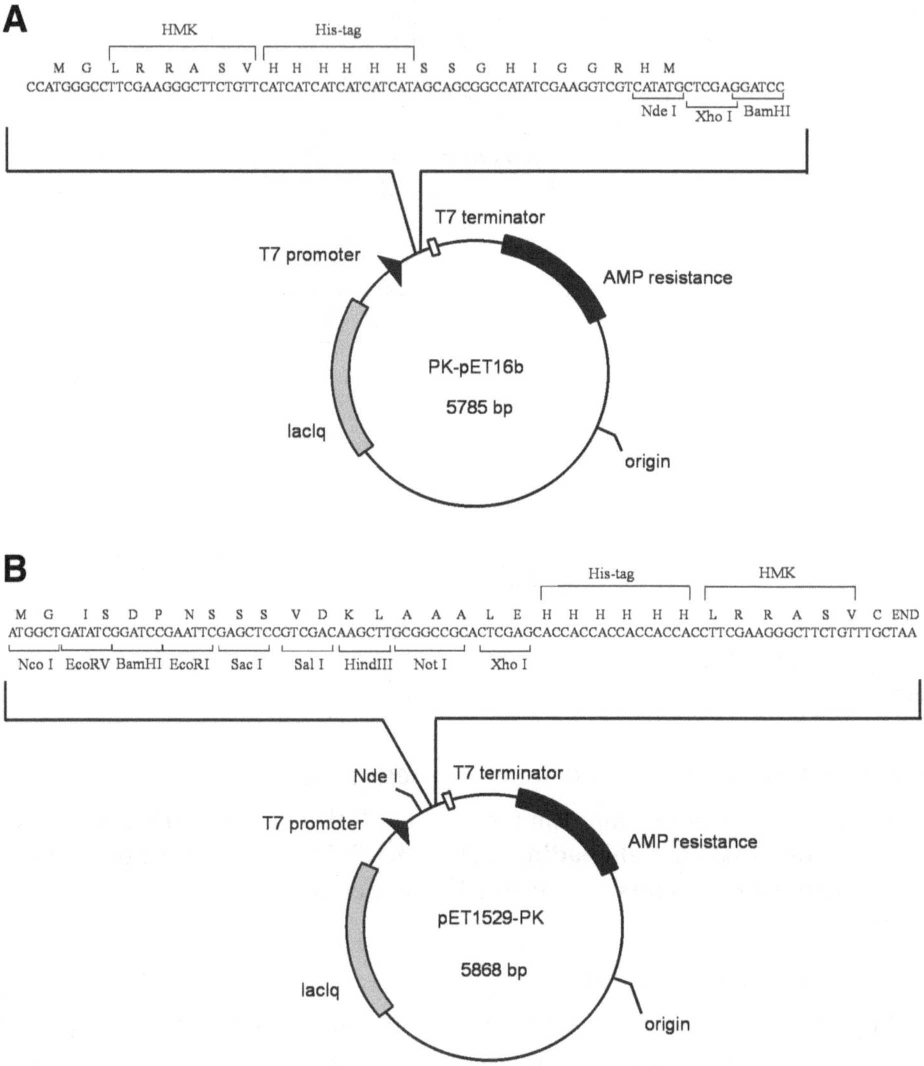

Fig. 2. Schematic drawings of **(A)** PK-pET16b and **(B)** pET1529-PK expression vectors. Features of these vectors include the ampicillin (AMP) resistance gene, lacIq repressor gene, pBR322 origin of DNA replication and T7 promoter and terminator. PK-pET16b was kindly provided by Dr. O'Donnell ***(12)***. Plasmid pET1529-PK was constructed to contain a pET15b background and pET29a multiple cloning site (i.e., the multiple cloning site of pET29a [sequences between the XbaI to Bpu1102I sites] was exchanged with the multiple cloning site of pET15b [sequences between the XbaI to Bpu1102I sites]). The HMK coding sequence was then inserted by recombinant PCR directly downstream of and in frame with the His-tag.

Expression vectors can be transformed into *E.coli* and induced to express recombinant GreB proteins. These proteins can be purified by Ni^{2+}-column chromatography according to manufacturer's instructions.

3.1.2. Activity Check for the Recombinant Forms of GreB

Purified HMK-His_6-GreB was shown to behave as wild-type GreB by several criteria. First, HMK-His_6-GreB was soluble when expressed in *E. coli*. Second, it was equally active in an in vitro transcription elongation assay. Finally, it bound RNA polymerase (Core) in a native gel-shift assay with an affinity essentially the same as that observed for wild-type GreB ***(9)*** (*see* **Note 2**).

3.2. Hydroxyl Radical Cleavage of the Protein in the Absence and Presence of Ligand

The next sections involve footprinting the end-labeled test protein using hydroxyl radicals generated by the Fenton reaction ***(13)***, followed by resolving and identifying the cleavage products in a tricine SDS-PAGE system developed for the resolution of small polypeptides ***(14)***.

3.2.1. End-Labeling Protein

1. End-label 40 μg of HMK-His_6-GreB with ^{32}P (*see* **Note 3**) in a 400 μL reaction (5 μ*M* final concentration). The reaction contains 100 units reconstituted HMK, 0.02 μ*M* [γ-^{32}P]ATP (6000 Ci/mmol), 20 m*M* Tris-HCl, pH 7.9, 0.5 *M* NaCl, and 12 m*M* $MgCl_2$. Incubate 15 min at 30°C. The following steps are necessary to remove free label.
2. Add 20 μL of Ni^{2+}–NTA agarose beads, preequilibrated in buffer A, to the reaction, gently mix, and incubate on ice for 15 min. Pellet beads and wash twice. Wash by adding 1 mL of buffer A, inverting the tube three times, then spinning it for 2-3 s in a table top centrifuge.
3. Resuspend beads in 200 μL of buffer B and incubate on ice for 10 min. Then centrifuge and discard pellet.
4. Add 300 μL of buffer A to the supernatant and apply to a Microcon 10 microconcentrator. Concentrate sample to 100 μL. Add 400 μL of buffer A and reconcentrate to reduce imidazole concentration to below 20 m*M*. The yield should be approx 6 μg (285 pmol) of end-labeled GreB in a 100 μL volume.

3.2.2. FE-EDTA Cleavage Reaction

1. In test tubes add 25 pmoles of HMK-His_6-GreB alone or with 25 pmol of Core. Adjust to a total volume of 14 μL/tube with footprinting reaction buffer (*see* **Note 4**). Incubate for 5 min at 25°C to allow for complex formation.
2. Add 2 μL of an iron–EDTA solution [25 μL of 40 m*M* ammonium iron (II) sulfate + 25 μL of 80 m*M* EDTA + 50 μL of footprinting reaction buffer, prepared just before use], 2 μL of 0.2 *M* sodium ascorbate, and 2 μL of 10 m*M* hydrogen peroxide.

3. Initiate reactions simultaneously by spotting the cleavage reagents to the inside lid of the test tubes followed by briefly centrifuging. The final concentrations of cleavage and protein components are: 1 m*M* $(NH_4)_2Fe(SO_4)_2$, 2 m*M* EDTA, 20 m*M* ascorbate, 1 m*M* H_2O_2, and 1.25–2.5 μ*M* of GreB and Core (*see* **Note 5**).
4. Incubate reactions for 20 min at 25°C and terminate with 10 μL of 3X loading buffer.

3.2.3. Tricine SDS-PAGE System

1. Prepare a tricine/SDS 16.5%T, 3%C polyacrylamide separating gel. Components for the gel are 17.5 mL of 33%T, 3%C polyacrylamide, 11.69 mL buffer C, 1.19 mL H_2O, and 4.62 mL 80% glycerol. Then prepare a 6% T, 3% C stacking gel (4.5 mL of 33%T, 3%C polyacrylamide, 6.2 mL buffer C, and 14.3 mL H_2O). The gel dimensions are 14 cm × 16 cm × 0.8 mm.
2. Set up gel in vertical electrophoresis apparatus and add anode and cathode buffers.
3. Load cleavage reactions (and molecular weight standards from **Subheading 3.2.4.**) onto the gel and run at 14 mAs for 20 min, switch to 20 mA for 16 h.
4. Dry gel for 2 h with a slow ramp up speed to 80°C.
5. Obtain digital image of the gel using a phosphorimager or equivalent instrument.

Figure 3A shows the range of fragments generated by hydroxyl radical cleavage of HMK-His_6-GreB free in solution. **Figures 4A** and **B** show the cleavage profiles of radical-cleaved HMK-His_6-GreB alone and in the presence of saturating amounts of Core.

3.2.4. Molecular Weight Standards

The hydroxyl-radical cleavage products are run on the same gel as a set of molecular weight markers and radical cleavage sites assigned by interpolation between the known cleavage sites of the markers. The molecular weight markers can be generated by residue-specific cleavage of denatured GreB using endoproteinases Lys-C, Glu-C, or Asp-N (specific for lysine, glutamate, and aspartate residues, respectively), or the chemical cleavage reagent cyanogen bromide (specific for methionine) (*see* **Notes 6** and **7**).

1. Endoproteinase reaction mixtures (total volume 30 μL) contain 1.25 μ*M* HMK-His_6-GreB, 8 *M* urea, 50 m*M* Tris-HCl, pH 9.0, and either 0.25 μg Lys-C, 0.5 μg Glu-C, or 0.5 μg Asp-N.
2. After a 15 min incubation at 25°C, terminate reactions by adding 10 μL of 3X loading buffer.
3. Reaction for methionine-specific cleavage contains (20 μL; pH adjusted to 2 with 1 *M* HCl): 1.25 μ*M* HMK-His_6-GreB, 400 m*M* CNBr, and 0.4% SDS. Terminate by lyophilization and addition of 30 μL of 1X loading buffer.
4. Load and run reactions on footprinting gel (**Fig. 3A**) (*see* **Note 8**).

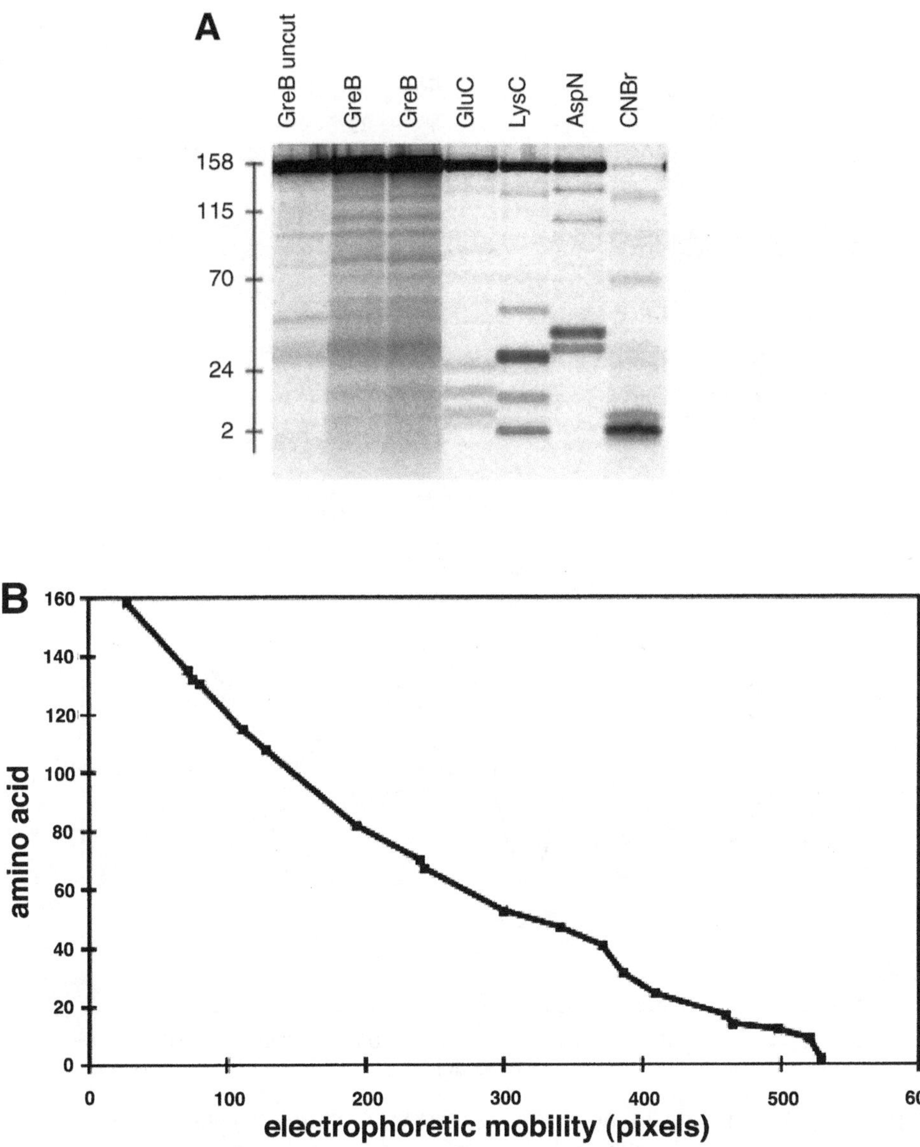

Fig. 3. Hydroxyl radical cleavage of GreB. **(A)** Phosphorimaged gel showing GreB fragments generated by cleavage of GreB with hydroxyl radicals (GreB lane) or proteases (lanes GluC, LysC, AspN, and CNBr) under single-hit conditions. The size of cleavage products (in amino acid residues, but not including the 22 non-native N-terminal residues) is indicated on the left. **(B)** Standard curve of cleavage site (amino acid residue) versus mobility for molecular weight markers obtained in A (protease treatments of GreB). The distance on the footprinting gel from uncut GreB to the molecular weight markers is expressed in terms of pixels.

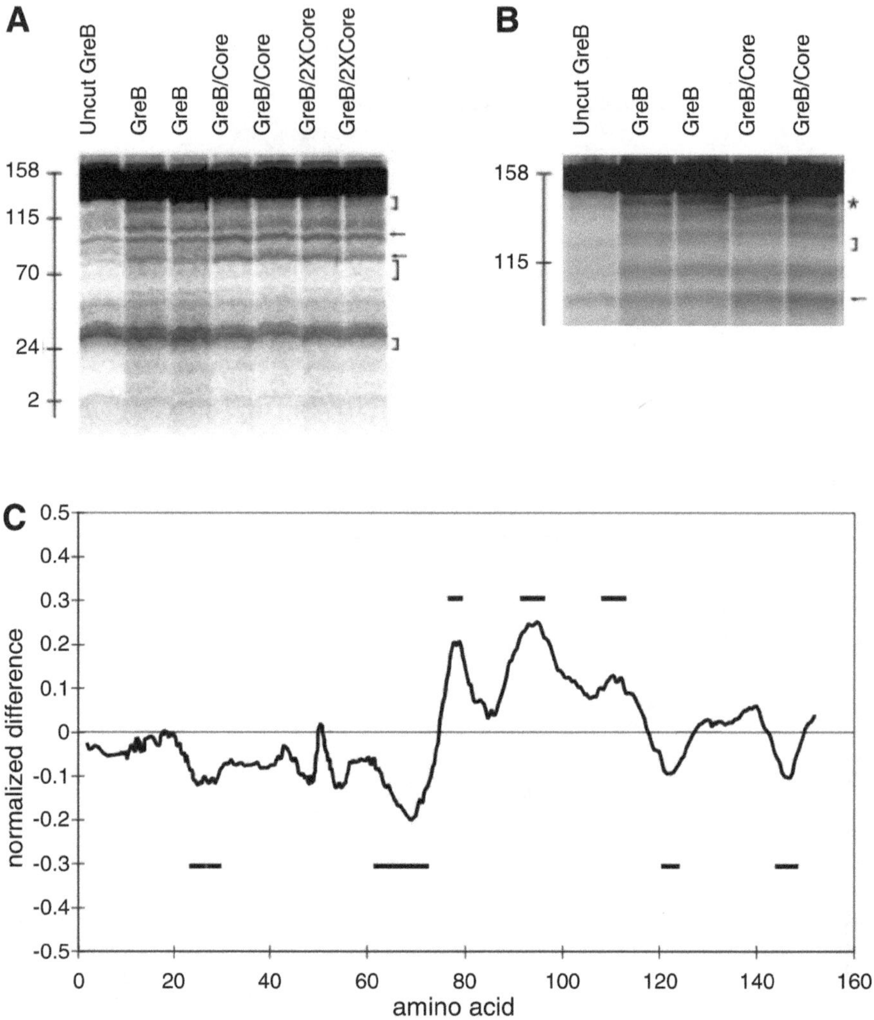

Fig. 4. Hydroxyl radical footprint of core RNAP on GreB. (**A**) phosphorimaged gel of hydroxyl-radical cleaved GreB and GreB–Core complex. Enhanced and protected areas are marked with arrows and brackets, respectively. (**B**) Top part of phosphorimaged gel from experiment using ^{33}P-labeled GreB. Asterisk marks additional protected area not seen in A due to masking by the intensity of the uncleaved product. (**C**) Difference plot showing normalized intensity difference, $(I_{complex} - I_{GreB})/(I_{complex} + I_{GreB})$, plotted against residue number ($I_{complex}$ and I_{GreB} are the corrected intensities for the GreB–Core complex and GreB, respectively). Statistically significant differences according to a Student's *t* test (confidence level of 0.99999) are denoted by black bars.

3.3. Data Analysis

The digital images of the footprinting gels shown in **Figs. 3A**, **4A**, and **4B** can be subjected to quantitative analysis and the results presented in the form of a difference plot (*see* **Note 9**). The next steps describe how to (a) generate intensity plots for each gel lane, (b) align them to each other to correct for lane-to-lane distortions, (c) subtract intensity values for free GreB from the GreB/Core complex, and (d) graph the differences versus amino acid number.

3.3.1. Alignment of Intensity Plots

To generate intensity plots, use the program IMAGEQUANT (Molecular Dynamics) to make a line down the center of each gel lane. Widen the line with the object attributes menu item to bracket the entire lane. Then use the Create graph command to integrate phosphorimager intensities across full lane widths and plot them vs electrophoretic mobilities (pixels).

To correct for lane-to-lane distortions, align intensity plots using the program ALIGN. ALIGN uses a linear interpolation to adjust peaks of one plot to homologous peaks in another. Chose one plot (gel lane) as a reference that all others will be aligned to. The lanes containing molecular weight standards can only and should be aligned from the band corresponding to the uncleaved protein.

3.3.2. Difference Plots

1. Import the aligned intensity scans, excluding portions corresponding to the uncleaved full-length protein, into a Microsoft Excel worksheet. Place pixel numbers in column A of the worksheet and the corresponding aligned intensity values for cleaved GreB and GreB:Core complex into columns B and C, respectively. In addition to obtaining the difference between GreB:Core and GreB plots, the following steps correct for gel loading and cleavage efficiencies between the samples.
2. Set column D to calculate the ratios of the intensity values for each pixel in B and C [i.e., GreB/(GreB:Core)].
3. Round off the values in column D to a single digit and place in column E. Then calculate the mode of the values in E.
4. Multiple the values in column C by the mode, then subtract the values in lane B and divide the difference by the additive values of B plus C multiplied by the mode $\{[(\text{Mode} \times C) - B]/[B + (\text{Mode} \times C)]\}$. Place these values in lane G.

The next series of steps converts pixel number to amino acid number. Knowledge of the residue specificity for each endoproteinase, as well as the primary amino acid sequence for GreB, allows one to determine how many bands (i.e., molecular weight standards) and their mobilities relative to each other would be generated under limited cleavage conditions.

5. Assign each of the bands of standards in the gel an amino acid number (where the endoproteinase or CNBr cuts) and pixel number (distance from the beginning of the line made to generate the intensity plots). Place the pixel and amino acid numbers into columns I and J, respectively. Pixel numbers start from lowest to highest and amino acid numbers highest (full-length) to lowest available marker.
6. Determine the log of J values and place in column K. Use the data in I and K to make the standard curve of log(M_r) vs relative mobility (**Fig. 3B**).
7. Find the slopes and intercepts for the lines between the data points of the standard curve and place in columns L and M, respectively.
8. The pixel values in column A can then be converted to amino acid number by the formula {10^ [(L × A) + M]}. Place interpolated amino acid values in column N and use N and G values to generate the difference plot (**Fig. 4C**).
9. Average difference plots from different gel lanes and from different gels (interpolated to a reference gel). Test the statistical significance of the differences between the (Core/GreB complex – GreB) experiments using a Student's t test for small samples with unequal variances.

Regions of statistically significant protection (negative values) and enhancement (positive values) of radical cleavage by the binding of core RNA polymerase are observed (**Fig. 4C**). Functional studies of the isolated Gre factor domains revealed that the main determinants for binding to RNAP lie within the C-terminal domain (amino acids 75–158) *(17)*. Therefore, surface-exposed amino acids within the GreB–CTD/Core interaction areas (from residues 116–127 and 134–148) were identified using the program GRASP *(18)* and then were individually substituted with alanine. Alanine substitution for residues 121 and 123 resulted in a decrease in core binding *(9)*. These results indicate that the protection from radical cleavage within this area of GreB most likely results from a direct interaction with core. (*see* **Note 10** for other applications of this procedure).

4. Notes

1. Before beginning any footprinting project, it is important to show that the wild-type test protein is not phosphorylated by the kinase at an internal "cryptic" site. If a cryptic site is present, phosphorylation conditions may be found that eliminate or reduce labeling to acceptably low levels compared to the usually robust labeling of the cognate site for the kinase.
2. It must be shown that the terminal modifications to the test protein do not disrupt folding or interactions with the ligand. How this is shown will depend on the binding and/or activity assays available in each specific case. A derivatized GreB with C-terminal modifications (GreB-His_6-HMK) was also generated but was not soluble when overexpressed and therefore was refolded from inclusion bodies. Although this refolded derivative was active in transcript elongation assays, dynamic light scattering showed it to be somewhat aggregated. The presence of aggregates or possibly a fraction of misfolded proteins in a preparation of the C-

terminal derivative could affect cleavage susceptibility to hydroxyl radicals and thereby complicate the analysis of cleavage profiles. Indeed, the hydroxyl radical cleavage profile for GreB-His_6-HMK was altered when compared with N-terminally modified GreB, and therefore the C-terminal derivative was not used in further studies. It is best to use refolded proteins in footprinting experiments only when an activity-based purification step is available (such as an affinity column using the ligand to be tested in the footprinting experiment).

3. ^{33}P can be used instead of ^{32}P. The lower energy beta particles arising from ^{33}P decay (0.248 MeV) compared with ^{32}P (1.71 MeV) do not penetrate and scatter as far into the detecting medium, resulting in sharper bands. However, ^{32}P gives a stronger signal, should a low amount of protein be used in the footprinting reaction ***(9)***.
4. Alternative buffers can be used but should not contain radical scavengers (e.g. Tris or glycerol) ***(3)***. $MgCl_2$ can be omitted if necessary to prevent Mg^{2+}-dependent activation of proteins during the reaction ***(9)***.
5. The concentration of the components must leave a large fraction of the molecules uncut, thus satisfying the need for single-hit conditions during footprinting ***(15)***.
6. Depending on the sequence of the test protein, other chemical cleavage reagents, such as BNPS skatole and 2-nitro-5-thiocyanobenzoic acid (specific for tryptophan and cysteine, respectively), may be used to generate molecular weight markers ***(16)***.
7. Gel mobility of proteins is not related in a simple way to the molecular weight, thus small protein fragments may show atypical mobility in SDS-PAGE. Therefore, calibration of gel mobility using molecular weight markers generated from the test protein itself is essential to accurately convert the gel mobilities into amino acid cleavage sites.
8. Products generated by radical cleavage of HMK-His_6-GreB very close to the N terminus of GreB were resolved on the gels (note products visible next to the last molecular weight marker in the LysC lane, which corresponds to cleavage at amino acid position 2 of GreB; **Fig. 3A**). It has been found that cleavage products shorter than about 25 amino acids in length are lost during electrophoresis ***(2)***. It is therefore advantageous that 22 non-native amino acids of HMK-His_6-GreB are on the N-terminal side of GreB, thereby giving the ability to detect protections or enhancements at that terminus. Such was the case with footprinting GreB on a transcription elongation complex ***(9)***. Therefore, adding non-native amino acids between the kinase site and the beginning of the protein can help resolve footprints at termini of test proteins.
9. Unlike the situation for DNA footprinting, the footprint on the test protein resulting from the bound ligand may not be immediately apparent from observation of the imaged gel. The reason for this is mainly because single amino acid resolution is not achieved by the SDS-PAGE analysis. Usually a cleavage site can be assigned to ± three to five amino acids using the gel dimensions given ***(3)***. Thus, careful quantitative analysis of the phosphorimager intensities is required for the footprint to be fully revealed.

10. While the above procedure focuses on a protein–protein interaction, proteins in complex with nucleic acids have also been footprinted with success ***(19,20)***. Given that nucleic acids are more susceptible than proteins to hydroxyl radical cleavage, care must be taken that the ligand and/or complex is not altered during the footprinting reaction. The footprinting procedure here has also been used to identify conformational changes in proteins in response to ligand ***(19)***.

References

1. Heyduk, E. and Heyduk, T. (1994) Mapping protein domains involved in macromolecular interactions: a novel protein footprinting approach. *Biochemistry* **33,** 9643–9650.
2. Heyduk, T., Heyduk, E., Severinov, K., Tang, H., and Ebright, R. H. (1996) Determinants of RNA polymerase α subunit for interaction with β, β', and σ subunits: hydroxyl-radical protein footprinting. *Proc. Natl. Acad. Sci. USA* **93,** 10,162–10,166.
3. Heyduk, T., Baichoo, N., and Heyduk, E. (2001) Hydroxyl radical footprinting of proteins using metal ion complexes. *Met. Ions. Biol. Syst.* **38,** 255–287.
4. Loizos, N. and Darst, S.A. (1998) Mapping protein-ligand interactions by footprinting, a radical idea. *Structure* **6,** 691–695.
5. Nagai, H. and Shimamoto, N. (1997) Regions of the *Escherichia coli* primary sigma factor σ70 that are involved in interaction with RNA polymerase core enzyme. *Genes to Cells* **2,** 725–734.
6. Wang. Y., Severinov, K., Loizos, N., et al. (1997) Determinants for *Escherichia coli* RNA polymerase assembly within the β subunit. *J. Mol. Biol.* **270,** 648–662.
7. Casaz, P. and Buck, M. (1999) Region I modifies DNA-binding domain conformation of sigma 54 within the holoenzyme. *J. Mol. Biol.* **285,** 507–514.
8. Colland, F., Orsini, G., Brody, E.N., Buc, H., and Kolb, A. (1998) The bacteriophage T4 AsiA protein: a molecular switch for sigma 70-dependent promoters. *Mol. Microbiol.* **27,** 819–829.
9. Loizos, N. and Darst, S.A. (1999) Mapping interactions of *Escherichia coli* GreB with RNA polymerase and ternary elongation complexes. *J. Biol. Chem.* **274,** 23,378–23,386.
10. Mossessova, E., Gulbis, M., and Goldberg, J. (1998) Structure of the guanine nucleotide exchange factor Sec7 domain of human Arno and analysis of the interaction with ARF GTPase. *Cell* **92,** 415–423.
11. Li, B.L., Langer, J.A., Schwartz, B., and Pestka, S. (1989) Creation of phosphorylation sites in proteins: construction of a phosphorylatable human interferon α. *Proc. Natl. Acad. Sci. USA* **86,** 558–562.
12. Kelman, Z. and O'Donnell, M. (1995) *Escherichia coli* expression vectors containing a protein kinase recognition motif, His_6-tag and hemagglutinin epitope. *Gene* **166,** 177–178.
13. Kim, K., Rhee, S.G., and Stadtman, E.R. (1985) Nonenzymatic cleavage of proteins by Reactive oxygen species generated by dithiothreitol and iron. *J. Biol. Chem.* **260,** 15,394–15,397.

14. Schagger, H. and von Jagow, G. (1987) Tricine-sodium dodecyl sulfate-polyacrylamide gel electrophoresis for the separation of proteins in the range from 1 to 100 kDa. *Anal. Biochem.* **166,** 368–379.
15. Brenowitz, M., Senear, D.F., Shea, M.A., and Ackers, G.A. (1986) Quantitative Dnase footprint titration: a method for studying protein-DNA interactions. *Methods Enzymol.* **130,** 132–181.
16. Borukhov, S., Lee, J., and Goldfarb, A. (1991) Mapping of a contact for the RNA 3' terminus in the largest subunit of RNA polymerase. *J. Biol. Chem.* **266,** 23,932–23,935.
17. Koulich, D., Nikiforov, V., and Borukhov, S. (1998) Distinct functions of N and C-terminal domains of GreA, an *Escherichia coli* transcript cleavage factor. *J. Mol. Biol.* **276,** 379–389.
18. Nicholls, A., Sharp, K.A., and Honig, B. (1991) Protein folding and association: insights from the interfacial and thermodynamic properties of hydrocarbons. *Proteins* **11,** 281–296.
19. Baichoo, N. and Heyduk, T. (1999) DNA-induced conformational changes in cyclic AMP receptor protein: detection and mapping by a protein footprinting technique using multiple chemical proteases. *J. Mol. Biol.* **90,** 37–48.
20. Schwanbeck, R., Manfioletti, G., and Wisniewski, J.R. (2000) Architecture of high mobility group protein I-C.DNA complex and its perturbtion upon phosphorylation by Cdc2 kinase. *J. Biol. Chem.* **275,** 1793–1801.

16

Use of Phage Display and Polyvalency to Design Inhibitors of Protein–Protein Interactions

Michael Mourez and R. John Collier

Abstract

We describe the synthesis of an inhibitor that interferes with critical protein–protein interactions occurring during the assembly of anthrax toxin. Using a phage display selection strategy, we isolated a peptide directed against the cell binding moiety of the toxin that was able to interfere with binding of the enzymatic moieties. Because the cell binding moiety of the toxin is a heptamer, the peptide can potentially bind up to seven equivalent sites. We synthesized a polyvalent molecule displaying multiple copies of this peptide and showed that it is a much more potent inhibitor than the free peptide. Because little structural knowledge of the interacting proteins was required to synthesize this inhibitor, we believe that this approach may prove useful in the design of inhibitors of protein–protein interactions in other systems.

Key Words

Inhibitor; polyvalent molecules; phage display; peptides; protein–protein interaction; anthrax; toxin.

1. Introduction

1.1. Peptides and Polyvalent Molecules as Inhibitors of Protein–Protein Interactions

Biologically active peptides often exert their functions by either mimicking or competing with the natural ligands of proteins. Consequently, there has been a lot of interest in genetic or structure-based methods to select or design peptides able to interact with a given protein target. The phage display technology is one of the most popular of these approaches because it is inexpensive and easy to perform. This technique uses peptide libraries genetically fused to a phage coat protein to select phages binding a specific target via the interaction of the displayed peptide *(1,2)*. Because it relies on a selection, phage display

From: *Methods in Molecular Biology, vol. 261: Protein–Protein Interactions: Methods and Protocols*
Edited by: H. Fu © Humana Press Inc., Totowa, NJ

has the added advantage of not requiring structural knowledge of the target. The use of phage display has had some stunning successes, like the selection of peptidic hormone receptor agonists *(3)*. In most cases, however, the peptides have rather weak affinities for their intended targets, precluding their use as therapeutics.

When a target offers multiple equivalent binding sites to a ligand, it can be advantageous to synthesize molecules displaying multiple copies of the ligand, which may foster the simultaneous interactions of multiple ligands with the target. Indeed, these polyvalent molecules can interact with their target far more efficiently than the monomeric ligand *(4)*. Polyvalent molecules designed in recent years have led to promising pharmaceuticals, for instance, polyvalent inhibitors of toxins displaying multiple copies of the carbohydrates that act as toxin receptors *(5)*.

In order to find a small-molecular-weight inhibitor of anthrax toxin, we decided to combine phage display technology and synthesis of polyvalent molecules. By this approach, we quickly developed a potent in vivo inhibitor of anthrax toxin that disrupts critical protein–protein interactions *(6)*.

1.2. Anthrax Toxin

Anthrax toxin is produced by *Bacillus anthracis*, the causative agent of anthrax *(7)*. The toxin is responsible for the major symptoms of the disease and, therefore, a specific inhibitor of its action might prove a valuable therapeutic. The intoxication mechanism has been well studied (**Fig. 1**). The toxin consists of a single receptor-binding moiety, termed protective antigen (PA), and two enzymatic moieties, termed edema factor (EF) and lethal factor (LF). After release from the bacteria as nontoxic monomers, these three proteins diffuse to the surface of mammalian cells and assemble into toxic, cell-bound complexes. Cleavage of PA into two fragments by a cell-surface protease enables the fragment that remains bound to the cell, PA_{63}, to heptamerize and bind EF and LF with high affinity. After internalization by receptor-mediated endocytosis, the complexes are trafficked to the endosome. There, at low pH, the PA moiety inserts into the membrane and mediates translocation of EF and LF to the cytosol. Interestingly, the N-terminal PA binding domain of LF, LFn, can be fused to recombinant polypeptides, and the resulting chimeras can be translocated in the same fashion. EF is an adenylate cyclase that perturbs water homeostasis in cells, causing edema. LF is a zinc-dependent protease that cleaves specifically mitogen-activated protein kinase kinases, triggering a poorly understood cascade of events culminating with the death of the host.

One way to prevent the action of anthrax toxin would be to interfere with assembly of PA, EF, and LF into toxic complexes. In this chapter, we will describe how we used phage display to identify a peptide that interfered with

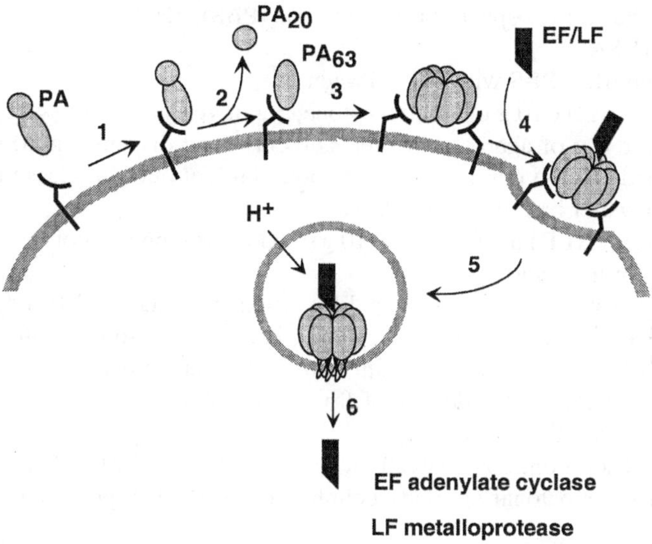

Fig. 1. Assembly of anthrax toxin complexes and their entry into cells. 1. Binding of PA to its receptor. 2. Proteolytic activation of PA and dissociation of PA_{20}. 3. Self-association of monomeric PA_{63} to form the heptameric prepore. 4. Binding of EF/LF to the prepore. 5. Endocytosis of the receptor:PA_{63}:ligand complex. 6. pH-dependent insertion of PA_{63} and translocation of the ligand. The polyvalent inhibitor synthesized by this protocol blocks step 4.

the binding of EF/LF to heptamers of PA_{63}, how we synthesized a molecule displaying multiple copies of this peptide so that the polyvalent molecule could bind multiple sites on the heptamers, and then how we performed an initial characterization of the activity of this polyvalent inhibitor (PVI). The approach we used could be applied to other systems where a target can present multiple equivalent binding sites to an inhibitory peptide. Our approach could also be adapted to other targets by linking multiple copies of different peptides binding to nonequivalent sites on to a unique backbone.

2. Materials

1. Ph.D.-12 Phage Display Library Kit (New England Biolabs): peptide phage display library, sequencing primers, and *Escherichia coli* ER2537.
2. M13KO7 Phage (New England Biolabs).
3. Purified proteins for the phage display selection: PA, PA_{63}, and LFn purified as before ***(8)***.
4. Maxisorp tubes with caps and Maxisorp plates (Nunc).

5. Blocking buffer: Phosphate-buffered Saline (PBS), pH 7.4 with 2% bovine serum albumin (BSA).
6. Washing buffer: PBS with 0.5% Tween 20.
7. Agarose top: Mix 10 g of bacto tryptone, 5 g of yeast extract, 5 g of NaCl, 1 g of $MgCl_2$, and 7 g of agarose per liter; autoclave and dispense in 50 mL aliquots. When needed, melt in microwave and pour 3 mL aliquots in glass tubes. Keep at 45°C in a water bath or heating block.
8. Luria–Bertani (LB) medium: Mix 10 g of bacto tryptone, 5 g of yeast extract, 5 g of NaCl per liter; autoclave.
9. IPTG/Xgal solution: 1.25 g isopropyl-β-D-thiogalactoside (IPTG) and 1 g 5-bromo-4-chloro-3-indolyl-β-D-galactoside (Xgal) in 25 mL *N,N*-dimethylformamide.
10. LB/IPTG/Xgal plates: LB medium with 15 g agarose per liter, autoclaved and cooled to 70°C prior to addition of 1 mL of the above Xgal/IPTG solution. Keep in the dark at 4°C.
11. PEG/NaCl solution: 20% polyethylene glycol 8000, 2.5 *M* NaCl, autoclaved.
12. Anti-M13 monoclonal antibody coupled to horseradish peroxidase (Pharmacia Biotech).
13. Solution of 3,3',5,5'-tetramethylbenzidine (TMB) liquid substrate for ELISA (Sigma) and sulfuric acid.
14. Poly *N*-acryloyloxy succinimide (PNAS), which can be custom-made and characterized in a chemistry laboratory according to published protocol ***(9)***. PNAS may also be obtained commercially (*see* for instance http://www.phoenixpeptide.com).
15. Custom synthetic peptide purified at greater than 95% purity. The N terminus of the peptide should be acetylated and its C terminus amidated.
16. Triethylamine, *N,N*-dimethylformamide, ammonium hydroxide.
17. Dialysis membrane with a 30-kDa molecular weight cut-off.
18. Specifically for the characterization of the anthrax toxin inhibitor:
    a. CHO cells and cell culture medium (HAM-F12 with 10% calf serum).
    b. LFn labeled with ^{35}S-methionine by in vitro coupled transcription and translation, as described previously ***(10)***.
    c. LFnDTA purified as described previously ***(11)***.
    d. Tritiated leucine and leucine free HAM-F12 media.
19. Lysis buffer: 100 m*M* NaCl, 20 m*M* sodium phosphate pH 7.4, 10 m*M* EDTA, 1% Triton X100.

3. Methods

3.1. Selection of PA_{63} Heptamer-Specific Peptides by Phage Display

PA_{63} heptamer, and not monomeric PA, has an interface to interact with EF/LF. By selecting members of a phage library that bind to PA_{63} heptamers and eliminating those that bind to the monomeric uncleaved PA, we hoped to enrich for phages that bind at or near the EF/LF binding site of PA_{63} heptamers. Such peptides might then be able to inhibit the interactions between PA_{63} heptamers and EF/LF. This section describes the three steps of

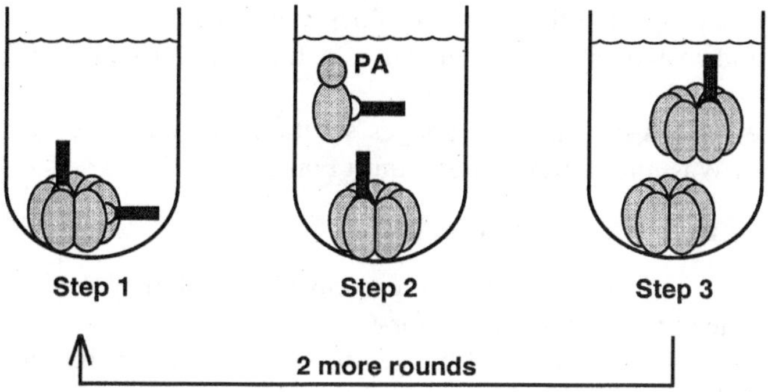

Fig. 2. Selection of bacteriophages binding to PA_{63} heptamer at or near the EF/LF site. Purified heptamer is coated on a plastic surface and a library of bacteriophages displaying 12-amino-acid peptides is allowed to bind the heptamer (Step 1). Purified intact PA is added to elute the phages that are bound to surfaces present both in intact PA and in the heptamer (Step 2), the remaining phages are eluted with purified heptamer (Step 3). The final eluate is used to perform another round of selection.

the protocol to select such phages (**Fig. 2**): (1) coat PA_{63} heptamers on a plastic surface and add a library of phages displaying random peptides; (2) wash the surface and add monomeric uncleaved PA to elute phages that bind to the surfaces not specific for the heptamer; and (3) add a solution of PA_{63} heptamers to recover phages that adsorbed to its specific surface (*see* **Note 1**). In this experiment we use a commercial library of M13 bacteriophages, Ph.D.-12, by New England Biolabs. The peptide library is made of 12-amino-acid random peptides fused to the N terminus of the phage coat protein pIII (*see* **Note 2**).

3.1.1. Binding to PA_{63} Heptamers

1. Add 1 mL of a purified PA_{63} heptamer solution at 2 μg/mL in PBS in a 5 mL Maxisorp tube (Tube 1) to coat the plastic surface with heptamer. Place the tube on a rotating wheel overnight in a cold room at 4°C.
2. The next day, add 5 mL of Blocking buffer to an uncoated Maxisorp tube (Tube 2) and incubate at 37°C for 2 h to block the plastic surface. Wash the empty tube briefly twice with Washing buffer using a squirt bottle. Add the phages, 1.5×10^{11} plaque forming unit (pfu) diluted in 1 mL of PBS–0.1% Tween 20, and incubate on a rotating wheel at room temperature for 1 h (*see* **Note 3**).
3. During the blocking of the uncoated tube (Tube 2), discard the heptamer solution of the coated tube (Tube 1), add 5 mL of Blocking buffer, and incubate at 37°C for 2 h.

4. Transfer the phage solution from the blocked uncoated tube (Tube 2) to the tube blocked and coated with heptamer (Tube 1). Incubate at room temperature for 1 h on a rotating wheel.
5. Carefully discard the unbound phages solution and wash the tube at least six times with Washing buffer using a squirt bottle.

3.1.2. Elution with Monomeric PA

1. Add 1 mL of a purified monomeric uncleaved PA solution at 15 μg/mL in PBS and incubate at room temperature for 1 h.
2. Carefully discard the unbound phages solution and wash the tube at least six times with Washing buffer using a squirt bottle.

3.1.3. Elution with PA_{63} Heptamers

1. Elute the remaining phages by adding 1 mL of a purified PA_{63} heptamer solution at 40 μg/mL in PBS to the tube and rotate at room temperature for 1 h.
2. Collect the eluted phages in a microfuge tube.
3. Determine the titer of the phages in the eluate:
   a. Prepare five 10-fold serial dilutions of the eluate in PBS.
   b. Add 10 μL of each dilution in a glass tube containing 3 mL of melted agarose top and mix with 200 μL of a mid-log-phase culture of ER2537. Immediately pour on a LB/Xgal/IPTG plate and let the agar top solidify. Incubate the plate at 37°C overnight.
   c. The *E. coli* strain ER2537, provided with the peptide library, is a rapidly growing F^+ strain carrying the mutation Δ(lacZ)M15. Because the M13 phages of the library contain the *lacZα* gene, phage plaques appear blue on a lawn of ER2537 growing on a plate of LB/Xgal/IPTG. Therefore, the next day count the blue plaques and calculate the titer of the eluate (the number of blue plaques multiplied by the dilution factor) to obtain a number of pfu per mL.
4. Amplify the eluate:
   a. Add 500 μL of the eluate to 25 mL of LB with 500 μL of an overnight culture of ER2537. Incubate at 37°C with vigorous shaking for a good aeration for 4–5 h.
   b. Spin down the bacterial cultures and recover the supernatant. Add 5 mL of PEG/NaCl solution and let the phage particles precipitate overnight at 4°C.
   c. Spin at 20,000*g* for 15 min, discard the supernatant, and resuspend the pellet in 1 mL of PBS. Transfer to microfuge tubes, spin briefly at maximum speed in a microfuge, and transfer 900 μL to a new microfuge tube.
   d. Add 150 μL of PEG/NaCl and incubate on ice for 1 h.
   e. Spin at maximum speed in a microfuge for 10 min, resuspend the pellet in 1 mL of PBS, spin again at maximum speed for 2 min, and transfer 900 μL to a new microfuge tube. This is the amplified eluate.
5. Titer the amplified eluate using the technique of **step 3** (*see* **Note 4**) and coat new tube to repeat the whole selection (**Subheadings 3.1.1.** to **3.1.3.**).

A total of three rounds are typically needed. Changes have to be introduced in the second and third round: Incubate the phages with the heptamer coated tube for 30 min in the second round and 5 min in the third instead of 1 h in the first. Elute with heptamer solution overnight for the second and third round instead of 1 h in the first. This results in the selection of phages having a higher affinity. The eluate of the third round need not be amplified, but it should be titered, and the phages from the plaques of the titer determination will be individually tested. Typically the titer of the first round eluate is 10^4–10^5 pfu/mL and the titer of the third round eluate 10^7–10^8 pfu/mL. The increase in the titer of the eluates means that more specific phages are retained at each round and is an indication that the selection is successful.

3.1.4. Characterization of Phage Eluates

The amplified eluate of each round should be tested for enrichment in the specific phages. This is typically done by doing a phage-ELISA, which assays for the phage ability to bind a purified protein coated on plastic.

1. Coat Maxisorp 96-well plate overnight at 4°C with 100 μL per well of a 10 μg/mL solution of purified protein in PBS. We coat PA_{63} heptamers as the intended target and uncleaved monomeric PA as the control protein toward which the phages should have no specific binding. The wells should be coated in duplicate.
2. Discard the protein solution, drain excess liquid on a paper towel. Block the plate for 2 h at 37°C with 200 μL per well of Blocking buffer.
3. Incubate the plate for 1 h at room temperature with 100 μL of a dilution of the phage eluate at 10^9 pfu/mL.
4. Wash at least six times with Washing buffer and then incubate for 1 h at room temperature with a monoclonal anti-M13 antibody coupled to horseradish peroxidase (HRP), diluted 5000-fold in PBS.
5. Repeat the wash step and reveal HRP activity by assaying the oxidation of 100 μL of a solution of TMB. Let the reaction develop for about 30 min and stop the reaction by addition of 100 μL of 0.2 *M* sulfuric acid. Measure the absorbance at 450 nm.

Round after round, the signal generated by phages binding specifically to PA_{63} heptamers and not monomeric PA should increase (**Fig. 3A**). If this is not the case, a fourth round of selection might be carried out. If no improvement is observed after the fourth selection, it usually indicates that the selection failed.

3.1.5. Characterization of Selected Individual Phages

After the last round of selection, the individual phages should be characterized. The phages should be sequenced in order to know what peptides have been selected. For each peptide, an ELISA should then be performed to assay the selectivity and affinity of the phage. The following protocol can be used to prepare phage solutions from an individual plaque:

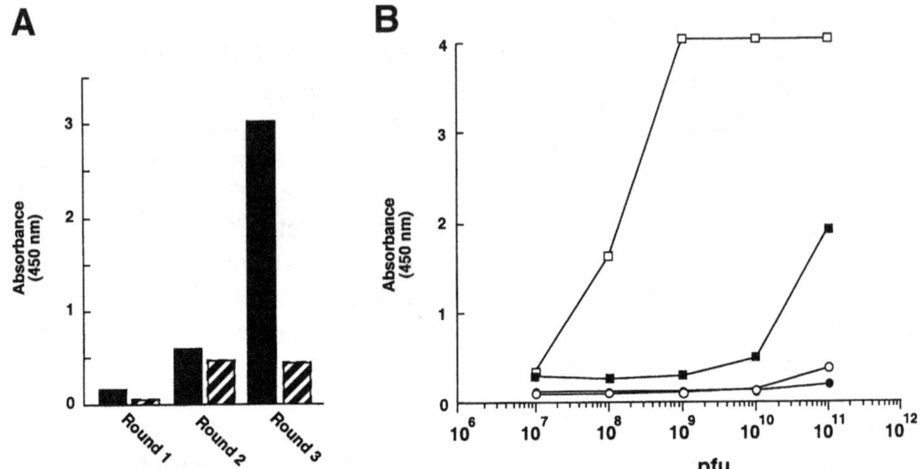

Fig. 3. **(A)** Phage-ELISA of the amplified eluates at the end of each round of the selection. The phages are allowed to bind to PA_{63} heptamer (solid bars) or uncleaved PA (striped bars) adsorbed on the surface of a polystyrene plate. Each well is coated overnight with 1 µg protein in PBS. **(B)** Phage-ELISA of dilutions of one selected phage displaying the HTSTYWWLDGAP peptide (squares) or of dilutions of M13KO7 phage (circles). Each well is coated overnight with 1 µg of PA_{63} heptamer (solid symbols) or uncleaved PA (open symbols) in PBS.

1. Using a sterile toothpick, stab an isolated blue plaque from one of the plates used to determine the titer of the last eluate. Transfer to a tube containing 2 mL of a 1:100 dilution of an overnight culture of ER2537 in fresh LB media.
2. Grow for 4–5 h at 37°C with vigorous shaking.
3. Pellet bacteria and transfer 1.5 mL of the supernatant containing the phages to a clean microfuge tube and add 250 µL of PEG/NaCl solution. Let the phage precipitate for 1 h on ice.
4. Spin down the precipitated phages at maximum speed for 10 min and resuspend the pellet in 1 mL of PBS. Briefly spin down at maximum speed and transfer 900 µL of the supernatant to a new microfuge tube. This is a solution of the individual phage.

The DNA of the phage can be obtained using a standard phenol–chloroform extraction followed by ethanol precipitation. The sequence of the N-terminal portion of the pIII protein can then be determined using the sequencing primer provided by New England Biolabs with the Ph.D.-12 library. If the selection is successful, a consensus motif is expected to appear from the alignment of the sequences of the peptides displayed by 10 or more of the selected individual phages. For instance, when we performed the selection described above, we obtained a phage displaying the peptide HTSTYWWLDGAP and another phage displaying the peptide HQLPYWWLSPG, suggesting the importance of the consensus core YWWL.

Phages corresponding to each unique selected sequence should be amplified using under the **Subheading 3.1.3., step 5**, of the selection procedure and assayed by doing phage-ELISA using the protocol described in the previous section. However, to evaluate the strength of the interaction between the displayed peptide and the target, one should perform the ELISA with serial dilutions of the individual phage solutions (**Fig. 3B**). Dilutions typically range between 10^{11} and 10^6 pfu per well, the stronger the interaction between peptide and target is, the less phage it takes to saturate the ELISA signal. An M13 control phage not displaying any peptide, like M13KO7, or a phage displaying an irrelevant peptide should be used as a control phage unable to bind the coated proteins. To test the hypothesis that the selected phages display peptides are able to compete with the enzymatic moieties for binding to PA_{63} heptamers, the ELISA can also be performed in the presence of an excess of purified natural ligand (we used 10 μM of LFn, for instance). If the hypothesis is true, then phage binding should be completely blocked by the addition of such an excess of competitor.

3.2. Making a Polyvalent Molecule

The selected dodecapeptide can then be synthesized and studied for its ability to prevent toxin assembly, but often the concentration required to inhibit 50% of the targeted protein–protein interaction will be high. Indeed, a 150 μ*M* concentration of the HTSTYWWLDGAP peptide was required to block 50% of anthrax toxin assembly (see below), precluding its usefulness as an inhibitor. Owing to the formation of PA_{63} heptamers, the selected peptide probably has multiple binding sites. We, therefore, decided to synthesize a derivative of polyacrylamide that had multiple, covalently linked copies of the peptide. To this end, we used a method devised by George M. Whitesides and colleagues (**Fig. 4**). This coupling method involves the reaction of a peptide amino group with an activated ester of the polyacrylamide-based backbone, poly-*N*-acryloyloxy succinimide (PNAS). To provide a reactive amino group, the peptide to couple should, therefore, be ordered with an additional C-terminal lysine (*see* **Note 5**).

1. Mix a solution of PNAS (20 mg, 118 μmol) in 1.5 mL of *N,N*-dimethyl formamide (DMF) with triethylamine (50 μL, 360 μmol) and a solution of the peptide (9.46 mg, 5.9 μmol) in DMF.
2. Stir the mixture overnight and quench the reaction by adding aqueous ammonium hydroxide (500 μL, 30% NH_3 by weight) (*see* **Note 6**). This will produce the polyvalent inhibitor (PVI). A control backbone polymer should also be synthesized by performing the same reaction without peptide.

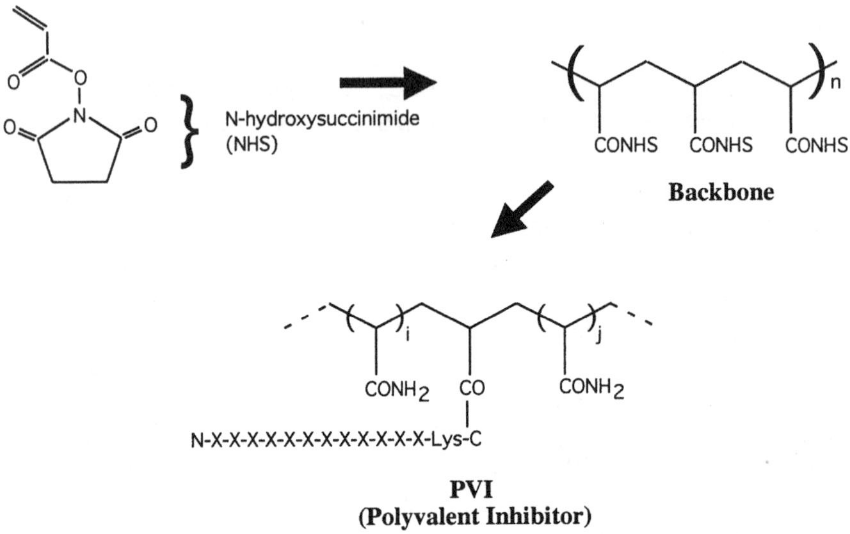

Fig. 4. Schematic description of the chemical synthesis of the polyvalent inhibitor (PVI). The backbone, poly-*N*-acryloyl succinimide (PNAS), is made by radical catalyzed polymerization of the *N*-hydroxysuccinimide (NHS) ester of acrylic acid, itself obtained by the reaction of acryloyl chloride with NHS. The NHS-activated esters of the backbone can randomly react with the ε amino group of a C-terminal lysine of the peptide (other residues are marked by X). Unreacted esters are then quenched by ammonia. The resulting molecule is the PVI.

3. Purify the PVI and backbone by exhaustive dialysis against distilled deionized water using a membrane with a 10-kDa molecular weight cut-off. Lyophilize the dialyzed molecules in a microfuge tube using a speedvac. Calculate the amount of the synthesized molecules by comparing the weight of the microfuge tube prior to and after lyophilization.
4. Resuspend the PVI and backbone in water at an identical concentration, typically 2–5 mg/mL.
5. Determine the concentration of peptide in PVI. Measure the absorbance of the polymer solution at 280 nm and compare it with the absorbance of a range of dilutions of monomeric peptide at known concentrations. The control polymer with no peptide should not have any absorption at this wavelength (*see* **Note 7**).

In our experiment, the peptide concentration of a 2.5 mg/mL PVI solution was determined to be 1.5 m*M*. The size of the polyacrylamide backbone was determined by gel filtration chromatography and shown to have an average molecular weight of 96.5 kDa. The ratio of peptide to acrylamide monomers was determined by NMR to be 1 to 40. Hence, our PVI is, on average, a molecule with 900 acrylamide monomers and 22 randomly grafted peptides.

3.3. Characterization of the Polyvalent Inhibitor

The biological activity of the peptide, backbone, or PVI is initially assayed by measuring the inhibition of the binding of radiolabeled LFn to PA_{63} heptamer on the surface of mammalian cells. As a further test of PVI activity, we look at its effect on the toxicity toward CHO cells of a mixture of PA and LFnDTA. LFnDTA, a fusion of the diphtheria toxin A chain (DTA) to the C terminus of LFn, binds to PA_{63} heptamers and enters cells by the same pathway as EF/LF. The DTA moiety then catalyzes ADP-ribosylation of elongation factor-2 within the cytosol and causes an inhibition of protein synthesis. We use this assay because the effect of DTA can be easily and precisely measured in many different cell types.

3.3.1. Inhibition of the Assembly of the Toxin

1. The day before the assay, seed CHO cells in a 24-well plate to confluence (200,000 cells per well). We usually perform the assay in duplicate and test the efficacy of peptide, underivatized backbone, and PVI. Seed two additional control wells, one that will contain no PA, to correct for background binding of LFn in the absence of PA, and one that will contain no inhibitor, to determine the maximum level of LFn binding.
2. On the day of the assay, chill the cells on ice for 10 min and wash once with 500 µL of cold PBS. Add 500 µL of purified "nicked PA" diluted in HAM medium buffered with 20 m*M* HEPES, pH 7.4 to a concentration of 2 µg/mL and incubate for 1 h. "Nicked PA" is purified PA that has been cleaved by trypsin, as described previously *(8)*. In the first control well add only HAM–HEPES buffer without PA.
3. Discard the PA solution and wash the cells once with 500 µL cold PBS. Add 250 µL of radiolabeled LFn diluted in HAM–HEPES buffer. Add 50 µL of various concentrations of peptide, PVI, or backbone, except in the second control well, where no inhibitor should be added. The concentration of PVI must be measured in concentration of linked peptide in order to make a direct comparison with free peptide significant. Use an initial dilution of the molecule so that the final peptide concentration is 10^{-6} *M* for the PVI and 10^{-3} *M* for the free peptide and make 10-fold dilutions of the molecules in water. Incubate on ice for 1 h.
4. Wash three times with 500 µL of cold PBS and resuspend the cells in 200 µL of lysis buffer. Count the radioactivity recovered in the cell lysate, corresponding to the amount of labeled LFn bound on cells. Subtract background binding of LFn independent of PA addition, which should be around 5–10 % of the amount bound in the presence of PA. **Figure 5A** shows a typical result of this experiment. This proves that the selected peptide is able to compete with LFn for the binding to PA_{63} heptamers on cells and that polyvalent display of this peptide greatly increases its biological efficacy, up to 7500-fold.

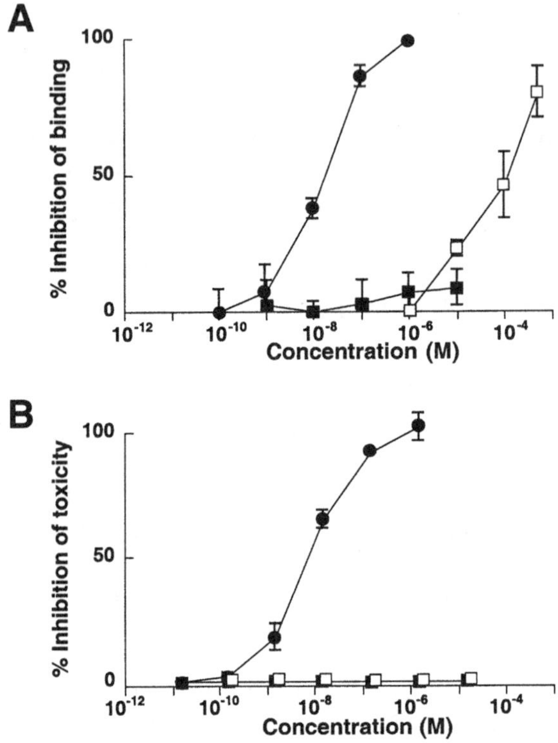

Fig. 5. Inhibition of toxin action in cell culture. The effects of various amounts of PVI (●), backbone (■), or monomeric peptide (□) were tested. The concentration of PVI is given as concentration of linked peptide, not the concentration of the whole PVI molecule. (**A**) Inhibition of toxin association. The association of toxin is measured as the binding of ^{35}S-labeled LFn on CHO cells incubated with PA. (**B**) Inhibition of toxin cytotoxicity. The cytotoxicity is assayed by measuring the ^{3}H-leucine uptake in cells incubated with PA and LFnDTA.

3.3.2. Inhibition of Toxicity

1. The day before the assay, seed CHO cells in a 96-well plate to confluence (20,000 cells per well). We usually perform the assay in duplicate and test the efficacy of peptide, backbone, and PVI. Prepare two wells for a control without toxin and two wells for a control without inhibitors.
2. Dilute purified PA and purified LFnDTA in cell culture medium to a final concentration of 10^{-9} *M* PA and 2×10^{-11} *M* LFnDTA. Prepare eight 10-fold dilutions of peptide, PVI, or backbone starting with a concentration of peptide of 10^{-5} *M*. Discard the old medium and add 100 µL of the PA–LFnDTA mix to each well along with 10 µL of the dilutions of each inhibitory molecule to be tested. In the control wells add either medium alone (no toxin control) or only the medium

containing PA and LFnDTA (no inhibitor control). Incubate for 4 h at 37°C in a cell incubator.

3. Discard the medium and add 100 μL of fresh leucine-free cell culture medium supplemented with 1 μCi/mL of tritiated leucine. Incubate for 1 h at 37°C in a cell incubator.
4. Discard the medium. Add 200 μL of cold 5% TCA to each well for 10 min, lysing cells and precipitating proteins. Repeat the precipitation step once.
5. Neutralize the TCA by adding 100 μL 0.1 *N* KOH per well. Incubate at room temperature for 10 min. Add 100 μL of HCl 0.2 *N* per well and mix thoroughly. Count the amount of radioactivity in each well, corresponding to the amount of leucine incorporated in newly synthesized proteins. Subtract the amount of radioactivity incorporated in the "no inhibitor" control wells; it should be 1–5 % of the amount incorporated in the "no toxin" control wells. **Figure 5B** shows the result of this experiment and proves that PVI is a potent inhibitor. This also confirms the results of the inhibition of LFn binding and the efficacy of polyvalent display.

4. Notes

1. There are other theoretical ways to do the selection we performed. One alternative is to bind the phages on monomeric PA, keep the unbound phages, and then bind them on PA_{63} heptamers. Another option is to bind the phages on PA_{63} heptamers and then elute by adding an excess of ligand, LFn for instance, in order to displace the phages specifically bound on the EF/LF binding site. These approaches did not work in our hands. Phages selected by the first method still bound to monomeric PA, and those selected by the second method bind LFn. Combinations of all those methods should still be considered when applying this approach to other systems.
2. There are two other phage libraries available from New England Biolabs, one displaying seven-amino-acid peptides and one displaying seven-amino-acid peptides constrained by a disulfide bridge made by two cysteines located before and after the randomized peptide. The seven-amino-acid libraries have the advantage of being "complete," meaning that statistically all seven-amino-acid sequences are represented in the library, whereas the 12-amino-acid library is not. Other random libraries can be obtained from laboratories working with phage display. A new library can also be constructed by random peptide sequences from one natural protein partner of the target. Finally, other phage display systems exist (display on pVIII protein, other filamentous phages, other phages like lambda or T7), and other display methods (bacterial surface display or ribosomal display).
3. Take great care when handling and discarding phage-contaminated material. Avoid spills and use filter tips. When washing tubes, drain excess liquid with a paper towel and use a new towel each time. Two dangers exist. First, phages can contaminate your lab cultures of *E. coli*. Second, if multiple parallel selections are performed, cross contamination of phages can occur between the tubes of the different selections.

4. High titers of filamentous phages such as a solution of amplified eluate can also be approximated by measuring the absorption spectra of the phages solution between 240 and 320 nm and using the following formula:

$$\text{Titer in pfu/mL} = \frac{\left(A_{269} - A_{320}\right) * 6 \times 10^{16}}{\text{number of nucleotide bases per phage chromosome}}$$

A full protocol and more phage display related material can be obtained from the Internet site of George P. Smith (http://www.biosci.missouri.edu/smithgp/).

5. The peptide we used did not have a primary amino group. If a selected peptide has an amino acid that can react with PNAS, then alternate chemistry must be considered or a peptide having another residue in place of the reactive one must be ordered. Another thing to consider when ordering the peptide is that the solubility of the free peptide and/or the PVI in water can be poor. Introducing a charged residue can sometimes relieve that problem.
6. The peptide to backbone ratio is the crucial variable in the reaction. Different ratios have to be tried to generate different types of products and the activity of the resulting molecules has to be tested. A mix of different peptides can also be used instead of just one. Finally, before quenching with ammonia the backbone can be modified by different chemicals, ethanolamine, for example, in order to change its hydrophobicity. This might become interesting when trying to enhance the solubility of the polyvalent molecule in water.
7. If there is access to a chemistry laboratory, the main characteristics of the PVI should be determined in order to have an idea of the structure of the resulting molecule. This mainly consists of determining the average length of the backbone and the average number of peptides coupled to the backbone. The average size of the polyacrylamide backbone can be evaluated using any gel filtration chromatography or light scattering. Determining the ratio of peptide to backbone is typically done by NMR, comparing the integration of the aromatic peaks to the integration for the hydrogen of the carbon α to the carbonyl group.

Acknowledgments

We thank D. Borden Lacy for helpful comments and critical reading of the manuscript. PNAS and PVI synthesis was performed by Ravi S. Kane, Steve Metallo, and Pascal Deschatelets in the laboratory of George M. Whitesides. One of us (R. J. C.) holds equity in AVANT Immunotherapeutics and PharmAthene, Inc.

References

1. Kay, B. K., Winter, J., and McCafferty, J., eds. (1996) *Phage Display of Peptides and Proteins. A Laboratory Manual.* Academic Press, San Diego, CA.
2. Zwick, M. B., Shen, J., and Scott, J. K. (1998) Phage-displayed peptide libraries. *Curr. Opin. Biotechnol.* **9(4),** 427–436.

3. Wrighton, N. C., Farrell, F. X., Chang, R., et al. (1996) Small peptides as potent mimetics of the protein hormone erythropoietin. *Science* **273(5274),** 458–464.
4. Mammen, M., Choi, S.-K., and Whitesides, G. M. (1998) Polyvalent interactions in biological systems: implications for design and use of multivalent ligands and inhibitors. *Angew. Chem. Int. Ed.* **37,** 2754–2794.
5. Kitov, P. I., Sadowska, J. M., Mulvey, G., et al. (2000) Shiga-like toxins are neutralized by tailored multivalent carbohydrate ligands. *Nature* **403(6770),** 669–672.
6. Mourez, M., Kane, R. S., Mogridge, J., et al. (2001) Designing a polyvalent inhibitor of anthrax toxin. *Nat. Biotechnol.* **19(10),** 958–961.
7. Brossier, F. and Mock, M. (2001) Toxins of *Bacillus anthracis*. *Toxicon* **39(11),** 1747–1755.
8. Miller, C. J., Elliott, J. L., and Collier, R. J. (1999) Anthrax protective antigen: prepore-to-pore conversion. *Biochemistry* **38(32),** 10,432–10,441.
9. Mammen, M., Dahmann, G., and Whitesides, G. M. (1995) Effective Inhibitors of hemagglutination by influenza virus synthesized from polymers having active ester groups. Insight into mechanism of inhibition. *J. Med. Chem.* **38,** 4179–4190.
10. Wesche, J., Elliott, J. L., Falnes, P. O., Olsnes, S., and Collier, R. J. (1998) Characterization of membrane translocation by anthrax protective antigen. *Biochemistry* **37(45),** 15,737–15,746.
11. Milne, J. C., Blanke, S. R., Hanna, P. C., and Collier, R. J. (1995) Protective antigen-binding domain of anthrax lethal factor mediates translocation of a heterologous protein fused to its amino- or carboxy-terminus. *Mol. Microbiol.* **15(4),** 661–666.

III

Detecting Protein–Protein Interactions in Heterologous Systems

17

A Bacterial Two-Hybrid System Based on Transcription Activation

Simon L. Dove and Ann Hochschild

Abstract

We describe the use of a bacterial two-hybrid system for the study of protein–protein interactions in *Escherichia coli*. This system is based on transcription activation and involves the synthesis of two fusion proteins within the bacterial cell whose interaction stimulates transcription of a reporter gene. Specifically, one of the fusion proteins can function as a transcription activator when its interaction partner is fused to a subunit of the bacterial RNA polymerase. This bacterial two-hybrid system has been used to study a number of interacting proteins from both prokaryotes and eukaryotes, and can be used to find interacting proteins from complex protein libraries.

Key Words

Bacterial two-hybrid system; protein–protein interactions; RNA polymerase; λcI.

1. Introduction

Two-hybrid systems are genetic approaches for detecting protein–protein interactions in vivo. Recently, a number of bacterial counterparts of the powerful yeast two-hybrid system ***(1,2)*** have been developed (reviewed in refs. ***3–6***), and this chapter will focus on one of these. In particular, we will describe the use of a transcription activation–based two-hybrid system for the study of protein–protein interactions in *Escherichia coli*. This bacterial system has been used to detect and analyze the interactions between a number of different proteins from both prokaryotes and eukaryotes ***(7–15)*** and has been adapted for the study of both phosphorylation-dependent ***(16)*** and small-molecule mediated ***(17)*** protein–protein interactions.

From: *Methods in Molecular Biology, vol. 261: Protein–Protein Interactions: Methods and Protocols*
Edited by: H. Fu © Humana Press Inc., Totowa, NJ

The bacterial two-hybrid system described here is based on the finding that any sufficiently strong interaction between two proteins can activate transcription in *E. coli* provided one of the interacting proteins is tethered to the DNA by a DNA-binding domain and the other is tethered to a subunit of the bacterial RNA polymerase (RNAP) *(7,8)*. To use this system, one of the protein domains to be tested (the bait) is fused to a sequence-specific DNA-binding protein, and the other protein under investigation (the prey) is fused to a subunit of the bacterial RNAP. Typically, the DNA-binding protein we use is the cI protein from bacteriophage λ (λcI) (*see* **Note 1**), whereas the bacterial RNAP subunit we use is α (*see* **Note 2**). Alternative DNA-binding proteins and RNAP subunits can also be employed *(8,18)* (*see* **Note 3**). Compatible plasmids directing the synthesis of the λcI and α fusion proteins are introduced into a suitable strain of *E. coli*, which contains a test promoter that drives the expression of a linked reporter gene. The test promoter used in this system contains a binding site for λcI (O_R2) in its upstream regulatory region (*see* **Fig. 1**). When induced, expression of the λcI fusion gene results in the binding of the corresponding λcI fusion protein to O_R2. Expression of the α fusion gene leads to the assembly of the resulting α fusion protein into RNAP. Interaction between the DNA-bound λcI fusion protein and the assembled α fusion protein stabilizes the binding of RNAP to the test promoter, thus activating transcription of the reporter genes (*see* **Fig. 1**). The strength of the protein–protein interaction between the two fusion proteins (the artificial activator and its artificial activation target) determines, in part, the magnitude of reporter gene activation *(7)*.

To facilitate the use of this bacterial two-hybrid system, two reporter genes are used: *bla*, which encodes β-lactamase, and *lacZ*, which encodes β-galactosidase. In cells containing this twin reporter construct, transcription initiating from the test promoter results in co-expression of the *bla* and *lacZ* genes (*see* **Fig. 1**). The *bla* gene confers resistance to carbenicillin, and *lacZ* gene expression can be measured either quantitatively, using a liquid β-galactosidase assay, or qualitatively by examining colony color on indicator medium containing the chromogenic substrate X-Gal (5-bromo-4-chloro-3-indolyl-β-D-thiogalactopyranoside). The use of selectable reporter genes, like *bla*, with this system should facilitate the selection of interacting proteins from complex protein libraries *(16,18)*.

2. Materials

2.1. Making Fusions to λcI

1. Plasmids pBT and pACλcI32 can be used for making fusions to the C-terminal end of λcI (*see* **Table 1**).

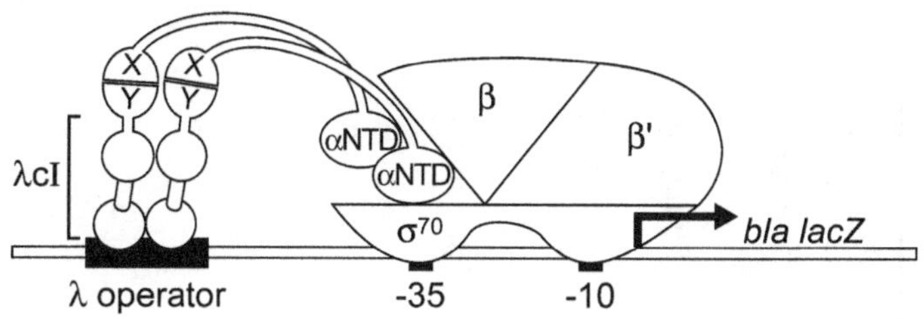

Fig. 1. Principle of bacterial two-hybrid system. Contact between protein domains X and Y fused, respectively, to the α-NTD and to λcI activates transcription from the test promoter. β, β', and σ^{70} are subunits of *E. coli* RNAP. The illustrated test promoter, p*lac*O_{R2}-62, contains a λ operator (O_R2) positioned 62 bp upstream from the transcription start site of the *lac* core promoter. The –10 and –35 elements that constitute the *lac* core promoter, and to which RNAP binds, are depicted as small black boxes. In reporter strain US3F'3.1 and the BacterioMatch reporter strain, the test promoter drives the expression of both the *bla* and *lacZ* reporter genes, as indicated.

2. Competent cells (*see* **Note 4** about making chemically competent cells): Strains XL1-Blue, XL1-Blue MRF' Kan, and JM109 can be used for propagating plasmids pBT and pACλcI32 (*see* **Table 2**).
3. Antibiotic stock solutions: 100 mg/mL carbenicillin, in 50% ethanol; 25 mg/mL chloramphenicol, in methanol; 50 mg/mL kanamycin, in distilled water, filter sterilized; 10 mg/mL tetracycline, in methanol. Antibiotic stock solutions should be stored at –20°C.
4. Luria–Bertani (LB)-agar plates (containing appropriate antibiotics): 10 g bacto tryptone, 5 g yeast extract, 10 g NaCl, and 15 g bacto agar in 1 L distilled water. Autoclave LB-agar to sterilize. Antibiotics should be added once the medium has cooled to below 50°C. LB-agar plates should be stored at 4°C.
5. LB: 10 g bacto tryptone, 5 g yeast extract, and 10 g NaCl in 1 L of distilled water. Autoclave to sterilize. Store at room temperature.

2.2. Making Fusions to the α *Subunit of* E. coli *RNA Polymerase*

1. Plasmids pTRG, pBRSTAR, and pBRαLN can be used for making fusions to the α subunit of *E. coli* RNAP (*see* **Table 1**).
2. Competent cells (*see* **Note 4** about making chemically competent cells): Strains XL1-Blue MRF' Kan and JM109 can be used for manipulating plasmids pTRG and pBRSTAR, whereas strains XL1-Blue, XL1-Blue MRF' Kan, and JM109 can be used to manipulate plasmid pBRαLN (*see* **Table 2**).
3. Antibiotic stock solutions (*see* **Subheading 2.1., item 3**).
4. LB-agar plates containing appropriate antibiotics (*see* **Subheading 2.1., item 4**).
5. LB (*see* **Subheading 2.1., item 5**).

Table 1
Plasmids for Use With Bacterial Two-Hybrid System

Plasmid	Description	Antibiotic Resistance	Useful cloning sites	Source/Ref
pBT	Bait plasmid, encodes λcI	Chloramphenicol (25 μg/mL)	*Not*I, *Eco*RI, *Sma*I, *Bam*HI, *Xho*I, *Bgl*II	Stratagene
pBT-LGF2	Positive control plasmid, encodes λcI-Gal4 fusion	Chloramphenicol (25 μg/mL)		Stratagene
pACλcI32	Bait plasmid, encodes λcI	Chloramphenicol (25 μg/mL)	*Not*I, *Bgl*II (*Bst*YI), *Asc*I, *Bst*YI	***3***
pTRG	Prey plasmid, encodes α-NTD	Tetracycline (10 μg/mL)	*Bam*HI, *Not*I, *Eco*RI, *Xho*I, *Spe*I	Stratagene
pTRG-Gal11P	Positive control plasmid, encodes α-Gal11P fusion	Tetracycline (10 μg/mL)		Stratagene
pBRSTAR	Prey plasmid, encodes α-NTD	Tetracycline (10 μg/mL)	*Not*I, *Bam*HI	***16***
pBRαLN	Prey plasmid, encodes α-NTD	Carbenicillin (50 μg/mL)	*Not*I, *Bam*HI	***3***
pBRα-Gal11P	Positive control plasmid, encodes α-Gal11P fusion	Carbenicillin (50 μg/mL)		***8***

Table 2
Bacterial Strains for Use With Bacterial Two-Hybrid System

Strain	Relevant genotype	Antibiotic resistance	Reporter genes	Source/Ref
KS1	F' *lacI^q^*	Kanamycin (50 μg/mL)	*lacZ*	***7***
BacterioMatch Reporter[a]	*recA*, F' *lacI^q^*	Kanamycin (50 μg/mL)	*bla*, *lacZ*	Stratagene
US3F'3.1	*recA*, F' *lacI^q^*	Kanamycin (50 μg/mL)	*bla*, *lacZ*	***16***
XL1-Blue[b]	*recA*, F' *lacI^q^*	Tetracycline (10 μg/mL)		Stratagene
XL1-Blue MRF' Kan[b]	*recA*, F' *lacI^q^*	Kanamycin (50 μg/mL)		Stratagene
JM109[b]	*recA*, F' *lacI^q^*			***24***

[a] Higher transformation efficiencies can be achieved with this strain than can be achieved with strain US3F'3.1.
[b] Strains used for propagation of plasmids used with the bacterial two-hybrid system.

2.3. Detecting Protein–Protein Interactions

2.3.1. Rapid Protocol for Transformation of Reporter Strains with Bait and Prey Plasmids

1. Competent cells of reporter strain KS1, or the BacterioMatch reporter strain (*see* **Note 4**).
2. LB-agar plates containing appropriate antibiotics (*see* **Subheading 2.1., item 4**).
3. Antibiotic stock solutions (*see* **Subheading 2.1., item 3**).

2.3.2. Quantifying Reporter Gene Expression in Liquid Cultures: The β-Galactosidase Assay

1. LB (*see* **Subheading 2.1., item 5**).
2. Antibiotic stock solutions (*see* **Subheading 2.1.**, item 3).
3. IPTG stock solution (100 m*M*): Dissolve 0.238 g isopropyl-β-D-thiogalactoside (IPTG) in 10 mL distilled water, filter sterilize, aliquot, and store at –20°C.
4. Z-buffer: 16.1 g $Na_2HPO_47H\ _2O$, 5.5 g $NaH_2PO_4H_2O0.75$ g KCl, 0.246 g $MgSO_47H\ _2O$, 2.7 mL β-mercaptoethanol, and distilled water to 1 L. Adjust pH to 7.0. Do not autoclave.
5. 0.1% SDS.
6. Chloroform.
7. ONPG solution: O-nitrophenyl-β-D-galactoside (ONPG; 4 mg/mL) in Z-buffer. Store at –20°C.
8. 1 *M* Na_2CO_3.

2.4. Screening Libraries in E. coli for Protein Interaction Partners

1. Antibiotic stock solutions (*see* **Subheading 2.1., item 3**).
2. X-Gal stock solution: 10 mg/mL X-Gal in di-methylformamide.
3. 150 mm LB-agar plates containing 50 μg/mL kanamycin, 10 μg/mL tetracycline, 25 μg/mL chloramphenicol, 20 μ*M* IPTG, 500–2000 μg/mL carbenicillin.
4. 100 mm LB-agar plates containing 50 μg/mL kanamycin, 10 μg/mL tetracycline, 25 μg/mL chloramphenicol, 20 μ*M* IPTG, 500–2000 μg/mL carbenicillin.
5. LB (*see* **Subheading 2.1., item 5**).
6. IPTG (*see* **Subheading 2.3.2., item 3**).

3. Methods

3.1. Making Fusions to λcI

Plasmids pBT and pACλcI32 can be used for making fusions to the C-terminus of λcI. These related plasmids are derived from the low-copy-number vector pACYC184 and confer resistance to chloramphenicol (25 μg/mL). Expression of the λcI fusion gene in both of these plasmids is under the control of the *lac*UV5 promoter and can be induced with IPTG. In the absence of IPTG, expression of the λcI fusion gene can be repressed by the Lac repressor (encoded by the *lac*I gene). Plasmids pBT and pACλcI32 differ from one another with respect to the multiple cloning sites that have been introduced at the end of the *cI* gene (to facilitate the fusion of bait proteins to the C-terminus of λcI; *see* **Table 1**). The bait (a protein of interest or domain thereof) can be fused to the C-terminus of λcI by cloning a suitably designed PCR product into pBT or pACλcI32 such that the gene encoding the bait is in-frame with the *cI* coding region (*see* **Note 5**). Plasmids pBT and pACλcI32 contain the *lac*UV5 promoter and therefore should be propagated only in strains of *E. coli* that contain *lacI*^{q} such as XL1-Blue, XL1-Blue MRF' Kan, and JM109 (*see* **Note 6**). Because of the low copy number of these plasmids, larger culture volumes should be used when isolating plasmid DNA. Bait plasmids should be constructed using standard molecular biology procedures and the materials listed under **Subheading 2.1.**

3.2. Making Fusions to the α *Subunit of* E. coli *RNA Polymerase*

Plasmids pTRG, pBRSTAR, and pBRαLN can be used for making fusions to the N-terminal domain of the α subunit of *E. coli* RNAP (α-NTD). Fusions are actually made to the C-terminus of the α linker; a long and flexible linker that normally connects the N- and C-terminal domains of α (*19,20*; *see* **Note 2**). Plasmids pTRG, pBRSTAR, and pBRαLN are derived from the low–medium-copy-number plasmid pBR322 and harbor a truncated version of the *rpoA* gene (encoding amino acids 1–248 of the α subunit of RNAP, corresponding to the α-NTD and linker) under the control of tandem *lpp* (a strong constitutive promoter) and *lac*UV5 promoters. Plasmids pTRG and pBRSTAR confer resistance to tetracycline (10 μg/mL), whereas plasmid pBRαLN confers resistance to carbenicillin (or ampicillin) (50–100 μg/mL). Plasmids pTRG and pBRSTAR differ from one another with respect to the multiple cloning sites that have been introduced at the end of the truncated *rpoA* gene to facilitate the fusion of prey proteins to the end of the α linker (*see* **Table 1**). The prey can be fused to the C-terminus of the α linker by cloning a suitably designed PCR product into these plasmids such that the gene encoding the prey is in-frame with the *rpoA* coding region (*see* **Note 5**). Because plasmids pTRG, pBRSTAR, and pBRαLN contain the *lac*UV5 promoter they should be propagated only in strains of *E. coli* that contain *lacI*^{q}. Plasmid pBRαLN can be propagated in strains XL1-Blue, XL1-Blue MRF' Kan, and JM109. Plasmids pTRG and pBRSTAR can be propagated in strains XL1-Blue MRF' Kan and JM109 (but not XL1-Blue because this strain already carries a resistance determinant for tetracycline on an F' episome). Prey plasmids should be constructed using standard molecular biology procedures and the materials listed in **Subheading 2.2.**

3.3. Detecting Protein–Protein Interactions

In order to test whether two proteins can interact with one another, the bait and prey plasmids directing the synthesis of the desired λcI fusion protein and the desired α fusion protein, respectively, are transformed together into a suitable reporter strain of *E. coli* along with the appropriate controls. Transformants are then assayed for β-galactosidase activity. Two proteins are said to interact with one another when the bait stimulates production of β-galactosidase only in the presence of the prey. A suitable positive control is provided by plasmids that direct the synthesis of the λcI-Gal4 and α-Gal11P fusion proteins (*see* **Table 1**). The λcI-Gal4 and α-Gal11P fusion proteins have been shown previously to interact with one another in *E. coli* and strongly stimulate production of β-galactosidase in an appropriate reporter strain *(8)*. Suitable negative controls comprise the bait plasmid transformed together with the parent vector of the prey plasmid (i.e., pTRG, pBRSTAR, or pBRαLN; *see* **Note 7**), and the prey plasmid transformed together with the parent vector of the bait plasmid (i.e., pBT or pACλcI32). Alternative negative controls comprise the bait plasmid transformed together with a plasmid that directs the synthesis of the α-Gal11P fusion protein (i.e., pTRG-Gal11P or pBRα-Gal11P; *see* **Note 7**), and the prey plasmid transformed together with a plasmid directing the synthesis of the λcI-Gal4 fusion protein (i.e., pBT-LGF2).

Several different reporter strains have been designed for use with this bacterial two-hybrid system (*see* **Table 2**). The authors routinely use reporter strain KS1, which is a derivative of *E. coli* strain MC1000 that harbors an F' episome that confers resistance to kanamycin and carries *lacI^q^* *(7)*. KS1 also harbors an imm 21 prophage on the chromosome that bears a *lacZ* reporter gene linked to the test promoter p*lac*O_R2-62 *(7)*. This test promoter (illustrated in **Fig. 1**) consists of the *lac* core promoter together with a λ operator (O_R2) positioned 62 bp upstream from the *lac* transcription start site. Reporter strain US3F'3.1 and the BacterioMatch reporter strain can also be used. These reporter strains contain the same F' episome carrying both *bla* and *lacZ* reporter genes under the control of the p*lac*O_R2-62 test promoter (*see* **Fig. 1**). In these strains the F' episome also bears *lacI*^{q} and a kanamycin resistance cassette. The US3F'3.1 and BacterioMatch reporter strains have lower apparent basal or unstimulated levels of l*acZ* expression (approx 15 Miller Units of β-galactosidase activity) compared to that of reporter strain KS1 (approx 60 Miller Units of β-galactosidase activity). Furthermore, it should be noted that higher transformation efficiencies can be achieved with the BacterioMatch reporter strain than with strain US3F'3.1, presumably as a result of differences in the genotypes of these two strains.

3.3.1. Rapid Protocol for Transformation of Reporter Strains with Bait and Prey Plasmids

1. Thaw chemically competent reporter strain cells on ice (*see* **Note 4** about making chemically competent cells).

2. Add approx 10 ng each of the bait and prey plasmids or control plasmids to a sterile 1.5 mL microcentrifuge tube and place on ice for 5 min.
3. Add chemically competent reporter strain cells (25 μL) directly to the DNA in each tube.
4. Incubate tubes on ice for 10 min, heat shock at 42°C for 2 min, and then incubate on ice for a further 2 min.
5. Spread each transformation mix directly onto an LB-agar plate containing the appropriate antibiotics (*see* **Note 8**).
6. Incubate plates at 37°C overnight.

This protocol does not include a step that allows for phenotypic outgrowth of antibiotic resistance determinants and so does not result in particularly high transformation efficiencies. For a transformation protocol that results in higher transformation efficiencies *see* **Note 9**.

3.3.2. Quantifying Reporter Gene Expression in Liquid Cultures: The β-Galactosidase Assay

This protocol is derived from that originally described by Miller ***(21)***.

1. Individual colonies from the transformation plates (see **Subheading 3.3.1.**) are inoculated into 3 mL of LB containing the appropriate antibiotics (*see* **Note 8**) and IPTG at a concentration of between 0 and 200 μ*M* (*see* **Note 10**).
2. Incubate cultures overnight (approx 16 h) at 37°C with aeration.
3. Inoculate 30 μL of each overnight culture into 3 mL of LB supplemented with antibiotics and IPTG at the same concentration used for the overnight culture.
4. Incubate cultures at 37°C with aeration until cultures reach an OD_{600} of 0.3–0.7 (*see* **Note 11**).
5. Place cultures on ice for 20 min.
6. Transfer 1 mL of the bacterial culture from each tube to a cuvet and record the OD_{600}.
   At this point, an aliquot of the bacterial culture may also be taken for Western blot analysis if desired (*see* **Note 12**).
7. Transfer 200 μL of each culture to a small glass test tube (the assay tube) containing 800 μL of Z-buffer. This should be performed in duplicate so there will be two assay tubes for each culture. Duplicate assay tubes that will serve as the blank should be prepared by adding 200 μL of LB to 800 μL of Z-buffer.
8. Add 30 μL 0.1% SDS to each assay tube.
9. Add 60 μL chloroform to each assay tube (*see* **Note 13**).
10. Vortex each assay tube for 6 s.
11. Place tubes and thawed ONPG stock solution in a 28°C water bath and allow them to equilibrate to 28°C by incubation for 10 min.
12. Start the assay by adding 200 μL of 4 mg/mL ONPG (allowed to equilibrate to 28°C) to each assay tube and, using a timer, record the time at which each assay was started.
13. Mix the contents of the assay tubes by shaking or gentle vortexing.

14. Stop the reactions by adding 500 μL of 1 *M* Na_2CO_3 when the tubes become sufficiently yellow (*see* **Note 14**) and record the time at which each assay was stopped.
15. Vortex the tubes briefly at a low setting, and let sit at room temperature or 4°C until the remaining reactions are complete.
17. Transfer 1 mL from each assay tube to a cuvet (*see* **Note 15**) and determine the optical density at 420 nm and 550 nm for each assay using the control as the blank.
18. Determine β-galactosidase activity (in Miller Units) for each assay using the following equation:

$$\text{Units} = 1000 \times [OD_{420} - (1.75 \times OD_{550}) / t \times v \times OD_{600}]$$

Note that t is the time of the reaction in minutes, v is the volume of the culture used in the assay in mL (i.e., 0.2), and the OD_{600} is that determined for the culture used in each assay ***(21)***.

3.3.3. Interpreting Results of β-galactosidase Assays

An interaction between the bait and prey proteins results in at least a several-fold increase in β-galactosidase activity above that seen with the appropriate negative controls. It is important to note that a certain amount of β-galactosidase activity will be apparent for the negative controls, reflecting basal transcription from the test promoter that drives expression of the l*acZ* reporter gene (*see* refs. ***7*** and ***8*** for example).

3.4. Screening Libraries in E. coli for Protein Interaction Partners

The bacterial two-hybrid system described here can be used to identify interacting bait and prey proteins from complex libraries in which only a few of the combined bait and prey proteins interact ***(16)***. Crucially, interaction between the bait and prey proteins in reporter strain cells containing the *bla* gene linked to the activateable p*lac*O_R2-62 test promoter results in cells that are resistant to higher concentrations of carbenicillin.

The bait should be made by fusing the protein of interest (or domain thereof) to the C-terminus of the λcI protein as described under **Subheading 3.1.** Prior to screening a library for proteins that can interact with the bait, it is advisable to confirm that the λcI fusion protein to be used as the bait can bind to a λ operator. This can be done using reporter strain S11-LAM1 ***(22)***, which carries an F' episome harboring a λ operator positioned between the –10 and –35 elements of a relatively strong test promoter. The test promoter in this reporter strain drives expression of a linked *lacZ* reporter gene and is repressed when the λ operator is occupied (*see* **Note 16**). The degree to which the test promoter is repressed in strain S11-LAM1 is therefore a measure of how efficiently a particular protein can bind the λ operator. Any λcI fusion protein that binds the λ operator poorly, or not at all, is unsuitable as a bait.

Alternatively, an immunity test can be used to confirm that the λcI fusion protein can bind to a λ operator. Thus, cells containing plasmid pBT or pACλcI32 encoding the λcI fusion protein should be immune to superinfecting λcI⁻ phage, even when grown in the absence of IPTG. This can be assayed by performing a simple cross-streak test *(3)*.

Libraries of α fusion proteins (prey libraries) can be made by cloning genomic DNA, cDNA, or PCR products into plasmids pTRG or pBRSTAR. However, the construction of these libraries is beyond the scope of this chapter. Certain cDNA libraries made using pTRG are commercially available from Stratagene. What follows is one protocol for screening an α fusion library, made in pTRG or pBRSTAR, for proteins that interact with a predetermined bait (*see* **Note 17**).

3.4.1. Library Screening

1. Add 1 μg of bait plasmid to a sterile 1.5 mL microcentrifuge tube and incubate tube on ice for 5 min (*see* **Note 18**).
2. Add 100 μL of chemically competent BacterioMatch reporter strain cells containing the prey library (cloned into pTRG or pBRSTAR) (*see* **Note 19**).
3. Incubate on ice for 30 min. Heat shock for 2 min at 42°C, then return tubes to ice for at least 2 min.
4. Add 1 mL LB to the transformation mix and incubate at 37°C for at least 1 h.
5. Add IPTG to the transformation mix to a final concentration of 20 μ*M* and incubate at 37°C for 1 hour.
6. Spread sufficient transformation mix to plate approx 10^7 transformants onto 150 mm LB-agar plates containing 50 μg/mL kanamycin, 10 μg/mL tetracycline, 25 μg/mL chloramphenicol, 20 μ*M* IPTG (*see* **Note 10**), 500–2000 μg/mL carbenicillin (*see* **Note 20**), and X-Gal (*see* **Note 21**).
7. Incubate plates overnight at 37°C (*see* **Note 22**). Positive clones will form blue colonies on these selection plates.

3.4.2. Analysis of Putative Interaction Partners

False positives can appear as blue colonies on selection plates (containing carbenicillin), but, unlike true positives, the corresponding colonies do not contain plasmids with interacting fusion proteins. In order to eliminate one class of false positives, any putative positive clone can first be restreaked onto 100 mm LB-agar plates containing 50 μg/mL kanamycin, 10 μg/mL tetracycline, 25 μg/mL chloramphenicol, 20 μ*M* IPTG (or concentration used in original selection), carbenicillin (concentration used in original selection), and X-Gal. A true positive should again give rise to blue colonies on these plates. Any putative positive that passes this test should be inoculated into 5 mL LB containing 50 μg/mL kanamycin and 10 μg/mL tetracycline and grown overnight with aeration at the appropriate growth temperature (37°C or 30°C) (*see*

Note 22). Plasmid DNA is then isolated from the overnight cultures and transformed into *E. coli* strain XL1-Blue MRF' Kan or strain JM109 so that the prey plasmid of interest (containing the tetracycline resistance gene) can be isolated from the bait plasmid and purified (transformants should be plated on LB-agar plates containing 10 μg/mL tetracycline). Once the desired prey plasmid has been isolated, it can be transformed together with the original bait plasmid, alongside the appropriate controls (*see* **Subheading 3.3.**), into a suitable reporter strain. Whether the prey plasmid directs the synthesis of an α fusion protein that can interact with the bait can then be assessed quantitatively by performing β-galactosidase assays on liquid cultures (*see* **Subheading 3.3.2.**) or qualitatively by plating transformants onto selection plates containing carbenicillin and X-Gal (*see* **Subheading 3.4.1.**). Any plasmid that encodes a prey capable of interacting with the bait can then be sequenced to determine the identity of the interaction partner (*see* **Note 23**).

4. Notes

1. The cI protein from bacteriophage λ (λcI) is a sequence-specific DNA-binding protein that binds its recognition site as a dimer (**23**). Although λcI can function as both an activator and a repressor of transcription (**23**), in this bacterial two-hybrid system it serves simply as a DNA-binding protein.
2. The bacterial two-hybrid system detailed here exploits the domain structure of the α subunit of *E. coli* RNAP. α is a homodimer, each subunit of which consists of two independently folded domains separated by a long flexible linker ***(19,20)***. The α subunit of RNAP is an essential protein and so the reporter strains used in this bacterial two-hybrid system contain wild-type α (expressed from the native *rpoA* gene on the *E. coli* chromosome). Consequently, any cell expressing a particular α fusion protein (or prey) will contain a population of RNAP molecules that contain either 2, 1, or 0 copies of the α fusion protein.
3. Sequence-specific DNA-binding proteins other than λcI, such as a derivative of the monomeric zinc-finger protein Zif268, have also been used in this system as DNA-binding domains to which to fuse proteins of interest ***(18)***. Furthermore, α is not the only subunit of RNAP that can be used to display heterologous protein domains. We have previously shown that the monomeric ω subunit of *E. coli* RNAP can also be used ***(8)***.
4. A number of protocols for making chemically competent cells exist and one is described here. An isolated colony of the desired strain is inoculated into 5 mL LB containing the appropriate antibiotic. The culture is incubated at 37°C overnight with aeration, 0.5 mL of the overnight culture is used to inoculate 200 mL LB (containing the appropriate antibiotic where required) in a 1 L conical flask; 3 mL of sterile 1 *M* $MgCl_2$ are also added to the 200 mL culture. The culture is then incubated at 37°C with aeration until an OD_{600} of approx 0.5 is reached. Cells are transferred to a large sterile centrifuge bottle and pelleted by centrifugation at 4°C. The cell pellet is then resuspended in 60 mL of cold solution A and the suspension

is incubated on ice for 20 min. (Solution A is made by combining 10 mL 1 *M* $MnCl_2$, 50 mL 1 *M* $CaCl_2$, 200 mL 50 m*M* MES, pH 6.3, and 740 mL distilled water. The solution is then filter sterilized.) Cells are then pelleted by centrifugation at 4°C and resuspended in 12 mL cold solution A containing 15% glycerol. Aliquot competent cells in 0.5–1.0 mL volumes in sterile precooled 1.5 mL microcentrifuge tubes and freeze on dry ice. Competent cells can be stored at –80°C.

5. In plasmids pBT and pACλcI32 a *Not*I site has been introduced at the end of the cI gene such that the 8 bp *Not*I site, together with an additional base pair (i.e., GCGGCCGCA), adds three alanine residues to the end of the *cI* gene, thus providing a short linker with which to attach a bait protein ***(3)***. Useful cloning sites at the end of the *cI* gene in plasmids pBT and pACλcI32 are listed in **Table 1**. In plasmids pTRG, pBRSTAR, and pBRαLN, a *Not*I site has been introduced into the *rpoA* gene after codon 248 such that the *Not*I site (together with an additional base pair) similarly adds three alanine residues to the end of the α linker. The *Not*I restriction sites in the bait and prey plasmids provide a convenient means to clone PCR products encoding the bait and prey, respectively. Useful cloning sites at the end of the truncated *rpoA* gene in plasmids pTRG, pBRSTAR, and pBRαLN are also listed in **Table 1**. Note that PCR products to be cloned into pACλcI32, pBRαLN, or pBRSTAR typically contain a *Not*I site at one end and a *Bam*HI site preceded by an in-frame stop codon at the other.
6. The *lacI^q^* allele is required to provide sufficiently high quantities of Lac repressor to keep the *lac*UV5 promoter efficiently repressed. In the absence of *lacI^q^* the strong *lac*UV5 promoter will be close to being fully derepressed and, as a result, undesirable mutations within the expression vectors could be selected.
7. Note that plasmids pBRαLN and pBRα-Gal11P should not be used in strain US3F'3.1 or the BacterioMatch reporter strain because they confer resistance to carbenicillin.
8. Antibiotics that select for the bait and prey plasmids should be used in conjunction with 50 μg/mL kanamycin to select for maintenance of the F' in each of the reporter strains.
9. Add approx 10 ng each of the bait and prey plasmids to a sterile 1.5 mL microcentrifuge tube and allow tube to cool on ice for 5 min. Add chemically competent reporter strain cells (100 μL) directly to the DNA in each tube. Incubate tubes on ice for 30 min, heat shock at 42°C for 2 min, and then incubate on ice for a further 2–3 min. Add 1 mL LB (do not use SOC) to each tube and incubate at 37°C for 1 h. Pellet cells by centrifugation in a microcentrifuge. Pour off most of the supernatant and resuspend the cell pellet in the remainder. Spread each transformation mix directly onto an LB-agar plate containing the appropriate antibiotics (*see* **Note 8**) and incubate plates at 37°C overnight.
10. To begin with, an IPTG concentration of 20 μ*M* is recommended; however, the optimal IPTG concentration (0–200 μ*M*) should be determined empirically.
11. Typically, cultures take between 1.5 and 4 h to reach mid-logarithmic phase; the exact time varies depending on the particular bait and target proteins under investigation and the concentration of IPTG being used to induce their synthesis.

12. An antibody against λcI is available commercially from Invitrogen, and can be used to determine the intracellular levels of λcI fusion proteins.
13. In order to pipet chloroform accurately, first pipet the chloroform up and down several times before adding it to the first tube, and do not change the tip.
14. The OD_{420} (the yellow color) of the stopped reaction should be between 0.6 and 0.9 (*see* ref. ***21***). Each reaction should be stopped when approximately the same intensity of yellow is reached, so that the reaction time is the main variable.
15. Care should be taken not to transfer any of the chloroform residing at the bottom of the assay tube to the cuvet as this will interfere with the OD readings.
16. Occupancy of the λ operator in reporter strain S11-LAM1 by λcI, or a λcI fusion protein, prevents RNAP from binding the test promoter, resulting in repression *(22)*. Plasmids pBT, pBT-LGF2, and pACλcI32 are suitable positive controls for use in this strain, whereas plasmid pACΔcI *(7)* is a suitable negative control for use in this strain. Note that λcI fusion proteins often do not bind the λ operator as efficiently as wild-type λcI; nevertheless, as long as they can repress transcription from this test promoter, they can be used in the two-hybrid assay.
17. It may be preferable to transform the prey library into competent cells of the reporter strain containing the bait plasmid. When making such competent cells, IPTG can be included in the growth medium so that synthesis of the λcI fusion protein is induced. This may improve the chance of finding an interaction partner.
18. The number of potential false positives that might occur as a result of mutations in the BacterioMatch reporter strain, for example, can be determined by performing a control transformation with 1 μg of the bait parent vector (pTRG or pACλcI32).
19. Chemically competent cells harboring the prey library should ideally have a transformation efficiency of at least 1×10^7/μg of DNA.
20. For an initial screen a concentration of 1000 μg/mL carbencillin is suggested. However, ideally a range of carbenicillin concentrations (500–2000 μg/mL) should be tested. Carbenicillin concentrations as low as 250 μg/mL can be used provided the number of transformants that are spread on a plate are limited to approx 10^5. It is also important to note that ampicillin should not be used instead of carbenicillin.
21. X-Gal can be added to the 150 mm LB-agar plate by first adding 200 μL of LB to the center of the plate followed by 160 μL of 10 mg/mL X-Gal. The X-Gal/LB mix is then spread onto the plate. The plate should then be allowed to dry for approx 40 min before any transformation mix is added.
22. The selection can be performed at either 37°C or 30°C. In *E. coli*, certain proteins are more soluble at lower growth temperatures.
23. An oligonucleotide (sequence 5'-GCAATGAGAGTTGTTCCGTTGTGG-3') that hybridizes to the end of the *cI* gene and reads toward the *Not*I site can be used for sequencing inserts in bait plasmids pBT and pACλcI32. An oligonucleotide (sequence 5'-GGTCATCGAAATGGAAACCAACG-3') that hybridizes to *rpoA* and reads toward the NotI site can be used for sequencing inserts in prey plasmids pTRG, pBRSTAR, and pBRαLN.

Acknowledgments

This chapter is adapted from Dove, S.L. (2002) Studying protein-protein interactions using a bacterial two-hybrid system. *Methods Mol. Biol.* **205,** 251–265. We thank Renate Hellmiss for artwork. Supported by NIH grants GM55637 and GM44025 to A. Hochschild and a grant from the Charles H. Hood Foundation, Inc., Boston, MA, to S. Dove.

References

1. Fields, S. and Song, O. (1989) A novel genetic system to detect protein-protein interactions. *Nature* **340,** 245–246.
2. Gyuris, J., Golemis, E. A., Chertkov, H., and Brent, R. (1993) Cdi1, a human G1 and S phase protein phosphatase that associates with Cdk2. *Cell* **75,** 791–803.
3. Hu, J. C., Kornacker, M. G., and Hochschild, A. (2000) *Escherichia coli* one- and two-hybrid systems for the analysis and identification of protein-protein interactions. *Methods* **20,** 80–94.
4. Ladant, D. and Karimova, G. (2000) Genetic systems for analyzing protein-protein interactions in bacteria. *Res. Microbiol.* **151,** 711–720.
5. Legrain, P. and Selig, L. (2000) Genome-wide protein interaction maps using two-hybrid systems. *FEBS Lett.* **480,** 32–36.
6. Hu, J. C. (2001) Model systems: Studying molecular recognition using bacterial n-hybrid systems. *Trends Microbiol.* **9,** 219–222.
7. Dove, S. L., Joung, J. K., and Hochschild, A. (1997) Activation of prokaryotic transcription through arbitrary protein-protein contacts. *Nature* **386,** 627–630.
8. Dove, S. L. and Hochschild, A. (1998) Conversion of the ω subunit of *Escherichia coli* RNA polymerase into a transcriptional activator or an activation target. *Genes Dev.* **12,** 745–754.
9. Dove, S. L., Huang, F. W., and Hochschild, A. (2000) Mechanism for a transcriptional activator that works at the isomerization step. *Proc. Natl. Acad. Sci. USA* **97,** 13,215–13,220.
10. Dove, S. L. and Hochschild, A. (2001) Bacterial two-hybrid analysis of interactions between region 4 of the σ^{70} subunit of RNA polymerase and the transcriptional regulators Rsd from *Escherichia coli* and AlgQ from *Pseudomonas aeruginosa*. *J. Bacteriol.* **183,** 6413–6421.
11. Blum, J. H., Dove, S. L., Hochschild, A., and Mekalanos, J. J. (2000) Isolation of peptide aptamers that inhibit intracellular processes. *Proc. Natl. Acad. Sci. USA* **97,** 2241–2246.
12. Tottey, S., Rondet, S. A. M., Borelly, G. P. M., Robinson, P. J., Rich, P. R., and Robinson, N. J. (2002) A copper metallochaperone for photosynthesis and respiration reveals metal-specific targets, interaction with an importer, and alternative sites for copper acquisition. *J. Biol. Chem.* **277,** 5490–5497.
13. Kuznedelov, K., Minakhin, L., Niedziela-Majka, A., et al. (2002) A role for interaction of the RNA polymerase flap domain with the σ subunit in promoter recognition. *Science* **295,** 855–857.

14. Pande, S., Makela, A., Dove, S. L., Nickels, B. E., Hochschild, A., and Hinton, D.M. (2002) The bacteriophage T4 transcription activator MotA interacts with the far-C-terminal region of the σ^{70} subunit of *Escherichia coli* RNA polymerase. *J. Bacteriol.* **184,** 3957–3964.
15. Nickels, B. E., Roberts, C. W., Sun, H., Roberts, J. W., and Hochschild, A. (2002) The σ^{70} subunit of RNA polymerase is contacted by the λQ antiterminator during early elongation. *Mol. Cell* **10,** 611–622.
16. Shaywitz, A. J., Dove, S. L., Kornhauser, J. M., Hochschild, A., and Greenberg, M. E. (2000) Magnitude of the CREB-dependent transcriptional response is determined by the strength of the interaction between the kinase-inducible domain of CREB and the KIX domain of CREB-binding protein. *Mol. Cell. Biol.* **20,** 9409–9422.
17. Althoff, E. A. and Cornish, V. W. (2002) A bacterial small-molecule three-hybrid system. *Angew. Chem. Int. Ed.* **41,** 2327–2330.
18. Joung, J. K., Ramm, E. I., and Pabo, C. O. (2000) A bacterial two-hybrid selection system for studying protein-DNA and protein-protein interactions. *Proc. Natl. Acad. Sci. USA* **97,** 7382–7387.
19. Blatter, E. E., Ross, W., Tang, H., Gourse, R. L., and Ebright, R. H. (1994) Domain organization of RNA polymerase α subunit: C-terminal 85 amino acids constitute a domain capable of dimerization and DNA-binding. *Cell* **78,** 889–896.
20. Jeon, Y. H., Yamazaki, T., Otomo, T., Ishihama, A., and Kyogoku, Y. (1997) Flexible linker in the RNA polymerase alpha subunit facilitates the independent motion of the C-terminal activator contact domain. *J. Mol. Biol.* **267,** 953–962.
21. Miller, J. H. (1972) *Experiments in Molecular Genetics.* Cold Spring Harbor Laboratory, Cold Spring Harbor, NY.
22. Whipple, F. W. (1998) Genetic analysis of prokaryotic and eukaryotic DNA-binding proteins in *Escherichia coli. Nucleic Acids Res.* **26,** 3700–3706.
23. Ptashne, M. (1992) *A Genetic Switch, Phage Lambda and Higher Organisms,* Cell Press, Blackwell, Cambridge, UK.
24. Yanisch-Perron, C., Vieira, J., and Messing, J. (1985) Improved M13 phage cloning vectors and host strains: nucleotide sequences of the M13mp18 and pUC19 vectors. *Gene* **33,** 103–119.

18

Using the Yeast Two-Hybrid System to Identify Interacting Proteins

John Miller and Igor Stagljar

Abstract

The yeast two-hybrid system is a powerful technique for studying protein–protein interactions. Two proteins are separately fused to the independent DNA-binding and transcriptional activation domains of the Gal4p transcription factor. If the proteins interact, they reconstitute a functional Gal4p that activates expression of reporter gene(s). In this way, two individual proteins may be tested for their ability to interact, and a transcriptional readout can be measured to detect this interaction. Furthermore, novel interacting partners can be found by screening a single protein or domain against a library of other proteins using this system. It is this latter feature—the ability to search for interacting proteins without any prior knowledge of the identity of such proteins—that is the most powerful application of the two-hybrid technique.

Key Words

Two-hybrid methodology; protocol; library screening.

1. Introduction

Among the myriad methods for examining protein–protein interactions, the yeast two-hybrid system is one of the easiest and most inexpensive to perform. All that is required is a protein(s) of interest, the ability to clone into a vector, yeast, and various yeast media. The technique works by capitalizing on the modular nature of a transcription factor, Gal4p, which contains separable DNA-binding and transcriptional-activation domains ***(1,2)***. The amino-terminal 147 residues ($Gal4_{1-147}$) of Gal4p code for the DNA-Binding Domain (BD) of the protein, while the last 114 residues ($Gal4_{768-881}$) encompass a strong transcription Activating Domain (AD). Fields and Song demonstrated that these separate domains of the Gal4p protein could stimulate transcription at a promoter that contained *cis* elements [upstream activating sequences (UAS)] recognized

From: *Methods in Molecular Biology, vol. 261: Protein–Protein Interactions: Methods and Protocols*
Edited by: H. Fu © Humana Press Inc., Totowa, NJ

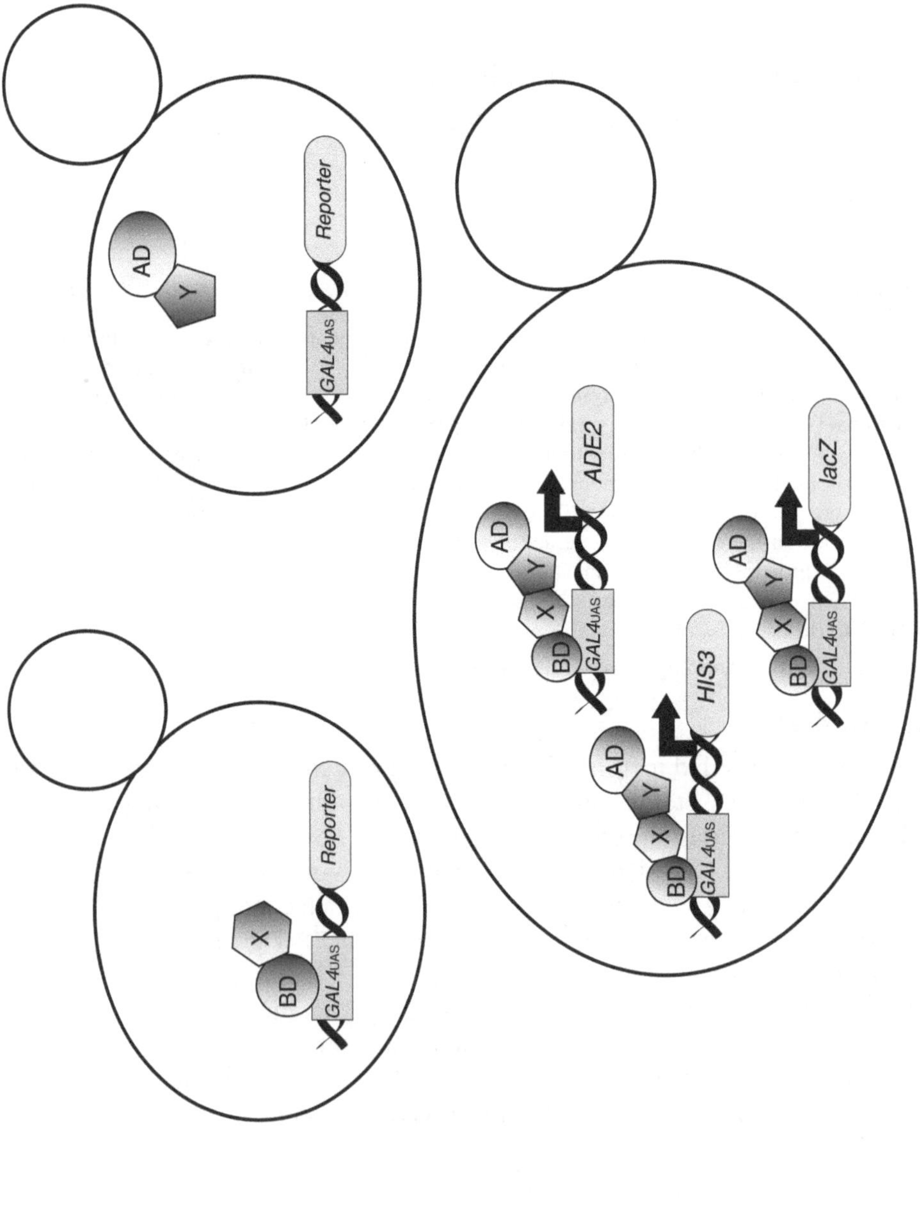

AD
Y
GAL4UAS
Reporter
BD
X
GAL4UAS
Reporter
AD
Y
X
BD
GAL4UAS
ADE2
AD
Y
X
BD
GAL4UAS
HIS3
AD
Y
X
BD
GAL4UAS
lacZ

by Gal4p if they were attached to two proteins that physically associate (**Fig. 1**) *(3)*. Therefore, the isolated BD and AD domains could potentially be fused to any two proteins, and if the proteins interact, then the transcription from genes bearing GAL4-UAS elements will report the interaction. BDs (in combination with the appropriate *cis* elements) and/or ADs from other transcription factors can be employed in a similar manner *(4,5)*.

A tremendously useful approach with the yeast two-hybrid is the searching of cDNA libraries for interacting partners of a protein of interest *(6)*. In this way, large numbers of proteins can be tested in pair-wise combinations with the protein of interest and those that interact can be selected for by GAL4-UAS sequences located upstream of reporter genes (*HIS3*, *ADE2*, etc.). In about 3–4 wk, a researcher can go from having a protein of interest a collection of candidate interacting partners that can be explored by further experimentation.

The interaction between two DNA-damage repair pathway components, Mec3p and Rad17p, will be used as an example for a yeast two-hybrid investigation of an interacting protein pair *(7)*. Additionally, the steps involved in carrying out a two-hybrid screen of an activation domain library for novel interacting partners will be described.

2. Materials

Unless otherwise noted, all chemicals are from Sigma.

1. Yeast Strain PJ694A: *Mat*α, *trp1-901, leu2-3, 112, ura3-52, his3-200, gal4*Δ, *gal80*Δ, *GAL2-ADE2, LYS2::GAL1-HIS3, met2::GAL7-lacZ* ***(8)***.
2. Yeast Strain PJ694α: *Mat*α, *trp1-901, leu2-3, 112, ura3-52, his3-200, gal4*Δ, *gal80*Δ, *GAL2-ADE2, LYS2::GAL1-HIS3, met2::GAL7-lacZ* ***(9)***.
3. pOAD: GAL4 activation domain fusion vector with *LEU2* marker ***(7,10)***.

Fig. 1. *(opposite page)* The yeast two-hybrid system. A yeast cell expressing protein X fused to the DNA-binding domain (BD) of Gal4p does not activate reporter genes despite binding of BD-X to the $GAL4_{UAS}$ sequences upstream of the reporter genes (**A**). Similarly, expression of protein Y fused to the transcriptional activation domain (AD) in a yeast cell is not sufficient to stimulate transcription of the reporter genes (**B**). However, upon co-expression of the BD-X and AD-Y in the same cell by either co-transformation or mating, if X and Y interact, then the binding of the BD to the $GAL4_{UAS}$ sequence recruits the AD to the reporter gene promoter, which in turn recruits the transcriptional machinery and thus activates transcription of the reporter genes: *HIS3* encoding an enzyme involved in histidine biosynthesis; *ADE2* encoding an enzyme involved in adenine biosynthesis; and *lacZ*, the gene for the bacterial β-galactosidase enzyme (**C**).

4. pOBD: GAL4 DNA-binding domain fusion vector with *TRP1* marker ***(7,10)***.
5. cDNA of gene of interest (or yeast genomic DNA).
6. Oligonucleotide primers.
7. Activation domain library from relevant organism/tissue source (store at –20°C).
8. YEPD medium: 1% yeast extract, 2% Bacto-peptone (Difco), 1.4% Meer agar (Difco). After autoclaving, add 2% dextrose (Fisher) (glucose)(4°C). Omit agar for liquid.
9. Synthetic dropout medium (SD): 0.67% yeast nitrogen base (Difco), 1.4% Meer agar. After autoclaving, add 1X amino acid mix and 2% glucose. Omit agar for liquid. Dropout medium has the above ingredients minus the appropriate supplement: No tryptophan (–Trp), no leucine (–Leu), no tryptophan/leucine (–Trp –Leu), no tryptophan/leucine/adenine (–Trp, –Leu, –Ade), or no tryptophan/leucine/histidine with added 3AT (–Trp, –Leu, –His + X m*M* 3AT) (4°C).
10. 3AT stock: Make 1 *M* 3-amino-1,2,4-triazole stock in ddH_2O (store at –20°C).
11. Yeast β-galactosidase Assay Kit (Pierce; cat. no. 75768) (–20°C) ***(12)***.
12. *E. coli* strains DH5α and MH4 ($leuB^-$) prepared for electroporation ***(17)***.
13. M9 Minimal medium plates (1 L): 750 mL ddH_2O, 5X M9 salts, 15 g Meer agar, water up to 980 mL; then 2 mL 1 *M* $MgSO_4$, 0.1 mL 1 *M* $CaCl_2$. After autoclaving, add 20 mL 20% glucose (4°C) ***(17)***.
14. 5X M9 salts (1 L): 64 g $Na_2HPO_4 7H_2O$, 15 g KH_2PO_4, 2.5 g NaCl, and 5 g NH_4Cl ***(17)***.
15. LB plates (1 L): 14 g Meer agar, 10 g tryptone, 5 g yeast extract, 5 g NaCl (Fisher), 1 mL 1 N NaOH. After autoclaving, add ampicillin (100 μg/mL: final concentration) (4°C).
16. PCR enzyme, restriction enzymes, and T4 DNA ligase.
17. Yeast 10X "Amino Acid" mix (per liter): 0.2 g adenine, 0.2 g arginine, 0.2 g histidine, 0.3 g isoleucine, 1 g leucine, 0.3 g lysine, 0.2 g methionine, 0.5 g phenylalanine, 2 g threonine, 0.2 g tryptophan, 0.3 g tyrosine, 0.2 g uracil, 1.5 g valine. Leave out appropriate components for dropout medium.
18. LiOAc-PEG solution: 40% PEG (Fisher), 100 m*M* lithium acetate (LiOAc), 10 m*M* Tris-HCl, 1 m*M* EDTA (TE). Make up fresh from 50% PEG 4000, 1 *M* LiOAc, and 1 *M* Tris-HCl, pH 7.5, 200 m*M* EDTA (TE). Vortex vigorously to get complete mixing.
19. Sheared Salmon Sperm DNA (Sigma) (5 mg/mL).
20. Qiagen Miniprep Kit and Qiagen Maxiprep Kit (Qiagen).
21. Qiagen Qiaquick Gel Extraction Kit (Qiagen).
22. 400–500 μm acid-washed glass beads (Sigma).
23. Replica stamp and velvets (wrapped in foil and autoclaved prior to use).

3. Methods

Described below are the (1) generation of the two-hybrid constructs, (2) assaying for an interaction, (3) screening of a library for the identification of new interacting partners, and (4) retesting of the newly identified protein partners.

3.1. Building Two-Hybrid Constructs

3.1.1. Generating Inserts by PCR

Insertion of *MEC3* and *RAD17* into both of the pOAD and pOBD vectors was carried out by homologous recombination in yeast (*see* **Note 1**). The primer sequences below are from ref. *7*.

1. By the polymerase chain reaction (PCR), the sequence {5'-CTATCTATTCG ATGATGAAGATACCCCACCAAACCCAAAAAAAGAGATC**GAATTCCA GCTGACCACC**-3'} was added to the 5'-end of each gene for insertion into pOAD. The sequence {5'-CTTGCGGGGTTTTTCAGTATCTACGATTCATAG ATCTCTGCAGGTCGACG**GATCCCCGGGAATTGCCATG**-3'} was added to the 3' end of the genes.
2. Insertion into pOBD was mediated by the sequence {5'-ATCGGAAGAGAG TAGTAACAAAGGTCAAAGACAGTTGACTGTATCGCCG**GAATTCCA GCTGACCACC**-3'} added onto the 5' end, and extending the 3' end of the genes with {5'-TCATAAATCATAAGAAATTCGCCCGGAATTAGCTTGGCTGCA GGTCGACG**GATCCCCGGGAATTGCCATG**-3'}.

3.1.2. Cloning by Homologous Recombination in Yeast

The PCR products of *MEC3* and *RAD17* were co-transformed into yeast along with pOAD or pOBD vectors that had been cut with the *Nco*I restriction enzyme and gel-purified. As a control, vector alone was also transformed to identify the amount of uncut plasmid present. Transformation was carried out by the lithium acetate/polyethelene glycol method as follows (modified from *11*):

1. A single yeast colony of strain PJ694A (for pOAD) or PJ694α for (pOBD) was inoculated into 5 mL of liquid YEPD and grown overnight at 30°C with shaking.
2. One milliliter of the saturated yeast culture was then pelleted in an Eppendorf tube for each transformation to be performed (three for each strain, i.e., *RAD17*, *MEC3*, and no insert).
3. The supernatant was aspirated off and 3 μL of sheared 5 mg/mL salmon sperm DNA that had been boiled 5 min and transferred immediately to ice was added to each tube.
4. Tubes were vortexed and 1 μL of gel-extracted, *Nco*I-cut pOAD was added to each tube of PJ694A, while 1 μL of gel-extracted, *Nco*I-cut pOBD was added to each tube of PJ694α.
5. Following vortexing, 5 μL of the appropriate PCR products of *MEC3* and *RAD17* were added to a single tube of the appropriate strain and all tubes were again vortexed.
6. 500 μL of freshly made LiOAc/PEG was added to each tube and mixing was done by pipetting the viscous solution up and down several times.
7. 57 μL of DMSO was then added, the tubes were vortexed, and then incubated at 30°C for 15 min.

8. Following this incubation, the transformations were heat-shocked by a 15-min incubation in a 42°C water bath.
9. The cells were collected by a 1-min centrifugation in a microfuge, and the supernatant was aspirated off. 500 μL of 1X TE was used to resuspend the cells, followed by an additional 1-min centrifugation.
10. 400 μL of the supernatant was removed, and the cells were resuspended in the remaining approx 100 μL and plated onto SD plates lacking the appropriate amino acid (–Leu for pOAD constructs, –Trp for pOBD). The plates were incubated upside down for 3 d at 30°C.

3.1.3. Confirming Insertion of the Gene into the Vector

A comparison of the number of colonies on plates where PCR product was added versus the vector alone control gives an estimate of the number of colonies that contain insert.

1. Five colonies of each of the BD fusions and each of the AD fusions of *MEC3* and *RAD17* were grown in 5 mL of the appropriate SD medium for plasmid isolation (*see* **Subheading 3.4.1.**).
2. The plasmids were electroporated into *E. coli* (DH5α), isolated from the overnight *E. coli* culture using the Qiagen Miniprep Kit, and sequenced.
3. In-frame fusions of each gene with the AD or BD were used for the two-hybrid assay (*see* **Note 2**).

3.2. Performing the Assay

3.2.1. Mating of Yeast Cells Expressing Hybrids

Individual colonies of each haploid (PJ694A) expressing an AD fusion were resuspended in a small volume of water and then 2 μL of each cell suspension was spotted onto a YEPD plate. Onto these spots the appropriate BD fusion yeast (PJ694α) suspensions were spotted and the plates were incubated for 24 h at 30°C.

The YEPD plates were then replica-plated using an autoclaved velvet onto diploid selective plates, i.e., SD/ –Trp –Leu. This plate selects for diploids because only the yeast cells containing both the *LEU2*-bearing pOAD and *TRP1*-carrying pOBD will be able to grow: the only cells with both plasmids will be those that have mated (PJ694A/α diploids). The diploid plates were grown for 3 d at 30°C.

3.2.2. Replica Plating to Change to Selective Media

Diploids were then replica-plated using a stamp and autoclaved velvet onto plates to assess reporter gene activation. These were (SD/ –Trp –Leu) –Ade, –His, and –His with 3AT (3 m*M*, 30 m*M*, 50 m*M*, and 100 m*M*, respectively) (*see* **Note 3**). Plates were allowed to grow for 3 d at 30°C (**Fig. 2**).

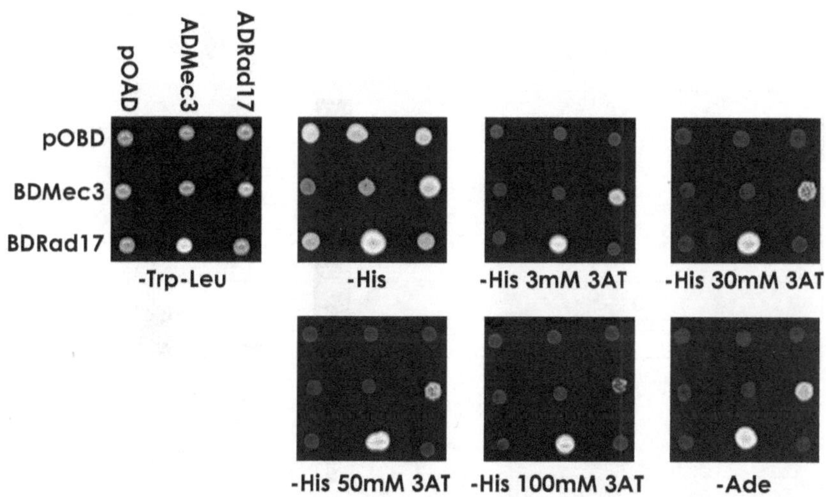

Fig. 2. The two-hybrid interaction of Mec3p and Rad17p demonstrated on selective plates. Each row contains a different BD construct and each column a different AD construct. Diploids containing the different AD and BD pairs were spotted onto SD plates lacking the indicated components and containing the indicated concentration of 3AT. Plates were incubated for 3 d at 30°C. Note that yeast containing the empty vector controls are able to grow on the SD/–His plate (*see* **Note 3**).

3.2.3. Assessing the Strength of Reporter Gene Transcription by a β-Galactosidase Assay

The Yeast β-Galactosidase Assay Kit from Pierce was used to carry out β-galactosidase assays on each of the diploids. The protocol that accompanies the kit *(12)* was followed (**Fig. 3**).

3.3. Screening an Activation Domain Library for Interactions

Most often researchers do not have a specific pair of proteins they are assaying for an interaction. In addition, if a researcher is interested in a small number of potentially interacting proteins, it is often desirable to examine the proposed interactions by a more direct biochemical technique. However, when an investigator has a favorite protein and wants to find out what other proteins it interacts with, the two-hybrid system is one of the best ways to start searching for these other proteins. To do this, they make use of a library of fusions to the Gal4p-AD and search with their protein of interest as a fusion to the Gal4p-BD. In this way it is possible to perform one experiment and identify many interacting partners for a particular protein.

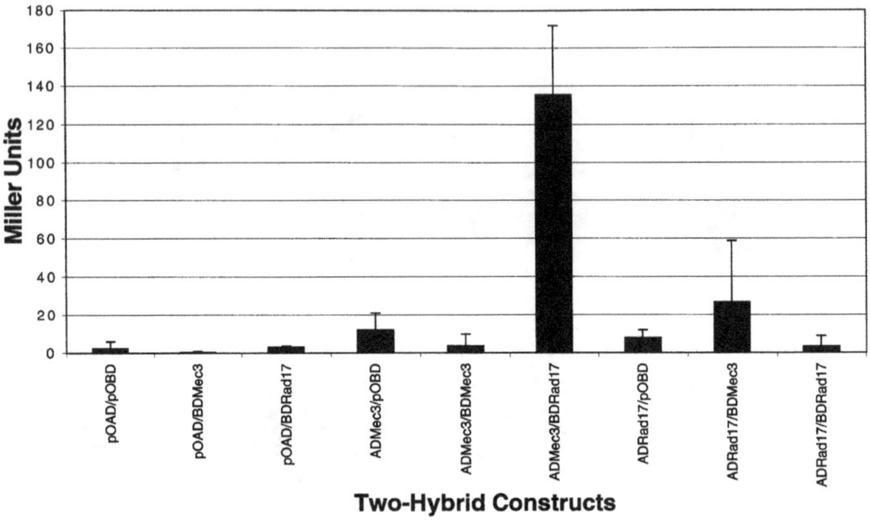

Fig. 3. β-galactosidase assays of Mec3p and Rad17p two-hybrid pairs. β-galactosidase activity was assayed using the Yeast β-Galactosidase Assay Kit (Pierce) ***(12)***. Three replicates of each experiment were carried out.

A critical test prior to screening a library is to make sure that your protein fused to the Gal4p-BD does not activate the reporter genes on its own:

1. Co-transform yeast with pOAD and the BD-fusion of your protein of interest in pOBD.
2. Streak these transformants onto SD/–Trp –Leu –His +3m*M* 3AT plates and look for growth.
3. Do a β-galactosidase assay comparing your BD-fusion/pOAD to cells containing pOBD/pOAD.

If your fusion is expressed in yeast and does not self-activate, then you are ready to search an AD library for interacting partners. If the reporter genes are activated by your BD-fusion alone, then you may be able to overcome this with higher concentrations of 3AT or by using less than the full-length protein as a BD-fusion (*see* **Note 4**).

3.3.1. Getting a Good Library

The best way to find an activation domain library that will work for you is to search the literature. Look for a library that's reported in multiple publications and pay attention to the quantity and quality of the identified proteins found: Do screens with this library provide multiple clones encoding the same protein? Are the identified clones full-length cDNAs or only fragments? The answer to this latter question shouldn't deter you from obtaining and using libraries with fragments in them as these are sometimes superior to full-length clones in that interactions are found in the two-hybrid with fragments, but testing of the full-length protein diminishes or eliminates the strength of the interaction. These of course must be scrutinized with further experiments to confirm the authenticity of the interaction, but examples of such results have been reported ***(13)***.

Additionally, the activation domain library should of course be from your organism of study if at all possible, and from the specific tissue(s) that are known to express your protein of interest. If these are not available, orthologues from a different organism can be tried against a library for the orthologous species. If something interesting is found, you can then go back to the original model organism and see if the proteins are interacting by two-hybrid or biochemical means in that system.

If no suitable libraries are available, you may wish to construct your own activation domain library either by doing it yourself or by paying to have one made [BD Biosciences Clontech offers this service ***(14)***]. These alternatives offer you the ability to have the exact library you want, but at the expense associated with creating it. One caveat is that such a library will have been untested prior to your use, and its quality cannot be established to the same degree as one used by multiple other users who've already published interacting proteins found in the library.

As this is probably your first foray into the two-hybrid system, we will not describe the undertaking of building your own activation domain library, but other sources can provide you with the methodologies involved ***(15)***. Instead, we will focus on screening the library with your protein fused to the Gal4p-BD. A flow-chart of library screening is included for clarity (**Fig. 4**).

3.3.2. Protocol for Screening

Depending on the form and amount of the library that you receive, you may have to amplify the library. Amplification is not desirable if it can be avoided because of the potential to lose rare clones in the process of growing up the library in *E. coli*. However, it is necessary if you are going to use the library for multiple screens. A good approximation is to use 50 μg of library DNA to transform a liter of yeast. More or less can be used, but should be based on the complexity of the library, i.e., how many different inserts are contained in the library.

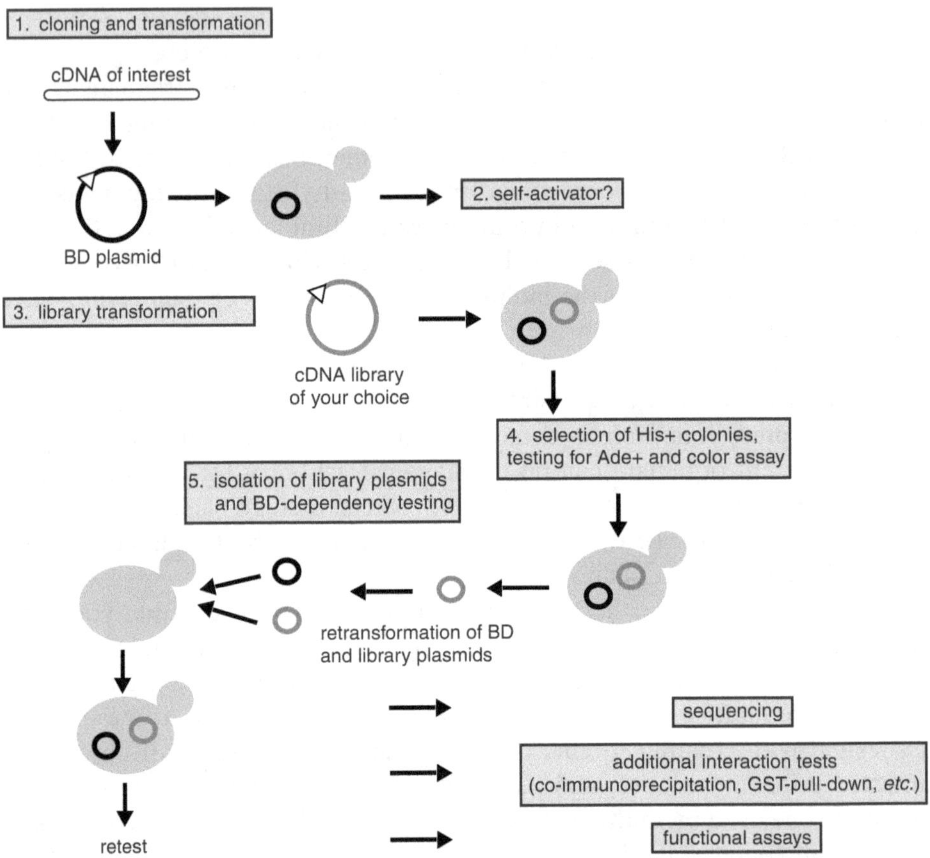

Fig. 4. Flow-chart for a yeast two-hybrid library screening. (1) The BD-fusion is transformed into yeast and (2) tested for reporter gene activation. (3) The AD-fusion library is transformed into cells bearing the BD-fusion and (4) the *HIS3* reporter gene is used to select for interaction positives. The other reporters are tested, and (5) plasmids are isolated and reintroduced into naive yeast cells for retesting of the reporter genes. Positive AD-fusions passing the retests are sequenced and further characterized.

You will want to obtain a multiple of transformants greater than this complexity (at least three times greater). For example, if the AD library you are using has 1×10^5 different inserts, if you get a million transformants (which should be possible from 50 μg of DNA and 1 L of yeast), you will have covered the library 10 times, or each insert is expected to be present in your transformation in 10 different colonies.

3.3.2.1. Amplifying the Library

1. If it has not been provided to you, then determine the number of colony forming units/microgram (cfu/µg) of DNA by electroporating *E. coli* (DH5α) with a small fraction of the library DNA (e.g., 10 ng) and plating the transformants onto multiple 100 mm Petri plates with LB plus ampicillin. Also make 1:10, 1:100, and 1:1000 dilutions and plate these to ensure an accurate count of the cfu/µg. Based on this number, you can decide how many 150 mm LB amp plates you will need in order to plate out individual colonies (approx 2000 colonies per plate) for the actual amplification in the next step.
2. Electroporate *E. coli* (Dh5α) with the amount of the library determined in **step 1** and grow the plates overnight at 37°C.
3. Scrape the cells off of the plates with 1 mL of LB per plate and a spreader. [We have also used cell culture spatulas (costar®Cell Lifter, cat. no. 3008) for this purpose, but care must be taken not to dig up too much agar from the plate.]
4. Weigh the cell pellet and perform a MaxiPrep (Qiagen) using volumes based on this weight to isolate the DNA.
5. Use a fluorimeter and/or spectrophotometer to measure the concentration of DNA.

3.3.2.2. Transforming Yeast (Adapted from refs. *14* and *16*)

1. Grow 1 L of PJ694α previously transformed with your bait protein fused to the BD in YEPD to mid-log phase (1×10^7 cells/mL or an absorbance of approx 0.5–0.6 at a wavelength of 660 nm).
2. Pellet the cells in two 500 mL centrifuge tubes by centrifugation at 1000*g* for 15 min.
3. Resuspend the cells in 500 mL of ddH_2O. Repeat centrifugation.
4. Resuspend the cells in 8 mL of TE/LiOAc and transfer to a 200 mL centrifuge tube with screw cap.
5. Prepare 100 mL PEG/LiOAc being sure to vortex thoroughly as PEG is extremely viscous.
6. Add 20 mg of sheared salmon sperm DNA (boiled for 5 min then placed immediately on ice). Mix by vortexing.
7. Add 50–500 µg of the AD library DNA to the yeast. Mix by vortexing.
8. Add 60 mL of PEG/LiOAc and vortex to mix.
9. Shake the tube at 200 rpm at 30°C for 30 min.
10. Add DMSO to a final concentration of 10% and mix by inverting.
11. Heat shock the cells in a 42°C water bath for 15 min with occasional swirling.
12. Chill cells on ice.
13. Collect cells by centrifugation in swinging bucket rotor for 5 min at 1000*g*.
14. Resuspend the transformed cells in 10 mL of TE.
15. Plate out 100 µL of a 1:10, 1:100, and 1:1000 dilutions onto SD/–Trp, –Leu plates to determine the transformation efficiency.

16. Plate the remaining volume onto 40 (250 μL/ plate) 150 mm petri plates with SD/–Trp –Leu –His + 3 m*M* 3AT (or the 3AT concentration you've determined for your BD-fusion) (*see* **Note 4**).
17. Incubate plates upside down at 30°C for up to 2 wk (*see* **Note 5**). Check for positives after 3 d.

3.3.3. What Positives Should be Further Pursued?

When positive colonies have grown on the plates, streak them onto a new plate containing SD/–Trp –Leu –His + 3 m*M* 3AT. Anything that doesn't grow anew should be discarded. You can then check the other reporter genes by replica plating onto SD/–Trp –Leu –Ade, and carrying out β-galactosidase assays (*see* **Note 6**). Anything that isn't activating all three reporter genes should most likely be excluded.

The –Ade plate will most likely give you several variations in your positives: (i) white, strong growth; (ii) pink, moderate growth; and (iii) red, little or no growth. Pursue the white positives first, because they represent the strongest positives. However, the *ADE2* reporter is the most stringent and weak interactions that are "real" by the *HIS3* and *lacZ* reporters might be worth pursuing if you don't have many positives or none are white on a –Ade plate.

3.4. Retesting Interaction Positives

With a narrowed-down number of candidates to study further, you can begin the retesting process. You'll first want to ensure that the reporter gene activation is due to the two-hybrid plasmids by isolating the plasmids and moving them into naive yeast. It is important to then test the AD-fusions against the empty pOBD vector as well as other BD-fusions to determine if your positives activate reporter genes nonspecifically with any BD-fusion. Lastly, interactions observed by the two-hybrid should be reproduced by testing the reciprocal fusions and other experiments.

3.4.1. Isolating Plasmids for Retesting

1. Inoculate 5 mL of SD/ –Trp –Leu liquid media with a positive from your screen that has shown activation of all of the reporter genes. Grow overnight at 30°C with shaking.
2. Centrifuge to collect cells in a swinging bucket rotor for 5 min at 1000*g*.
3. Resuspend the pellet in 1 mL of water and transfer to an Eppendorf tube. Collect cells with a quick spin and aspirate off the supernatant.
4. Follow the Qiagen Miniprep protocol for bacterial plasmid isolation except upon addition of 250 μL of P1 buffer, also add approx 100 μL of 400–500 μm acid-washed glass beads. Vortex for 7 min. Continue with protocol as normal.
5. Electroporate *E. coli* strain MH4 (*leuB*⁻) prepared as described in *(17)* with 1.5 μL of the isolated DNA.

6. Plate half of the cells onto LB +ampicillin, and the other half onto an M9 minimal plate (*see* **Note 7**).
7. If using LB + ampicillin colonies, replica-plate or streak to M9 plates to select for bacteria containing the *LEU2* (AD-fusion) plasmids.
8. Miniprep the plasmid DNA and identify inserts by sequencing with primers located in the AD vector. We use the oligo [5'- TACCACTACAATGGAT GATGTATATAAC] to sequence the AD-protein junction, and the oligo [5'- GATGCACAGTTGAAGTGAACTTGCGG] to sequence the 3' end of the gene.
9. AD-fusions with in-frame inserts should be retested by assaying for an interaction with the BD-fusion used to first identify them by transforming (PJ694A) as in **Subheading 3.1.2.** and performing the two-hybrid as in **Subheading 3.2.** (*see* **Note 1**).

3.4.2. Testing Activation Domain Fusions Against Lamin

An important caveat for any positive found with a single bait plasmid is that the positive AD-fusion might be a false-positive. What is a "false" positive? The simple answer is that it is an AD-fusion that activates the reporter genes when combined with any (or at least a large number of) BD-fusions tested. The possible mechanisms for false-activation are complicated and not of interest to the researcher trying to identify a new interaction. So, how can a false-positive be identified? By testing multiple BD-fusions against it.

1. Transform PJ694A with the AD-fusion DNAs from **Subheading 3.4.1., step 9**.
2. Transform PJ694α with pOBD and at least one BD-fusion other than the one you did the screen with (BD-lamin is a good tester).
3. Follow the mating strategy for performing a two-hybrid as in **Subheading 3.2.** and examine the reporter genes for activation.
4. Discard AD-fusions that interact with unrelated BD-fusions as likely false-positives.
5. Study further those AD-fusions that are specific to your BD-fusion.

3.4.3. Independently Confirming Interactions

With interacting pairs of proteins in hand you'll want to move into your system to do experiments to confirm the importance of the interactions for your organism. This is the best way to confirm interactions, but depending on the difficulty and/or expense involved, it may be prudent to acquire independent confirmation of the interaction(s) you've found by other means. Biochemical methods for validating two-hybrid interactions are advised, including co-immunoprecipitation *(17)* and GST-pulldown *(17)*, both of which are described in this volume.

4. Notes

1. We make use of the ResGen (Invitrogen) yeast open reading frame (ORF) collection *(7)* in which every yeast ORF has the same 20 nucleotides on its 5' ends and a different 20 nucleotide sequence on its 3' end (shown in bold). These allow a single set of primers to be used to amplify any of the ORFs. Such a strategy expedites the construction of different fusions into the pOBD and pOAD vectors. However, if only a few two-hybrid constructs need to be made, cloning by restriction digests and ligation can be carried out using sites encoded in the ResGen 20mers that are contained head to tail in the vectors.
2. The commercially available antibodies we have tried that are raised against the Gal4p AD and BD don't work well to detect expression from the pOAD and pOBD vectors in our hands. Presumably this is due to the lower expression level of these CEN-based vectors relative to multicopy 2μ plasmids used for the two-hybrid in many other labs. If antibodies to the protein(s) of interest are available, then a Western blot would be another way to confirm in-frame insertion.
3. As shown in **Fig. 2**, PJ694A/α grow on –His alone due to leakiness of the *HIS3* reporter in these strains. To overcome this, 3-amino-1,2,4-triazole (3AT) is added to the medium because it is a stoichiometric inhibitor of the *HIS3* gene product. Thus, more 3AT included in the media means more transcription from the *HIS3* gene is required for growth. The strength of a two-hybrid signal can be qualitatively estimated from the concentrations of 3AT that yeast containing the two-hybrid pair is resistant to.
4. If your bait protein appears to be activating the reporter genes by itself, you can try growing the yeast with your BD-fusion on higher concentrations of 3AT until you find the point at which they are no longer resistant. Perform your screen at this concentration of 3AT. Alternatively, delete portions/domains of your protein with the purpose of removing the amino acids that are activating transcription and test the deletions for reporter gene activation prior to screening.
5. When storing plates for such a long time in the 30°C incubator, be sure to keep pans of clean water in the incubator to provide humidity. You can also incubate the plates in the plastic sleeves that the Petri plates came in to prevent excessive moisture loss.
6. A commonly used qualitative assay for β-galactosidase activity is the filter-lift. Unfortunately, this is not useful in the PJ694A/α strains because the strain itself is β-galactosidase positive by this assay. However, the quantitative liquid assay works well and the advantage of the three reporter genes outweighs the need for a quick β-galactosidase activity test present in other two-hybrid reporter strains.
7. The growth of the MH4 strain on the minimal plate will only be possible if it has been transformed with a *LEU2*-containing plasmid (the AD-fusion plasmid). Thus, you can separate the BD and AD plasmids by passaging through these *E. coli*. However, growth is slow such that 2 d are usually required to get colonies, so we also plate onto LB with ampicillin and after one day replica plate to M9 minimal medium to get good-sized colonies.

Acknowledgments

I. Stagljar is supported by Zücher Krebsliga, Gebert-Rü Stiftung, Walter Honegger Stiftung, Bonizzi-Theler Stiftung, EMDO Stiftung, Stiftung fü medizinische Forschung, Kommission fü Technische Inovation (KTI, Nr. 5343.2 SUS), and the Swiss National Science Foundation (31-58798.99).

References

1. Keegan, L., Gill, G., and Ptashne, M. (1986) Separation of DNA binding from the transcription-activating function of a eukaryotic regulatory protein. *Science* **231,** 699–704.
2. Ma, J. and Ptashne, M. (1987) Deletion analysis of GAL4 defines two transcriptional activating segments. *Cell* **48,** 847–853.
3. Fields, S. and Song, O. (1989) A novel genetic system to detect protein-protein interactions. *Nature* **340,** 245–246.
4. Gyuris, J., Golemis, E., Chertkov, H., and Brent, R. (1993) Cdi1, a human G1 and S phase protein phosphatase that associates with Cdk2. *Cell* **75,** 791–803.
5. Ma, J. and Ptashne, M. (1987) A new class of yeast transcriptional activators. *Cell* **51,** 113–119.
6. Chien, C. T., Bartel, P. L., Sternglanz, R., and Fields, S. (1991) The two-hybrid system: a method to identify and clone genes for proteins that interact with a protein of interest. *Proc. Natl. Acad. Sci. USA* **88,** 9578–9582.
7. Hudson, J.R. Jr., Dawson, E.P., Rushing, K.L., et al. (1997) The complete set of predicted genes from *Saccharomyces cerevisiae* in a readily usable form. *Genome Res.* **7,** 1169–1173.
8. James, P., Halladay, J., and Craig, E. A. (1996) Genomic libraries and a host strain designed for highly efficient two-hybrid selection in yeast. *Genetics* **144,** 1425–1436.
9. Uetz, P., Giot, L., Cagney, G., et al. (2000) A comprehensive analysis of protein-protein interactions in *Saccharomyces cerevisiae. Nature* **403,** 623–627.
10. Bartel, P. L., Roecklein, J. A., SenGupta, D., and Fields, S. (1996) A protein linkage map of *Escherichia coli* bacteriophage T7. *Nat Genet.* **12,** 72–77.
11. Soni, R., Carmichael, J. P., and Murray, J.A. (1993) Parameters affecting lithium acetate-mediated transformation of *Saccharomyces cerevisiae* and development of a rapid and simplified procedure. *Curr. Genet.* **24,** 455–459.
12. Pierce Chemical Co., P.O. Box 117, Rockford IL 61105. http://www.piercenet.com.
13. Flajolet, M., Rotondo, G., Daviet, L., et al. (2000) A genomic approach of the hepatitis C virus generates a protein interaction map. *Gene* **242,** 369–379.
14. BD Biosciences Clontech, 1020 E Meadow Circle, Palo Alto, CA 94303. MATCHMAKER Library User Manual (PT1020-1) version PR2562. http://www.clontech.com.
15. Zhu, L., Gunn, D., and Kuchibhatla, S. (1997) *The Yeast Two-Hybrid System* (Bartel, P. L. and Fields, S., eds.) Oxford University Press, New York, NY, pp. 73–96.

16. Bai, C. and Elledge, S.J. in ref. ***15***, pp. 11–28.
17. Sambrook, J. and Russell, D.W. (2001) *Molecular Cloning, A Laboratory Manual*, 3rd Ed. Cold Spring Harbor Laboratory Press, Cold Spring Harbor, NY.

19

Analysis of Protein–Protein Interactions Utilizing Dual Bait Yeast Two-Hybrid System

Ilya G. Serebriiskii and Elena Kotova

Abstract

To characterize a protein's function, it is often advantageous to identify other proteins with which it interacts. The yeast two-hybrid system is one of the most versatile methods available for detection and characterization of protein–protein interactions, and in the recent years it has become a mature and robust technology. A further improvement to this technique is the ability to examine and distinguish more than one interaction simultaneously. This is achieved in the Dual Bait, which has successfully been used to detect proteins and peptides that target specific motifs in larger proteins, to facilitate rapid identification of specific interactors from a pool of putative interacting proteins obtained in a library screen, and to score specific drug-mediated disruption of protein–protein interaction.

Key Words

Yeast two-hybrid system; protein–protein interactions; LexA; cI; specificity.

1. Introduction

In the 10 yr since it was first described *(1)*, the yeast two-hybrid system has become a mature and robust technology. It has allowed for the identification of hundreds of new proteins and novel protein interactions. The occurrence of false positives and negatives, although not completely eliminated, has been minimized to the point that a two-hybrid interaction can provide an important clue about protein function. The development of comprehensive high-throughput two-hybrid systems has provided genome-wide protein linkage maps. A further improvement to this technology is the ability to examine and distinguish more than one interaction simultaneously. This is achieved in the two bait systems *(4,5)*, represented well by the Dual Bait system *(6)*. Derived from

From: *Methods in Molecular Biology, vol. 261: Protein–Protein Interactions: Methods and Protocols*
Edited by: H. Fu © Humana Press Inc., Totowa, NJ

the classic Interaction Trap *(7)* two-hybrid system, the Dual Bait can simultaneously assess the interaction of two distinct baits with one interactive partner. In this system (shown in schematic form in **Fig. 1**), one protein of interest is expressed as a fusion to the DNA-binding protein LexA (Bait 1), while a second is expressed as a fusion to the DNA-binding protein cI (Bait 2). Strains of *Saccharomyces cerevisiae*, designed for screening of these two baits, carry four separate reporter genes: *LacZ* and *LEU2* are both transcriptionally responsive to an operator for L*exA (lexAop-LacZ* and *lexAop-LEU2)*, and *GusA* and *LYS2* are both transcriptionally responsive to an operator for cI (*cIop-GusA* and *cIop-LYS2*). A final component of the Dual Bait system is a plasmid expressing an activation-domain (AD) fused protein, where the protein is either a defined protein interactor or a cDNA library. Selective interaction of the AD-fusion with one or the other of the two baits is scored by observing a transcriptional activation (e.g., if AD-fusion preferentially interacts with Bait 1, then activation of *LacZ* = *LEU2* >> *LYS2* = *GusA* and vice versa).

In order to demonstrate the utility of the Dual Bait system, we have utilized the Ras superfamily of proteins. Members of this superfamily include a large number of proteins that share extensive sequence homology. A particular challenge for any technique used to characterize protein interactions, especially those of the Ras superfamily, is to devise a means of distinguishing those interactions specific for individual members from interactions that do not discriminate between family members. We have utilized the Ras superfamily of proteins to provide an initial validation of the Dual Bait approach. We analyzed the selective interactions of Raf, Ral-GDS, and Krit1(AD-fused components) with related Ras family GTPases, Ras (cI-fused Bait 2), and Krev-1 (LexA-fused Bait 1). It has been shown that Raf interacts preferentially with Ras; Krit1 with Krev-1; and RalGDS with both Ras and Krev-1. The Dual Bait system successfully identified correct interactions among these proteins, and was able to select specific high-affinity interacting pairs from seeded pools of low-affinity interacting proteins ***(6)***, suggesting these reagents would be useful for a more demanding library screening and mutational applications. These reagents are used as a set of controls in the protocol described below.

In another example of the Dual Bait system use, the small GTPases Rac and Cdc42 were studied. An important effector for these proteins is p21-activated kinase (Pak1). Pak1 contains a p21 binding domain (PBD), which appears to be used similarly to bind Rac and Cdc42. To delineate which pathways originating with Rac or Cdc42 involve Pak, we have randomly mutagenized the PBD segment of the kinase Pak1, creating an AD fusion library of mutant variants. Rac and Cdc42 were simultaneously used as baits to screen a library of more than 2000 randomly mutagenized Pak proteins for the ability to differentially bind the two GTPases. We have successfully isolated mutants that spe-

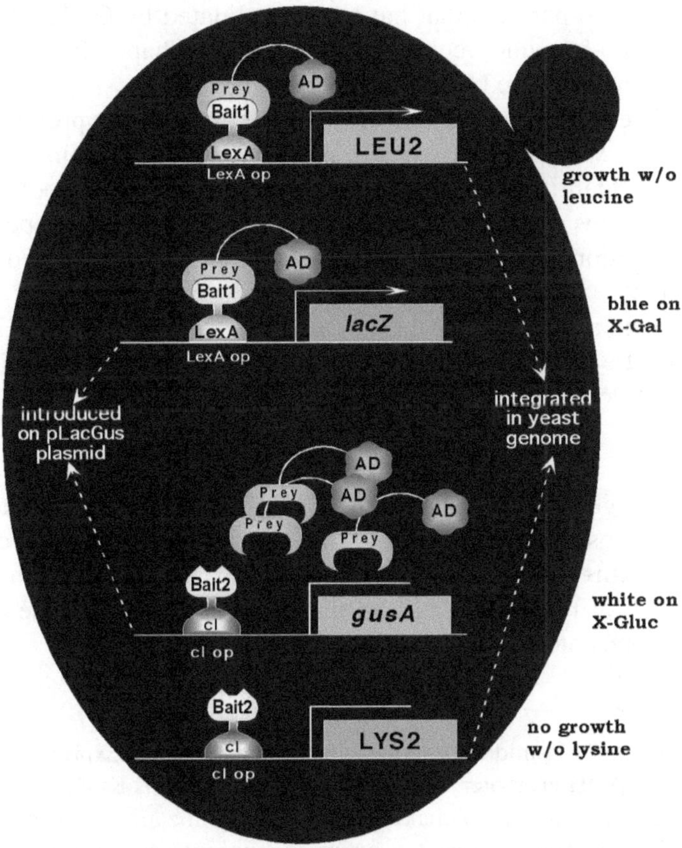

Fig. 1. Schematic of Dual Bait System. An activation domain-fused protein (Prey) interacts with a LexA-fused protein (Bait1) to drive transcription of *lexAop*-responsive *LEU2* and *lacZ* reporters but does not interact with a cI-fused bait and, thus, does not turn on transcription of *cIop*-responsive *LYS2* and *gusA* reporters. Note, as drawn here, cI-fused bait is representing a nonspecific partner. The system can also be configured so prey interacts with both baits. AD, activation domain.

cifically bind one, but not both of the GTPases and have used these mutants to distinguish Pak signaling pathways within the cell *(8)*.

The reagents described here have been also used effectively to study interactions between known sets of proteins, and have been utilized in a number of library screening applications.

In these screens the two baits were used either to simultaneously screen for interactors with a bait of interest versus a nonspecific control or to probe a wild-type versus mutated form of the same bait. These screens have yielded specific interactive partners that have been validated by GST-pulldown (*see* Chapter 13) or co-immunoprecipitation assays (*see* Chapter 23), or are known through other means to be biologically valid.

Some other examples of the use of this system, to detect proteins and peptides that target well-defined specific motifs in larger protein structures, to facilitate rapid identification of specific interactors from a pool of putative interacting proteins obtained in a library screen, and to score specific drug-mediated disruption of protein–protein interaction, are summarized in a recent publication ***(19)***.

The following method can be used to investigate specific protein interaction problems that are difficult to clarify by other means. This protocol is a derivation of the classic Interaction Trap two-hybrid system.

2. Materials

Dual Bait reagents, developed in the E. Golemis Lab at Fox Chase Cancer Center, is the most recent addition to the set of Interaction Trap–compatible reagents (*see* **Note 1**). All the protocols for the Dual Bait system utilize an overlapping set of reagents. Thus, all materials necessary for the three basic protocols are presented here.

2.1. Plasmids

1. pMW103: *HIS3* plasmid for making LexA fusion protein. Expression is from the constitutive ADH promoter. Bacterial selective marker is Km^R.
2. pJG4-5: *TRP1* plasmid for making a nuclear localization sequence–activation domain–hemaglutinin epitope tag fusion to a unique protein or a cDNA library. Expression is from the *GAL1* galactose-inducible promoter. Bacterial selective marker is Ap^R.
3. pLacGus: *URA3* plasmid containing eight LexA operators upstream of the *LacZ* reporter gene, and three cI operators upstream of *GusA* reporter gene. Bacterial selective marker is Km^R.
4. pGBS10: $G418^R$ plasmid for making cI fusion protein. Expression is from the constitutive ADH promoter. Bacterial selective marker is Km^R.
5. pSH17-4: *HIS3* plasmid encoding LexA-GAL4, a strong positive control for activation. Bacterial selective marker is Ap^R.
6. pEG202-hsRPB7: *HIS3* plasmid encoding LexA-hsRPB7, a weak positive control for activation. Bacterial selective marker is Ap^R.
7. pEG202-Ras: *HIS3* plasmid encoding LexA-Ras, a negative control for activation and positive control for interaction. Bacterial selective marker is Ap^R.
8. pGBS9-Krit: $G418^R$ plasmid encoding cI-Krit1, a weak positive control for activation. Bacterial selective marker is Km^R.
9. pGBS9-VP16: $G418^R$ plasmid encoding cI-VP16 fusion, a strong positive control for activation. Bacterial selective marker is Km^R.

10. pGBS10-Krev: *G418* plasmid encoding cI-Krev1, a negative control for activation and positive control for interaction. Bacterial selective marker is Km^R.
11. pJG4-5:Raf: library plasmid encoding a positive control for interaction with Ras.
12. pJG4-5:Krit1: library plasmid encoding a positive control for interaction with Krev1.
13. pYesTrp:RalGDS: *TRP1* library plasmid encoding a positive control for interaction with both Ras and Krev1
14. pJK202, pGilda, pNLexA (optional) : plasmids related to pEG202 that incorporate a nuclear localization sequence into the LexA-fusion construct, are expressed from the inducible *GAL1* promoter, or fuse LexA to the carboxy-terminal end of the test protein, respectively.
15. pGBS9, pGMS12 (optional) : plasmids related to pGBS10 that express cI- fusion construct at lower levels or from the inducible *GAL1* promoter, respectively.

2.2. Strains

1. Yeast strain SKY 48: *MATα ura3 trp1 his3 3LexAop-leu2 3cIop-lys2.*
2. Yeast strain SKY 191: *MATα ura3 trp1 his3 1LexAop-leu2 3cIop-lys2.*
3. Yeast strain SKY473: *MATa ura3 trp1 his3 2LexAop-leu2 3cIop-lys2.*
4. *Escherichia coli* strain KC8: *pyrF leuB600 trpC hisB463.*

2.3. Reagents for LiOAc Transformation of Yeast

1. LiOAc solution: 10 m*M* Tris-HCl, pH 8.0, 1 m*M* EDTA, 0.1 *M* lithium acetate. Sterile filtered.
2. LiOAc/PEG solution: 10 m*M* Tris-HCl, pH 8.0, 1 m*M* EDTA, 0.1 *M* lithium acetate, 40% PEG4000. Sterile filtered.

2.4. PCR and Miniprepration of DNA From Yeast

1. STES lysis solution: 100 m*M* NaCl, 10 m*M* Tris-HCl, pH 8.0, 1 m*M* EDTA, 0.1%SDS.
2. Acid washed sterile glass beads: 0.15–0.45 mm diameter (e.g., Sigma cat. no. G-1145).
3. TE: 10 m*M* Tris-HCl, pH 8.0, 1 m*M* EDTA.
4. 1:50 β-glucuronidase type HP-2 (crude solution from *H. pomatia*; Sigma).
5. 50 m*M* Tris-HCl, pH 7.5, 10 m*M* EDTA, 0.3% (v/v) 2-mercaptoethanol (freshly prepared).

2.5. X-gal/ X-Gluc Overlay Assays

1% low-melting agarose in 100 m*M* K_2HPO_4, pH 7.0; add X-Gal or X-Gluc solutions to 0.25 mg/mL when cooled to approx 60°C.

2.6. Plates for Growing Bacteria (100 mm)

1. Luria–Bertani (LB) supplemented with ampicillin (50 μg/mL).

2. KC8 plates for selecting library plasmids:
   a. 1 L of H_2O containing 15 g agar, 1 g $(NH_4)_2SO_4$, 4.5 g KH_2PO_4, 10.5 g K_2HPO_4, and 0.5 g sodium citrate2H $_2$O; autoclave, cool to 50°C.
   b. Add 1 mL sterile filtered 1 *M* $MgSO_4$7H $_2$O, 10 mL sterile-filtered 20% glucose, and 5 mL each of 40 µg/mL sterile-filtered stocks of L-histidine, L-leucine, and uracil. Pour plates.

2.7. General Directions for Defined Minimal Yeast Medium

1. All minimal yeast media, liquid and solid plates, are based on the following three ingredients, which are sterilized by autoclaving for 15–20 min: Per liter, 6.7 g Yeast Nitrogen Base-amino acids (Difco cat. no. 0919-15), 20 g glucose (or 20 g galactose and 10 g raffinose), 2 g appropriate nutrient "dropout" mix (*see* below). For plates, 20 g Difco bacto-agar (Difco cat. no. 0140-01) are also added.
2. A complete minimal (CM) nutrient mix includes the following: Adenine (2.5 g), L-arginine (1.2 g), L-aspartic acid (6.0 g), L-glutamic acid (6.0 g), L-histidine (1.2 g), L-isoleucine (1.2 g), L-leucine (3.6 g), L-lysine (1.8 g), L-methionine (1.2 g), L-phenylalanine (3.0 g), L-serine (22.0 g), L-threonine (12.0 g), L-tryptophan (2.4 g), L-tyrosine (1.8 g), L-valine (9.0 g), uracil (1.2 g). This amount of nutrient mix is sufficient to make 40 L of medium.
3. Dropout medium: Leaving out one or more nutrients from the complete minimal nutrient mix, termed "dropout medium," it is used to select for yeast carrying a plasmid with corresponding nutritional marker. Thus, "dropout medium" lacking uracil (denoted -Ura in the following recipes) would select for the presence of plasmid with the URA3 marker. Premade dropout mixes are available from some commercial suppliers.

2.8. Plates for Growing Yeast

G418 should be added, where appropriate, to the final concentration of 200 µg/mL in YPD medium and to 350 µg/mL in minimal medium.

1. Defined complete minimal medium dropout plates (100 mm) with glucose as a carbon source (Glu/CM): –Trp; –Ura-His; –Ura-His-Trp; –Ura-His-Trp-Leu; –Ura-His-Trp-Lys.
2. Defined minimal dropout plates (100 mm) with galactose and raffinose as carbon sources (Gal-Raff/CM): –Ura-His-Trp-Leu; –Ura-His-Trp-Lys.
3. YPD plates (100 mm; rich medium; per liter): 10 g yeast extract, 20 g peptone, 20 g glucose, 20 g Difco Bactoagar. Autoclave for approx 18 min. Pour approx 40 plates.
4. Dropout plates for library transformations (240 × 240 mm): Defined minimal dropout plates with glucose as a carbon source (–Trp). Each plate requires approx 250 mL of medium.

2.9. Liquid Medium for Growing Yeast

1. Defined minimal dropout media with glucose as a carbon source (Glu/CM): –Ura-His; –Trp.

2. YPD (per liter): 10 g yeast extract, 20 g peptone, 20 g glucose. Autoclave for approx 15 min.

2.10. Primers

1. For LexA-fusion plasmids: Forward primer for confirming reading frame: 5' CGT CAG CAG AGC TTC ACC ATT G.
2. For cI-fusion plasmids: Forward primer for confirming reading frame: 5' ATG ATC CCA TGC AAT GAG AG.
3. For JG4-5 plasmid: Forward primer, FP1: 5' CTG AGT GGA GAT GCC TCC. Reverse primer, FP2: 5' CTG GCA AGG TAG ACA AGC CG.
4. pYESTrp plasmid: Forward primer: FP1 can be used, or 5' GATGTTAA CGATACCAGCC (Invitrogen-recommended). Reverse primer: 5' GCG TGA ATG TAA GCG TGA C.

2.11. Miscellaneous

1. Sterile glass beads: 3–4 mm (cat. no. 3000, Thomas Scientific 5663L19 or Fisher cat. no. 11-312A).
2. Sterile glycerol solution for freezing transformants: 65% sterile glycerol, 0.1 *M* $MgSO_4$, 25 m*M*, Tris-HCl, pH 8.0.
3. Insert grid from a rack of pipette tips (Rainin RT series, 200 µL capacity).
4. A metal frogger (Dankar Scientific cat. no. MC48).
5. A plastic replicator (Bel-Blotter, Bel-Art Products cat. no. 378776-0002 or Fisher cat. no. 1371213).
6. 2X Laemmli sample buffer: 0.125 *M* Tris-HCl (pH 6.8), 4% (w/v) SDS, 20% (v/v) glycerol, 10% (v/v) 2-mercaptoethanol, and 0.002% (w/v) bromophenol blue.

3. Methods

The following procedures are used (i) to make and characterize plasmids that express bait proteins, (ii) to screen cDNA libraries for interacting proteins using characterized bait plasmids, and (iii) to confirm positive interactions (**Fig. 2**).

3.1. Expression of Bait Proteins

A prerequisite for screening for interacting proteins (an interactor hunt) is the construction of plasmids that expresse the proteins of interest (Bait) as fusions to the bacterial protein LexA and the lambda phage protein cI. These plasmids are transformed *(9)* into a yeast reporter strain to assess the suitability of the bait proteins for library screening. Comparison to established controls allows determination of appropriately synthesized baits that are not by themselves transcriptionally active and are nontoxic. If any of these conditions is not met, strategies for modifying bait or screening conditions are suggested (**Table 1**). To minimize the chance of false results or other difficulties, it is a good idea to move rapidly through the suggested characterization steps before

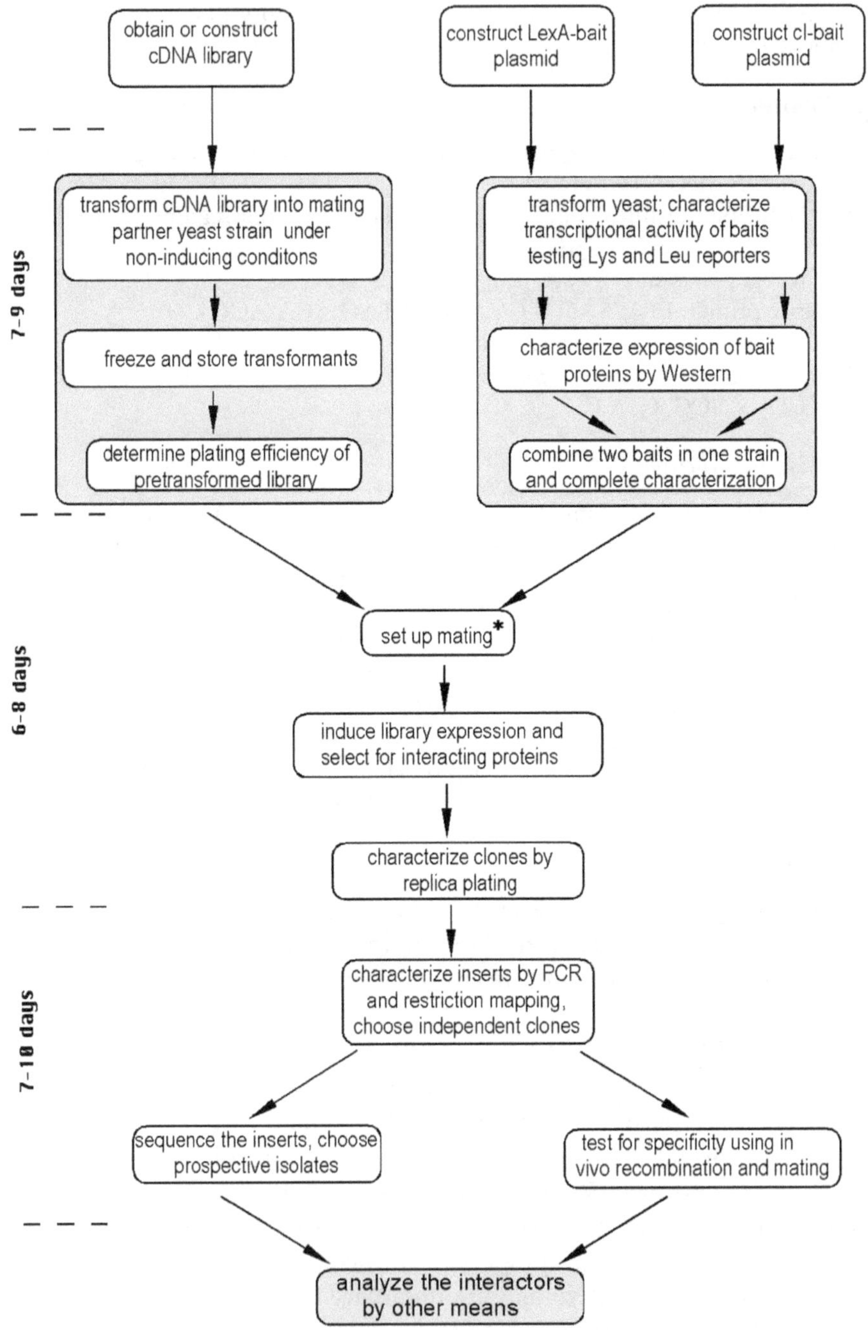

Fig. 2. Flow chart of the two-hybrid screen done by interaction mating (from the website in **Note 1**). Day 7–10 stage allows flexibility, shown in greater detail in **Fig. 5** (*see* text for details).

Table 1
Troubleshooting and Alternative Plasmids

A. Possible Modifications to Enhance LexA-Fused Bait Performance in Specific Applications (18)

Response / **Bait Problem**	truncate/modify bait	Use more stringent strain/reporter combination	fuse to nuclear localization sequence **pJK202**	Put LexA-fused protein under GAL1-inducible promoter **pGilda**	Fuse LexA to the carboxy-terminal end of the bait **pNLexA**	Integrate bait, reduce concentration **pEG202I**
strongly activating	+	-	-	-	-	-
weakly activating	+	+	-	-	+?	+?
low expression level or not transported to the nucleus	-	-	+	-	-	-
Continuous expression of LexA-fusion is toxic to yeast	+	-	-	+	+?	+?
Bait protein requires unblocked N-terminal end for function	-	-	-	-	+	-
Bait protein expressed at high levels, unstable or interacts promiscuously	+?	-	-	+?	+?	+
Potential New Problem*	It may be necessary to subdivide bait into two or three overlapping constructs, each of which must be tested independently	Use of very stringent interaction strains may eliminate detection of biological relevant interactions		Can no longer use Gal-dependence of reporter phenotype to indicate cDNA-dependent interaction	Generally, LexA poorly tolerates attachment of the N-terminal fusion domain. Only ~60% of constructs are expressed correctly.	Reduced Bait protein concentration may lead to reduced assay sensitivity

+: would usually help; +?: may help; - : will not help.

*: All of the alternative bait expression vectors have an Ap^R selection marker. If using them without modification, the investigator may need to use an *E. coli* KC8 strain to isolate the Ap^R-library plasmid after a library screening. *(continued)*

Table 1 ***(Continued)***
Troubleshooting and Alternative Plasmids

B. Possible Modifications to Enhance cI-Fused Bait Performance in Specific Applications (from the website in **Note 1**)

Response / Bait Problem	truncate / modify bait	Use more stringent strain/reporter combination	Put cI-fused protein under GAL1-inducible promoter **pGMS12**	Reduce concentration **pGBS9**
strongly activating	+	-	-	-
weakly activating	+	+	-	+?
Continuous expression of cI-fusion is toxic to yeast	+	-	+	+?
Bait protein expressed at high levels	+?	-	+?	+
Potential New Problem*	It may be necessary to subdivide bait into two or three overlapping constructs, each of which must be tested independently	Use of very stringent interaction strains may eliminate detection of biological relevant interactions	Can no longer use Gal-dependence of reporter phenotype to indicate cDNA-dependent interaction	Reduced Bait protein concentration may lead to reduced assay sensitivity

+: would usually help; +?: may help; - : will not help.

*: All of the alternative bait expression vectors have an Ap^R selection marker. If using them without modification, the investigator may need to use an *E. coli* KC8 strain to isolate the Ap^R-library plasmid after a library screening.

undertaking a library screening. Use freshly growing yeast cells whenever possible. Although plasmids can be retained for an extended period of time in yeast maintained at 4°C on stock plates, variable protein expression and transcriptional activation results will be more likely to occur in yeast left for approx 10–14 d at 4°C.

3.1.1. Construction of Plasmids for cI–Bait Fusion Protein and Yeast Transformation

1. Subclone the DNA encoding the protein of interest into the polylinker of pGBS10 (**Fig. 3**) to enable synthesis of an in-frame protein fusion to cI (*see* **Notes 2** and **3**).
2. Select a colony of yeast strain SKY191 and grow a 5-mL culture in liquid YPD medium overnight at 30°C with shaking (*see* **Note 4**).
3. Dilute the overnight culture into 50–60 mL of YPD liquid medium such that the diluted culture has an $OD_{600\ nm}$ of approx 0.15. Incubate at 30°C on an orbital shaker until this culture has reached an $OD_{600\ nm}$ of 0.5–0.7.

A **LexA-fusion vectors' polylinkers:**

pEG202 (aka pLexA, displayBait)

```
                          SalI        NotI          SalI
  EcoRI       BamHI             NcoI*        XhoI
GAA TTC CCG GGG ATC CGT CGA CCA TGG CGG CCG CTC GAG TCG AC
```

pLexZeo

```
EcoRI                                   ApaI            NotI         SalI
                SacI          PvuII          KpnI             XhoI          PstI
GAA TTC AAG CTT GAG CTC AGA TCT CAG CTG GGC CCG GTA CCG CGG CCG CTC GAG TCG ACC T
E   F   E   L   R   S   Q   L   G   P   V   P   R   P   L   E   S   T   C
```

B **cI-fusion vectors' polylinkers: description of AB variants**

A = AAT frame

```
                       BglII        ApaI            NotI          SalI
 EcoRI           SacI         PvuII         KpnI             XhoI
G AAT TCA AGC TTG AGC TCA GAT CTC AGC TGG GCC CGG TAC CGC GGC CGC TCG AGT CGA Cct
  N   S   S   L   S   S   D   L   S   W   A   R   Y   R   G   R   S   S   R   P
```

B = GAA frame

```
          EcoRI         BglII         ApaI            NotI          SalI
                  SacI          PvuII         KpnI             XhoI
G AAT Ttg GAA TTC GAG CTC AGA TCT CAG CTG GGC CCG GTA CCG CGG CCG CTC GAG TCG ACC TGC
  N   L   E   F   E   L   R   S   Q   L   G   P   V   P   R   P   L   E   S   T   C
```

C **AD-fusion (library) vectors' polylinkers:**

pJG4-5 (aka pB42AD, displayTarget)

```
                                      EcoRI           XhoI
ATG GGT GCT CCT CCA AAA AAG AAG ...  CCC GAA TTC GGC CGA CTC GAG AAG CTT ...
M   G   A   P   P   K   K   K   ...  P   E   F   G   R   L   E   K   L   ...
```

pYesTrp2

```
                              KpnI        BamHI
                        HinDIII     SacI
ATG GGT AAG CCT ...     AAG CTT GGT ACC GAG CTC GGA TCC ACT AGT AAC GGC
M   G   K   P           K   L   G   T   E   L   G   S   T   S   N   G

                EcoRI                             NotI          SphI
    BstXI                              BstXI                XhoI
CGC CAG TGT GCT GGA ATT CTG CAG ATA TCC ATC ACA CTG GCG GCC GCT CGA GGC ATG C
R   Q   C   A   G   I   L   Q   I   S   I   T   L   A   A   A   R   G   M   H
```

Fig. 3. Polylinkers of two-hybrid basic vectors. Maps and sequences are available on the Golemis lab Website (*see* **Note 1**). Only restriction sites that are available for subcloning are shown. Those shown in bold are unique. pEG202 is the same as pMW103 except that it has an Ap^R marker.

4. Transfer 50 mL of culture to a sterile 50-mL Falcon tube, and centrifuge for 5 min at 1000–1500*g* at room temperature. Gently resuspend the pellet in 5 mL of sterile water.
5. Re-centrifuge the cells for 5 min at 1000–1500*g* and resuspend the yeast cells in 0.5 mL of LiOAc solution. This volume of competent yeast cells is sufficient for 10 transformations.
6. Add 50 μL of competent yeast cells to microfuge tubes containing 1 μg of freshly sheared, denatured salmon sperm DNA and the following cI plasmids (100–500 ng each):
   a. pGBS10-Bait2 (test for activation).
   b. pGBS9-Krit (weak positive control for activation).
   c. pGBS9-VP16 (strong positive control for activation).
   d. pGBS10-Krev1 (negative control for activation).
7. To each tube, add 300 μL of sterile LiOAc/PEG solution. Mix by gently inverting the tubes several times (do not vortex). Place the tubes at 30°C for 30–60 min.
8. To each tube, add 40 μL of dimethylsulfoxide, and again mix by inversion. Place the tubes at 42°C for 10 min.
9. Centrifuge the cells for 20 s at 10,000–15,000*g*. Pour off the supernatant and resuspend the yeast in 0.5 mL of YPD. Incubate for 6 h to overnight at 30°C (*see* **Note 5**).
10. Plate each transformation mixture on YPD G418 plates, and maintain at 30°C for 2 d to select for transformed yeast colonies containing desired plasmids.

3.1.2. Construction of Plasmids for LexA–Bait Fusion Protein and Yeast Transformation

1. Subclone the DNA encoding the protein of interest into the polylinker of pMW103 (**Fig. 3**) to enable synthesis of an in-frame protein fusion to LexA (*see* **Notes 2** and **3**).
2. Select a colony of yeast strain SKY48 and prepare competent yeast as described above (**Subheading 3.1.1., steps 2–5**).
3. Add 50 μL of competent yeast cells to microfuge tubes containing 1 μg of freshly sheared, denatured salmon sperm DNA and the following combinations of LexA fusion and *lexAop-LacZ* plasmids (100–500 ng each):
   a. pMW103-Bait1 + pLacGus (test for activation).
   b. pEG202-hsRPB7 + pLacGus (weak positive control for activation).
   c. pSH17-4 + pLacGus (strong positive control for activation).
   d. pEG202-Ras + pLacGus (negative control for activation).
4. Continue transformation as described above (**Subheading 3.1.1., steps 7–9**), except in **step 9**, where yeast should be resuspend in sterile water rather than in YPD, and eliminating the overnight incubation.
5. Plate each transformation mixture on Glu/CM –Ura-His dropout plates, and maintain at 30°C for 2–3 d to select for transformed yeast colonies containing desired plasmids (*see* **Note 6**).

3.1.3. Assessing Bait Activation of Reporters Using Replica Technique

For each combination of plasmids, assay at least six independent colonies for activation phenotype of auxotrophic and colorimetric reporters (*see* **Note 7**). Assessment of transcriptional activation requires the transfer of yeast from master plates to several selective media. This transfer can be accomplished simply by using a sterile toothpick to move cells from individual patches on the master plate to each of the selective media. However, in cases in which large numbers of colonies and combinations of bait and prey are to be examined, and particularly in genomic-scale applications, it is useful to use a transfer technique that facilitates high-throughput analysis. The following technique, based on microtiter plates, is an example of such an approach.

1. Add approx 50 μL of sterile water to each well of one-half (wells A1–H6) of a 96-well microtiter plate (e.g., using a syringe-based repeater). Place an insert grid from a rack of micropipet tips over the top of the microtiter plate and attach it with tape. The holes in the insert grid should be placed exactly over the wells of the microtiter plate. Placing insert is not essential but will stabilize the tips in the plate and allow simultaneous removal, speeding the replica process.
2. Using sterile plastic micropipet tips, pick six yeast colonies (1–2 mm diameter) from each of the transformation plates *a–d* (**Subheading 3.1.1., step 6** and **Subheading 3.1.2., step 3**). Leave the tips supported in a near-vertical position by the insert grid until all the colonies have been picked.
3. Swirl the plate gently to mix the yeast into suspension, remove the sealing tape, and lift the insert grid, thereby removing all the tips at once.
4. Use a replicator to plate yeast suspensions (*see* **Note 8**). Each spoke will leave a drop approximately equal to a 3-μL volume on the following plates:
   i. To assess activation of cI-fused bait, use YPD G418 (a new master plate), Glu/CM –Lys, and Gal-Raff/CM –Lys plates (for LYS2 reporter). Incubate the plates at 30°C for up to 4 d and save the YPD G418 master plate at 4°C.
   ii. To assess activation of LexA-fused bait, use Glu/CM –Ura-His (a new master plate), Gal-Raff/CM –Ura-His-Leu (for LEU2 reporter), and Gal-Raff/CM –Ura-His (for LacZ reporter in **step 6**). Incubate the plates at 30°C for up to 4 d, and save the Glu/CM –Ura-His master plate at 4°C.
5. For yeast with LexA-fused bait, growth of the strong positive control should be detectable within 1–2 d on Gal-Raff/CM –Ura-His-Leu plates. Yeast containing the weak positive control should exhibit growth within 4 d. Yeast containing the negative control should not grow. If the yeast containing the bait under test shows no growth in this period, it is probably suitable for library screening (*see* **Notes 6** and **10**).
6. For detecting the expression of the LacZ reporter ***(10)***, overlay the Gal-Raff/CM –Ura-His plates with X-Gal agarose in about 24–30 h after the plating as follows (*see* **Note 9**):
   i. Gently overlay each plate with chloroform. Pipet slowly in from the side so as not to smear colonies. Leave colonies completely covered for 5 min. Caution:

$CHCl_3$ is quite toxic and should neither be inhaled nor allowed to come into contact with skin. Wear gloves and work in a chemical hood. Try to minimize the amount of $CHCl_3$ used, just enough to cover the colonies. Avoid prolonged contact of the $CHCl_3$ with the plate, as the plastic will melt.

ii. Briefly rinse the plates with another approx 5 mL of chloroform (optional). Drain and let dry, uncovered, for another 5 min at 37°C or for 10 min in the hood.

iii. Overlay the plate with approx 10 mL of X-Gal-agarose, making sure that all yeast spots are completely covered. Plates will be chilled after $CHCl_3$ evaporation, so it will be difficult to spread less than 7 mL of top agarose.

iv. Incubate plates at 30°C and monitor for color changes. It is generally useful to check the plates after 20 min, and again after 1–3 h. In assessing the transformants from reactions **a–d** in **Subheading 3.1.2.**, strongly activating baits will be detectable as dark blue colonies in 20–60 min, whereas negative controls should remain as faint blue or white colonies. An optimal bait would either mimic the negative control or only develop faint blue color (*see* **Note 9**).

7. For yeast with cI-fused baits: yeast containing positive controls should exhibit growth within 2–4 d on Glu/CM -Lys and Gal-Raff/CM -Lys plates. Yeast containing the negative control should not grow (*see* **Note 10**).

Ideally, all six colonies assayed representing the same transformation would have essentially the same phenotype. For a small number of baits, this is not the case. The most typical deviation is that of six colonies assayed for a new bait, some fraction appear to be inactive while the remaining fraction display some degree of blueness and growth. Yeast cells with inactive baits appear white in colorimetric assay and do not grow on auxotrophic selection medium. Do NOT select the white, nongrowing colonies as starting point in a library screen; generally, these colonies synthesize little or no bait protein, as can be assayed by Western blot.

3.1.4. Detection of Bait Protein Expression

One excellent confirmation that a bait is correctly expressed would be its specific interaction with a known partner expressed as an AD fusion protein. In the absence of such confirmation, Western analysis of yeast lysates containing LexA or cI-fused baits is helpful in characterization of the bait's expression level and size (*see* **Note 11**).

1. From the appropriate master plate (**Subheading 3.1.2., step 5**), inoculate at least two primary bait/reporter transformants for each bait to be tested into liquid medium (same as medium on the master plate). Include pGBS10-Krev1 and pEG202-Ras transformants as a positive and negative controls for protein expression. Grow overnight cultures on an orbital shaker at 30°C. Dilute the saturated cultures into fresh tubes containing 2 mL of the same medium to an $OD_{600\ nm}$ of approx 0.15, and grow at 30°C (*see* **Note 12**).

2. After the $OD_{600\ nm}$ of the cultures reaches 0.45–0.7 after about 4–6 h, harvest 1.5 mL of each culture by centrifuging at 13,000*g* for 3–5 min (*see* **Note 13**).
3. Add 50 µL of 2X Laemmli sample buffer to each pellet and rapidly vortex to resuspend pellet. Boil samples for 5 min for immediate assay, or freeze at –70°C for subsequent use. Such samples will be stable for at least 4–6 month.
4. After heating, chill the samples on ice and centrifuge for 30 s at 13,000*g* to pellet large cell debris. Load 10–25 µL of each sample onto a SDS polyacrylamide gel.
5. Prepare a Western blot and screen LexA fusions using an antibody to LexA, and cI fusions using an antibody to cI ***(11,12)***, allowing comparison of expression levels of the bait protein under test with other standard bait proteins, e.g., provided controls (*see* **Note 14**).
6. Note which colonies on the master plate express bait appropriately and use one of these colonies as a founder to propagate for second bait transformation/library mating.

3.1.5. Combining the Two Baits Together and Finalizing Bait Characterization

1. Select a colony each of yeast strain SKY191 expressing pGBS10-Krit and pGBS10-Krev1, and two colonies expressing pGBS10-Bait2. Grow 2 mL of each culture in liquid YPD-G418 medium overnight at 30°C in a shaking incubator. Prepare competent cells as described above.
2. Prepare four tubes containing 1 µg of freshly sheared, denatured salmon sperm DNA and approx 250 ng of pLacGus plasmid. Into these tubes, mix 50 µL of competent yeast cells with the LexA fusion and reporter plasmids (100–500 ng each), to obtain the following baits/reporter combinations:
   a. pMW103-Bait1 + pLacGus + pGBS10-Bait2 (clone 1).
   b. pMW103-Bait1 + pLacGus + pGBS10-Bait2 (clone 2).
   c. pEG202-Ras + pLacGus + pGBS10-Krev1.
   d. pEG202-hsRPB7 + pLacGus + pGBS9-Krit.
3. To each tube, add 300 µL of sterile LiOAc/PEG solution . Mix by gently inverting the tubes several times. Place the tubes at 30°C for 30–60 min.
4. To each tube, add 40 µL of dimethylsulfoxide, and again mix by inversion. Place the tubes at 42°C for 10 min.
5. Centrifuge the cells for 20 s at 10,000–15,000*g*. Pour off the supernatant and resuspend the yeast in 0.15 mL of sterile water. Plate each transformation mixture on Glu/CM –Ura-His G418 dropout plates, and maintain at 30°C for 2–4 d to select for yeast colonies containing transformed plasmids (*see* **Note 15**).
6. Replica-plate 12 colonies of each transformants on the following set of plates:

   Glu/CM –Ura-His G418 (a new master plate).
   Gal-Raff/CM –Ura-His G418 (for X-Gal overlay assay).
   Gal-Raff/CM –Ura-His G418 (for X-Gluc overlay assay).
   Gal-Raff/CM –Ura-His-Leu.
   Gal-Raff/CM –Ura-His-Lys.

Incubate the plates at 30°C for up to 4 d, and save the Glu/CM -Ura-His master plate at 4°C.

7. It is recommended to reconfirm expression of the baits by Western blotting. Confirmed bait-containing strains will be used for library screening in **Subheading 3.2.3.**

3.2. Screening of cDNA Libraries for Interacting Proteins (Fig. 4)

A partial list of available libraries compatible with the Interaction Trap can be found at the website in **Note 1**. Currently, the most convenient source of libraries suitable for the Interaction Trap is commercial; the broadest selection is found at Origene. If one wishes to make one's own library, it should be cloned in a vector such as pJG4-5 or a related vector such as pYesTrp2 (Invitrogen). The polylinker sequence at the site of cDNA insertion is shown in **Fig. 3**.

The following protocols are designed with the goal of saturation screening of a cDNA library derived from a genome of mammalian complexity. Fewer plates will be required for screens with libraries derived from organisms with less complex genomes, and researchers should scale back accordingly. A protocol is provided for mating in ref. ***13*** the library against the bait of interest. An advantage of this approach is that if the investigator wishes to use the same library to screen multiple baits, only a single large-scale transformation is required (*see* **Note 16**).

It is generally a good idea to additionally mate new bait strains with a negative control strain. The control strain is the same strain used for the library but containing the library vector with no cDNA insert. This control will provide a clear estimate of the frequency of cDNA-independent false positives, which is important to know when deciding how many positives to pick and characterize. In a subsequent characterization of potential interactors, a positive control is useful. While you set up your library transformation or interaction mating, it is useful to use the same technique to get a set of three proteins, expressed in library plasmids, in the same strain. Normally, these positive controls will interact with either or both of the baits of the control strain SKY191(pEG202-Ras + pLacGus + pGBS10-Krev1) obtained in **Subheading 3.1.5., step 2c**.

3.2.1. Transforming the cDNA Library

1. Select a colony of yeast strain SKY473 and grow a 20-mL culture in liquid YPD medium overnight at 30°C in an orbital shaker (*see* **Note 4**).
2. Dilute the 20-mL overnight culture into approx 300 mL of YPD liquid medium such that the diluted culture has an $OD_{600\ nm}$ of 0.15. Incubate at 30°C on an orbital shaker until the culture has reached an $OD_{600\ nm}$ of 0.5–0.7.
3. Subdivide the culture among six sterile 50-mL tubes, and centrifuge at 1000–1500*g* for 5 min at room temperature. Gently resuspend each pellet in 5 mL of

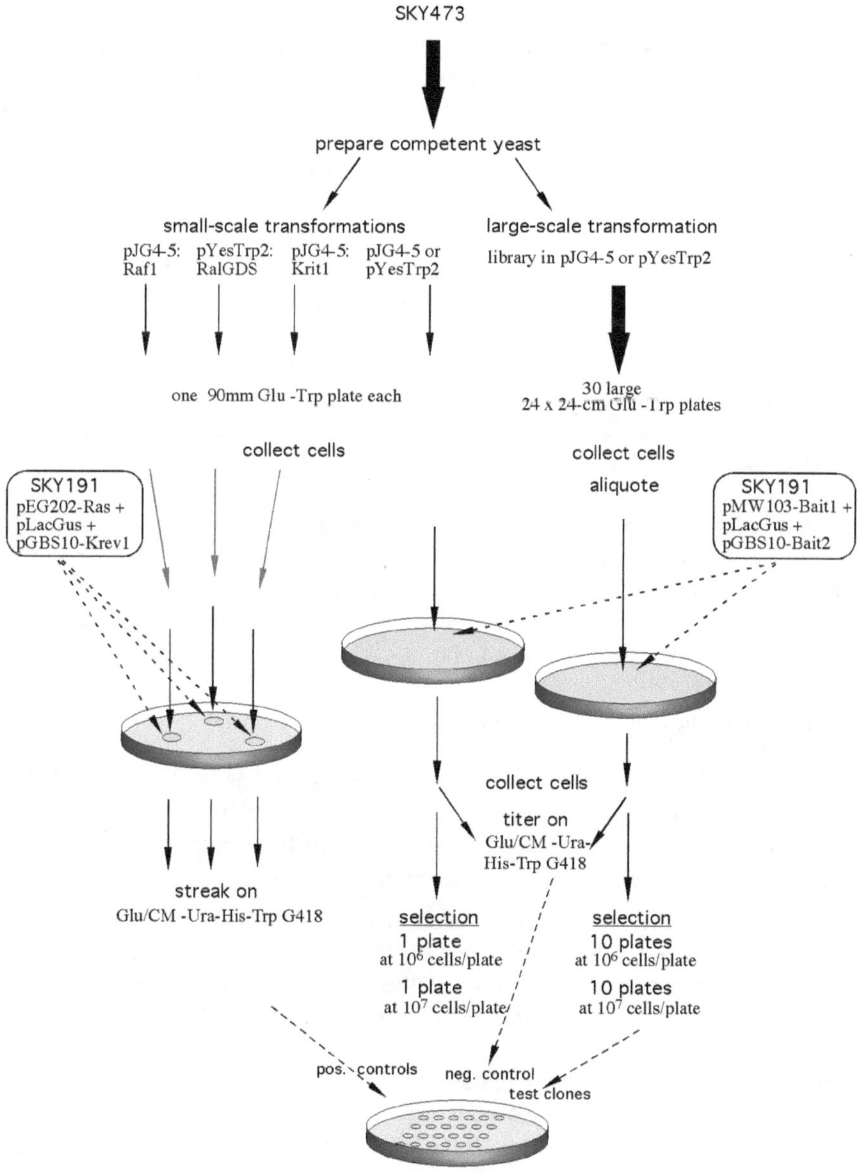

Fig. 4. Detailed library screening flow chart (from the website in **Note 1**; *see* text for details).

sterile water, and combine all the slurries into a single tube. Add sterile water and mix.

4. Recentrifuge the cells at 1000–1500*g* for 5 min at room temperature. Pour off the water and resuspend the yeast in 1.5 mL of LiOAc solution.

5. Mix 30 μg of library DNA and 1.5 mg of freshly denatured sheared salmon sperm DNA. Add the DNA mix to the yeast. Mix gently and dispense 60-μL aliquots of DNA/yeast suspension into 30 microfuge tubes (*see* **Note 17**).
6. To each tube, add 300 μL of sterile LiOAc/PEG solution. Mix by gently inverting the tubes several times. Do not vortex. Place the tubes at 30°C for 30–60 min.
7. To each tube, add 40 μL of dimethylsulfoxide, and again mix by inversion. Place the tubes at 42°C for 10 min.
8. Pipet the contents of each tube onto a separate 240 × 240 mm Glu-Trp dropout plate, and spread the cells evenly using 12–24 sterile glass beads. Invert the plates without discarding glass beads (*see* **Note 18**) and incubate at 30°C until all colonies appear (3–4 d).
9. Select two representative transformation plates. Draw a 23 × 23 mm square (1% of the plate bottom surface) over an average density spot, count the colonies in each grid section, and recalculate for the whole transformation. A good transformation performed according to this protocol should yield approx 20,000–40,000 colonies per plate.
10. Use small aliquots of competent yeast from **step 4** to transform the empty library plasmid (pJG4-5 or pYesTrp2), pJG4-5:Raf1, pJG4-5:Krit1, and pYesTrp2: RalGDS. Plate on 100 mm plate, and collect the transformed cells as for the library (protocol outlined below), scaling down accordingly. The control strain with an empty library plasmid can be safely reamplified in liquid medium.

3.2.2. Harvesting and Pooling Primary Transformants

In the next step, a homogenized slurry is prepared (*see* **Note 19**) from the pool of primary transformants (approx 3×10^5–10^6 colonies), aliquoted, and frozen. Each of these aliquots is representative of the complete set of primary transformants and can be used in subsequent mating.

1. Pour 10 mL of sterile water onto each of five 240 × 240 mm plates containing transformants. Stack the five plates on top of each other. Holding on tightly, shake the stack horizontally until all the colonies are resuspended (1–2 min). Using a sterile pipet, collect the yeast slurry from each plate by tilting the plates, and pool in a sterile 50-mL tube.
2. Repeat for further sets of five plates of transformants, resulting in a total of up to 150 mL of suspension (*see* **Note 20**).
3. Fill each tube containing yeast to the top with sterile TE or water, and vortex/invert to suspend the cells. Centrifuge the tubes at 1000–1500*g* for 5 min at room temperature, and discard the supernatants. Repeat this step. After the second wash, the cumulative pellet volume should be approx 25 mL of cells derived from up to 10^6 transformants.
4. Resuspend each packed cell pellet in 1 vol of glycerol solution. Combine the contents of the three tubes and mix thoroughly. Disperse as 0.2- to 1.0-mL aliquots in sterile Eppendorf tubes and freeze at –70°C. The aliquots are stable for at least 1 y.

3.2.3. Mating the Bait Strain and the Pretransformed Library Strain

After the bait strain has been made and characterized, and the library strain has been transformed and frozen in aliquots, the next step is to mate the two strains. To mate the two strains, the bait strain is grown in liquid culture and mixed with a thawed aliquot of the pretransformed library strain. The mixture is plated onto rich media and grown overnight. During this time individual cells of the bait strain will fuse with individual cells of the library strain to form diploid cells. The diploids, along with unmated haploids, are collected and plated on media to select for interactors (**Subheading 3.2.4.**). In practice, the diploid/haploid mixture is generally frozen in a few aliquots to allow titering and repeated platings at various dilutions. Mating with the negative control strain (*see* **Subheading 3.2.1.**) should be performed at the same time as the library mating, and both matings can be treated identically in the next step, selecting interactors.

1. Start a 30-mL Glu/CM –Ura-His G418 liquid culture of the bait strain (pMW103-Bait1 + pLacGus + pGBS10-Bait2) from the master plate (**Subheading 3.1.5.**; *see* **Note 21**). Grow with shaking at 30°C to mid- to late-log phase ($OD_{600\ nm}$ = 1.0–2.0).
2. Collect the cells by centrifuging at 1000*g* for 5 min at room temperature. Resuspend the cell pellet in 1 mL of sterile water and transfer to a sterile 1.5-mL microfuge tube. This will yield yeast suspension of about 1×10^9 cells/mL.
3. Thaw an aliquot of the pretransformed library strain and an aliquot of the negative control strain at room temperature. Mix approx 2×10^8 cells of the bait strain (approx 200 µL) with approx 10^8 colony-forming units (CFU) of the pretransformed library strain (*see* **Subheading 3.2.2.**) on a single 100-mm diameter YPD plate and incubate at 30°C for 12–15 h. In parallel, set up mating with the negative control strain.
4. Add 1.5–2 mL of sterile water to the surface of each YPD plate and suspend the cells using sterile glass beads. Transfer the suspension to a sterile tube and vortex gently for 2 min. Collect the cells by centrifugation at 1000*g* for 5 min and resuspend in 1 vol of sterile glycerol solution. Distribute into 200 µL aliquots and freeze at –80°C (*see* **Note 22**). Another option is to leave one aliquot unfrozen and proceed directly to plating on selective medium.
5. Titer the mated cells by plating serial dilutions on Glu/CM –Trp-His-Ura plates (unmated haploids will not grow on this medium). Count the colonies that grow after 2–3 d, and determine the titer of the frozen mated cells (*see* **Note 23**).
6. In parallel with **steps 1–4** above, grow up approx 1.5 mL of control bait strain (pEG202-Ras + pLacGus + pGBS10-Krev1) in Glu/CM –Ura-His G418. Also, grow up three control prey strains (*see* **Subheading 3.2.1., step 10**) in Glu/CM-Trp. On a YPD plate, make three spots of control bait strain by placing a drop (approx 5 µL) of the liquid culture on its surface. Without waiting for the liquid to soak in, add 5 µL of one of the three control prey strains to the same spots.

Incubate overnight, and then streak all three matings onto Glu/CM –Ura-His-Trp G418 plates.

3.2.4. Screening for Interacting Proteins

This section describes how interactors are selected by plating the mated cells onto auxotrophic selection plates. It is important to know how many viable diploids were plated onto these selection plates in order to gain a sense of how much of the library has been screened and to determine the false-positive frequency. This information is provided by the titer (colony-forming units per milliliter) of the frozen mated cells (*see* **Subheading 3.2.3., step 6**).

Dual Bait allows selection for the interaction with LexA-fused bait (on Gal-Raff/CM –Ura-His-Trp-Leu G418 plates), or for the interaction with cI-fused bait (on Gal-Raff/CM –Ura-His-Trp-Lys plates), or for interaction with both at the same time (on Gal-Raff/CM –Ura-His-Trp-Leu-Lys plates) (*see* **Note 24**). Negative selection with one interaction but not the second one is theoretically possible, but impractical. Depending on the purpose of your screen, you will plate mated cells on the appropriate selective plates. If your second bait is to be used mainly to increase the specificity of your primary screen, plate your mating on Leu selection plates only. If you screen two independent baits, plate on two separate sets of Leu selection and Lys selection plates. Finally, if you are trying to identify proteins that simultaneously interact with both baits, we suggest plating a fraction of your mating on Leu selection plates and another fraction on double selection (-Leu-Lys) plates.

1. Thaw an aliquot of the mated yeast cells. Dilute 100 µL into 10 mL of Gal-Raff/CM –Ura-His-Trp G418 liquid dropout medium and incubate with shaking at 30°C for 5 h. If the frozen culture was not previously titered, plate serial dilutions onto Glu/CM –Ura-His-Trp G418 plates.
2. On the assumption that a culture at $OD_{600\ nm} = 1.0$ contains 1×10^7 cells/mL, plate 10^6 cells on five 100-mm plates with the appropriate auxotrophic selection medium. Plate 10^7 cells on each of five additional plates with the same medium. Plating 10^7 cells/plate allows screening more diploids on fewer number of plates, but may or may not result in higher levels of background growth.
3. Incubate for up to 6 d at 30°C (*see* **Note 25**). Depending on the individual bait used, good candidates for positive interactors will generally produce LEU^+ colonies over this time period, with the most common appearance of colonies at 2–4 d. LYS^+ colonies typically form at 3–5 d (*see* **Note 26**).
4. Inspect the plates on a daily basis. Mark the location of colonies visible on d 1 with dots of a given color on the plate. Each day, mark new colonies with different colors. At d 4 or 5, streak colonies in a microtiter plate format onto a solid master plate (Glu/CM –Ura-His-Trp G418), in which colonies are grouped by day of appearance (*see* **Note 27**). If many apparent positives appear, pick separate master plates for colonies arising on d 2, 3, and 4, respectively (*see* **Note 28**).

5. Include the positive control colonies (from mating with the control baits strain) on each of the master plates. Also, include a few colonies from the titer plate (**step 1** above); because they contain randomly chosen library plasmids, the phenotype of these colonies is most likely be negative.
6. Incubate the master plates at 30°C until patches/colonies form.

3.2.5. Confirmation of Positive Interactions

The following steps test for galactose-inducible transcriptional activation of both the auxotrophic and colorimetric reporters. Simultaneous activation of both reporters in a galactose-specific manner generally indicates that the transcriptional phenotype is attributable to expression of library-encoded proteins, rather than derived from mutation of the yeast.

1. Invert a frogger on a flat surface and place a master plate upside down on the spokes, making sure that the spokes and colonies are properly aligned. Remove the plate and insert the frogger into a microtiter plate containing 50 μL of sterile water in each well. Let the plate sit for 5–10 min, shaking from time to time to resuspend the cells left on the spokes. When all yeast are resuspended, print on the following plates (*see* **Note 8**):
   a. Master plate: Glu/CM –Ura-His-Trp G418
   b. Test for activation of LEU2: Gal-Raff/CM (–Ura-His-Trp)-Leu, Glu/CM (–Ura-His-Trp)-Leu
   c. Test for activation of LYS2: Gal-Raff/CM (–Ura-His-Trp)-Lys, Glu/CM (–Ura-His-Trp)- Lys
   d. Two sets of plates, for *LacZ* and *GusA* activation: Glu/CM –Ura-His-Trp G418, Gal-Raff /CM –Ura-His-Trp G418

   (*see* **Notes 24** and **29**).
2. Repeat for each master plate (from **Subheading 3.2.4., step 6**).
3. Incubate the plates at 30°C for 3–4 d. After 20–30 h, take out all –Ura-His-Trp plates. Retain one Glu/CM –Ura-His-Trp G418 plate as a fresh master plate, and overlay the remaining two sets with X-Gal or X-Gluc agarose, as described in **Subheading 3.1.3., step 6**. Score growth on the –Leu and –Lys plates 48–72 h after plating.
4. For interpretation of the results, refer to **Table 2**.

3.3. DNA Isolation and Secondary Confirmation of Positive Interactions

Execution of the above protocols for a given bait will result in the isolation of between zero to hundreds of potential "positive" interactors (*see* **Note 30**). These positives must be evaluated for reproducible phenotype and specific interaction with the bait used to select them, using a strategy outlined in **Fig. 5**.

Table 2
Interpretation of Primary Isolates' Behavior (from the Website in **Note 1**)

A. Interpretation of Interaction Phenotype for Each Bait

Observed phenotype				Interpretation		Recommendation
Auxotrophic. reporter		Colorimetric reporter				
Glu	Gal	Glu	Gal	Traditional	Optimistic	
-	+	-	+	Very good sign		Work with those clones first
(+)	+	(+)	+	Bait is mutated or its expression is upregulated to give a high background of transcriptional activation.	• GAL1 promoter is slightly leaky • both proteins are very stable • interaction occurs with high affinity	Take one clone for confirmation of interaction, store the rest.
-	+	-	-	Yeast mutation occurred that favors growth or transcriptional activation on galactose medium	Some bait-interactor combinations are known to preferentially activate one reporter versus another	If all other candidates winnow out, check these clones.
All other phenotypes				Contamination/plasmid rearrangements/ mutations	Something really new	Trash

Each feature marked as positive should be also Gal-dependent, as in Table 2A, top row. Comparing the phenotype of the clones of interest to this set of controls should help to assess whether the isolated prey interacts with one or both baits.

(continued)

Table 2 *(Continued)*
Interpretation of Primary Isolates' Behavior
(from the Website in **Note 1**)

B. Expected Phenotype for Control Set of Interactions

Bait	Prey	LEU2	lacZ	LYS2	gusA	**Explanation**
cI-Krev + LexA-Ras	Raf1	+	+	-	-	interaction with LexA-fused bait only
	RalGDS	+	+	+	+	interaction with both baits
	Krit1	-	-	+	+	interaction with cI-fused bait only
LexA-bait1 + cI-bait2	random	-	-	-	-	no interaction with any bait

Each feature marked as positive should be also Gal-dependent, as in Table 2A, top row. Comparing the phenotype of the clones of interest to this set of controls should help to assess whether the isolated prey interacts with one or both baits.

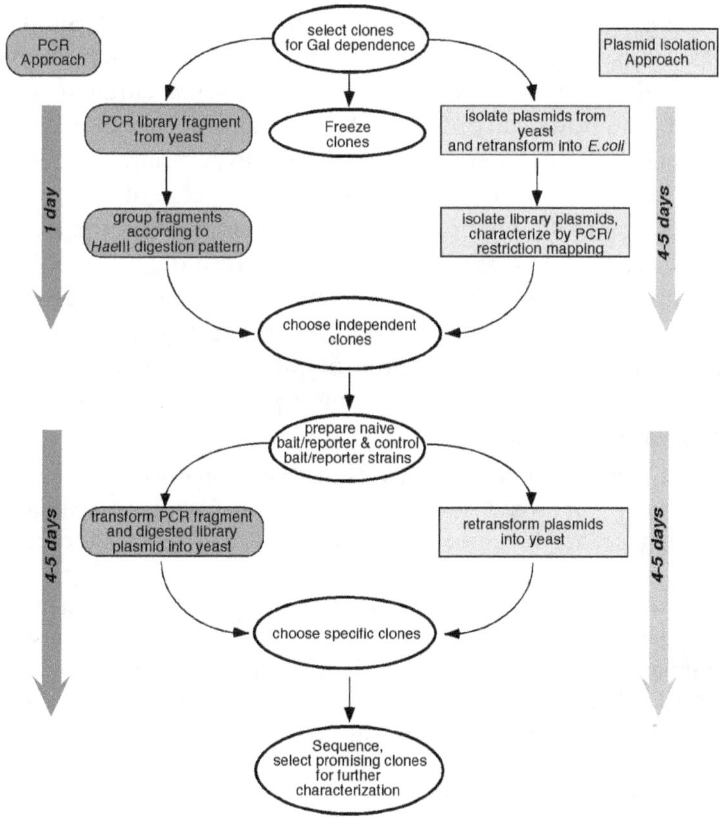

Fig. 5. Detailed flow chart for characterization and secondary confirmation of primary positives (*(18) see* text for details).

If a large number of positives are obtained, these subsequent characterizations require prioritization. In this case, select up to approx 24–48 independent colonies with robust phenotype for the first round of characterization, while maintaining a master plate of additional positives at 4°C. This first analysis set will be tested for specificity, and screened by PCR/restriction analysis and/or sequencing to determine if clusters of frequently isolated cDNAs are obtained. Such clusters are generally a good indication for a specific interaction. Two strategies for analyzing positives are provided below, and are summarized in the flow chart in **Fig. 5**. While both utilize similar methods, the order with which techniques are applied differ. The choice between strategies depends on whether the individual investigator would rather spend time and money doing bulk yeast plasmid recovery, or bulk PCR. The PCR protocol is generally 1–3 d faster.

3.3.1. PCR Approach: Rapid Screen for Interaction Trap Positives

A major strength of the PCR protocol is that it will identify redundant clones prior to plasmid isolation and bacterial transformation, which in some cases greatly reduces the amount of work required. However, accurate records should be maintained as to how many of each class of cDNAs are obtained and if any ambiguity is present as to whether a particular cDNA is part of a set or unique. The protocol outlined below includes enzymatic treatment to generate crude yeast lysates (**steps 1–3**), which is later used as template for the PCR reaction (**step 4**). PCR product can be obtained directly from the yeast colonies even without beta-glucuronidase treatment, e.g., by introducing a 10-min –94°C step at the beginning of the PCR program. However, yeast lysates obtained in this protocol also can be used as a source of plasmid for electroporation into *E. coli*, instead of the more time-consuming plasmid recovery protocol outlined in **Subheading 3.3.2.**

1. Starting from the Glu/CM –Ura-His-Trp G418 master plate, resuspend yeast in 25 μL of beta-glucuronidase solution in a 96-well microtiter plate. Seal the wells using tape, and incubate on a horizontal shaker at 37°C for 1.5–3.5 h (*see* **Note 31**).
2. Remove the tape, add about 25 μL of glass beads (0.15–0.45 mm) to each well, and seal again. Attach the microtiter plate to a vortex with a flat top surface (e.g., using rubber bands) and mix vigorously for 5 min.
3. Add 100 μL of sterile distilled water to each well. Take 0.8–2 μL as a template for each PCR reaction. Reseal the plate with tape, and keep the rest frozen at –70°C.
4. PCR.
   i. Use primers specific for the library plasmid used (*see* **Subheading 2.10.**). Perform a PCR amplification in approx 30 μL volume as follows: (a) 2 min at 94°C; (b) 45 s at 94°C; (c) 45 s at 56°C; (d) 45 s at 72°C, 31 cycles of **b–d** (*see* **Note 32**).

Table 3
Recommended Controls for PCR and Interpretation of Results *(18)*

Template	Possible outcomes			
Plasmid*	-	+	+	+
Yeast pPrey-control **	-	-	-	+/-
Plasmid + Yeast pPrey-control	-	+	-	+/-
Clone 1…to n	-	-	-	+/-
Interpretation	Bad mastermix/ Wrong settings/ Faulty amplifier	Not enough template	Lysed yeast inhibited PCR	Too much yeast: Uneven template load
Recommen-dation	Double-check/ Repeat	Add more template/ improve lysis	Add less template	Adjust template load/re-PCR from obtained bands

* PCR from the empty vector yields a product of approx 130 bp for JG4-5 (FP1 and FP2 primers); approx 185 bp for YesTrp (YesTrp forward and reverse primers).

** Approximate fragment sizes are approx 1.6 kb for Krit1, approx 2 kb for Raf1, 0.4 kb for RalGDS.

ii. In parallel, set up PCR reactions from the following control templates: Empty library plasmid (diluted!) and yeast from the positive control colonies (**Subheading 3.2.1., step 10**), treated along with experimental clones as above (*see* **Note 33**); and the same amount of diluted library plasmid and positive control yeast, mixed together. For analysis of results, *see* **Table 3**.

5. Take 10 μL of the PCR product for the following *Hae*III digestion in **step 6** and use the remainder of the PCR product (about 20 μL) for analysis on a 0.7% agarose gel. Identify PCR fragments that appear to be of the same size. Put gel in a refrigerator until you are ready to isolate fragments (*see* **Note 34**).
6. Perform a restriction digest of 10 μL of the PCR product with *Hae*III in a total volume of 20 μL. Rearrange the loading order according to the results obtained with nondigested PCR (**step 5**), and load the digestion products on a 1.5% agarose gel. Run out the DNA a sufficient distance to get good resolution of DNA products in the 200–1000 bp size range. This will generally yield distinctive and unambiguous groups of inserts, confirming whether multiple isolates of a small number of cDNAs have been obtained.

7. Purify fragments from the agarose gel using standard molecular biology techniques. In cases where a very large number of isolates representing a small number of cDNA classes have been obtained, the investigator may choose to directly sequence the PCR product (*see* **Note 35**).
8. The next step is to determine whether isolated cDNAs reproduce interaction phenotypes specifically with the bait(s) of interest and to exclude library-encoded cDNAs that either interact with the baits in a nonspecific manner or result in nonspecific transcriptional activation. This can be done using a PCR-recombination approach in a single step ***(13)***. Confirmed specific positive clones can be isolated through conventional plasmid purification.
   i. Digest an empty library plasmid with two enzymes producing incompatible ends in the polylinker region (e.g., *Eco*RI and *Xho*I) (*see* **Note 36**).
   ii. Perform PCR from positive control plasmid(s) using the same primers as before (*see* **Note 33**) and purify the PCR product if necessary.
   iii. Transform SKY191 containing pEG202-Ras + pLacGus + pGBS10-Krev1 (*see* **Note 21**) with (a) digested library plasmid, (b) digested library plasmid (50–100 ng) and control PCR product (0.5–1 μg), and (c) uncut library plasmid. Save the digested library plasmid and the pPrey-control PCR product for **step 13** (*see* **Note 37**).
   iv. Plate the transformations on Glu/CM –Ura, –His, –Trp dropout plates and incubate at 30°C until colonies grow (2–3 d). Store at 4°C (*see* **Note 38**). When transformed together, the PCR amplified cDNA fragment from pPrey-control PCR product and the digested library plasmid will undergo homologous recombination in vivo in up to 97% of the transformants that acquired both vector and insert ***(15)***. This is due to the identity between the cDNA PCR fragment and the plasmid at the priming sites. If transformation efficiency in (**iii.b**) is better than in (**iii.a**) by 5–20-fold, you can proceed to the next steps. (**iii.c**) is a positive control for the transformation.
   v. Using same ratios as in **step iii.b** above, transform digested library vector in combination with selected PCR products (again, include positive control(s) from **step ii**) to strains containing the following: (a) pMW103-Bait1 + pLacGus + pGBS10-Bait2; (b) pEG202-Ras + pLacGus + pGBS10-Krev1.
   vi. Plate each transformation mix on Glu/CM –Ura, –His, –Trp G418 dropout plates and incubate at 30°C until colonies grow (2–3 d).
   vii. Prepare a master plate for each library plasmid being tested. Each plate should contain at least 10 colonies of the transformed yeast derived from **step v**.
   viii. Test for coloration and for auxotrophic requirements exactly as described in **Subheading 3.2.5.** True positives should show an interaction phenotype with **a**, but not with **b**. Clones transformed with control PCR product will provide both positive and negative controls: **a** should be negative while **b** should be positive when assayed for both color and growth on the corresponding plates.
   ix. Proceed with sequencing and biological characterization. Most often, PCR provides ample source of DNA for all subsequent cloning. If needed, transform selected positives into *E. coli* by electroporation ***(12)***, using 1–2 μL of

the beta-glucuronidase treated frozen yeast (**step 3**), and isolate plasmid DNA from Ap^R colonies.

3.3.2. Plasmid Isolation Approach: Isolation of Plasmids and Transfer to Bacteria

The following option is suggested as an alternative to the basic PCR protocol. This protocol can also be scaled up if a large number of colonies are to be assayed. It is based on lysing the cells with glass beads and phenol–chloroform extraction of the DNA followed by plasmid isolation and plasmid retransformation into yeast. A number of kits for yeast minipreps are commercially available, e.g., from Clontech and others. Some companies provide service to isolate plasmids from yeast cells, transform, and amplify the plasmid in *E. coli* to produce a sequencing template, e.g., Qbiogene (http://www.qbiogene.com/services/two-hybrid.html).

1. Grow colonies with the appropriate phenotype in 2 mL of Glu/CM –Trp overnight at 30°C. It is advisable to omit the –Ura-His G418 selection in this case to enrich for library plasmids at the expense of the bait and reporter plasmids.
2. Centrifuge 1 mL of each culture at 13,000*g* for 1 min. Resuspend each pellet in 200 μL of STES lysis buffer. To each add approx 100 μL of 0.45-mm-diameter sterile glass beads (**Subheading 2.4.**) and vortex vigorously for 1 min.
3. Extract the mixtures first with buffer-saturated phenol without removing the glass beads and then with phenol/chloroform. Transfer each aqueous phase to a fresh microfuge tube.
4. Precipitate the plasmid DNA with 2 vol of 100% ethanol and resuspend each pelleted DNA in 5–10 μL of TE.
5. If only the library plasmid carries the Ap^R selection marker (e.g., other plasmids carry Km^R selective marker), transform the plasmids into *E. coli* DH5α cells using electroporation and select on medium containing ampicillin. Only bacteria that have taken up a library plasmid will grow (*see* **Note 39**).
6. Select at least two bacterial clones for each yeast clone and prepare a small quantity of plasmid DNA *(12)* from each bacterial clone (*see* **Note 40**).
7. Follow **Subheading 3.3.1.** from **step 8-v** to the end essentially as described, except transform with purified library plasmids, instead of a mixture of PCR product plus digested library vector.

3.4. Follow-Up for Library Screening

Following completion of the above specificity tests, the next step is to proceed to biological characterization of the interaction in the appropriate organism for the bait. Such characterization will be necessarily bait-specific, and should serve to further eliminate interactions of dubious physiological relevance. *Of note*, a database of common false positives, along with salient discussions, is available at the website listed in **Note 1**.

4. Notes

1. The original set of Interaction Trap reagents represent the work of many contributors: R.Brent, J. Kamens, S. Hanes, J. Gyuris, R. Finley, E. Golemis, I. York, M. Sainz , S. Nottwehr, D. Shaywitz, and others. A lot of work has been done at Glaxo in Research Triangle Park and at Invitrogen. A complete table with all reagents compatible with Interaction Trap and Dual Bait systems is available at http://www.fccc.edu/research/labs/golemis/interactiontrapinwork.html. Many of these reagents are available commercially, and also can be acquired by request from the Golemis Laboratory at Fox Chase Cancer Center, (215) 728-3885 phone, (215) 728-3616 fax, IG_Serebriiskii@fccc.edu.
2. Standard molecular biology techniques or alternative cloning strategies, e.g., in vivo recombination or GATEWAY cloning (Invitrogen), can be used. In any case, it is a good idea to include a translational stop sequence at the carboxyl-terminal end of the bait sequence. It is important to keep in mind that the assay depends on the ability of the bait to enter the nucleus, and requires the bait to be a transcriptional NON-activator. Hence, obvious membrane localization motifs and/or transcriptional activation domains should be removed. Using two-hybrid systems to find associating partners for proteins that are normally extracellular, even though they have worked in a few cases, should be regarded as extremely high risk.
3. A number of modified versions of the plasmid exist that contain additional sites, altered reading frame, and alternate antibiotic resistance markers (*see* **Table 1**, **Fig. 3**, and the website in **Note 1** for details).
4. It is important to use a fresh colony (thawed from –70°C and streaked to single colony less than approx 7 d previously) and maintain sterile conditions throughout all subsequent procedures.
5. Incubation in YPD for 6 h to overnight at 30°C prior to plating allows expression of antibiotic resistance proteins and significantly increases transformation efficiency for plasmids with G418 and Zeo markers.
6. An efficient transformation would yield approx 10^4 transformants per microgram of DNA when two plasmids are being simultaneously transformed. Therefore, this experiment also provides a good chance to assess transformation efficiency, which will be of much higher importance by the time of library transformation. Thus, if only a very small number of colonies are obtained, or colonies are not apparent within 3–4 d, it would imply that transformation is very inefficient. In this case all solutions/media/conditions must be double-checked or prepared fresh, and transformation be repeated. Sheared salmon sperm DNA (sssDNA), which is most often used as carrier, must be of very high quality; use of a poor quality preparation can drop transformation frequencies one to two orders of magnitude. SssDNA is available commercially from a number of companies or can be homemade easily ***(9)***. If very few transformants containing the bait plasmid appear compared to the controls, or yeast expressing the bait protein grow noticeably more poorly than control cells, or if colony population appears much more heterogeneous than control (e.g., presents a mix of large and small colonies), this would suggest that the bait

protein is somewhat toxic to the host (*see* **Table 1** for suggested modifications).

7. This is important, because for some baits, protein expression level is heterogeneous between independent colonies, with accompanying heterogeneity of apparent ability to activate transcription of the two reporters.
8. A replicator/frogger for the transfer of multiple colonies can be purchased or easily homemade. It is important that all of the spokes have a flat surface and that spoke ends are leveled. A metal frogger can be sterilized by autoclaving or by alcohol/flaming. A plastic replicator must be cut in half to fit to a standard 90 mm Petri plate, and it can be sterilized by autoclaving or rinsing with alcohol. The replicator should have 48 spokes in a 6 × 8 configuration. When making prints on a plate, dip the replicator in the wells of the microtiter plate, then put it on the surface of the solidified medium. Tilt slightly in circular movement, then lift replicator and put it back in the microtiter plate with the correct orientation! Make sure all the drops left on the surface are of approximately the same size. If only one or two drops are missing, it is easy to correct by dropping approx 3 μL of yeast suspension on the missing spots from the corresponding wells. If many drops are missing, make sure that all the spokes of the replicator are in good contact with liquid in the microtiter plate. It may be necessary to cut off the side protrusions on the edge spokes of the plastic replicator and redo the whole plate. Continue replicating by shuttling back and forth between microtiter and media plates. Let the liquid absorb to the agar before putting the plates upside down in the incubator. Alternative techniques to assessing *LacZ* reporter activation are available at the website in **Note 1**.
9. The technique described here is much more sensitive than a standard X-Gal plate assay and can be done within 24 h of plating on appropriate medium. It is generally preferred in high-throughput analysis.
10. At this step, the ability of LexA-fused bait to activate transcription is tested on both LEU2 and LacZ reporters, while potential cI self-activation is only tested for LYS2 reporter. Auxotrophic reporters are, however, the most important for the library screening, because they allow the direct selection for interaction phenotype. In addition, there is normally a good correlation between activation of the two reporters, so it is very unlikely, that the bait which is not activating LYS2 will significantly activate GusA. Therefore, if no activation is detected on -Lys plates, one should proceed further; if bait causes growth on -Lys plates, it should be modified. The ability of the cI-fused bait to activate GusA reporter will be tested in **Subheading 3.1.5.**
11. In addition to the simplified technique described below, a number of more elaborate and time-consuming protocols are available (e.g., *see* Clontech's *Yeast Protocols Handbook*, available as PDF file at http://www.clontech.com/clontech/Manuals/PDF/PT3024-1.pdf).
12. Many fusion proteins exhibit sharp decreases in detectable levels of protein with the onset of stationary phase. Therefore, use of the saturated cultures is not recommended for this assay.

13. Cultures growing in YPD G418 medium would typically reach the desired $OD_{600\ nm}$ faster than those growing in minimal medium.
14. A high percentage of the colonies not appropriately expressing the bait protein, although containing the bait plasmid, may be indicative that the bait is toxic to the yeast.
15. An investigator who is proficient in yeast transformations may wish to cotransform all baits and reporter plasmids into a single strain simultaneously. If the resultant strain behaves favorably, this single transformation can save time. It should be cautioned, however, that simultaneous transformation with three plasmids can be problematic for first-time users. In addition, the step-wise procedure that we recommend allows verification of one bait while the second is being constructed or modified.
16. A traditional alternative to the mating—directly transforming the library into yeast containing the bait—is even less practical for Dual Bait system. Such direct transformation in the bait strain requires media not only selective for library plasmid, but also maintaining selective pressure to keep both baits and reporter. Because in any currently available combination of bait plasmids at least one selection marker is antibiotic resistance (G418 or Zeo), antibiotic should be added to about 8 L of medium required to pool 30 large plates. This will make screening much more expensive.
17. A good library transformation efficiency should be approx 10^5 transformants per microgram of library DNA for a single transformation. Transformation of yeast in multiple small aliquots in parallel helps reduce the likelihood of contamination. Furthermore, it frequently results in significantly better transformation efficiency than that obtained by using larger volumes in a smaller number of tubes. Finally, do not use excess transforming library DNA per aliquot of competent yeast because cells may take up multiple library plasmids, complicating subsequent analysis. Under the conditions described here, less than 10% of yeast will contain two or more library plasmids.
18. Although it is possible to throw away the beads after spreading, it is acceptable and efficient to keep the glass beads on the lids while incubating the plates. Glass beads will be needed to harvest the library transformants . Contamination is much less likely to occur on the glass beads than on the plates themselves.
19. Thoroughly inspect the plates visually prior to the collecting transformants. If visible molds or other contaminants are observed on the plates, carefully excise them and a region around them using a sterile razor blade prior to beginning harvest of library transformants.
20. This technique also minimizes the time the plates are open and thus avoids contamination from airborne molds and bacteria. It is more important to ensure the same wash-off rate for all plates than to collect as much yeast as possible. About one third of the yeast slurry will be left on the plates. However, a second wash can greatly improve the yield. After the first wash, add 10 more mL of water, shake again, and transfer the slurry to the next *unwashed* plate. Optionally, the 24 × 24 cm plates can be reused many times after removing the

remaining agar, washing, and alcohol/UV sterilization. As these plates are quite expensive, this is a useful point of economy.

21. The bait and reporter plasmids should have been transformed into the yeast less than approx 7–10 d prior to mating with pretransformed library or transforming with library plasmids.
22. In general, for yeast frozen for less than 1 yr, viability will be greater than 90%. Refreezing a thawed aliquot results in the loss of viability.
23. Titering can also be done later in parallel with selection step in **Subheading 3.2.4.**
24. We omit antibiotic for the plates selecting for cI–Bait interactors to speed up growth of yeast colonies. Selecting for interacting bait–prey combination also automatically maintains pressure for the presence of bait plasmid. If, however, contamination occurred and is G418-sensitive, antibiotic should be included in the medium and selection repeated. Also see **Note 25**.
25. If contamination occurred at an earlier step and results in the growth of many (>500) colonies per plate, this will interfere with screening. In the case of bacterial contamination, the hunt may be salvaged by adding tetracycline (15 μg/mL) to the selective plates and repeating library induction and plating. If contamination is fungal, there is little to be done. Mating, or even library transformation, must be repeated.
26. Compare selection plates seeded with lower and higher densities. The number of colonies should be roughly proportional to the seeding density and there should be no background growth. If disproportionally more colonies, or a lawn, appear on the more densely seeded plates, this is background due to cross-feeding. In this case, a higher number of plates seeded at lower density should be used. Calculate how many plates at acceptable cell density are necessary for full representation of the desired number of diploids. If needed, repeat induction and plating from another frozen mating aliquot.
27. If colonies do not arise within the first week after plating, colonies appearing at later time points are not likely to represent *bona fide* positives. True interactors tend to come up in a window of time specific for a given bait with false positives clustering at a different time point. Hence, pregrouping by date of growth facilitates the decision of which clones to analyze first.
28. The number of candidate colonies to pick and characterize should be based on the number of cDNA-independent false positives that arise on the same selection plates for the control mating. The higher the frequency of false-positives, the more colonies that should be picked to find rare true-positives. Because the frequency of true-positives will be unknown at this step, the goal will be to pick through all of the false-positives that are expected in the number of library transformants being screened. For example, if the number of library transformants was 10^6, the goal will be to pick through the number of false positives expected in 10^6 diploids. If the cDNA-independent false positive frequency is 1 Leu^+ colony in 10^4 cfu plated, it will be necessary to pick at least 100 Leu^+ colonies to find a true-positive that exists at a frequency of 1 in 10^6.

29. In general, test plates for auxotrophic reporter characterization lacking only leucine or lysine would automatically keep selective pressure for the presence of the prey and the corresponding bait plasmids. Using plates with fewer dropped-out components would slightly accelerate the growth, and the potential loss of other plasmids would not influence the results of the assay on these plates. At the investigator's discretion, however, –Leu and –Lys plates can be substituted for –Ura-His-Trp-Leu or –Ura-His-Trp-Lys
30. In some cases no positives are obtained from library screens. Reasons for this might include inappropriate library source, an inadequate number of screened colonies (<500,000), a bait that in spite of production at high levels is nevertheless incorrectly folded or post-translationally modified, or alternatively, a bait that does not interact with its partners with a sufficiently high affinity to be detected. Be also aware of such simple explanations as a wrong batch of plates. In such cases, it may be worth trying screens again with a different variant of bait, screening strain, and/or library, although success is not guaranteed. It is rarely if ever profitable to continue to rescreen the same bait/strain/library combination through >3–5 $\times 10^6$ primary transformants.
31. Transfer approximately the volume of one middle-sized yeast colony (2–3 μL packed pellet). Do NOT take more, or quality of isolated DNA will suffer. The master plate does not need to be absolutely fresh. Plates that have been stored for 5 d at 4°C have been successfully used. If setting up multiple reactions from the same master plate, a multicolony replicator can be used (*see* **Subheading 3.2.5., step 1**).
32. Modified versions of this protocol with extended elongation times were also found to work. The variant given above has amplified fragments of as much as 1.8 kb in a pretty fair quantity.
33. If the library being screened is based on pJG4-5 plasmid and primers specific for this plasmid are used in PCR master mix, only clones containing Raf1 and Krit-1 plasmids would produce products. For a pYesTrp-based library, take RalGDS clone as a positive control.
34. Sometimes a single yeast cell will contain two or more different library plasmids. If this happens, it will be immediately revealed by PCR. In this case two bands can be separately isolated from the gel and reamplified. Also, after bacterial transformation an increased number of clones should be checked to avoid the loss of the "real" interactor.
35. Only the forward primer, FP1, works well in sequencing of PCR fragments . The reverse primer will only work in sequencing from purified plasmid. In general, the TA-rich nature of the ADH terminator sequences downstream of the polylinker in the pJG4-5 vector makes it difficult to design high-quality primers in this region.
36. Gel analysis produces little information on the completeness of digestion, because it is not possible to distinguish between plasimid species cut by one and two enzymes. Purification of the digested plasmid is not necessary.
37. This control experiment is an indicator of the degree of digestion of the library plasmid. The background level of colonies transformed with undigested empty library plasmid (**a**) should be minimal. In case the background is high, make sure

that the digestion of the empty library plasmid is completeby increasing the digestion incubation time or the restriction enzyme concentration.

38. The fraction of the correct clones can be assessed by replica-plating 12–24 clones to check their phenotype (as in **Subheading 3.2.5.**). Normally, it should be between 85 and 95%.
39. If using specialized LexA-fusion bait plasmid with Ap^R selection marker in combination with Ap^R-library plasmid, it will be necessary to use an *E. coli* trp^- strain to select for the library plasmid. A library plasmid can be specifically isolated by the ability of the yeast *TRP1* gene to complement the *E. coli trpC* mutation. (a) Electroporate ***(12)*** 1 μL of each plasmid DNA into *E. coli* KC8 (*pyrF leuB600 trpC hisB463*) and plate on LB/ampicillin. Incubate overnight at 37°C. (b) Restreak or replica plate colonies from the LB/ampicillin plates to bacterial defined minimal medium KC8 plates supplemented with uracil, histidine, and leucine but lacking tryptophan. Colonies that grow under these conditions contain the library plasmid, since the *TRP1* gene contained on this plasmid efficiently complements the bacterial *trpC9830* mutation. It is also feasible to plate *E. coli* KC8 transformants directly onto bacterial minimal medium, although it may take 2 d for colonies to grow. It is advisable to use standard molecular biology techniques to identify redundant clones prior to verification of the interaction specificity.

References

1. Fields, S. and Song, O. (1989) A novel genetic system to detect protein-protein interaction. *Nature* **340,** 245–246.
2. Ito, T. et al. (2001) A comprehensive two-hybrid analysis to explore the yeast protein interactome. *Proc. Natl. Acad. Sci. USA* **98,** 4569–4574.
3. Schwikowski, B., Uetz, P., and Fields, S. (2000) A network of protein-protein interactions in yeast. *Nat. Biotechnol.* **18,** 1257–1261.
4. Inouye, C., Dhillon, N., Durfee, T., Zambryski, P. C., and Thorner, J. (1997) Mutational analysis of STE5 in the yeast Saccharomyces cerevisiae: application of a differential interaction trap assay for examining protein-protein interactions. *Genetics* **147,** 479–492
5. Xu, C. W., Mendelsohn, A. R., and Brent, R. (1997) Cells that register logical relationships among proteins. *Proc. Natl. Acad. Sci. USA* **94,** 12,473–12,478
6. Serebriiskii, I., Khazak, V., and Golemis, E. A. (1999) A two-hybrid dual bait system to discriminate specificity of protein interactions. *J. Biol. Chem.* **274,** 17,080–17,087.
7. Gyuris, J., Golemis, E. A., Chertkov, H., and Brent, R. (1993) Cdi1, a human G1 and S phase protein phosphatase that associates with Cdk2. *Cell* **75,** 791–803.
8. Reeder, M. K., Serebriiskii, I. G., Golemis, E. A., and Chernoff, J. (2001) Analysis of small GTPase signaling pathways using Pak1 mutants that selectively couple to Cdc42. *J. Biol. Chem.* **276,** 40,606–40,613.
9. Schiestl, R. H. and Gietz, R. D. (1989) High efficiency transformation of intact yeast cells using single stranded nucleic acids as a carrier. *Curr. Genet.* **16,** 339–346.

10. Duttweiler, H. M. (1996) A highly sensitive and non-lethal beta-galactosidase plate assay for yeast. *TIG* **12,** 340–341.
11. Harlow, E. and Lane, D. (1988). *Antibodies: a Laboratory Manual.* Cold Spring Harbor Laboratory, Cold Spring Harbor, NY.
12. Sambrook, J., Fritsch, E. F., and Maniatis, T. (1989). *Molecular Cloning: a Laboratory Manual.* Cold Spring Harbor Laboratory, Cold Spring Harbor, NY.
13. Finley, R. and Brent, R. (1994) Interaction mating reveals binary and ternary connections between Drosophila cell cycle regulators. *Proc. Nat. Acad. Sci. USA* **91,** 12,980–12,984.
14. http://www.fermentas.com/TechInfo/PCR/DNAamplProtocol.html.
15. Petermann, R., Mossier, B. M., Aryee, D. N., and Kovar, H. (1998) A recombination based method to rapidly assess specificity of two-hybrid clones in yeast. *Nucleic Acids Res.* **26,** 2252–2253.
16. Oldenburg, K. R., Vo, K. T., Michaelis, S., and Paddon, C. (1997) Recombination-mediated PCR-directed plasmid construction *in vivo* in yeast. *Nucleic Acids Res.* **25,** 451–452.
17. DeMarini, D. J., Creasy, C. L., Lu, Q., et al. (2001) Oligonucleotide-mediated, PCR-independent cloning by homologous recombination. *Biotechniques* **30,** 520–523.
18. Serebriiskii, I., Toby, G., Finley, R.L., and Golemis, E.A.. (2001).Genomic analysis utilizing the yeast two-hybrid system, in *Genomic Protocols* (Starkey, M. ed.), Humana, Totowa, NJ, pp. 415–454.
19. Serebriiskii, I. G., Mitina, O., Pugacheva, E. N., et al. (2002) Detection of peptides, proteins, and drugs that selectively interact with protein targets. *Genome Res.* **12,** 1785–1791.

20

The Split-Ubiquitin Membrane-Based Yeast Two-Hybrid System

Safia Thaminy, John Miller, and Igor Stagljar

Abstract

Protein–protein interactions are essential in almost all biological processes, extending from the formation of cellular macromolecular structures and enzymatic complexes to the regulation of signal transduction pathways. It is assumed that approximately one-third of all proteins in eukaryotic cells are membrane associated. Because of their hydrophobic nature, the analysis of membrane–protein interactions is difficult to be studied in a conventional two-hybrid assay. We described here a new genetic method for in vivo detection of membrane–protein interactions in the budding yeast *Saccharomyces cerevisiae*. The system uses the split-ubiquitin approach based on the detection of the in vivo processing of a reconstituted split ubiquitin. On interaction of X and Y proteins, ubiquitin reconstitution occurs and leads to the proteolytic cleavage and subsequent release of a transcription factor that triggers the activation of a reporter system enabeling easy detection. In this manner, and in contrast to the conventional yeast-two hybrid system in which interactions occur in the nucleus, the membrane-based yeast two-hybrid system represents an in vivo system that detects interactions between membrane proteins in their natural environment.

Keys Words

Yeast two-hybrid system; protein–protein interaction; split-ubiquitin; membrane proteins; screening.

1. Introduction

1.1. Limitations and Problems of the Yeast Two-Hybrid System

Protein–protein interactions are essential for many biological processes, including replication, transcription, splicing, secretion, signal transduction, and metabolism, to name a few. Thus, one central task in the study of a protein is to determine its interaction partners. Although the yeast two-hybrid system is a useful and powerful genetic technique for the in vivo analysis of

From: *Methods in Molecular Biology, vol. 261: Protein–Protein Interactions: Methods and Protocols*
Edited by: H. Fu © Humana Press Inc., Totowa, NJ

protein–protein interactions *(1)*, its application is limited. First, hybrid proteins generated in the two-hybrid assay are targeted to the nucleus. Thus, proteins that cannot fold correctly in the nucleus are excluded from the method. For example, integral membrane proteins, which exist in the lipid bilayer, are excluded because they are unlikely to be able to enter the nucleus, or will be misfolded if they do. Second, interactions that are dependent on post-translational modifications taking place within the endoplasmic reticulum, such as glycosylation and disulfide-bond formation, will not be testable with the two-hybrid system as these modifications do not occur in the nucleus. Third, DNA–binding protein hybrids that activate transcription in the absence of a specific activation-domain hybrid partner cannot be studied by the two-hybrid method. Although the region required for inappropriate activation sometimes could be deleted, such an approach may also affect binding of the protein to its cellular partners. Finally, if the protein–protein interaction is mediated by an amino-terminal domain of either protein, the standard orientation of the hybrid constructs may not work because the transcription factor domain blocks accessibility.

1.2. Detection of Protein–Protein Interaction Using the Membrane-Based Yeast-Two Hybrid System

The split-ubiquitin system represents an alternative assay for the in vivo analysis of protein interactions. It was developed in 1994 for the detection of interactions between soluble proteins *(2)*. The system was then modified (*see* **Fig. 1**) and shown to work with membrane proteins *(3–7)* and as a screening system *(8,9)*. Similar to the two-hybrid system, where a transcription factor is reassembled upon interaction of two test proteins, the split-ubiquitin system consists of two fragments of ubiquitin that are brought together by interacting proteins. Ubiquitin is a small protein of 76 amino acids, which participates in many cellular pathways involving the degradation of proteins *(10)*. Specifically, the C-terminus of ubiquitin gets linked via an amide bond to proteins to be degraded by a complex system consisting of several enzymes. Ubiquitin-specific proteases (UBPs) recognize the ubiquitin part of these protein fusions and cleave the covalent bond between ubiquitin and the protein. The protein is then degraded by the proteasome *(11)*.

For the detection of interactions using split-ubiquitin, ubiquitin is expressed as two separate fragments in different plasmids: the N-terminal fragment of ubiquitin (amino acids 1–37, Nub*I*) and the C-terminal fragment (amino acids 35–76, Cub). A reporter protein is attached to the C-terminus of Cub, and upon co-expression with Nub*I* both ubiquitin pieces reassemble, forming a split-ubiquitin heterodimer that results in release of the reporter protein. This is because split-ubiquitin is a substrate for UBPs, which cleave the reporter protein from the C-terminus of Cub, allowing it to enter the nucleus and activate reporter genes. In the split-ubiquitin system, the N-terminal fragment Nub*I* was modified by a point mutation (Nub*G*), which does not spontaneously associate with Cub. Nub*G* and Cub-reporter can reassemble only upon interaction of the fusion proteins attached to them *(2)*.

Because the formation of the split-ubiquitin complex does not require a special localization of the proteins, the system has been modified to study interactions between membrane proteins. In the membrane-based yeast two-hybrid system, one protein of interest, the bait X, is fused to the Cub domain followed by an artificial transcription factor (TF), and the other, the prey Y, is fused to Nub*G*. To detect a potential interaction between these two proteins X and Y, the two plasmids are introduced into the yeast reporter strain L40. Interaction of both proteins results in the assembly of split-ubiquitin and the proteolytic release of TF, allowing TF to enter the nucleus leading to the activation of two reporter genes, *lacZ* and *His3*. Thus, interactions can be directly monitored by growing the yeast on a selective medium lacking histidine and by performing an X-Gal test for the expression of the β-galactosidase enzyme (*see* **Fig. 1**).

In this chapter, we describe the methodology for the in vivo detection of interactions either between a membrane protein and a soluble protein, or between two membrane proteins.

2. Materials

1. L40 yeast reporter strain: *MATα trp-1 leu2 his3 LYS2::lexA-HIS3 URA3::lexA-lacZ* ***(12)***.
2. "Bait" expression plasmid, pX-Cub-TF or pTF-Cub-X: expresses the bait protein as a Cub–TF fusion protein. The plasmid contains the LEU2 selection marker and ampicillin-resistance gene.

Fig. 1. *(pages 300, 301)* The membrane-based yeast two-hybrid system. A transmembrane bait protein X is fused to the Cub domain followed by an artificial transcription factor (TF) to generate a chimeric protein X-Cub-TF. The prey Y, which can be a cytosolic or membrane protein, is fused to the NubG domain. (**A**) Interaction between X and Y results in the formation of a split-ubiquitin heterodimer. The heterodimer is recognized and cleaved by the UBPs (scissors), liberating TF, which enters the nucleus by diffusion and binds to lexA-binding sites upstream of the *lacZ* and *HIS3* reporter genes resulting in β-galactosidase activity and histidine synthesis, respectively. Yeast cells containing the interacting X and Y proteins can be identified because they are blue in the presence of X-Gal and grow on agar plates lacking histidine. (**B**) If there is no interaction between X and Y, the split-ubiquitin heterodimer will not be reconstitute, resulting in the absence of cleavage of TF and activation of the reporter genes.

A

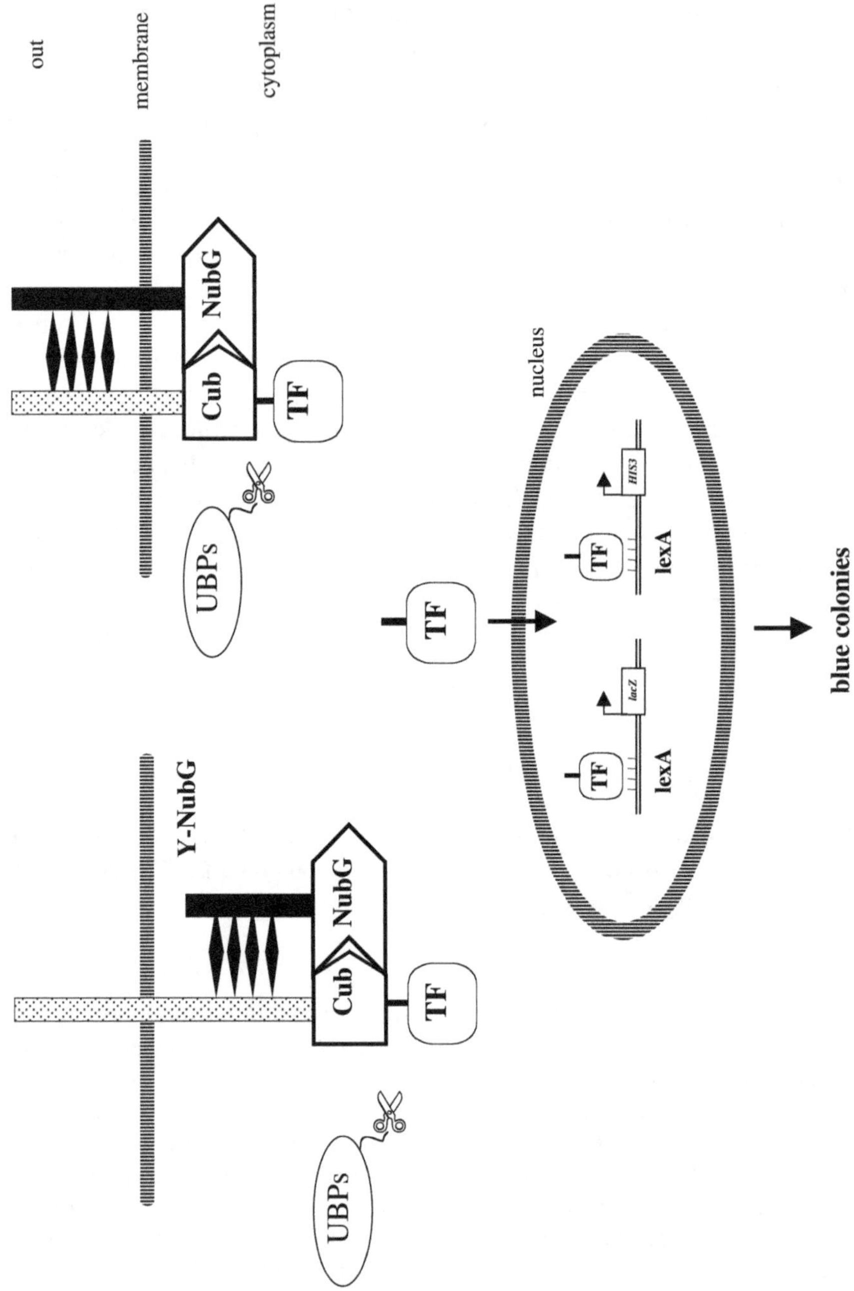

X-Cub-TF
Y-NubG
Cub
NubG
TF
UBPs
out
membrane
cytoplasm
nucleus
TF
lexA
lacZ
HIS3
blue colonies
HIS3+

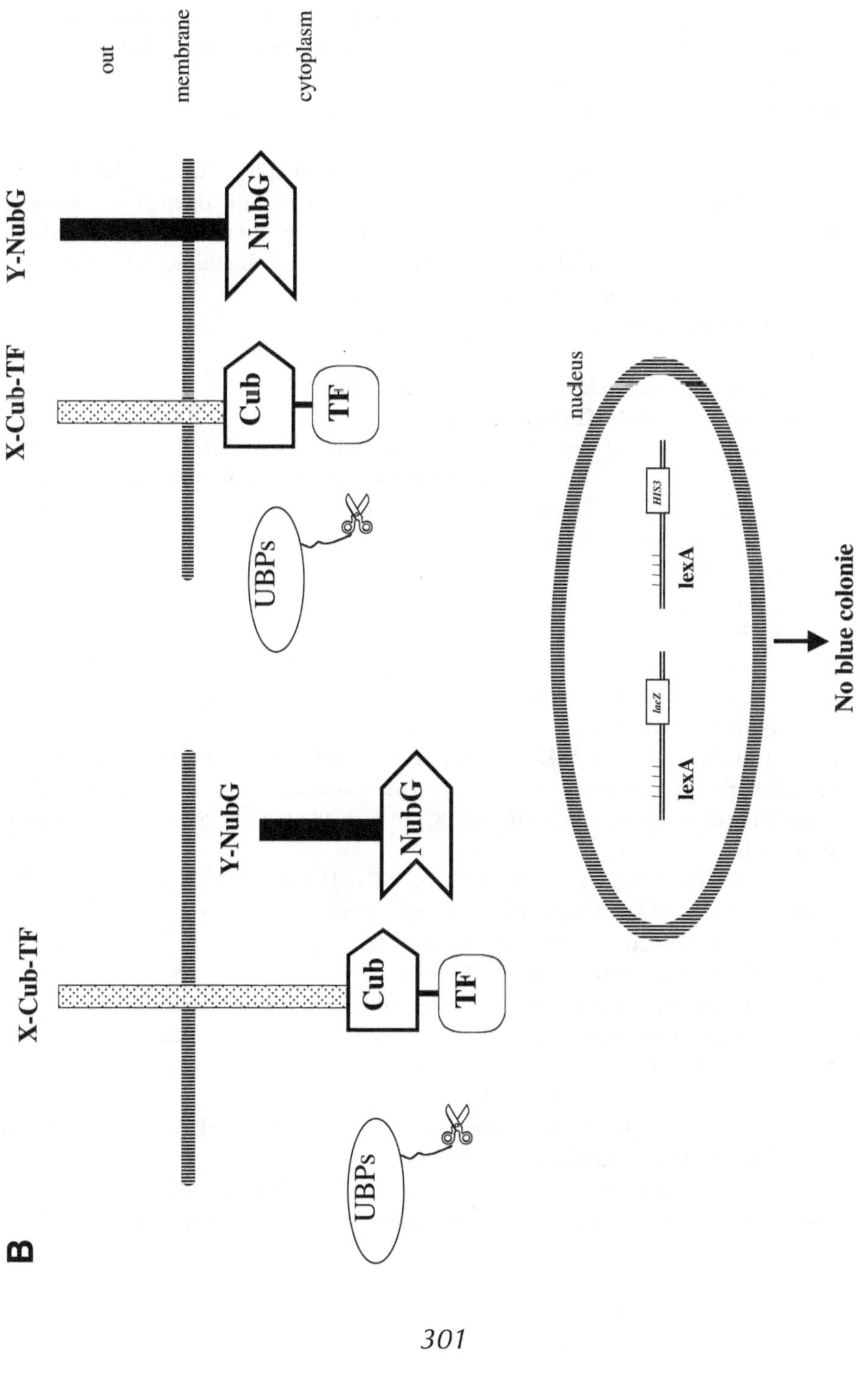
B
out
membrane
cytoplasm
X-Cub-TF
Cub
TF
Y-NubG
NubG
UBPs
X-Cub-TF
Cub
TF
Y-NubG
NubG
UBPs
nucleus
lexA
lacZ
lexA
HIS3
No blue colonie
HIS3-

3. "Prey" expression plasmid, pY-HA-NubG or pNubG-HA-Y: expresses the prey protein as a Nub*G* fusion protein. The plasmid contains the TRP1 selection marker and ampicillin-resistance gene.
4. YPD medium: 1% Bacto-yeast extract, 2% Bacto-peptone, 2% glucose, 2% Bacto-agar.
5. Drop-out medium without leucine, tryptophan, or histidine (CLONTECH).
6. SD (synthetic drop-out) medium: 0.67% Bacto-yeast nitrogen base (without amino acids), 2% glucose, 2% Bacto-agar, 0.1% drop-out mix. SD -Trp medium is complemented with 100 mg/L leucine and 20 mg/L histidine. SD -Trp -Leu medium is complemented with 20 mg/L histidine.
7. PEG (polyethylene glycol) 4000 (50 % w/v).
8. 1 *M* LiOAc.
9. 10 mg/mL single-stranded carrier DNA (Clontech).
10. 500 μm glass beads (SIGMA).
11. Lysis buffer: 100 m*M* Tris-HCl, pH 7.5, 200 m*M* NaCl, 20 % glycerol, 5 m*M* EDTA pH 8.0, 1 m*M* PMSF, 14 nM β-mercaptoethanol, 0.5 m*M* leupeptin, 0.5 m*M* pepstatin, and 0.5 m*M* bestatin at 4°C.
12. Polyallomer microfuge tubes.
13. Rolling wheel.
14. Rocker platform.
15. Shaker, 30°C.
16. Heat block, 42°C.
17. SDS-PAGE (sodium dodecyl sulfate-polyacrylamide gel electrophoresis) equipment.
18. Electroblotter equipment.
19. Sodium dodecyl sulfate (SDS) sample buffer: 0.0625 *M* Tris-HCl pH 6.8, 2% SDS, 5 % β-mercaptoethanol, 10% glycerol, 8 *M* urea, 0.025 % bromphenol blue.
20. Transfer buffer: 25 m*M* Tris-HCl pH 8.3, 192 m*M* glycine, 20% (v/v) methanol, store at 4°C.
21. Washing buffer: 10m*M* Tris-HCl pH 7.5, 150 m*M* NaCl, 0.05% (w/v) Tween-20.
22. Blocking buffer: 5% (w/v) nonfat dry milk in washing buffer.
23. Rabbit VP16 polyclonal antibody (Clontech).
24. Mouse HA monoclonal antibody (BabCO).
25. Secondary antibody conjugated to horseradish Peroxidase (HRP).
26. Enhanced chemiluminescence (Super Signal West Pico Substrate, Pierce).
27. X-ray films (Fuji).
28. 3MM Whatman filter paper, sterile.
29. 5-Bromo-4-chloro-β-D-galactopyranoside (X-Gal) 10 mg/mL freshly dissolved in *N*,*N*-dimethylformamide).
30. TBS buffer (tris-buffered saline): 0.8% NaCl, 0.02% KCl, 0.3 % Tris base pH 7.4.
31. X-Gal buffer: 0.5% agarose in 1X TBS. After cooling down the medium to 50°C, add 1 mL of X-Gal per 1 L 1X TBS.
32. Agarose.
33. Ampicillin.
34. Restriction enzymes, T7 DNA ligase.

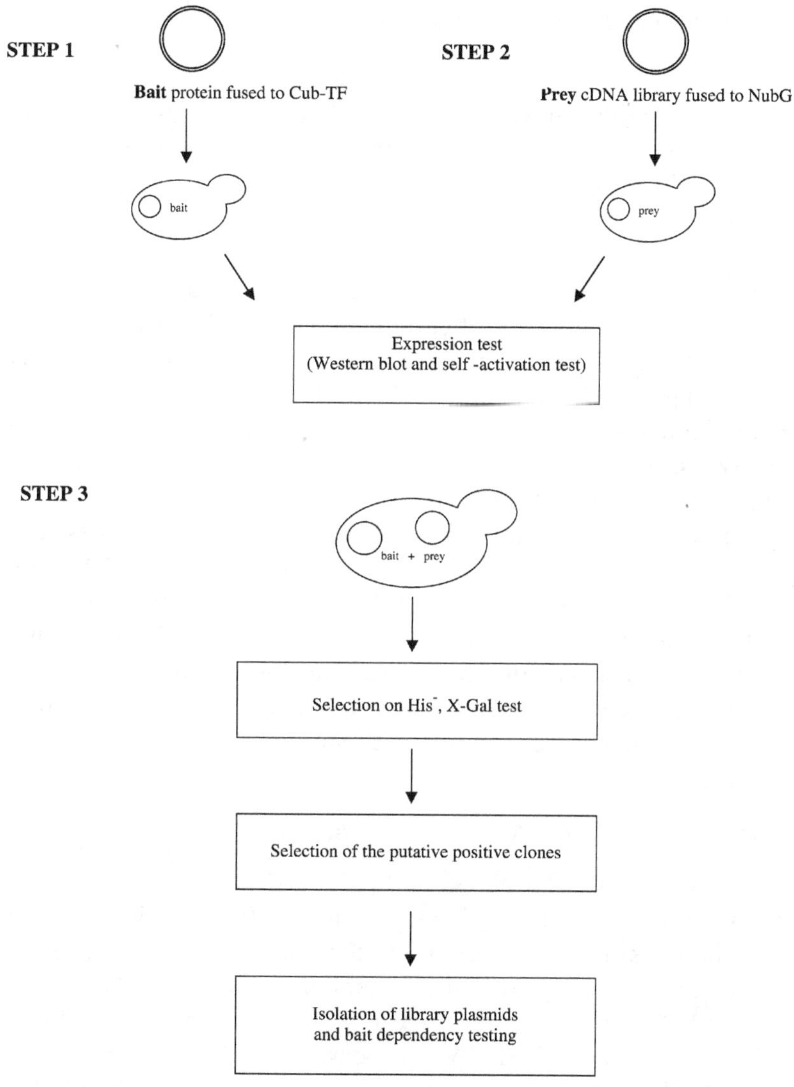

Fig. 2. Flow-chart representing the membrane-based yeast two-hybrid procedure.

3. Methods

The methods described below outline (1) the expression of full-length heterologous membrane proteins in yeast, (2) the reconstitution of known protein–protein interactions and (3) a library screening to detect novel interacting partners using the membrane-based yeast two-hybrid system (*see* **Fig. 2**).

3.1. Expression of Heterologous Proteins in Yeast

3.1.1. Cloning

This section describes how to insert the cDNAs of interest into the bait and prey vectors, transform yeast to express these fusion proteins, and test them for reporter gene activation.

3.1.1.1. pX-Cub-TF/pTF-Cub-X Bait Vectors

The bait vectors pX-Cub-TF and pTF-Cub-X are centromeric vectors that contain an ARSH4 origin of replication and the CEN6 locus to allow a low copy number (one to two plasmids per cell). It consists of the weak cytochrome-c oxidase (*CYC1*) promoter ***(12)***, which provides low-level expression of the fusion (*see* **Note 1**), a multiple cloning site (MCS), and a Cub–TF cassette (*see* **Note 2**). Both pX-Cub-TF and pTF-Cub-X are bearing the *LEU2* selectable marker (*see* **Fig. 3A**).

The gene of interest is amplified by PCR using oligonucleotides flanked by the sequence of a restriction enzyme also present in the MCS of the bait vector (*Xba*I, *Spe*I, *Pst*I, *Hind*III for pX-Cub-TF and *Nde*I, *Nhe*I, *Pst*I, *Spe*I, *Xba*I, *Sac*II for pTF-Cub-X. The PCR fragment and the pY-Cub-TF/ pTF-Cub-Y bait plasmids are then digested with this appropriate enzyme and ligated. The plasmid DNA is then transformed into *Escherichia coli* and checked for the presence of the insert and correct orientation using restriction enzymes and DNA sequencing.

3.1.1.2. pNubG-HA-Y/pY-HA-NubG Prey Vectors

The prey vectors pY-HA-NubG and pNubG-HA-Y contain the 2-μm origin of replication, allowing a high copy number (10–30 copies per cell). They consist of the constitutive *ADH1* promoter, the Nub*G* domain, and the HA epitope tag. The MCS is positioned in such a way that the foreign gene can be introduced as either an N-terminal or C-terminal fusion to Nub*G*. Both pX-HA-NubG and pNubG-HA-Y are bearing the *TRP1* selectable marker (*see* **Fig. 3B**). The gene of interest is fused in frame to the Nub*G* cassette using the unique restriction sites *Nde*I, *Nco*I, *Sma*I, *BamH*I for pY-HA-NubG and *Nco*I, *Sma*I, *BamH*I, *Pst*I for pNubG-HA-Y.

3.1.1.3. Building Libraries in the pNubG-HA-X and pX-HA-NubG Vectors

The pNubG-HA-Y and pY-HA-NubG libraries are built by cloning the DNA inserts in the unique "X" site of the MCS according to the traditional Gubler–Hoffman method ***(14)***. The construction of the library is performed by a standard protocol ***(15)***. To construct a cDNA library, the mRNA is extracted from cells or tissue expressing the gene of interest and the cDNA is synthesized by the reverse transcriptase using oligo-dT primers. Oligonucleotide linkers containing a single-restriction endonuclease recognition sequence are attached to the double-stranded cDNA product in a way that the inserts could be easily inserted into the "Y" site of the MCS of the prey plasmid. Finally, the cDNA is fractionated by gel filtration through Sepharose CL-4B columns, ligated to the pNubG-HA-Y and pY-HA-NubG vectors and propagated into *E. coli*.

3.1.2. Yeast Transformation

After cloning, the bait and the prey plasmids are first introduced alone into the yeast reporter strain and the expression of the proteins is checked by Western blot. Second, a filter assay is performed on the transformed yeast colonies to check if the corresponding proteins X-Cub-TF/ TF-Cub-X and Y-HA-NubG/ NubG-HA-Y are able to self-activate.

1. Grow L40 yeast cells in 30 mL of YPD medium at 30°C, 200 rpm to a concentration of approx 2×10^7 cells/mL (corresponding to an $OD_{600\ nm}$ ~1).
2. Centrifuge at 1700*g* for 5 min in sterile 50 mL Falcon tube.
3. Resuspend the cells in 40 mL of sterile distilled water by vortexing.
4. Centrifuge at 1700*g* for 5 min.
5. Remove the supernatant and add 1 mL of cold sterile distilled water.
6. Aliquot 100 µL of yeast cells in cold Eppendorf tubes.
7. Add in order: 240 µL PEG 4000 (50 % w/v), 36 µL 1 *M* LiOAc, 5 µL of 10 mg/mL single-stranded carrier DNA, and plasmid DNA (pX-Cub-TF/pTF-Cub-X or pY-HA-NubG/pNubG-HA-Y) (between 5 and 10 µg).
8. Vortex the tubes vigorously. Heat shock 25 min at 42°C.
9. Collect the cells by a 1 min centrifugation at 1000*g*.
10. Carefully remove the supernatant and resuspend the cells in 100 µL of sterile distilled water. Spread the resuspended cells on a selective plate, SD -Leu for the L40 cells transformed with the bait plasmid (pX-Cub-TF/pTF-Cub-X) and SD -Trp for those transformed with the prey plasmid (pY-HA-NubG/pNubG-HA-Y). Transformants should appear after 3 or 4 d at 30°C, depending on the strain and the selectable marker.

3.1.3. Protein Extraction

Depending on the expected localization of the expressed protein, cytoplasmic and membrane extracts can be made of a single transformed yeast colony. Streak out three or four individual colonies on a new selective plate and incubate for 2–3 d at 30°C.

1. Inoculate each colony into 50 mL appropriate selective medium at 30°C, 200 rpm. Grow to a concentration of approx 2×10^7 cells/mL ($OD_{600\ nm}$ ~1).
2. Centrifuge at 1700*g* for 5 min in 50 mL Falcon tube.
3. Discard the supernatant and add 1 mL cold lysis buffer and resuspend by vortexing.

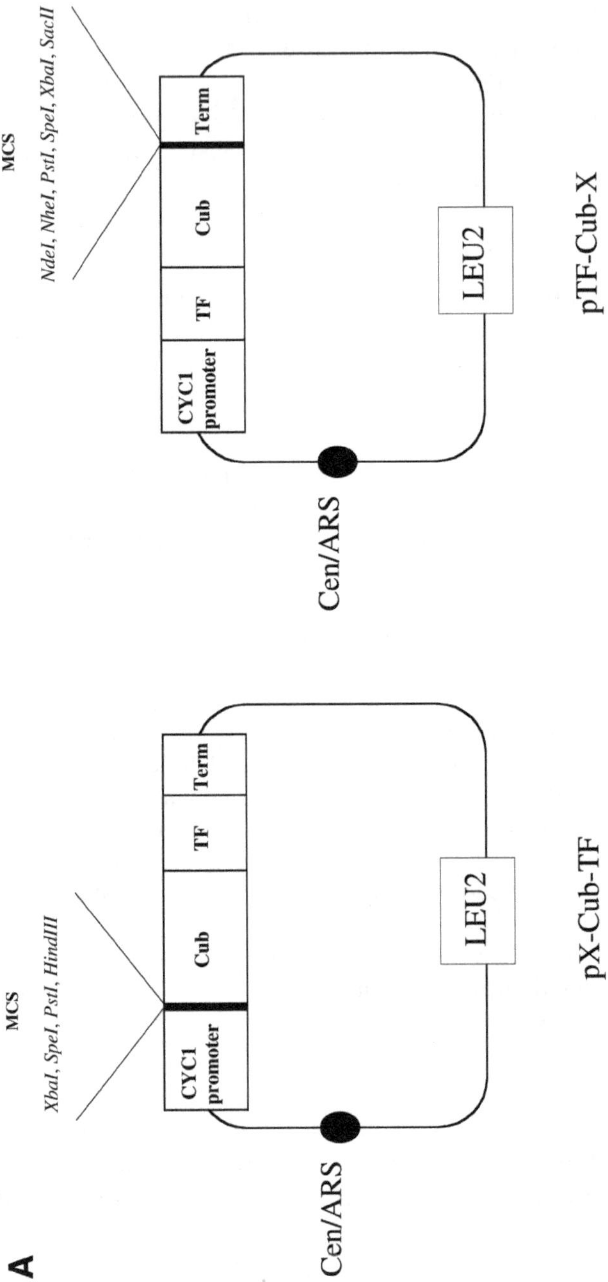
A
MCS
XbaI, SpeI, PstI, HindIII
CYC1 promoter
Cub
TF
Term
LEU2
Cen/ARS
pX-Cub-TF
MCS
NdeI, NheI, PstI, SpeI, XbaI, SacII
CYC1 promoter
TF
Cub
Term
LEU2
Cen/ARS
pTF-Cub-X

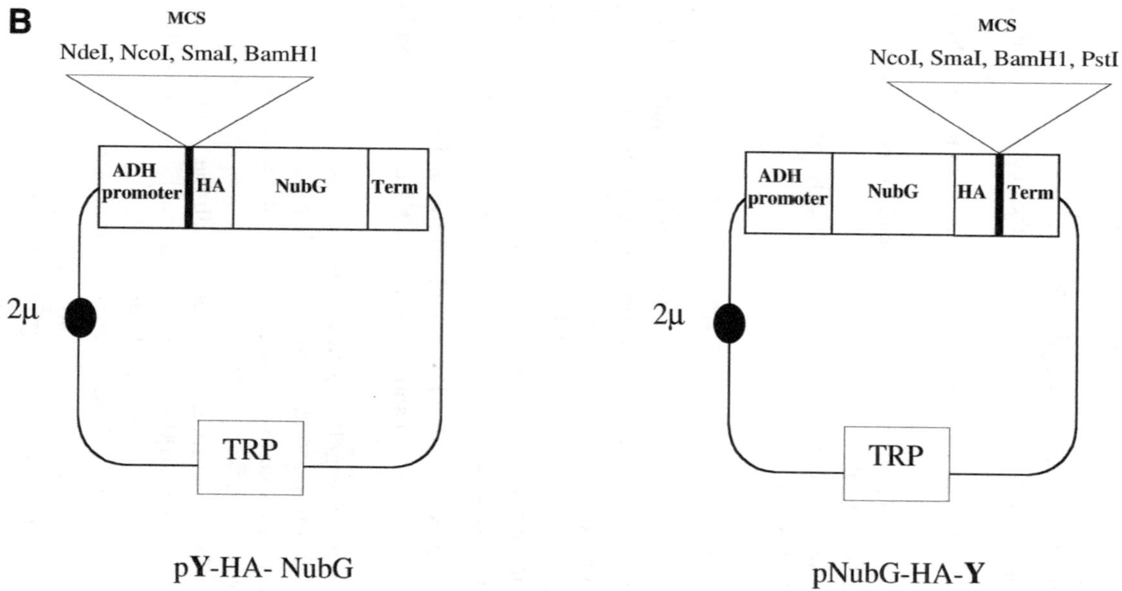

Fig. 3. The membrane-based yeast two-hybrid system vectors. **(A)** The bait vectors pX-Cub-TF and pTF-Cub-X are made by inserting the Cub-TF/TF-Cub cassette into the MCS of p415CYC1 ***(20)***. The foreign gene sequence is introduced into the MCS either N-terminally (*Xba*I, *Spe*I, *Pst*I, *Hind*III restriction sites) or C-terminally (*Nde*I, *Nhe*I, *Pst*I, *Spe*I, *Xba*I, *Sac*II restriction sites) to Cub. **(B)** The prey vectors pY-HA-NubG and pNubG-HA-Y were constructed by putting the Nub*G* cassette into the MCS of p414ADH ***(20)***. The foreign gene sequence is introduced into the MCS either N-terminally (*Nde*I, *Nco*I, *Sma*I, *BamH*I restriction sites) or C-terminally (*Nco*I, *Sma*I, *BamH*I, *Pst*I restriction sites) to Nub*G*.

4. Transfer the cells into a 2 mL Eppendorf tube and fill the tube with 300 μL of glass beads.
5. Vortex vigorously for 30 s intervals, cooling for 1 min on ice between vortex steps. Repeat seven times.
6. Centrifuge the extract in Eppendorf centrifuge at 700*g* for 20 min, 4°C.
7. Carefully transfer the supernatant fraction into a new tube and spin down again for 10 min.
8. Transfer the supernatant into polyallomer microfuge tubes and spin down at 150,000*g* in an ultracentrifuge for 2 h, 4°C.
9. Collect the supernatant (cytosolic fraction) and resuspend the pellet (membrane fraction) by pipetting in 100 μL of the lysis buffer.
10. Take an aliquot for measuring the protein concentration, freeze the extracts in liquid nitrogen and store tubes at –80°C.

3.1.4. Western Blot Analysis

To check the expression of the fusion proteins, the protein extracts are loaded on a SDS-PAGE gel and transferred to a nitrocellulose membrane (see Chapter 12). The blots are probed with the monoclonal HA antibody (raised against the HA epitope tag of the Nub*G* cassette) or the VP16 polyclonal antibody (raised against the VP16 protein present in the Cub–TF cassette). Extracts from yeast transformed with the empty bait and prey vectors are also loaded as controls.

1. Mix 30 μg of protein samples with the SDS- sample buffer and boil for 3 min.
2. Load the samples on a SDS-PAGE gel and carry out electrophoresis according to a standard protocol ***(15)***.
3. Transfer the proteins from the gel to a nitrocellulose membrane using an electroblotter apparatus according to the manufacturer's protocol.
4. Block nonspecific binding by soaking the membrane in the blocking buffer for 1 h at room temperature on a rocker platform.
5. Incubate the membrane for 1 h at room temperature with the primary antibody diluted in the washing buffer (1:1000 dilution for the rabbit polyclonal VP16 antibody and the mouse HA monoclonal antibody).
6. Wash the membrane three times for 5 min each with the washing buffer.
7. Incubate the membrane for 1 h with the horseradish peroxidase conjugated secondary antibody (dilution in the washing buffer as recommended by the supplier).
8. Wash three times for 5 min each with the washing buffer and one additional time for 5 min with the washing buffer without Tween-20.
9. Develop the blot following the Chemiluminescence HRP substrate instruction (Pierce) (*see* **Note 3**).
10. Expose the blot to X-ray film for 1–30 min and develop the film according to the manufacturer's instructions.

3.1.5. Filter Assay for Detection of β-Galactosidase Activity

After checking for expression of the protein fusions, yeast colonies transformed with the pX-Cub-TF/pTF-Cub-X bait plasmids alone can be tested for their ability to self-activate the *lacZ* reporter gene.

1. For each single transformation, streak out three or four individual colonies on a new selective plate and incubate for 2–3 d at 30°C.
2. Put 3MM Whatman filter paper cut to fit the size of the plate onto yeast for 10 min. This will allow the yeast to stick to the filter.
3. Using forceps, carefully remove the Whatman paper from the plate and transfer the filter into liquid nitrogen for 5 min.
4. Put the filter into a Petri dish, the printed face up, and let it thaw for few minutes at room temperature.
5. Overlay the filter with 0.5% (w/v) agarose in 1X TBS buffer containing 10 mg/L X-Gal. Pour X-Gal-containing agarose over the filter. Let it polymerize at room temperature for 30 min. Put the plates in an incubator at 30°C until a blue color can be observed (between 30 min and several hours). If the proteins are self-activating, two possibilities can be taken into consideration (*see* **Note 4**) and potentially remedied.

3.2. Detection of Protein Interactions Using the Membrane-Based Yeast Two-Hybrid System

After verifying that the fusion proteins are expressed in yeast and do not self-activate the *lacZ* reporter gene, a co-transformation of yeast with both bait and prey plasmids is performed to check for a potential interaction.

3.2.1. A Pairwise Interaction Test

The L40 reporter strain is transformed as described previously in **Subheading 3.1.2.**, but with both the bait and the prey plasmids being transformed into the same cell. Transformants are first selected on SD -Leu -Trp to verify the presence of both plasmids and then systematically streaked out on SD -Leu -Trp -His to test for the expression of the *HIS3* reporter gene (*see* **Note 5**). The *lacZ* reporter should also be assayed by the filter lift assay as described in **Subheading 3.1.5.**

3.2.2. Library Screening

In a library screening approach, a corresponding genomic or cDNA library fused to Nub*G* is introduced into the yeast strain expressing a particular membrane bait protein of interest.

1. Grow the L40 yeast strain expressing the X-Cub-TF/TF-Cub-X fusions overnight in 200 mL of the selective medium SD -Leu at 30°C, 200 rpm.

2. Dilute the overnight culture into 1 L of YPD medium to a starting $OD_{600\ nm}$ of approx 0.3.
3. When the culture reaches an $OD_{600\ nm}$ of between 0.8 and 1.0, centrifuge the culture at 1700*g* for 15 min, 4°C.
4. Resuspend the cells in 250 mL of sterile distilled water by vortexing.
5. Centrifuge at 1700*g* for 15 min, 4°C.
6. Remove the supernatant and add 20 mL of cold sterile distilled water.
7. Split the cells into four sterile 50 mL Falcon tubes and centrifuge at 1000*g* for 5 min.
8. Carefully discard the supernatant and add 8 mL of cold sterile distilled water to each tube.
9. Vortex and add 12 mL PEG 4000 (50 % w/v), 1.8 mL of 1 *M* LiOAc, 0.25 mL of 10 mg/mL single-stranded carrier DNA, and the library (between 50 and 100 μg) to each tube.
10. Fill the tubes to 18 mL with cold sterile distilled water and vortex vigorously for 5 min.
11. Place the tubes on a rolling wheel for 30 min at room temperature.
12. Heat shock the cells at 42°C while shaking at 200 rpm for 30 min.
13. Collect the cells by centrifugation at 1000*g* for 5 min, 4°C.
14. Carefully remove the supernatant and resuspend the cells in 10 mL of cold sterile distilled water.
15. Spread 100 μL of resuspended cells on SD -Leu -Trp -His (*see* **Note 5**). To check the transformation efficiency, spread 100 μL of dilute cells (10^{-4} to 10^{-1} serial dilutions) on SD -Leu -Trp.

3.2.3. Filter Assay for Detection of β-Galactosidase Activity

The X-Gal filter test is performed as described in **Subheading 3.1.5.** on any His positive colonies.

4. Notes

1. The bait vectors pX-Cub-TF/pTf-Cub-X contain the weak CYC1 promoter because higher levels of overexpression of heterologous proteins can activate the reporter genes by themselves. In this plasmid, most of the *UAS2* sequence of the *CYC1* gene has been deleted to generate a weak and noninducible promoter ***(13)***. We have had good results with this expression level for a variety of proteins.
2. Depending on protein topology (type I or type II transmembrane protein), the membrane protein of interest X can be fused either N-terminally (X-Cub-TF) or C-terminally (TF-Cub-X) to the Cub–TF cassette.
3. Because of the low expression of the X-Cub-TF/TF-Cub-X fusion proteins, the detection of the protein signal in western blot analyses can be enhanced by using a Super Signal West Dura Extended Duration Substrate (Pierce).
4. A potentially important consideration for the membrane-based yeast two-hybrid system is whether to include a yeast signal peptide to direct the heterologous

protein to the secretory pathway. To improve the targetting of heterologous transmembrane proteins to the endoplasmic reticulum, a yeast signal peptide such as that from Ste2p (first 14 amino acids MSDAAPSLSNLFYD) ***(15)*** or Mat α (MFα) (first 85 amino acids MRFPSIFTAVLFAASSALAAPVNTTTEDET AQIPAEAVIGYSDLEGDFDVAVLPFSNSTNNGLLFINTTIASIAAKEEGVSLDKR to generate KEEGVSLDKR) ***(17)*** should be used. In some cases, however, the heterologous protein may be targeted to the membrane without addition of a yeast signal peptide (Thaminy et al., personal communication; ***18***). Another important consideration for the membrane-based yeast two-hybrid system is the choice of the yeast strain, because expression of heterologous proteins could lead to a proteolytic degradation by the yeast machinery. This degradation can be avoided by using the yeast-protease-deficient strain YPH420 ***(19)***.

5. As a negative control for the co-transformation, the pY-IIA-NubG/pNubG-HA-X plasmid should be transformed into the yeast expressing the bait protein, and the pX-Cub-TF/pTF-Cub-X should be transformed into the yeast expressing the prey protein or the library before attempting the screen. If yeast colonies grow on the selective medium (SD -Leu -Trp -His), this background growth can potentially be suppressed by using 3-AT (3-amino-1,2,4-triazole), a competitive inhibitor of the yeast His3p protein. A titration of increasing concentrations of 3-AT in the selective medium (5, 10, 15, 20, 30, 40, and 50 m*M*) should be tested. The lowest concentration at which the strain is no longer able to grow is then used in plates to perform the library screen.

Acknowledgments

We are grateful to Arnoldo Anthony for critical reading of the manuscript. The I.S. group is supported by Zücher Krebsliga, Gebert-RüfStiftung, Walter Honegger Stiftung, Bonizzi-Theler Stiftung, EMDO Stiftung, Stiftung fü medizinische Forschung, Kommission fü Technische Inovation (KTI, Nr. 5343.2 SUS), and Swiss National Science Foundation (31-58798.99).

References

1. Bartel, P. L. and Fields, S. (1995) Analyzing protein-protein interactions using two-hybrid system. *Methods Enzymol.* **254,** 241–263.
2. Johnsson, N. and Varshavsky, A. (1994) Split ubiquitin as a sensor of protein interactions in vivo. *Proc. Natl. Acad. Sci. USA* **91,** 10,340–10,344.
3. Stagljar, I., Korostensky, C., Johnsson, N., and te Heesen, S. (1998) A genetic system based on split-ubiquitin for the analysis of interactions between membrane proteins in vivo. *Proc. Natl. Acad. Sci. USA* **95,** 5187–5192.
4. Wittke, S., Lewke, N., Muller, S., and Johnsson, N. (1999) Probing the molecular environment of membrane proteins in vivo. *Mol. Biol. Cell* **10,** 2519–2530.
5. Cervantes, S., Gonzalez-Duarte, R., and Marfany, G. (2001) Homodimerization of presenilin N-terminal fragments is affected by mutations linked to Alzheimer's disease. *FEBS Lett.* **505,** 81–86.

6. Massaad, M. J. and Herscovics, A. (2001) Interaction of the endoplasmic reticulum alpha 1,2-mannosidase Mns1p with Rer1p using the split-ubiquitin system. *J. Cell Sci.* **114,** 4629–4635.
7. Reinders, A., Schulze, W., Thaminy, S., Stagljar, I., Frommer, W. and Ward, J. (2002) Intra- and intermolecular interactions in sucrose transporters at the plasma membrane detected by the split-ubiquitin system and functional assays. *Structure* **10,** 1–20.
8. Thaminy, S, Auerbach, D, Arnoldo, A, Stagljar, I. (2003) Identification of novel ErbB3 interacting factors using the split-ubiquitin membrane yeast two-hybrid system. *Genome Res.* **13,** 1744–1753.
9. Wang, B., Nguyen, M., Breckenridge. D. G., Stojanovic, M., Clemons, P. A., Kuppig, S., Shore, G. C. (2003) Uncleaved BAP31 in association with A4 protein at the endoplasmic reticulum is an inhibitor of Fas-initiated release of cytochrome c from mitochondria. *J Biol Chem* **278,** 14461–14468.
10. Varshavsky, A. (1997) The N-end rule pathway of protein degradation. *Genes Cells* **2,** 13–28.
11. Varshavsky, A. (1997) The ubiquitin system. *Trends Biochem. Sci.* **22,** 383–387.
12. Vojtek, A. B., Hollenberg, S. M., and Cooper, J. A. (1993) Mammalian Ras interacts directly with the serine/threonine kinase Raf. *Cell* **74,** 205–214.
13. Guarente, L., Lalonde, B., Gifford, P., and Alani, E. (1984) Distinctly regulated tandem upstream activation sites mediate catabolite repression of the CYC1 gene of *S. cerevisiae. Cell* **36,** 503–511.
14. Gubler, U. and Hoffman, B. J. (1983) A simple and very efficient method for generating cDNA libraries. *Gene* **25,** 263–269.
15. Sambrook, J. and Russell, D. W. (2001) *Molecular Cloning, A Laboratory Manual,* 3rd Ed. Cold Spring Harbor Laboratory Press, Cold Spring Harbor, New York, NY.
16. King, K., Dohlman, H. G., Thorner, J., Caron, M. G., and Lefkowitz, R. J. (1990) Control of yeast mating signal transduction by a mammalian beta 2-adrenergic receptor and Gs alpha subunit. *Science* **250,** 121–123.
17. Brake, A. J. (1990) Alpha-factor leader-directed secretion of heterologous proteins from yeast. *Methods Enzymol.* **185,** 408–421.
18. Tu, J., Golde, D. W., Vera, J. C., and Heaney, M. L. (1998) Expression of the human GM-CSF receptor alpha subunit in *Saccharomyces cerevisiae. Cytokines Cell. Mol. Ther.* **4,** 147–151.
19. Nigro, J. M., Sikorski, R., Reed, S. I., and Vogelstein, B. (1992) Human p53 and CDC2Hs genes combine to inhibit the proliferation of *Saccharomyces cerevisiae. Mol. Cell Biol.* **12,** 1357–1365.
20. Mumberg, D., Muller, R., and Funk, M. (1995) Yeast vectors for the controlled expression of heterologous proteins in different genetic backgrounds. *Gene* **156,** 119–122.

21

Reverse Two-Hybrid Techniques in the Yeast *Saccharomyces cerevisiae*

Matthew A. Bennett, Jack F. Shern, and Richard A. Kahn

Abstract

Use of the yeast two-hybrid system has provided definition to many previously uncharacterized pathways through the identification and characterization of novel protein–protein interactions. The two-hybrid system uses the bi-functional nature of transcription factors, such as the yeast enhancer Gal4, to allow protein–protein interactions to be monitored through changes in transcription of reporter genes. Once a positive interaction has been identified, either of the interacting proteins can mutate, either by site-specific or randomly introduced changes, to produce proteins with a decreased ability to interact. Mutants generated using this strategy are very powerful reagents in tests of the biological significance of the interaction and in defining the residues involved in the interaction. Such techniques are termed reverse two-hybrid methods. We describe a reverse two-hybrid method that generates loss-of-interaction mutations of the catalytic subunit of the *Escherichia coli* heat-labile toxin (LTA1) with decreased binding to the active (GTP-bound) form of human ARF3, its protein cofactor.

Key Words

Reverse two-hybrid; two-hybrid; protein interaction; loss-of-interaction mutation.

1. Introduction

Protein interactions define key biochemical and regulatory networks in the cell. The identification and characterization of protein–protein interactions through the use of two-hybrid assays have provided definition to many previously uncharacterized pathways. Yeast two-hybrid systems use the bi-functional nature of transcription factors, e.g., Gal4 or LexA, to assay protein interactions in live cells *(1,2)*. The DNA binding and transcriptional activation domains of the transcription factor are each expressed as fusion proteins and promote transcription only when joined by the binding of the other portion of

From: *Methods in Molecular Biology, vol. 261: Protein–Protein Interactions: Methods and Protocols*
Edited by: H. Fu © Humana Press Inc., Totowa, NJ

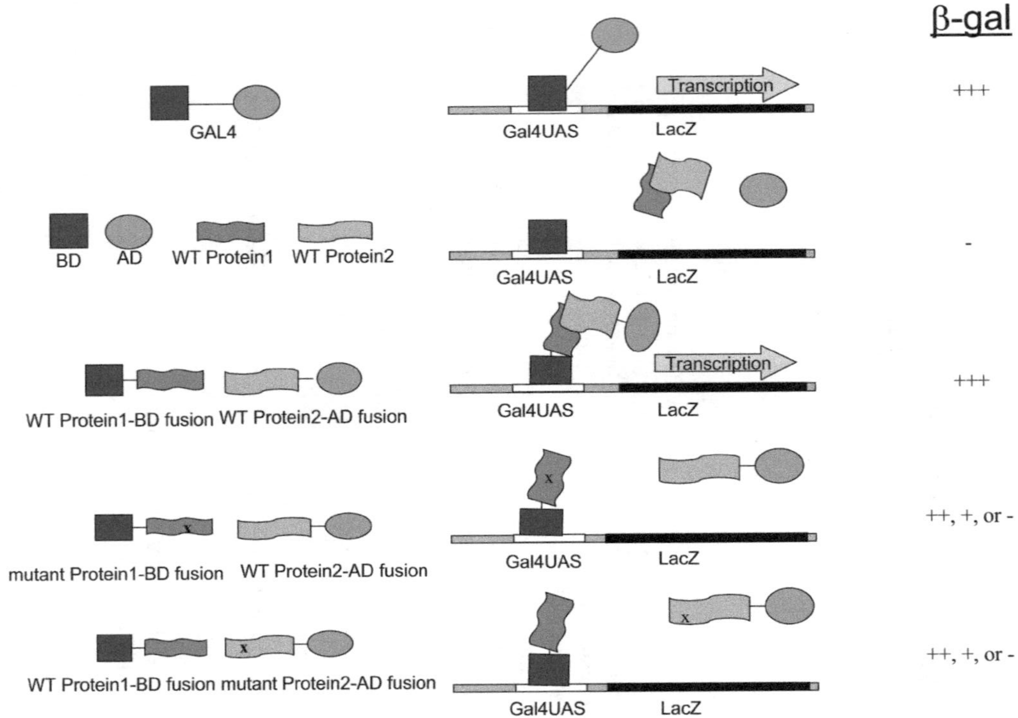

Fig. 1: Expression of reporter genes (e.g., *LacZ*) in two-hybrid systems is dependent on the interaction of two fusion proteins that can be lost as a result of a point mutation. The Gal4 protein, two interacting proteins of interest, and resulting fusion proteins are shown pictorially on the left. An "x" in a protein represents a point mutation. Interaction between the two fusion proteins generates a transactivator capable of promoting transcription of reporter genes (*LacZ*) downstream of the Gal4 upstream activating sequence (UAS). Point mutations in either fusion protein that decrease binding result in decreases in or elimination of reporter gene expression. The level of expression is shown on the right as the amount of β-galactosidase (β-gal) activity.

the fusion proteins (*see* **Fig. 1** and **Note 1**). Screens of two-hybrid libraries have led to the identification of a large number of protein binding partners *(3,4)* and proteome based screens *(5,6)* have increased the number of putative interactions markedly.

After a two-hybrid interaction is identified, either of the interacting proteins can be mutated to decrease the binding affinity. The generation and use of such mutants is collectively referred to as "reverse two-hybrid" methods. Such loss-of-interaction mutants can become essential reagents to test the functional importance of specific protein interactions in live cells, to obtain a low-resolution map of the binding interface, or to generate pairs of mutants that interact independently of endogenous proteins. These approaches are particularly attractive when one of the proteins under study binds multiple protein partners as mutagenesis and two-hybrid screens can identify specific loss of binding mutants that retain binding to other partners. In this chapter, we describe a reverse two-hybrid method for generating mutations of the catalytic subunit of the *Escherichia coli* heat-labile toxin (LTA1) with reduced binding to the activated (GTP bound) form of human ARF3, [Q71L]ARF3 *(7,8)*.

The main features of this approach include PCR-based random mutagenesis over a defined region of any coding region and identification of full-length mutants, expressed to the same levels as wild-type protein by immunoblot analysis using an epitope tag on the fusion proteins. The conditions used here have yielded approximately one change per 250 bp amplified. PCR allows the extent of mutagenesis to be controlled by modifying the concentration of nucleotides or manganese in the PCR reaction *(9)*. Although the use of selectable markers fused to the C-terminus of the mutated protein allows for genetic selection of full-length proteins *(10–12)*, which is important to high-throughput screens and can speed up this rate-limiting step in reverse two-hybrid screens, we prefer the use of immunoblots to allow estimates of protein expression as well as confirmation that the protein is full length. Using the methods described below we have been able to generate hundreds of colonies with reduced β-galactosidase activity in less than 1 wk and typically 25–50% of those were found to result from point mutations in full length proteins, suitable for further analysis. Thus, tens of mutants can be obtained, their interactions with multiple effectors characterized, and sequences determined in less than 1 mo.

The loss of an interaction is limited in interpretations due to the negative nature of the data. However, loss of one binding partner with retention of others provides a powerful probe of activities in live cells. In addition, after the generation of loss-of-interaction mutants in one protein it is even easier to then mutate the other binding partner and screen for mutants that regain the interaction. Such pairs of mutants can provide a formal proof of specific functionalities in live cells. Such mutants may have the added benefit of not binding the wild-type protein and can allow cell studies to be carried out independently of interference or complications presented by the presence of endogenous proteins. As with any technique, the data and conclusions from the use of reverse two-hybrid techniques are strengthened by confirmation using independent methods.

2. Materials

1. pACT2 vector ***(13)*** (Genbank accession no. U29899): a 2μ plasmid carrying the selectable *LEU2* gene and the *ADH1* promoter driving expression of the activation domain (AD) of Gal4p [Gal4 Region II ***(1)***, residues 768–881], followed by the HA epitope, sites for insertion of the open reading frame of your protein, and the *ADH1* terminator.
2. pBG4D vector ***(14)***: a 2μ plasmid carrying the selectable *TRP3* marker and the Gal4 DNA binding domain (BD) followed by the HA epitope, designed to add the Gal4-BD to the 3' end of any inserted open reading frame, with expression of the resulting fusion protein driven by the *ADH1* promoter. pBG4D was made by Robert M. Brasas by subcloning the *Sac*I/*Hind*III fragment from the pAS1-CYH plasmid into the plasmid D133.
3. Yeast strain Y190 (*MAT α gal4 gal80 his3 trp1-901 ade2-101 ura3-52 leu2-3,112 + URA3::GAL→lacZ, LYS2::GAL(UAS)-HIS3 cyh*^{R}) ***(13)*** kindly provided by Steve Elledge.
4. Whatman filter paper no. 1.
5. Nitrocellulose filters (Schleicher & Schuell Protran, 82 mm).
6. Monoclonal HA antibody 12CA5 (Boehringer), diluted 1:1000 (final = 1 μg/mL) for immunoblots.
7. X-gal (5-bromo-4-chloro-3-indolyl-β-D-galactopyranoside): 100 mg/mL stock solution in *N,N*-dimethylformamide (DMF).
8. Low dATP nucleotide mix: 2.5 m*M* dCTP, 2.5 m*M* dGTP, 2.5m*M* dTTP, and 0.625 m*M* dATP in water.
9. *Taq* polymerase.
10. Synthetic (SD) minimal, complete, and dropout media. Minimal SD medium: 1.5 g yeast nitrogen base lacking amino acids (Difco), 20 g glucose, and 5 g $(NH_4)_2SO_4$ per liter. Complete SD medium: 20 mg adenine, 20 mg histidine, 20 mg methionine, 20 mg tryptophan, 20 mg uracil, 30 mg leucine, and 30 mg lysine per liter are added, and each is omitted as needed to generate the dropout media, e.g., SD -leu -trp.
11. SD plates, made with the same components as SD media plus 20 g agar (Difco) per liter.
12. SD -leu/cycloheximide: SD -leu containing 2.5 μg/mL cycloheximide.
13. Agarose gel and DNA sequencing equipment.
14. SDS-PAGE (sodium dodecylsulfate polyacrylamide gel electrophoresis) and protein electrotransfer equipment for immunoblotting.
15. Z-Buffer: 60 m*M* Na_2HPO_4, 40 m*M* NaH_2PO_4, 10 m*M* KCl, 2 m*M* $MgSO_4$, pH 7.0.

16. Oligonucleotide primers for PCR mutagenesis: #827 (sense primer): GCGTATAACGCGTTTGGAATC, binds approx 70 bp 5' of the pACT2 MCS. #828 (antisense primer): GAGATGGTGCACGATGCACAG, binds approx 70 bp 3' of the pACT2 MCS.
17. Sequencing oligonucleotides: #1042 (sense primer): GCTTACCCATAC GATGTTCCA, binds within the pACT2 MCS at the 5' end. #1043 (antisense primer): TGAACTTGCGGGGTTTTTCAG, binds within the pACT2 MCS at the 3' end.
18. DNA purification kits (e.g., Qiagen mini prep, QiaQuick gel extraction, Wizard PCR purification kit).
19. pAB157: a plasmid encoding [Q71L]ARF3 with the Gal4-BD at the C-terminus by insertion into the *Bam*HI site of pBG4D, to direct expression of [Q71L]ARF3-BD.
20. pXJ12: a plasmid carrying the open reading frame of the catalytic (A1) subunit of the *E. coli* heat labile toxin (LTA1) cloned into the *Nco*I and *Bam*HI sites of pACT2, putting it in frame with the Gal4-AD, to direct expression of AD-LTA1 (*see* **Fig. 2**).
21. YAB457: the yeast strain, derived from Y190, carrying the pAB157 plasmid and thus expressing [Q71L]ARF3-BD.
22. SOS: 3.33 g/L yeast extract, 6.66 g/L dextrose, 6.66 g/L peptone, and 6.5 m*M* $CaCl_2$.
23. TE: 10 m*M* Tris-HCl, pH 8.0 and 1 m*M* EDTA, pH 8.0.
24. 0.1 *M* lithium acetate in TE buffer.
25. LiOAc/PEG solution: 40% (w/v) PEG 3350 suspended in 0.1 *M* lithium acetate in TE buffer.
26. Sonicated herring sperm DNA, 10 mg/mL.
27. STES: 500 m*M* NaCl, 200 m*M* Tris-HCl, pH 7.6, 10 m*M* EDTA, and 1% SDS.
28. Phenol/chloroform: 50% phenol and 50% chloroform.
29. Chloroform.
30. Polymerase chain reaction (PCR) buffer (1X): 10 m*M* Tris-HCl, pH 8.3, 50 m*M* KCl, 0.001% gelatin, which is made from a 10X solution (Sigma).

3. Methods

The methods listed in this section describe: (1) PCR-introduced, random mutagenesis of LTA1, (2) co-transfection of mutated LTA1 and gapped pACT2 plasmid into YAB457 to yield yeast colonies expressing mutant LTA-AD and [Q71L]ARF3 fusion proteins, (3) assaying for loss of interaction via a colony β-galactosidase assay, (4) rescuing the plasmids expressing mutant LTA1, and (5) characterizing the loss-of-interaction LTA1 mutants. These methods begin after a two-hybrid interaction has been identified, e.g., between the protein products of the plasmids pXJ12 (AD-LTA1) and pAB157 ([Q71L]ARF3-BD).

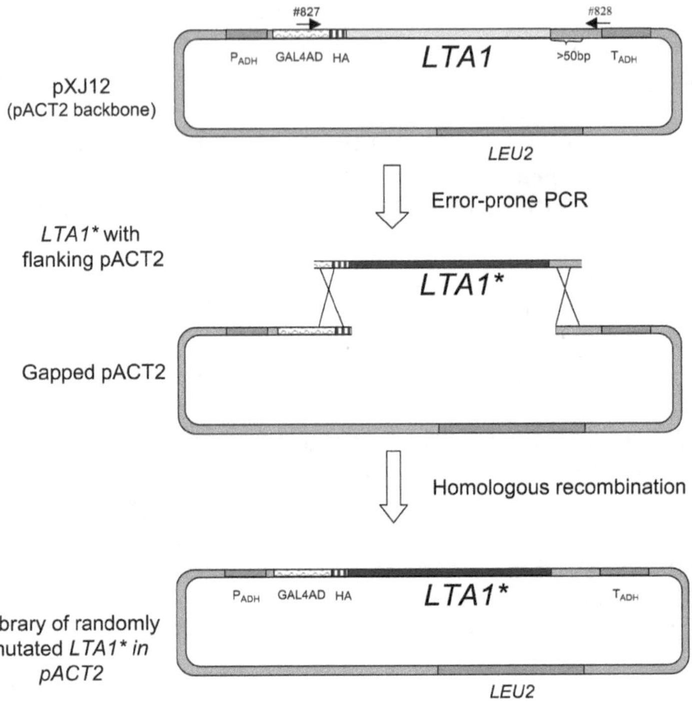

Fig. 2. Reverse two-hybrid screens use random PCR mutagenesis and gap repair in yeast to generate a large library of mutants, ready for screening. Amplification of the coding region of *LTA1* in pXJ12 is performed with primers that anneal ≥50 bp outside of the region targeted for mutagenesis to yield a product with ends homologous to those in the gapped plasmid, shown below. The error rate of the polymerase is increased by carrying out the PCR under conditions of reduced stringency. Note that any region of DNA can be targeted for mutagenesis. Co-transformation of the mutated PCR product with gapped pACT2 plasmid, prepared by restriction digestion, into yeast allows for repair of the plasmid by homologous recombination. When transformed into a yeast strain expressing the [Q71L]ARF3-BD fusion proteins, the resulting transformants can be assayed directly for loss of interaction using the colony β-galactosidase assay.

3.1. Generation of Mutant Substrates for Plasmids

3.1.1. Mutagenesis of LTA1

This method exploits the lack of proofreading by *Taq* polymerase and ability to replicate DNA with low fidelity and generate random changes into any region of DNA, in this case the LTA1-encoding portion of the AD-LTA1 plasmid (*see* **Fig. 2** and **Note 2**).

Mutagenesis of LTA1 was accomplished through PCR amplification of pXJ12 with primers #827 and #828 under conditions that promote polymerase-induced errors (*see* **Fig. 2**). These error-prone conditions included the addition of $MnCl_2$ and lowering the concentration of dATP, relative to other nucleotides, as described previously ***(15,16)***. The 50 µL PCR reaction consisted of the following: 1.5 m*M* $MgCl_2$, 0.05 m*M* $MnCl_2$, 0.25 m*M* each of dCTP, dGTP, and TTP, 0.0625 m*M* dATP, 2 µ*M* primer #827, 2 µ*M* primer #828, 1 µL *Taq* polymerase (0.05 U/µL), 25 ng pXJ12 in 1X PCR Buffer.

PCR amplification/mutagenesis was performed in 25 cycles using an annealing temperature of 55°C (15 s) and an extension temperature of 72°C (1 min for products <1 kb), with a denaturation step (94°C, 15 s) before each cycle. The product was purified with the Wizard PCR Clcanup kit (Promega) according to the manufacturer's directions. The result is the purified PCR product containing randomly introduced mutations in the LTA1 coding region and approx 80 bp of flanking pACT2-derived DNA upstream and downstream of the open reading frame. Yields from PCR under these conditions are lower than normal, but should produce plenty of product for later steps.

3.1.2. Gapped plasmid (pACT2) Production

The gapped plasmid was created by digesting pXJ12 with *Nco*I and *Xho*I (*see* **Note 3**). The resulting linearized, 8 kb vector backbone was purified from a 1% agarose gel with the QiaQuick gel extraction kit (Qiagen).

3.1.3. Estimating Vector and PCR Product Concentration

Concentrations of the purified, gapped (pACT2-derived) plasmid and PCR products were estimated by comparing the ethidium bromide fluorescence intensity from each sample in a 1% agarose gel to that of a known amount of a λ*Hind*III DNA marker of similar size.

3.2. Co-transformation of Gapped Plasmid and PCR Product

The insertion of the mutagenized open reading frame into the gapped plasmid was accomplished by taking advantage of the high level of homologous recombination in yeast cells. Co-transformation of yeast with gapped plasmid and PCR products that contain regions of identity of ≥ 50 bp at each end is approx 50% as efficient as transformation with the circular plasmid alone (*see* **Fig. 2** and **Note 4**).

1. Inoculate 50 mL SD -trp with YAB457 and grow at 30C to $OD_{600\ nm}$ ~0.6.
2. Harvest the cells by centrifugation at 1000*g* for 5 min.
3. Resuspend the cells in TE buffer and collect cells again by centrifugation.
4. Wash the cells in 0.1 *M* lithium acetate in TE and collect by centrifugation.
5. Resuspend the cells in 0.5mL 0.1 *M* lithium acetate/TE.

6. Aliquot 100 µL of cell suspension into each of five sterile microfuge tubes.
7. Add 5 µL freshly boiled sonicated herring sperm DNA and 0.7 mL of LiOAc/PEG solution to each tube. Single-stranded carrier DNA improves transformation frequency.
8. Add the transforming DNA (gapped plasmid and PCR products) to each tube. Equal molar amounts of gapped plasmid and PCR products are used. The amount of DNA will vary with the experimentally determined transformation frequency, designed to yield 400–600 transformants per 100 mm plate, but will probably be in the range of 0.1–1 µg per transformation (*see* **Notes 4** and **5**).
9. Incubate each transformation reaction at 30C for 30 min.
10. Transfer tubes to a 42°C water bath and heat shock for 15 min.
11. Add 600 µL of SOS to each transformation tube.
12. Plate an appropriate volume (100 µL) of the transformed cell suspension on SD -trp -leu plates and incubate at 28°C until small, distinct colonies become visible (approx 3 days).

3.3. Assaying Transformants for Loss-of-Interaction

The X-gal filter assay was used to measure levels of β-galactosidase activity in yeast colonies as an indicator of protein interactions *(17)*. The two fusion proteins (often referred to as bait and prey) interact, thereby activating transcription of the *lacZ* gene and production of β-galactosidase, which catalyzes the hydrolysis of X-gal to yield a blue product. Both the time of incubation and the darkness of the blue product formed on the filter were monitored and compared to the parent plasmid to allow initial determination of loss-of-interaction mutations.

1. Place a nitrocellulose filter on each plate for 30 s until wet, to replica the yeast colonies onto the filter.
2. Drop the filter into liquid nitrogen and carefully remove after 20 s. Avoid breaking the brittle filters while frozen.
3. Transfer the frozen nitrocellulose filter, with the yeast side facing up, to the lid of a 100 mm dish on which has been placed Whatman filter paper wetted with X-gal in Z buffer (1.5 mL of Z-buffer with 15 µL 100 mg/mL X-gal solution to yield a final X-gal concentration of 1 mg/mL).
4. Cover with the empty dish to prevent dehydration and incubate the filter at 30°C. Compare the blueness of colonies that develop on the co-transformation plates with that of the control plates at different times. The time will vary with the bait–prey combination. Severe changes (white colonies) are often later found to result from the introduction of premature stop codons but can also produce a mutant with the most dramatic difference in binding. It is a good idea to initially select groups of colonies with small, intermediate, and large differences in activity.
5. Colonies from the experimental plates, which turned less blue or remained white in comparison to the (parental) positive control plate, were streaked onto SD -trp -leu plates to select for retention of both plasmids. After colonies appear, they

may be re-screened using histidine auxotrophy for selection. We have found this to be a more sensitive assay than the X-gal filter assay.

3.4. Recovering Loss-of-Interaction Mutant Plasmids and Re-testing

Plasmids were recovered from yeast using a modification of the standard glass bead method *(18)*, also known as a "smash and grab" preparation. The modification simply adds the use of commercial plasmid purification columns to further remove contaminants that inhibit transformation. The use of plasmid purification columns is not required or used in all laboratories, but we have found it increases success rates, sometimes markedly.

1. Streak colonies with diminished β-galactosidase activity onto SD -leu cycloheximide plates to select for loss of the [Q71L]ARF3 BD expressing plasmid. This step is not required but often facilitates the isolation of the desired (AD-LTA1 mutant) plasmid from strains carrying two different plasmids (*see* **Note 6**).
2. Incubate the plates for ≥3 d at 28°C.
3. Inoculate 20 mL of SD -leu media with colonies from the SD -leu cycloheximide plates and grow each culture to an $OD_{600\ nm}$ ~ 1.
4. Collect cells by centrifugation at 1000*g* for 10 min.
5. Wash each pellet in 1mL of STES buffer and collect again.
6. Resuspend in 100 μL STES and add an equal volume of glass beads (*see* **Note 7**).
7. Vortex each tube for 2 min.
8. Add 100 μL STES and 200 μL phenol/chloroform and vortex for 5–30 min. Times will vary with the vortexer used.
9. Pellet cells and debris and separate phases in a microfuge at maximal speed (approx 14,000*g*) for 10 min, and transfer the aqueous phase to a new tube.
10. Extract with 200 μL phenol/chloroform, and then with chloroform alone.
11. Further purify the plasmid in the aqueous phase by using a commercial plasmid miniprep kit, according to manufacturer's directions. Then use the product to transform competent bacterial (e.g., DH5α) cells.
12. Purify plasmids from transformed DH5α using any commercial plasmid miniprep kit, according to manufacturer's instructions. Check by restriction digestion analysis to confirm the correct plasmid has been obtained.
13. Transform purified plasmids into YAB457 and assay transformants for interaction with [Q71L]ARF3 using the X-gal filter assay. The color developed at different times (15, 30, 60, and 180 min) was compared to that of the YAB457-derived strain expressing wild-type LTA1. Activities were scored visually on a three plus scale, with three plus equal to that of wild-type LTA1 (*see* **Note 8**).

3.5. Characterizing the Noninteracting Proteins

Multiple factors can be responsible for a loss of interaction, some of which are far less informative than a point mutation in a critical amino acid residue. Two types of uninteresting changes—premature stop codons and decreased levels of protein expression—were screened for by immunoblotting cell lysates with an HA antibody to determine the presence and relative abundance of the epitope tag in the LTA1-AD fusion proteins. The levels of protein expressed in each strain expressing mutant LTA1 were compared to wild-type LTA1 (*see* **Note 9**). Protein preparations from total yeast cell lysates were obtained using a modification of Horvath and Riezman ***(19)***.

1. Grow five OD of each yeast strain in selective liquid medium.
2. Collect cells by centrifugation at 1000*g* for 5 min.
3. Wash cells with water and collect again at 1000*g* for 5 min.
4. Resuspend each pellet in 20 μL of 1X SDS sample buffer and boil at 95°C for 5 min.
5. Add one cell volume of acid-washed glass beads and vortex 3 min at maximum speed.
6. Add 120 μL sample buffer and boil for 5 min.
7. Pellet cells and debris by centrifugation in a microfuge for 15 min at maximal setting. Transfer supernatant to another tube and boil for 5 min.
8. Load 7.5 μL of the sample on a 12% SDS gel and resolve proteins using standard methods.
9. Transfer the resolved proteins to nitrocellulose at 60 V for 2 h and immunoblot with the 12CA5 (HA) antibody using standard methods.

If the mutant protein library was transfected into an ARF3-expressing strain, then you should detect two bands in immunoblots, at sizes corresponding to the AD-LTA1 and ARF3-BD fusion proteins. Y190-derived strains also express a protein of approx 50 kDa that is bound by the 12CA5 antibody, so it is a good idea to use Y190 as a control, at least initially. A shift to a smaller size or the absence of (or clearly reduced) immunoreactivity is suggestive of premature termination or alteration in the level of expression, respectively. Plasmids encoding full-length LTA1 mutants that were expressed at approximately the same level as wild-type LTA1 were then sequenced with pACT2 sequencing primers #1259 and #1260 to determine any changes ***(7)***.

4. Notes

1. The two-hybrid system described here was that developed in the Elledge lab and uses a Gal4-based reporter system ***(13)***. Y190 is a strain of *S. cerevisiae* engineered to express the *HIS3* and *lacZ* gene products as reporters, in response to the presence of a functional Gal4 transactivator. Y190 is able to detect protein–protein interactions by making two Gal4 hybrid proteins involving two different domains of Gal4: one protein is fused with the Gal4's activation domain to generate a Gal4–AD fusion protein, while the other protein is fused with Gal4's DNA binding domain to yield a Gal4–BD fusion protein. If, when expressed in Y190, the two fusion proteins fail to interact, the yeast fail to activate transcription at the *GAL4* promoter and hence of the two reporters. However, if the two

proteins do interact, a functional Gal4 transactivator is created. Thus, interaction can be observed by selecting and assaying for histidine prototrophy and β-galactosidase activity.

Other two-hybrid systems have been described that use other Gal4-based plasmids ***(20)***, that use more than two reporters ***(21)***, that use other transcriptional activators (e.g., LexA; *see* other chapters in this section of the volume for details; also for reviews *see* refs. ***22,23***) . With only slight alterations, the steps outlined in this chapter can be applied to other such systems to perform a reverse two-hybrid screen.

2. The lack of proofreading by *Taq* polymerase facilitates the introduction of mutations, most often A-T and A-G mutations ***(9,24)***. Alternate DNA polymerases (e.g., Stratagene's Mutazyme) designed to introduce a broader variety of mutations can be used.
3. Gapped plasmid was generated from pXJ12 (as opposed to empty pACT2) to ensure that any uncut plasmid present in the gapped plasmid isolation would encode a protein that would retain interaction with [Q71L]ARF3 and remain blue in the X-gal filter lift assay.
4. Transformation efficiency was determined by transforming YAB457 with undigested pXJ12. We estimated the efficiency for gapped plasmid repair in the co-transformation reaction to be half that of intact plasmid when transforming with an equimolar amount of gapped plasmid and PCR product. The volume of cell suspension resulting from the co-transformation reaction plated on each SD -trp -leu plate was adjusted to yield between 400 and 600 transformants per 100 mm plate.
5. Up to four control transformation reactions can be performed, including (1) transformation of uncut pXJ12 plasmid to determine general transformation efficiency, transformation of (2) no DNA, (3) the gapped plasmid alone, or (4) the PCR product alone, provide estimates of the background levels of transformations that will appear on the experimental co-transformation plates. Although not required, these controls are strongly encouraged the first time reverse two-hybrid techniques are used in the lab.
6. pAS1-CYH and pBG4D both contain the Cyh^S gene that causes sensitivity to cycloheximide. By growing cells containing pACT2 and pBG4D-based plasmids on cycloheximide-containing media, cells not expressing pBG4D-based plasmids are selected and allow for easier isolation of pACT2-LTA1 plasmids.
7. This and the next step are critical and are the ones most likely to cause low recovery of plasmids. Too few beads and the cells simply rotate in the tube and too many beads and a slurry forms that also decreases shearing forces that produce cell lysis.
8. As an alternative to the filter lift assay, a liquid culture β-galactosidase activity assay can be used to generate more quantifiable measure of activity ***(25)***. Briefly, duplicate strains of yeast were grown in 5 mL of selective medium overnight at 30°C. The next day, 5 mL of fresh selective medium were added to 20–50 μL of the overnight culture and grown to mid log phase ($OD_{600} = 0.3 - 0.7$). Cells were

collected by centrifugation and resuspended in 1 mL Z-Buffer and put on ice. Cells (50–100 µL) were mixed with Z buffer to a volume of 1 mL, and a drop of 0.1% SDS and 2 drops of chloroform were added to the sample with a Pasteur pipet. The mixtures were vortexed for 15 s and incubated for 15 min at 30°C. *O*-nitrophenol-α-D-galactopyranoside (ONPG, 0.8 mg) was added, and the solution was vortexed for 5 s and incubated at 30°C until a medium yellow color had developed in the positive controls. The reaction was stopped by adding 0.5 mL of 1 *M* Na_2CO_3, and the time was recorded. OD_{420} and OD_{550} were determined after the cell debris was removed by centrifugation. Units of activity were determined using the equation:

$$U=1000\ (OD_{420}-OD_{550}) / (T \times V \times OD_{600})$$

where V is the volume of the culture used in the assay (µL), T is the time of the reaction (min), OD_{600} is the cell density at the start of the assay, OD_{420} represents the combination of absorbance by ONPG and the light scattering by cell debris, and OD_{550} represents the light scattering by cell debris.

9. In addition to providing the Gal4 activation domain, pACT2 also adds an HA-epitope after the Gal4–AD and before your protein. Other plasmids encoding the Gal4 DNA binding domain, such as pBG4D or pAS1-CYH, also include the HA tag to their fusion protein products and therefore will also appear when blotting with an HA antibody. When comparing levels of protein expression by immunoblotting, it is necessary to determine protein concentrations of each sample and load equivalent protein onto the gel as differences in cell lysis and protein recovery can occur. Note that not all proteins are expressed in yeast to a level that can be detected by immunoblotting but can still be used in two-hybrid and reverse two-hybrid methods.

References

1. Fields, S. and Song, O. (1989) A novel genetic system to detect protein-protein interactions. *Nature* **340,** 245–246.
2. Chien, C. T., Bartel, P. L., Sternglanz, R., and Fields, S. (1991) The two-hybrid system: a method to identify and clone genes for proteins that interact with a protein of interest. *Proc. Natl. Acad. Sci. USA* **88,** 9578–9582.
3. Boman, A. L., Zhang, C., Zhu, X., and Kahn, R. A. (2000) A family of ADP-ribosylation factor effectors that can alter membrane transport through the trans-Golgi. *Mol. Biol. Cell* **11,** 1241–1255.
4. Van Valkenburgh, H., Shern, J. F., Sharer, J. D., Zhu, X., and Kahn, R. A. (2001) ADP-ribosylation factors (ARFs) and ARF-like 1 (ARL1) have both specific and shared effectors: characterizing ARL1-binding proteins. *J. Biol. Chem.* **276,** 22,826–22,837.
5. Uetz, P., Giot, L., Cagney, G., et al. (2000) A comprehensive analysis of protein-protein interactions in *Saccharomyces cerevisiae*. *Nature* **403,** 623–627.
6. Ito, T., Chiba, T., Ozawa, R., Yoshida, M., Hattori, M., and Sakaki, Y. (2001) A comprehensive two-hybrid analysis to explore the yeast protein interactome. *Proc. Natl. Acad. Sci. USA* **98,** 4569–4574.

7. Zhu, X., Kim, E., Boman, A. L., Hodel, A., Cieplak, W., and Kahn, R. A. (2001) ARF binds the C-terminal region of the *Escherichia coli* heat-labile toxin (LTA1) and competes for the binding of LTA2. *Biochemistry* **40,** 4560–4568.
8. Zhu, X. and Kahn, R. A. (2001) The *Escherichia coli* heat labile toxin binds to Golgi membranes and alters Golgi and cell morphologies using ADP-ribosylation factor- dependent processes. *J. Biol. Chem.* **276,** 25,014–25,021.
9. Cadwell, R. C. and Joyce, G. F. (1992) Randomization of genes by PCR mutagenesis. *PCR Methods Appl.* **2,** 28–33.
10. Leanna, C. A. and Hannink, M. (1996) The reverse two-hybrid system: a genetic scheme for selection against specific protein/protein interactions. *Nucleic Acids Res.* **24,** 3341–3347.
11. Vidal, M., Brachmann, R. K., Fattaey, A., Harlow, E., and Boeke, J. D. (1996) Reverse two-hybrid and one-hybrid systems to detect dissociation of protein-protein and DNA-protein interactions. *Proc. Natl. Acad. Sci. USA* **93,** 10,315–10,320.
12. Puthalakath, H., Strasser, A., and Huang, D. C. (2001) Rapid selection against truncation mutants in yeast reverse two-hybrid screens. *Biotechniques* **30**, 984–988.
13. Durfee, T., Becherer, K., Chen, P. L., et al. (1993) The retinoblastoma protein associates with the protein phosphatase type 1 catalytic subunit. *Genes Dev.* **7,** 555–569.
14. Boman, A. L., Kuai, J., Zhu, X., Chen, J., Kuriyama, R. and Kahn, R. A. (1999) Arf proteins bind to mitotic kinesin-like protein 1 (MKLP1) in a GTP-dependent fashion. *Cell Motil. Cytoskeleton.* **44,** 119–132.
15. Muhlrad, D., Hunter, R., and Parker, R. (1992) A rapid method for localized mutagenesis of yeast genes. *Yeast* **8,** 79–82.
16. Kuai, J., Boman, A. L., Arnold, R. S., Zhu, X., and Kahn, R. A. (2000) Effects of activated ADP-ribosylation factors on Golgi morphology require neither activation of phospholipase D1 nor recruitment of coatomer. *J. Biol. Chem.* **275,** 4022–4032.
17. Bai, C. and Elledge, S. J. (1996) Gene identification using the yeast two-hybrid system. *Methods Enzymol.* **273,** 331–347.
18. Rose, M. D., Winston, F., and Hieter, P. (1990) *Laboratory Course Manual for Methods in Yeast Genetics*. Cold Spring Harbor Laboratory, Cold Spring Harbor, New York.
19. Horvath, A. and Riezman, H. (1994) Rapid protein extraction from *Saccharomyces cerevisiae*. *Yeast* **10,** 1305–1310.
20. Bartel, P., Chien, C. T., Sternglanz, R., and Fields, S. (1993) Elimination of false positives that arise in using the two-hybrid system. *Biotechniques* **14,** 920–924.
21. James, P., Halladay, J., and Craig, E. A. (1996) Genomic libraries and a host strain designed for highly efficient two- hybrid selection in yeast. *Genetics* **144,** 1425–1436.
22. Brent, R. and Finley, R. L., Jr. (1997) Understanding gene and allele function with two-hybrid methods. *Annu. Rev. Genet.* **31,** 663–704.
23. Vidal, M. and Legrain, P. (1999) Yeast forward and reverse ħ'-hybrid systems. *Nucleic Acids Res.* **27,** 919–929.

24. Shafikhani, S., Siegel, R. A., Ferrari, E., and Schellenberger, V. (1997) Generation of large libraries of random mutants in *Bacillus subtilis* by PCR-based plasmid multimerization. *Biotechniques* **23,** 304–310.
25. Guarente, L. (1983) Yeast promoters and lacZ fusions designed to study expression of cloned genes in yeast. *Methods Enzymol.* **101,** 181–191.

22

Mammalian Two-Hybrid Assay for Detecting Protein–Protein Interactions In Vivo

Jae Woon Lee and Soo-Kyung Lee

Abstract

Mammalian two-hybrid assay is a convenient, powerful tool to investigate protein–protein interactions in vivo. In particular, this method has a major advantage over the better known yeast version in that one can study interactions between mammalian proteins that may not fold correctly in yeast or that require post-translational modification or external stimulation that are not present in yeast.

Key Words

Chimera; Gal4; VP16; interaction; transcription.

1. Introduction

Mammalian two-hybrid assay is analogous to the better-known yeast two-hybrid assay *(1,2)*. In principle, both two-hybrid assays are based on the fact that many eukaryotic transcriptional activators consist of two physically and functionally separable modules *(3)*: a DNA-binding domain (DBD; depicted as "D" in **Fig. 1**) that specifically binds to a promoter/enhancer element and a transcriptional activation domain (TAD; depicted as "A" in **Fig. 1**) that directs RNA polymerase II to transcribe the gene downstream of the DNA-binding site. Successful transcriptional activation requires both domains, as in the case of the native transcription factors (**Fig. 1-I**). In contrast, expression of two domains separately will not lead to activation of the reporter gene (**Fig. 1-II**) unless the TAD is tethered to the DBD bound to the promoter.

From: *Methods in Molecular Biology, vol. 261: Protein–Protein Interactions: Methods and Protocols*
Edited by: H. Fu © Humana Press Inc., Totowa, NJ

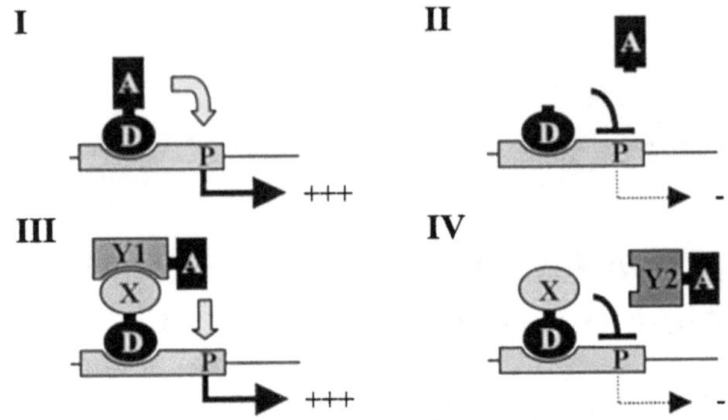

Fig. 1. Principles of mammalian two-hybrid assays. "D" and "A" denote DNA-binding and transcriptional activation domains, respectively. Most transcription factors consist of "D" and "A" in a single protein **(I)**. Lack of "A" results in the loss of transcriptional activity **(II)**. "D" and "A" can be brought together by two interacting proteins (X and Y1) that are fused to "D" and "A", respectively **(III)**. However, A'-fusion proteins that do not interact with X, such as Y2, do not restore the transcriptional activity of D-X fusion protein **(IV)**.

In two-hybrid assays, that tether is the protein–protein interaction between two interacting proteins (X and Y1 in **Fig. 1-III**) that are independently expressed as protein fusions to the TAD and DBD, respectively. In the mammalian two-hybrid assays, the DBD of yeast transcriptional activator Gal4 *(4)* and an acidic TAD located within the 78 carboxyl-terminal amino acids of the strong viral activator VP16 *(5)* are commonly used. A major advantage of mammalian two-hybrid assay over the yeast version *(1,2)* is that protein–protein interactions are studied in vivo in mammalian cell lines. Thus, it enables one to study the interactions between mammalian proteins that may not fold correctly in yeast or that require post-translational modification (e.g., phosphorylation) or external stimulation not present in yeast.

2. Materials

1. *Escherichia coli* competent cells (DH5α or other appropriate strain; Invitrogen cat. no. 18265017).
2. Alkaline phosphatase, calf intestinal (CIAP; New England Biolab, cat. no. M0290).
3. Restriction enzymes and T4 DNA ligase (New England Biolabs).
4. Ligation buffer (1X): 50 m*M* Tris-HCl, pH 7.5, 10 m*M* $MgCl_2$, 10 m*M* dithiothreitol, 1 m*M* ATP, 25 mg/mL bovine serum albumin.
5. Agarose gel equipment.

6. Luria–Bertani (LB) medium with and without ampicillin and agar plate with ampicillin (50 μg/mL; LB broth base, Invitrogen, cat. no. 12780029).
7. pCMX-Gal4-N Gal4 DBD fusion cloning vector (***10***; alternatively pM vector from Clontech, cat. no. K1618-1).
8. pCMX-VP16-N VP16 TAD fusion cloning vector (***10***; alternatively pVP16 vector, Clontech, cat. no. K1618-1).
9. pFR-Luc reporter plasmid (Stratagene, cat. no. 219050).
10. pCMX-Gal4-VP16 positive control vector ***(11)***.
11. pCMX-Gal4-ASC-2 positive control vector ***(11)***.
12. pCMX-VP16-RAR positive control vector ***(11)***.
13. Appropriate growth medium for mammalian cell in culture, for example, DMEM (Dulbecco's modified Eagle's medium; Invitrogen, cat. no. 11995065).
14. FBS (fetal bovine serum; Invitrogen, cat. no. 10082139).
15. D-PBS (Dulbecco's phosphate-buffered saline; Invitrogen, cat. no. 14080055).
16. 1X trypsin-EDTA (Invitrogen, cat. no. 25200056).
17. All-*trans*-retinoic acid (0.1–1 μ*M*; Sigma-Aldrich, cat. no. R2625).
18. Lipofectamine 2000 (Invitrogen, cat. no. 11668-019) or other appropriate transfection reagent, and OPTI-MEM I (Invitrogen, cat. no. 31985; necessary in case of using Lipofectamine).
19. Mammalian cells (e.g., HeLa, CHO, CV-1, COS, 293, and NIH3T3).
20. Luciferase assay kit (cell lysis buffer; Promega, cat. no. E1531 and luciferase assay buffer; Promega, cat. no. E1483).
21. Tissue culture dishes (24-wells; Costar, #3524).
22. Luminometer (MicroLumat Plus LB 96V, Berthold Technologies).

3. Methods

The methods described below outline (1) the construction of the chimera expression plasmids, (2) the transfection of plasmids into mammalian cell line, and (3) the measurement of luciferase activities.

3.1. Expression Plasmids

The construction of plasmids encoding two hybrid proteins is described. This includes Gal4 DBD-fusion vector pCMX-Gal4-N and VP16 TAD-fusion vector pCMX-VP16-N (kind gifts of Dr. Ron Evans). DNA encoding the bait protein is prepared for insertion into pCMX-Gal4-N and DNA encoding the target protein is prepared for insertion into pCMX-VP16-N either by restriction enzyme digestion or PCR amplification. DNA encoding the bait or the target protein must be expressed in the same reading frame as the Gal4 DBD and the VP16 TAD, respectively. In particular, an appropriate restriction site can be introduced at the end of the gene of interest by using a PCR primer, which incorporates the desired restriction enzyme site at the desired place and reading frame. In the multiple cloning sites (MCS) of both vectors there are nine unique sites for convenient cloning experiments (**Figs. 2A** and **B**).

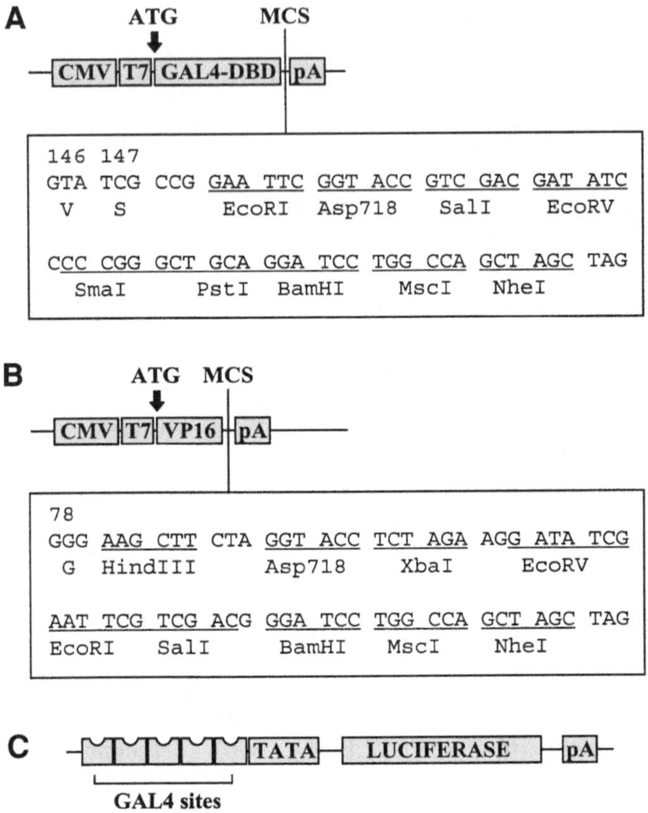

Fig. 2. Schematic representation of two fusion vectors and reporter. **(A)** pCMX-Gal4-N contains cytomegalovirus (CMV) promoter, T7 promoter, coding sequences for Gal4-DBD, multiple cloning sites (MCS), and polyadenylation sites (pA). **(B)** pCMX-VP16-N contains cytomegalovirus (CMV) promoter, T7 promoter, coding sequences for VP16, multiple cloning sites (MCS), and polyadenylation sites (pA). **(C)** pFR-Luc reporter plasmid (Stratagene, cat. no. 219050) contains five consecutive Gal4 binding sites, TATA box, coding sequences for luciferase and polyadenylation sites (pA) (*see* **Notes 6** and **7** for additional considerations).

3.1.1. Cloning

The vectors are digested with restriction endonucleases and dephosphorylated with CIAP prior to ligating with the insert DNA. For example, 1 μg of pCMX-Gal4-N was digested for 1 h at 37°C by 1 unit of *Eco*RI restriction enzyme in 20 μL containing 50 m*M* NaCl, 100 m*M* Tris-HCl, 10 m*M* $MgCl_2$, 0.025% Triton X-100, pH 7.5. At the end of the reaction, 0.5 unit of CIAP was added, and the reaction mixture was incubated for 1 h at 37°C. The ligation was executed by incubating *Eco*RI-digested pCMX-Gal4-N and *Eco*RI-digested insert (i.e., ASC-2) with T4 ligase (see below). If more than one restriction enzyme is used, the background (i.e., religation of vector) can be reduced further by electrophoresing the DNA on an agarose gel and recovering the desired plasmid band through electroelution, leaving behind the small fragment that appears between the two restriction enzyme sites. A variety of other methods, including adsorption onto glass beads and using low-melting agarose gels ***(6)***, could be utilized to extract and purify DNA. After purification and ethanol precipitation, the DNAs are resuspended in TE buffer.

3.1.2. Ligating the Insert

For ligation , the ideal insert-to-vector molar ratio of DNA is variable. In general, a reasonable starting point is a 2:1 insert-to-vector ratio. Incubate the reactions containing vector, insert, T4 DNA ligase, and the ligation buffer for 2 h at room temperature (22°C) or overnight at 4–16°C, as described previously ***(6)***. For blunt-end ligation, it helps to reduce the concentration of ATP to 5 m*M* and incubate the reactions overnight at 12–14°C.

3.1.3. Transformation

Thaw 100 μL of an aliquot of competent bacterial cells removed from the –80°C freezer and store on ice. Add 2 μL of the ligation reaction and leave the tube on ice for 30 min. Incubate the tube to 42°C water bath for 90 s and rapidly transfer to an ice bath to allow cells cool down for 2 min. Add 1 mL of LB medium and incubate the tube at 37°C for 30 min to 1 h. Plate the transformed cells on LB-ampicillin agar plate and incubate it at 37°C overnight.

3.1.4. Verifying the Presence and Reading Frame of the Insert

Isolated colonies are selected for miniprep (such as alkaline prep or boiling method; ***6***) analysis to identify transformed colonies containing the pCMX-Gal4-N or the pCMX-VP16-N vector with the DNA insert. The nucleotide sequence of the DNA insert should be determined to verify that the DNA insert is in-frame with the Gal4 DBD or the VP16-TAD and that the DNA inserts do not contain mutations. In addition, proper expression of the Gal4 DBD–bait fusion protein can be verified by Western blot analysis with an antibody that cross-reacts with either the protein expressed from the DNA insert or the Gal4 DBD. Similarly, the expression of the VP16 TAD–target protein can be verified by Western analysis with an antibody that detects either the protein expressed from the DNA insert or the VP16 TAD. For instance, we verify the expression of VP16/RAR in cells cotransfected with pCMX-VP16-RAR by Western analyses with an antibody raised against RAR.

3.2. Transient Transfection

The basic set of plasmid DNAs for mammalian two-hybrid assay is a reporter containing multiple copies of Gal4-binding sites (UAS), Gal4–bait DNA, VP16–target DNA, and a second reporter for normalizing transfection efficiency (*see* below). Plasmids are cotransfected into mammalian cells by any standard transfection method (e.g., calcium phosphate, DEAE-dextran, liposome reagents, electroporation, or viral transduction). We typically use Lipofectamine 2000 (Invitrogen) for mammalian transfections. The efficiency of transfection for different cell lines may vary by several orders of magnitude. A method that works well for one host cell line may not be adequate for another. Therefore, when working with a cell line for the first time, it's recommended to compare the efficiencies of several different transfection protocols. After a method of transfection is chosen, it may be necessary to optimize parameters such as cell density, the amount and purity of the DNA, media conditions, and transfection time (*see* **Note 1**). Once optimized, these parameters should be kept constant to obtain reproducible results. With each method, luciferase activity can be detected 36–48 h after transfection, depending on the host cell line used. For qualitative data, it's recommended to perform duplicate or triplicate transfections and average the results. It is also recommended to normalize for transfection efficiency by cotransfecting a constant amount of a second reporter under the control of a constitutive promoter. The values obtained in each sample for the primary reporter are then normalized to the values obtained for the second reporter in the same sample.

3.2.1. Reporter Plasmid

Various reporter plasmids responsive with upstream Gal4 binding sites are available, including pGal4-Luc. We routinely use pFR-Luc (Stratagene; San Diego, CA), which contains the entire coding sequence of *Photinus pyralis* (firefly) luciferase downstream of a basic promoter element (TATA box) and joined to five tandem repeats of the 17-bp Gal4 binding element (**Fig. 2C**).

3.2.2. Growing Mammalian Cells

The following protocol is designed for adherent cell lines such as CHO, CV-1, NIH 3T3, HeLa, and 293. Optimization of media and culture conditions may be required for other cell lines.

1. Thaw and seed frozen cell stocks in complete medium in a tissue culture flask.
2. Split the cells when they just become confluent.
3. Subculture the cells at an initial density of approx 1×10^5 to 2×10^5 cells/mL every 2–3 d.

4. Seed 0.5–1.0 × 10^5 cells in 1 mL of complete medium in each well of a 24-well tissue culture dish.
5. Incubate the cells at 37°C in a CO_2 incubator for overnight.

3.2.3. Preparing DNA Mixture for Transfection

Combine the plasmids to be cotransfected in a sterile tube. As each assay is run in duplicate or triplicate, the amount of plasmid DNA in each tube should be sufficient for two to three transfections (*see* **Notes 2**, **3**, and **5**). A range of DNA concentrations (usually 200–1000 ng/well) should be used in the pilot experiment to determine the optimal DNA concentration for a particular cell line.

3.3.4. Transfecting the Cells

A number of transfection methods, including calcium phosphate precipitation and lipid-mediated transfection, may be used. Transfection efficiencies vary between cell lines and different experimental conditions. Transfection procedures should be optimized according to manufacturer's instructions. The following guideline is for Lipofectamine 2000 that we usually use.

1. Dilute DNA in 50 μL of OPTI-MEM I.
2. Dilute 2 μL of Lipofectamine into 50 μL OPTI-MEM I and incubate for 5 min at room temperature.
3. Combine the diluted DNA with the diluted Lipofectamine and incubate for 20 min at room temperature.
4. Add the 100 μL of DNA–Lipofectamine complexes to each well and mix gently.
5. Incubate the cells at 37°C in a CO_2 inbubator for 36–48 h.

3.3. Luciferase Assays

1. Remove the medium from the cells 36–48 h after transfection in 24-well plate and carefully wash the cells twice with 0.5 mL of 1X PBS buffer.
2. Remove as much PBS as possible from the wells with a Pasteur pipet. Add 100 μL 1X cell lysis buffer to the wells and swirl the dishes gently to ensure uniform coverage of the cells.
3. Incubate the dishes for 15 min at room temperature. Swirl the dish gently midway through the incubation.
4. Assay for luciferase activity directly from the wells within 2 h (*see* **step 6**). If cell lysates are sticky, perform the centrifugation first and take supernatant to avoid inaccurate pipetting.
5. To store for later analysis, transfer the solutions from each well into a separate microcentrifuge tube. Spin the samples in a microcentrifuge at full speed. Store the supernatant at –80°C. It should be noted that each freeze–thaw cycle results in a significant loss of luciferase activity (as much as 50%).

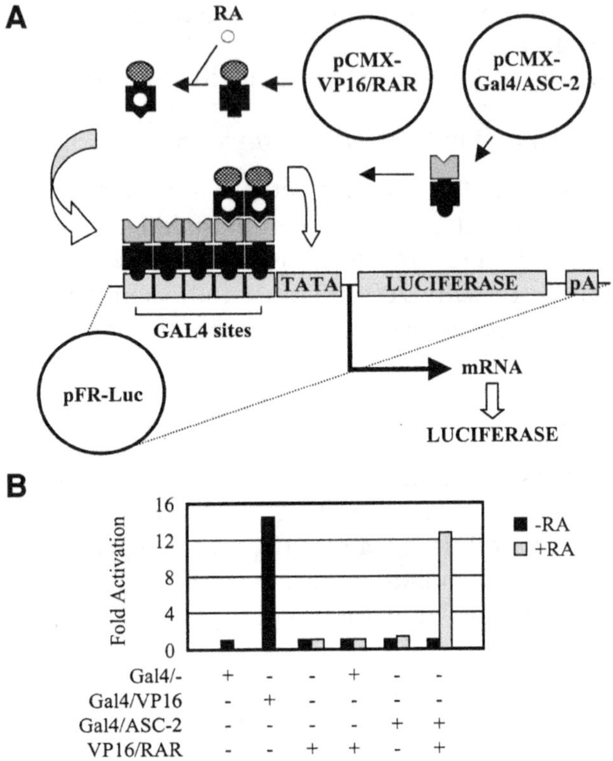

Fig. 3. **(A)** VP16/RAR derived from pCMX-VP16/RAR undergoes a conformational change upon binding all-*trans*-retinoic acid (RA), which interacts with Gal4/ASC-2 derived from pCMX-Gal4/ASC-2 (*see* **Note 8**). These protein–protein interactions bring transactivation domain VP16 to pFR-Luc reporter, which results in production of mRNA and protein for luciferase. **(B)** Luciferase activity was significantly increased in the presence of RA only when Gal4/ASC-2 and VP16/RAR were coexpressed. As a positive control, Gal4 fusion to VP16 was used (*see* **Note 9** for other applications and **Note 10** for complications).

6. Dispense 5–20 µL of cell extract (from **step 4** or **step 5**; *see* **Note 4**) to 96-well plate and place the plate into the sample chamber of luminometer equipped with an automatic injector and equilibrated with 1X luciferase assay buffer (100 µL).
7. Measure the light emitted from the reaction with a luminometer using an integration time of 10–30 s.
8. Luciferase activity may be expressed in relative light units (RLU) as detected by the luminometer from the sample (*see* **Note 4** and **5**). The activity may also be expressed as RLU/well, RLU/number of cells, or RLU/mg of total cellular protein (*see* **Fig. 3** for an example).

4. Notes

1. The DNA used for transfections must be of high quality (i.e., double cesium chloride banded). The plasmid DNA should have an $OD_{260/280}$ of approx 1.8–2.0 and be endotoxin free.
2. The ideal bait-to-target molar ratio of DNA is variable. In general, a reasonable starting point is to use the same amount of Gal4 DBD bait and VP16 TAD target DNA, and try various ratios of bait-to-target DNA for optimal results.
3. Many proteins often contain a cryptic autonomous transcriptional activation function, although they do not have any role in transcription. This could result in an unacceptably higher background. In this case, express this protein as VP16 TAD fusion, not as Gal4 DBD fusion.
4. When the luciferase activity is too low to calculate, try to increase the concentration of cell lysates in the assay and/or use more pFR-Luc reporter plasmid for transfection.
5. Excess amount of pCMX-Gal4-N and pCMX-VP16-N constructs should not be used to avoid the low increase of the luciferase activity over the background.
6. Gal4 DBD from pCMX-Gal4-N contains its own nuclear localization signal (NLS) ***(9)***. However, VP16 TAD from pCMX-VP16-N lacks any such signal. Thus, one should make sure that the gene of interest being inserted into pCMX-VP16-N vector brings in its own NLS or is engineered to have one during cloning. This becomes particularly critical when one works with nonnuclear protein or fragment of nuclear protein that may not be correctly localized into the nucleus due to the deletion of its inherent NLS. There are VP16 fusion vectors with a nuclear localization signal already engineered into them, such as pVP16 (Clontech; Palo Alto, CA).
7. It is recommended to test protein–protein interaction in both directions (i.e., exchange of the inserts between Gal4 DBD and VP16 TAD), due to the possible inhibitory effects of the fused domains. One can also express their gene of interest in front of Gal4 DBD and VP16 TAD to avoid such inhibitory effects.
8. pCMX-Gal4-VP16 encoding a Gal4 fusion protein to VP16 can be used as a positive control (**Fig. 3B**). In addition, we routinely use pCMX-VP16/RAR and pCMX-Gal4/ASC-2 in a control experiment to validate the mammalian two hybrid system *(7)*. Retinoic acid receptor (RAR) undergoes a conformational change upon binding retinoic acid (RA) *(8)* and a transcriptional coactivator ASC-2 *(7)* directly recognizes this structural change (**Fig. 3A**). Thus, protein–protein interactions between Gal4/ASC-2 and VP16/RAR is strictly RA-dependent, as demonstrated in **Fig. 3B**.
9. Once an interaction between two proteins has been detected in mammalian cells, the mammalian two-hybrid assay can be used to functionally analyze the interaction using deletional or site-directed mutagenesis. The mammalian two-hybrid assay is also useful for confirming the relevance of protein–protein interactions identified via yeast two-hybrid–based screenings ***(1,2)***. Such confirmation eliminates the possibility of a false-positive that is an artifact in yeast cells.

10. Protein–protein interactions in mammalian two-hybrid tests can be mediated by as yet uncharacterized endogenous adaptor proteins in cells. Thus, protein–protein interaction assays using purified proteins should be performed to verify such interactions.

Acknowledgments

The authors thank Dr. Ron Evans for plasmids pCMX-Gal4-N and pCMX-VP16-N. This work was supported by a research grant from 21C Frontier Functional Proteomics Project from Korean Ministry and Technology (to Jae Woon Lee).

References

1. Fields, S. and Song, O. (1989) A novel genetic system to detect protein-protein interactions. *Nature* **340,** 245–246.
2. Zervos, A. S., Gyuris, J., and Brent, R. (1993) Mxi1, a protein that specifically interacts with Max to bind Myc-Max recognition sites. *Cell* **72,** 223–232.
3. Ma, J. and Ptashne, M. (1987) A new class of yeast transcriptional activators. *Cell* **51,** 113–119.
4. Giniger, E., Varnum, S. M., and Ptashne, M. (1985) Specific DNA binding of GAL4, a positive regulatory protein of yeast. *Cell* **40,** 767–774.
5. Friedman, A. D., Triezenberg, S. J., and McKnight, S. L. (1988) Expression of a truncated viral trans-activator selectively impedes lytic infection by its cognate virus. *Nature* **335,** 452–454.
6. Ausubel, F. M., Brent, R., Kingston, R. E., et al. (eds.) (1995) *Current Protocols in Molecular Biology*. Greene Assoc., New York, NY.
7. Lee, S. K., Jung, S. Y., Kim, Y. S., Na, S. Y., Lee, Y. C., and Lee, J. W. (2001) Two distinct nuclear receptor-interaction domains and CREB-binding protein-dependent transactivation function of activating signal cointegrator-2. *Mol. Endocrinol.* **15,** 241–254.
8. Mangelsdorf, D. J. and Evans, R. M (1995) The RXR heterodimers and orphan receptors. *Cell* **83,** 841–850.
9. Silver, P. A., Keegan, L. P., and Ptashine, M. (1984) Amino terminus of the yeast GAL4 gene product is sufficient for nuclear localization. *Proc. Natl. Acad. Sci. USA* **81,** 5951–5955.
10. Sadowski, I. and Ptashne, M. (1989) A vector for expressing GAL4(1-147) fusions in mammalian cells. *Nucleic Acids Res.* **17,** 7539.
11. Lee, S., Anzick, S. L., Choi, J., et al. (1999) A nuclear factor, ASC-2, as a cancer-amplified transcriptional coactivator essential for ligand-dependent transactivation by nuclear receptors in vivo. *J. Biol. Chem.* **274,** 34,283–34,293.

23

Co-Immunoprecipitation from Transfected Cells

Shane C. Masters

Abstract

One of the most commonly used methods for determining whether two proteins can interact is co-immunoprecipitation. Co-immunoprecipitation relies on the ability of an antibody to stably and specifically bind complexes containing a bait protein. The antibody provides a means of immobilizing these complexes on a solid matrix, which in the protocol presented here is accomplished through interaction with Protein A, so that irrelevant proteins can be washed away. The presence of target proteins in the bait complexes is determined by Western blot. Because of the biochemical diversity of protein–protein interactions, it is not possible to describe a single set of conditions that will work for every immunoprecipitation experiment. Instead, the goal of this chapter is to provide practical starting conditions for co-immunoprecipitation assays and to describe potential modifications to the procedure so that conditions can be optimized.

Key Words

Immunoprecipitation; protein–protein interaction.

1. Introduction

Immunoprecipitation is one of the most commonly used methods for examining protein–protein interactions. The technique possesses several advantages that have made it a valuable component of many studies of protein binding. In the form presented here, immunoprecipitation uses proteins expressed in mammalian cells, which allows them to be post-translationally modified in ways not possible when using prokaryotic sources. For example, 14-3-3 proteins cannot bind recombinant Bad expressed in *Escherichia coli* without an additional serine phosphorylation step, but these two proteins can be co-immunoprecipitated from cells without special treatment ***(1,2)***. Immunoprecipitation can isolate multiprotein complexes, so that, when a complex mixture of pro-

From: *Methods in Molecular Biology, vol. 261: Protein–Protein Interactions: Methods and Protocols*
Edited by: H. Fu © Humana Press Inc., Totowa, NJ

teins and other cofactors (cell lysate) is used, it is possible to study indirect protein–protein interactions. This is an important point, as the roles of scaffolding proteins such as Gravin/AKAP250, which can bind the β_2-adrenergic receptor, PKA, PKC, and PP2B ***(3–5)***, and multiprotein regulatory complexes (i.e., Src, PKA, and the Slowpoke potassium channel; ***6***) are becoming increasingly appreciated. Because a wash step is included in the protocol, it is not necessary to purify the proteins of interest. This helps make immunoprecipitation relatively quick and inexpensive compared to many other methods. A related advantage is that there are few exotic reagents or devices needed to carry out immunoprecipitation, and most labs will be able to start using the technique with a minimal start-up investment in time and money.

There are some limitations to immunoprecipitation. Using complex mixtures instead of purified proteins allows examination of indirect protein–protein interactions and obviates the need for lengthy purification procedures, but it also makes it difficult to conclude that the two proteins under study bind each other directly (*see* ref. ***7*** for an example of how this issue can be addressed). Also, immunoprecipitation as it is normally performed does not provide quantitative data regarding the affinity or stoichiometry of an interaction. The variability among proteins and the myriad factors that affect protein structure and interaction leads to another potential problem, that of determining the experimental conditions (i.e., cell line, buffer components, and so on) for immunoprecipitation. This chapter provides reasonable starting conditions, but it is often necessary to empirically optimize the method for a particular interaction.

The procedure for immunoprecipitation is outlined in **Fig. 1**. In the first step, cells are transfected with plasmids coding for a bait protein and its potential ligand, the target protein. These cells are lysed with a gentle detergent treatment, creating a lysate containing bait–target complexes along with many irrelevant proteins. A capture antibody that specifically recognizes the bait protein is added to the lysates, forming a new antibody–bait–target complex. The antibody is then used as a handle to immobilize the proteins on inert Sepharose beads that are covalently coupled to Protein A, which stably binds the constant regions of many types of antibodies. At this point, those proteins not immobilized on beads are removed by a series of washes. Finally, the bait-protein complexes are eluted from the beads and dissociated by boiling in SDS. The presence or absence of the target protein, which is the endpoint of the assay, is evaluated by Western blot.

2. Materials

1. HEK293 cell line (American Type Culture Collection, Manassas, VA).
2. Materials and equipment for mammalian cell culture.

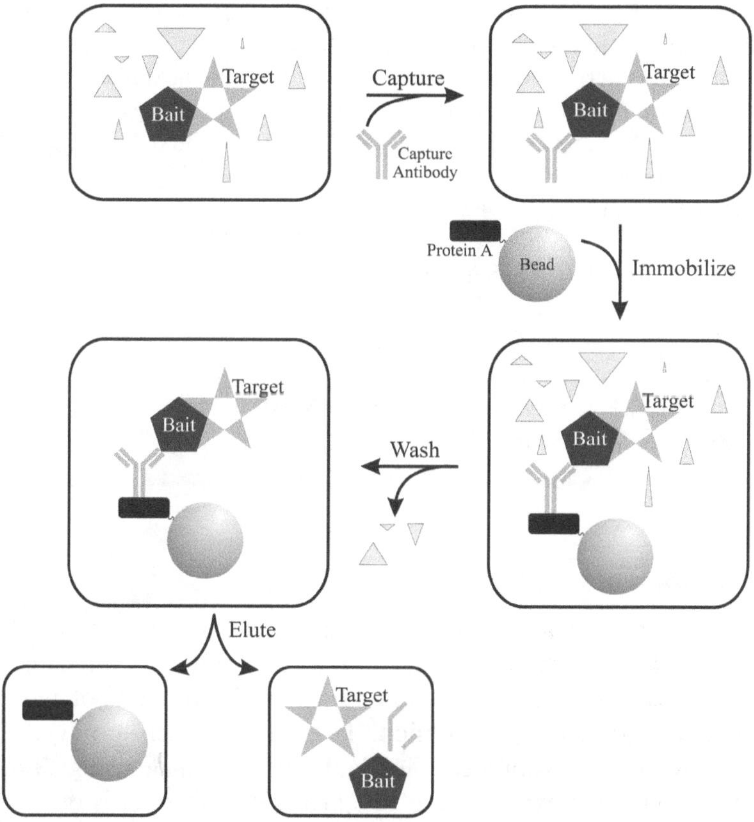

Fig. 1. Schematic diagram of immunoprecipitation. Cell lysates, which serve as a source of bait/target protein complexes, contain many irrelevant proteins (small triangles). The bait complexes are captured using a specific antibody. The antibody–bait–target assembly is immobilized using Protein A that is covalently attached to Sepharose beads. The beads are then washed free of irrelevant proteins, and the antibody, bait, and target are eluted by boiling.

3. Trypsin/EDTA: 0.25% Trypsin / 0.1% EDTA in HBSS (Mediatech, Herndon, VA).
4. Phosphate-buffered saline (PBS): 130 m*M* NaCl and 20 m*M* sodium phosphate, pH 7.5.
5. Lysis buffer: 150 m*M* NaCl, 10 m*M* HEPES, pH 7.5, 0.2% Igepal CA-630 or Nonidet P-40, 5 m*M* sodium fluoride, 5 m*M* sodium pyrophosphate, 2 m*M* sodium orthovanadate, 10 mg/L aprotinin, 10 mg/L leupeptin, and 1 m*M* phenylmethylsulfonyl fluoride, at 4°C (*see* **Note 1**).
6. Capture antibody.

7. Protein A Sepharose 4 Fast Flow (PAS; Amersham Biosciences, Piscataway, NJ) or Protein G Sepharose 4 Fast Flow (*see* **Note 2**).
8. Rocking platform (i.e., Adams Nutator Mixer, Becton-Dickinson, Franklin Lakes, NJ) or other tube mixing device.
9. Wash buffer: 150 m*M* NaCl, 10 m*M* HEPES, pH 7.5, and 0.2% Igepal CA-630 or Nonidet P-40, at 4°C.
10. 50 m*M* HEPES, pH 7.5, at 4°C.
11. Nonreducing 2X SDS buffer: 120 m*M* Tris-HCl, pH 6.8, 3.3% SDS, 10% glycerol, and 40 μg/mL bromphenol blue.
12. Reducing 2X SDS buffer: nonreducing 2X SDS buffer with 200 m*M* dithiothreitol.
13. SDS-PAGE and Western blotting equipment.
14. Primary antibody for Western blot.
15. Secondary (HRP conjugated) antibody for Western blot.
16. ECL reagents (Amersham Biosciences).
17. Film and film developing equipment.

3. Methods

3.1. Preparation of Cell Lysates

The initial part of the protocol describes the steps necessary to obtain the crude mixture that will serve as the source of bait-protein complexes. Cells in culture are transfected with plasmids coding for the proteins of interest, and time is allowed for protein expression. The cells are then removed from the culture dish, washed, and counted. Finally, lysates are made using a mild detergent treatment, and the expression of the proteins of interest is determined by Western blot.

3.1.1. Selection of Samples

1. When starting to experiment with a new immunoprecipitation, one useful strategy is to express the bait and target proteins as fusions to two different affinity tags (**Table 1**). Because of the widespread use of these tags, there are good antibodies available that have reasonable chances of working well either for Western blot or as capture antibodies. This approach also reduces the number of expensive reagents that must be purchased, as the same antibody can be used for multiple experiments.
2. An important consideration for immunoprecipitation is whether the observed interaction between two proteins is specific. Negative controls should be included to help address this issue. At a minimum, untransfected cells, as well as cells transfected singly with the bait and target proteins, should be tested in addition to the double-transfected cells. If the interacting domains of the two proteins can be predicted, mutant proteins can also be examined. Another useful control is to test the reciprocal immunoprecipitation (i.e., immunoprecipitate the original target protein and look for co-precipitated bait).

Table 1
Some Commonly Used Affinity Tags[a]

Tag name	Tag sequence	Antibody name	Antibody vendor
EGFP	(protein)	A.v. Peptide antibody	BD Biosciences Clontech (Palo Alto, CA)
FLAG	DYKDDDDK	Anti-FLAG (M2)	Sigma-Aldrich (St. Louis, MO)
HA	YPYDVPDYA	12CA5	Roche Applied Science (Indianapolis, IN)
myc	EQKLISEEDL	9E10	Santa Cruz Biotechnology (Santa Cruz, CA)

[a]The antibodies listed here are commonly cited in the scientific literature and have been used by the author. This list is not intended to be exclusive, and other tags or antibodies may be equally useful.

3.1.2. Transfection

1. Assemble mammalian expression vectors coding for the two proteins that you would like to examine for interaction, as well as any mutants or tagged variants, using standard molecular biology techniques (*see* **Note 3**).
2. Transiently transfect two 35 mm dishes (or two wells of a six-well plate) of HEK293 cells with DNA constructs coding for the bait and target proteins (*see* **Note 4**). HEK293 cells can be transfected easily by most popular methods, such as calcium phosphate *(8,9)* or lipid-mediated (i.e., FuGENE6, Roche Applied Science, Indianapolis, IN) protocols.
3. Grow the cells 24–48 h to allow time for protein expression. At the time of harvest, cells should be 80–100% confluent if possible.

3.1.3. Collection of Cells

1. From this point of the method on, it is not necessary to maintain sterility. However, the cells, culture media, and cell lysates are still considered biohazardous materials and should be handled with appropriate care.
2. Collect the culture medium from the transfected cells into a centrifuge tube.
3. Add 1 mL PBS to each dish of cells and gently swirl. Collect the PBS and pool with the culture medium.
4. Add 300 μL trypsin/EDTA to each dish of cells. Incubate at 37°C until the cells detach from the dish. This occurs very rapidly for HEK293 cells. Pipet the cells up and down to break up clumps, then transfer the cells to the centrifuge tube containing culture medium and PBS.

5. Pellet the cells by centrifugation (500*g*, 5 min). Aspirate the supernatant and discard.
6. Add 2 mL PBS and gently resuspend the cells by vortexing at low speed.
7. Take a 20 µL sample of the cell suspension, to be used later for counting, and place on ice.

3.1.4. Cell Lysis

1. The cells and lysates should be maintained at 4°C from this point on.
2. Pellet the cells by centrifugation (500*g*, 5 min, 4°C). Aspirate the supernatant and discard.
3. Add 600 µL lysis buffer to the cell pellet (i.e., 300 µL per 35 mm dish) and vortex or pipet up and down to resuspend. Some foaming of the buffer may occur. Place the mixture on ice for 30 min to allow time for lysis.
4. While the lysates are being made, count the cells in the sample taken in **Subheading 3.1.3., step 7**. For two 35 mm dishes of nearly confluent HEK293 cells harvested as described above, roughly 1×10^6 cells/mL can be expected (*see* **Note 5**).
5. Remove unlysed cells, cell nuclei, and debris from the lysates by centrifugation (14,000*g*, 10 min, 4°C). Transfer the supernatant to a clean tube.
6. Very often, it is possible to store the lysates at –70°C for long periods of time, in unit of use aliquots to minimize freeze/thaw cycles. However, this treatment could potentially disrupt some protein interactions. If freezing of the lysates is not desirable, the Western blots described in **Subheading 3.1.5.** can be done concurrently with those of **Subheading 3.3.2.**, with the risk that time and reagents used for immunoprecipitation will be wasted if the proteins of interest are not present in the lysates.

3.1.5. Assessment of Protein Expression

1. Run a sample of lysate equivalent to 3×10^4 cells on SDS-PAGE and transfer proteins to a membrane using standard techniques.
2. Probe the membrane with antibodies recognizing the bait and target proteins. If the presence of all proteins of interest cannot be verified by Western blot, it is not advisable to continue the experiment. In addition, each lysate used in an experiment should contain similar amounts of the desired proteins. Problems at this step are usually solved by repeating the transfection, possibly modifying the amount of DNA used, the time allowed for protein expression, or the cell line.

3.2. Immunoprecipitation

This part of the method describes the isolation and purification of bait-protein complexes from the lysates. The goal of the first step, preclearing, is to remove from the lysates proteins that bind directly to the Protein A Sepharose matrix (PAS). Lysates are incubated with PAS in the absence of antibody and then the PAS, with bound proteins, is discarded (*see* **Note 6**). The desired proteins, along with their interacting partners, are then bound to the capture antibody before precipitation with a fresh aliquot of PAS. Finally, the proteins remaining in solution are removed by a series of washes.

3.2.1. Preclear Lysates

1. Before use, the PAS should be washed into lysis buffer. For each sample, about 110 μL of the stock 50% PAS suspension will be needed (*see* **Note 7**). Pellet the PAS by brief centrifugation (5–10 s in a microcentrifuge using the pulse setting) and remove the supernatant with a pipet or using a gentle vacuum (*see* **Note 8**). Add a volume of lysis buffer equal to the pellet volume and resuspend the beads by tapping the tube. Repeat this procedure two more times to yield a suspension that contains 50% PAS in lysis buffer.
2. Dilute a volume of lysate equivalent to 8×10^5 cells to 300 μL in a 1.5-mL microcentrifuge tube. Add 50 μL washed 50% PAS.
3. Gently rock the PAS/lysate mixture at 4°C for 1 h. The level of agitation should be sufficient to keep the PAS in suspension, but the production of bubbles should be avoided.
4. Pellet the PAS by centrifugation (14,000*g*, 1 min, 4°C). Transfer the supernatant (cleared lysate) to a clean tube. Avoid transfer of the used PAS, which can be discarded.

3.2.2. Formation of Bait Protein-Antibody Complex

1. During immunoprecipitation, proteins are in a native, folded conformation, while they are denatured during Western blot. Thus, the performance and specificity of an antibody for immunoprecipitation cannot be predicted from Western blot data. It is not necessary that the capture antibody show only a single band, or even that it recognize its target protein, on Western blots. The scientific literature, along with the "suggested applications" ratings provided by antibody vendors, are usually sufficient to allow an informed selection for capture antibody.
2. Add 1–5 μg of capture antibody (*see* **Note 7**) to the cleared lysate.
3. Gently rock the samples at 4°C for 1.5 h to allow the capture antibody to bind the bait-protein complexes.

3.2.3. Antibody Precipitation

1. Add 50 μL washed 50% PAS to the lysate/antibody mixture.
2. Gently rock the samples at 4°C for 1.5 h so that the beads remain in suspension.

3.2.4. Washing

1. Pellet the PAS, which now carries the capture antibody and associated protein complexes, by brief centrifugation (5–10 s in a microcentrifuge). Remove the supernatant by pipeting or gentle aspiration and discard. Because the PAS pellet is easily disturbed, a small amount (approx 10–20 μL) of supernatant should be left over the beads.

2. Add 500 µL wash buffer (at 4°C) to the PAS. Gently resuspend the PAS beads in the wash buffer by tapping or inverting the tube. Pellet the PAS and discard the wash buffer as in **step 1**. Repeat twice for a total of three washes, working quickly to maintain the temperature near 4°C throughout the procedure.
3. Add 500 µL 50 m*M* HEPES, pH 7.5, to the PAS and resuspend. Pellet the PAS as in **step 1**, but this time remove the supernatant as completely as possible. The low ionic strength of the HEPES buffer helps prevent gel artifacts during SDS-PAGE.

3.3. Analysis of Precipitated Proteins

In this section of the protocol, the purified bait complexes are denatured and dissociated from the PAS matrix by boiling in SDS. The resulting protein mixture is resolved by SDS-PAGE and tested for the presence of target proteins by Western blot.

3.3.1. Protein Elution

1. Pick an elution buffer. If the bait or target proteins are 20–30 or 50–60 kDa, choose the nonreducing 2X SDS sample buffer (*see* **Note 9**); otherwise, use reducing 2X SDS sample buffer.
2. Add 25 µL elution buffer to the washed PAS beads. Briefly vortex to resuspend. Incubate the beads at 100°C in a tightly capped tube for 5 min. Before starting the experiment, you may wish to test whether the tubes you intend to use can withstand boiling for 5 min without popping open, because this can occasionally lead to significant sample loss.
3. Pellet the PAS by centrifugation (14,000*g*, 5 min, room temperature). It is important that this centrifugation be done at room temperature to prevent SDS precipitation. Transfer 20 µL of the supernatant to a clean tube, being careful not to take the PAS. The used PAS can be discarded.
4. If desired, the sample may be stored indefinitely at –20°C or less before proceeding with the protocol. After thawing, the sample should again be incubated at 100°C before SDS-PAGE.

3.3.2. SDS-PAGE and Western Blot

1. Separate the samples by SDS-PAGE and transfer proteins to a membrane using standard techniques (*see* Chapter 12).
2. Probe membranes for desired target protein. If at all possible, for the Western blot use primary antibodies from a different animal species than the one the capture antibody was generated in (*see* **Note 9**). Also probe for the immunoprecipitated bait protein to determine whether approximately equal amounts are present.

3.3.3. Data Interpretation and Troubleshooting

Ideally, you will see equal levels of bait protein in all samples where it was transfected and less or none in the other samples, depending on the endogenous levels and whether the capture antibody was directed against the bait itself or a tag. The target protein should appear only for samples that showed significant precipitation of the bait, and should be reduced or absent in samples lacking transfected target protein. These results indicate that the target and bait proteins are capable of interacting in transfected cells. It is important to note that this does not mean that the proteins bind each other directly, because there are many other proteins present that may mediate bait–target interaction. Many in vitro techniques in Part II (Chapters 3–16) of this volume can be used to address this issue. Also, such data obviously cannot prove that the proteins interact in vivo, given the potential problems related to protein overexpression and the use of cell lines. Co-localization experiments (i.e., using confocal microscopy; *see* Chapter 27) and immunoprecipitation from untransfected cells or animal tissues (*see* **Note 3**) are some commonly used approaches that can help resolve this important issue.

If there is a significant amount of target protein in the samples that do not show bait-protein precipitation ("nonspecific binding"), the background must be reduced before a conclusion can be reached regarding the bait–target interaction. This problem can be addressed by raising the stringency of the washes. Increasing the amounts of salt (to disrupt ionic interactions) or detergent (to disrupt hydrophobic interactions) in the buffer for some or all of the wash steps is a common and effective practice. The concentration of sodium chloride in the wash buffer can be raised, up to 500 m*M* or more, or multiple detergents can be included in the buffer at various concentrations, as in RIPA buffer (150 m*M* NaCl, 10 m*M* HEPES, pH 7.5, 1% Triton X-100, 0.1% sodium dodecyl sulfate, 1% sodium deoxycholate). Also, the PAS can be incubated in wash buffer with gentle agitation for a short time (5–15 min) at each wash step. Increasing the number of washes may also help, although probably not as much as the other two suggestions. Be aware that overly stringent washing could disrupt target/bait or capture antibody/bait binding in addition to nonspecific interactions.

It is possible that the target protein will not be seen in the bait-protein immunoprecipitates. This could indicate that the two proteins do not interact in this system. However, if the bait protein can be precipitated and the target protein is expressed, there are several technical issues to be considered before concluding that bait–target binding cannot be observed. Some of these issues, as well as potential solutions, are:

1. Does the capture antibody interfere with bait/target interaction? (Try a different capture antibody; move any tags to a different location; try reciprocal immunoprecipitation.)
2. Does one of the proteins need to be post-translationally modified to induce interaction? (Try treating the cells with agents that activate kinases, inhibit phos-

phatases, and so on; improve inhibitor coverage in the lysis buffer; try a different cell line.)

3. Is the Western blot sensitive enough to detect low levels of co-precipitated target protein? (Try a different antibody for Western blot; use a more-sensitive detection method for the Western blot; scale up the amount of lysate, antibody, and PAS used.)
4. Are endogenous proteins competing for the same interaction epitope on the bait protein? (Try to increase expression of bait and target proteins; try a different cell line; try reciprocal immunoprecipitation.)

4. Notes

1. The sodium fluoride, orthovanadate, and pyrophosphate are used here as phosphatase inhibitors, while the aprotinin, leupeptin, and phenylmethylsulfonyl fluoride are present to block proteases. The inhibitors used in this buffer all display some degree of toxicity and should be handled carefully. These inhibitors may not be sufficient to prevent enzymatic modification of all proteins of interest, and they can be removed, replaced, or supplemented as necessary.
2. Protein A is a staphylococcal protein that binds to the constant region of antibody molecules, with an affinity that depends on the species and isotype of the antibody. A similar molecule, Protein G, is available that has a different spectrum of affinities for different antibody types. A useful rule of thumb is to use Protein A for rabbit capture antibodies and Protein G for mouse capture antibodies. Throughout this chapter, PAS will be used to mean either Protein A Sepharose or Protein G Sepharose.
3. Instead of using transfected cells, it is possible to perform immunoprecipitation from cell lines or tissues expressing their endogenous complement of proteins. This approach has the major advantages of avoiding overexpression of proteins, promoting more native subcellular localization and post-translational modification, and eliminating foreign linker and tag sequences. However, it can be considerably more difficult to perform these experiments because of the requirement for cells that express sufficient levels of both proteins of interest, the need for several high-quality capture antibodies, and often the lack of good negative controls. The procedures listed in this chapter can be used as a starting point, but multiple sources of cell lysates, multiple-capture antibodies, and species- and isotype-matched irrelevant capture antibodies should be tested to help determine whether the proteins of interest interact specifically. In addition, if the protein interaction is thought to be affected by drug treatment or cell culture conditions, this should be tested in the immunoprecipitation assay.
4. HEK293 cells are used in this protocol because they are easy to culture and can be transfected with high efficiency. Other cell lines can be used if desired.
5. Instead of counting cells, it is possible to perform a protein assay on the lysate. This is particularly convenient if large numbers of samples are being processed. For the immunoprecipitation, use roughly one-half of the lysate of a sample containing an "average" amount of protein and enough lysate to give equivalent amounts of protein for the other samples.

6. The relatively inert nature of the PAS, along with specific detection of target proteins by Western blot, means that in many cases the preclearing procedure is unnecessary, and it is often omitted in order to save time and expense. However, it is probably wise to include this step at least the first time that a particular immunoprecipitation is attempted.
7. The amounts of capture antibody and PAS listed in the protocol are quite high relative to the amount of cell lysate. These values were chosen to maximize the precipitation of the bait protein. In addition, the large amount of PAS makes it easier to see the pellet and minimizes the impact of minor bead losses. However, these reagents can be very expensive. It is certainly possible in many cases to use less antibody and PAS; the optimal amounts will vary for each system and should be determined empirically.
8. There are a few techniques for handling PAS that should be mentioned. Before removing beads from the stock bottle, let them settle and verify that there are approximately equal volumes of settled beads and storage buffer, because the buffer has a tendency to evaporate. Add more storage buffer (described in the package insert) if necessary. When pipeting the suspension, use tips with a relatively large orifice or cut off the ends with a razor blade. The PAS can be damaged easily by centrifugation, so be sure to either use low speeds or, as indicated here, very short spins. The PAS will not form a solid pellet; do not accidentally aspirate the beads. During washes, do not let the beads dry out. Finally, prepare only enough PAS for one experiment at a time, as long-term storage in lysis buffer may lower binding capacity for antibody.
9. The IgG antibodies most commonly used for immunoprecipitation and Western blot are heterotetramers composed of two heavy chains (approx 50 kDa each) and two light chains (approx 25 kDa each) that are held together by disulfide bonds. When boiling the PAS to collect the protein complexes of interest, most of the antibody will co-elute, resulting in a large amount of antibody contamination in the final sample. If the elution buffer contains reducing agents, the antibody will be present mostly as 50 and 25 kDa species, and if not, it will appear mainly at approximately 150 kDa with reduced but still significant amounts of 50 and 25 kDa proteins. This can become a problem when probing the Western blot, because the secondary antibody for Western is usually a peroxidase-conjugated anti-IgG. If the capture and primary Western antibodies are made in the same species, the intensity of the signal from the co-eluted capture antibody can be extremely high, thus antibodies from different species should be used. Even so, proteins with molecular weights similar to the antibody components can be obscured, as there will usually be some cross reactivity of the Western secondary antibody toward IgG of different species. If this problem cannot be overcome by the addition or removal of reducing agents from the elution buffer, two other potential solutions are to run a longer gel to physically separate the protein of interest from the capture antibody or to use a capture antibody that is covalently linked to a bead matrix.

Acknowledgment

I am grateful to Dr. Haian Fu for his helpful advice and editing of this work.

References

1. Zha, J., Harada, H., Yang, E., Jockel, J., and Korsmeyer, S. J. (1996) Serine phosphorylation of death agonist BAD in response to survival factor results in binding to 14-3-3 not BCL-X(L). *Cell* **87(4),** 619–628.
2. Masters, S. C., Yang, H., Datta, S. R., Greenberg, M. E., and Fu, H. (2001) 14-3-3 inhibits Bad-induced cell death through interaction with serine-136. *Mol. Pharmacol.* **60(6),** 1325–1331.
3. Nauert, J. B., Klauck, T. M., Langeberg, L. K., and Scott, J. D. (1997) Gravin, an autoantigen recognized by serum from myasthenia gravis patients, is a kinase scaffold protein. *Curr. Biol.* **7(1),** 52–62.
4. Shih, M., Lin, F., Scott, J. D., Wang, H. Y., and Malbon, C. C. (1999) Dynamic complexes of beta2-adrenergic receptors with protein kinases and phosphatases and the role of gravin. *J. Biol. Chem.* **274(3),** 1588–1595.
5. Lin, F., Wang, H., and Malbon, C. C. (2000) Gravin-mediated formation of signaling complexes in beta 2-adrenergic receptor desensitization and resensitization. *J. Biol. Chem.* **275(25),** 19,025–19,034.
6. Wang, J., Zhou, Y., Wen, H., and Levitan, I. B. (1999) Simultaneous binding of two protein kinases to a calcium-dependent potassium channel. *J. Neurosci.* **19(10),** RC4.
7. Chen, J., Fujii, K., Zhang, L., Roberts, T., and Fu, H. (2001) Raf-1 promotes cell survival by antagonizing apoptosis signal-regulating kinase 1 through a MEK-ERK independent mechanism. *Proc. Natl. Acad. Sci. USA* **98(14),** 7783–7788.
8. Chen, C. and Okayama, H. (1987). High-efficiency transformation of mammalian cells by plasmid DNA. *Mol. Cell. Biol.* **7(8),** 2745–2752.
9. Chen, C. A. and Okayama, H. (1988). Calcium phosphate-mediated gene transfer: a highly efficient transfection system for stably transforming cells with plasmid DNA. *Biotechniques* **6(7),** 632–638.

IV

Probing Protein–Protein Interactions in Living Cells

24

Microscopic Analysis of Fluorescence Resonance Energy Transfer (FRET)

Brian Herman, R. Venkata Krishnan, and Victoria E. Centonze

Abstract

We describe here detailed practical procedures for implementing various approaches of fluorescence resonance energy transfer (FRET) in microscope-based measurements. A comprehensive theoretical formalism is developed and the different experimental procedures are outlined. A step-by-step protocol is provided for preparing the specimens for FRET measurements, data acquisition procedures, analysis, and quantification. Particular emphasis is given to exemplify the FRET applications in the study of protein–protein interactions.

Key Words

Microscopy; FRET; FLIM; protein interactions.

1. Introduction

The ability to spatially resolve ever-smaller structures defines any advancement in microscopy. Resolution is defined as the smallest distance between any two points (or structures) so as to be able to distinguish them clearly in space, the numerical estimate of which is given by Rayleigh's criterion:

$$d = 0.6\lambda/\mathrm{NA} \tag{1}$$

As can be seen, spatial resolution (or the minimum resolving distance, d) varies directly with the wavelength (λ) of the light used and inversely with the numerical aperture (NA) of the objective lens used. Although electron microscopes can offer greater resolution than optical imaging using visible wavelengths of light, this form of microscopy is not useful for live cell applications because of the non-native imaging environment (vacuum) and other stringent specimen preparation conditions that are required. Additionally, one cannot

From: *Methods in Molecular Biology, vol. 261: Protein–Protein Interactions: Methods and Protocols*
Edited by: H. Fu © Humana Press Inc., Totowa, NJ

increase NA indefinitely as this would lead to impractically small working distances between the specimen and the objective. Thus, biological imaging applications are limited to visible light and maximal NA of approx 1.5–1.6. Considering these limitations, the best possible resolution achievable by any light microscope is approx 250 nm. Modern confocal and multiphoton microscopes do have improved axial resolution compared to their wide-field counterparts; however, they still suffer the same limitations of lateral resolution *(1)*. Although there have been some recent developments that have increased the spatial resolution of the optical microscope beyond this limit, no optical microscope allows observations of objects at the molecular level *(2)*. This limitation has made it difficult to study a number of physiological and metabolic processes in cells that result from interaction between macromolecules on a spatial scale of a few nanometers. Fluorescence (Forster) resonance energy transfer (FRET) holds an unique niche in modern fluorescence microscopy because it provides the means by which one can probe inter- and intra-molecular interactions in the 1–10 nm range *(3–6)*. FRET microscopy methods do not improve the resolution *per se* over that of conventional light microscopy methods but only extend the detection limit to a few nanometers. In other words, FRET microscopy yields quantitative information about the molecular interactions occurring in the range of 1–10 nm that is otherwise impossible to obtain using other light microscopy methods. For a biologist, this implies that it is possible to extend light microscopy from being a mere visualization tool to a precise quantitative tool—while addressing the various issues of macromolecular interactions on a molecular level. This chapter attempts to provide the novice in this field with practical notes for implementing the different FRET approaches for cell biological applications.

1.1. Theoretical Formalism

Quantum mechanically, there exists a certain probability that an excited donor can transfer a part of its excitation energy nonradiatively to a neighboring molecule (via direct dipole–dipole interactions) in addition to other classical radiative decay processes *(7)*. These two decay paths differ in their time scales: radiative fluorescence decay occurs typically over a few nanoseconds after excitation, whereas the nonradiative energy transfer process is 100–1000 times faster than the radiative decay. FRET is one such nonradiative decay process and requires stringent conditions in terms of the spatial as well as spectral relationship between the donor and acceptor (*see* **Note 1**). Energy transfer efficiency (E) serves as a quantitative measure of the FRET process. It is therefore natural that all the available experimental approaches are designed to measure E in various circumstances. The energy transfer efficiency depends on the relative orientation and separation (R) between the donor and acceptor dipoles

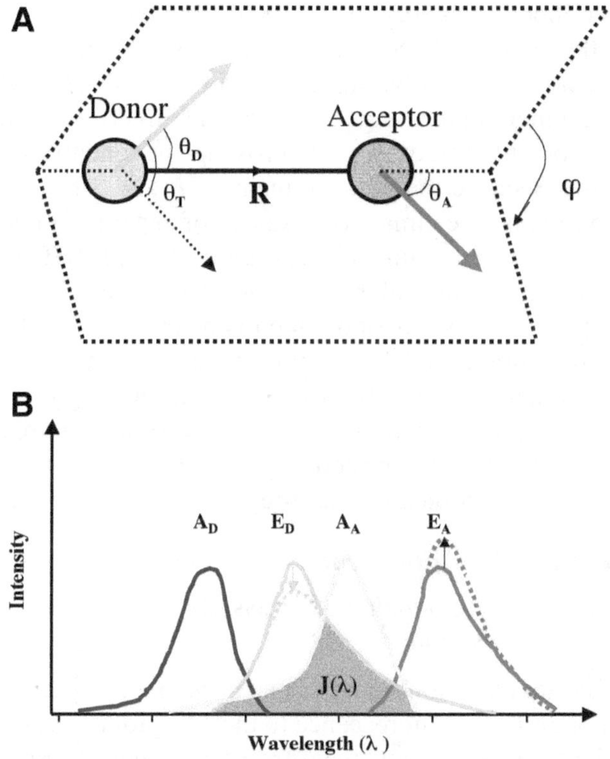

Fig. 1. **(A)** A schematic representation of FRET mechanism. The angles and distance of separation are explained in **Note 1**. **(B)** Schematic of energy transfer in terms of donor/acceptor spectra.

(**Fig. 1A**) in addition to the overlap between the donor emission and the acceptor absorption spectra (**Fig. 1B**). This relationship is given by

$$E = \frac{R_0^6}{R_0^6 + R^6} \tag{2}$$

By definition, R_0 ("Forster distance") is the distance of separation at which the energy transfer efficiency is 50%, that is to say, when $R = R_0$ there exist equal probabilities that nonradiative energy transfer and radiative deactivation can occur from the donor excited state. See **Note 1** for a detailed explanation of individual parameters in the above Forster equation. What gives the FRET phenomenon its sensitivity and specificity is the sixth-power dependence of transfer efficiency on the distance separating the donor–acceptor (*see* **Note 2**).

Because the likelihood of energy transfer varies by the sixth power of the distance between the donor and acceptor, even very small (angstrom level) separations can be measured reliably. In practice, this provides the effectiveness of the FRET measurement in the range 1–10 nm and effectively extends the measurement resolution of the fluorescent microscope beyond that of conventional light microscopy. Because the measurement of E and a few other constants (*see* **Note 1**) can yield an estimate of distance of separation between the pair of fluorescent molecules that undergo energy transfer, FRET process is often referred to as "spectroscopic ruler" ***(8)***. The phenomenon of energy transfer also contains molecular information which is different from that revealed by solvent-relaxation, fluorescence quenching, and so on, which provide information about the interaction of the fluorophore with the neighboring molecules in the surrounding solvent shell ***(7)***. There are various manifestations of the FRET process, and the following section gives a brief account of the different approaches that can be adopted to measure the energy transfer efficiency.

1.2. FRET Experimental Approaches

1.2.1. Donor Quenching/Sensitized Emission of the Acceptor Fluorescence

Because the donor emission fluorescence is quenched upon energy transfer, the ratio of donor emission fluorescence to the acceptor emission fluorescence intensity can be used as a good measure of the energy transfer efficiency (**Fig. 1B**). It is possible to quantify these ratios in individual cells. The most conventional form of observing the energy transfer event is to excite the donor and observe the fluorescence from both the donor and acceptor ***(3)***. The transfer efficiency in this case can be expressed as

$$E = 1 - \left(I_{\mathrm{DA}}/I_{\mathrm{D}}\right) \tag{3}$$

where I_{DA} is the donor fluorescence in the presence of acceptor (energy transfer condition) and I_{D} is the donor fluorescence when there is no energy transfer. This method can be implemented in both wide-field and confocal microscopes (*see* **Fig 2** and **Note 3**).

1.2.2. Donor Dequenching Upon Acceptor Photobleaching

If the acceptor molecules are sufficiently photolabile, then another way of monitoring FRET is to observe the increase (or the recovery) in donor fluorescence when the acceptors are selectively photobleached. Energy transfer efficiency can be now calculated from the following equation:

$$E = 1 - \left(I_{\mathrm{D(pre)}}/I_{\mathrm{D(post)}}\right) \tag{4}$$

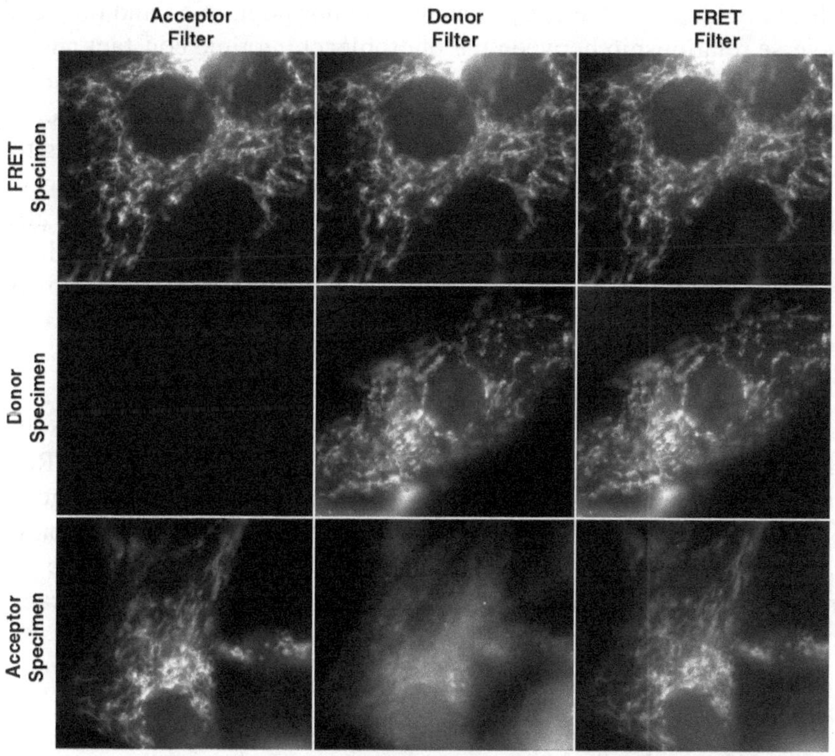

Fig. 2. FRET measurement by sensitized emission approach. The nine-panel image shows donor-only, acceptor-only, and FRET specimens taken with the three different filter sets marked in the figure. Baby hamster kidney (BHK) cells were transfected with mitochondrially targeted protein constructs: **(A)** mCFP: *Donor only*; **(B)** mYFP: *Acceptor only*; and **(C)** mCGY: *FRET specimen*. The measurements were carried out in Olympus IX-70 inverted microscope with appropriate excitation and emission filters as described in **Table 1**. The data acquisition and analysis are carried out as described in **Subheading 3.3.**

where $I_{D(pre)}$ and $I_{D(post)}$ are the donor intensities in pre- and postbleaching conditions. An advantage of this method is that one can use laser scanning confocal or multiphoton microscopy to visualize even small regions in cells and to selectively photobleach the acceptor in these small regions. In this way, one can achieve a high spatial resolution in the FRET measurement. Jovin and Jovin have introduced an alternate way of obtaining energy transfer efficiency by photobleaching the donor instead of the acceptor *(9)*.

Photobleaching depletes the excited state of donor population and this leads to an inverse relationship between the photobleaching time constant and the fluorescence lifetime of the donor. The FRET process results in a decrease in the donor lifetime and hence a corresponding increase in the photobleaching time constant. Energy transfer efficiency can be calculated by measuring this bleaching time constant for the donor in the presence and absence of acceptors. However, as many biological experiments require keeping cells alive under long periods of observation, the use of either acceptor or donor photobleaching approaches can result in irreversible photobleaching of the cells (*see* **Fig. 3** and **Note 4**).

1.2.3. Concentration-Dependent Depolarization of Fluorescence

The transition dipoles of most fluorophores are anisotropic. When excited with plane-polarized light, their emission also becomes plane polarized. Rotational diffusion of the fluorophores during the excited-state lifetime is generally found to be the dominant cause of depolarization ***(7)***. Depending on the relative magnitudes of the rotational diffusion time (ϕ) and the intrinsic lifetime (τ) of the fluorophore, the change in anisotropy (r) of a fluorophore can be calculated by the Perrin equation ***(7)***:

$$r = \frac{r_0}{1 + (\tau / \phi)} \tag{5}$$

where r_0 (=2/5) is the limiting anisotropy for a randomized ensemble of fluorophores and using single-photon excitation. One of the consequences of FRET is that, if the donor is excited with plane-polarized light, then following an energy transfer event, the acceptor fluorescence is depolarized. The corresponding energy transfer efficiency can be calculated by the following equation ***(10)***:

$$E = 1 - (r / r^0) \tag{6}$$

where r and r^0 are the measured anisotropies of the acceptor with and without energy transfer, respectively. An advantage of this method is that one can use this to measure homo-FRET (i.e., where the same fluorophore serves as both donor and acceptor) when the fluorophore used has a sufficient overlap between its own excitation and emission spectra (i.e., very small Stokes' shift) (***11–14***; *see* **Note 5**).

1.2.4. FLIM : Reduction of Donor Lifetime Upon Energy Transfer

Another important manifestation of a FRET event is that the donor lifetime decreases upon the energy transfer ***(7)***. This can be exploited experimentally by measuring the fluorescence lifetime of the donor in the presence and absence of the acceptor by one of the many time-domain or frequency-

Table 1
Typical FRET Pairs and the Corresponding Filter Combinations[a]

	Donor imaging path			Acceptor imaging path		
FRET Pair	Excitation filter	Dichroic	Emission filter	Excitation filter	Dichoric	Emission filter
FITC–TRITC	480/40	505 LP	535/40	545/30	565 LP	610/60
FITC–Cy3	480/40	505 LP	535/40	540/25	565 LP	605/55
Cy3–Cy5	540/25	565 LP	605/55	640/20	660 LP	680/30
BFP–GFP	390/20	420 LP	460/50	475/40	460 LP	535/50
GFP–Cy3	475/40	460 LP	505/40	540/25	565 LP	605/55
CFP–YFP	436/20	455 LP	480/40	500/20	515 LP	535/30
CFP–DsRed	436/20	455 LP	480/40	546/12	560 LP	620/20

[a] All the values are expressed as center wavelength (nm) with the spread on either side of the peak value. Donor-only and acceptor-only images are acquired separately with donor and acceptor filter sets, respectively. FRET images are acquired with donor excitation filter, donor dichroic and acceptor emission filter. These filter combinations were adapted from the catalog of filter sets published by Chroma Techonology Corporation.

domain fluorescence lifetime imaging microscopy (FLIM) methods ***(15–17)***. The energy transfer efficiency can be calculated by

$$E = 1 - \left(\tau_{DA}/\tau_{D}\right) \tag{7}$$

where τ_{DA} and τ_{D} are the donor lifetimes in the presence and absence of acceptor. The main advantage of this technique is that the FRET signal depends only on the excited-state reactions and not on the donor concentration or light path length. Furthermore, this method does not suffer from problems due to cross-talk between the donor and acceptor spectra or the need to precisely control the concentration of the donor and acceptor (*see* **Note 6**).

In the following sections, we give a generalized outline of equipments and procedures for implementing the various FRET approaches described above. As the field of FRET microscopy is rapidly progressing in a variety of applications, it is not possible to cover all these specific details in this chapter. It is therefore recommended to take care of the specific requirements (besides the general guidelines given below) for cell culture and handling before implementing the FRET measurements. For the sake of exemplification, we choose a specific FRET pair, namely, cyan fluorescent protein–yellow fluorescent protein (CFP-YFP) and typical equipments that we use routinely in our laboratory. However, the general theme of **Materials** and **Methods** holds good for any combination of FRET pairs and is valid for all the different FRET approaches described in this section.

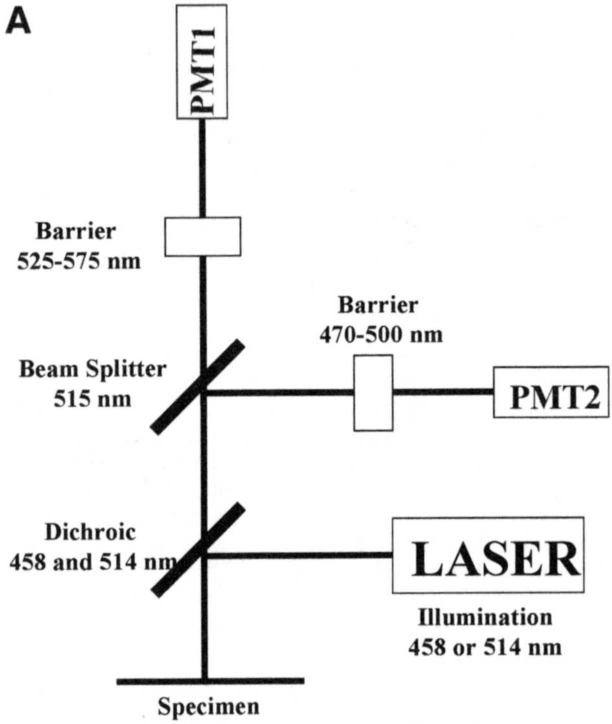

Fig. 3. (**A**) Typical imaging path in the laser scanning confocal microscope for the specific FRET pair CFP-YFP. Laser excitation (458 nm for CFP and 514 nm for YFP) from the source is routed through the dichroic to the specimen. The fluorescence emission from the specimen passes through the dichroic to the beam splitter. Fluorescence signal below 515 nm is reflected toward the CFP emission filter (470–500 nm bandpass) and detected by the photomultiplier tube (PMT2) in the donor (CFP) channel. The rest of the signal (>515 nm) passes through the beam splitter and the YFP emission filter (525–575 nm bandpass) and eventually gets recorded in the acceptor (YFP) channel. For FRET imaging, the specimen is excited with donor excitation wavelength (458 nm) and the signal is detected in the acceptor channel. (**B**) FRET measurement by Acceptor photobleaching approach. The FRET sample (mCGY) sample is the same as in **Fig.2**. The measurements were carried out in Zeiss-LSM510 confocal microscope with 458 nm and 514 nm laser lines for exciting CFP and YFP, respectively. HT515LP dichroic was used in the excitation path and the fluorescence emission from CFP and YFP were collected through the bandpass filters 470-500 nm and 525–575 nm respectively. The data acquisition and analysis are carried out as described in **Subheading 3.3.** The upper panel shows the images of the FRET sample in donor- and acceptor-channels—both before and after photobleaching by 514 nm laser line. Many regions were marked in the mitochondria and an average of fluorescence intensity in these regions is calculated and is given in the lower panel of the figure; 20% FRET efficiency was obtained in these conditions as calculated.

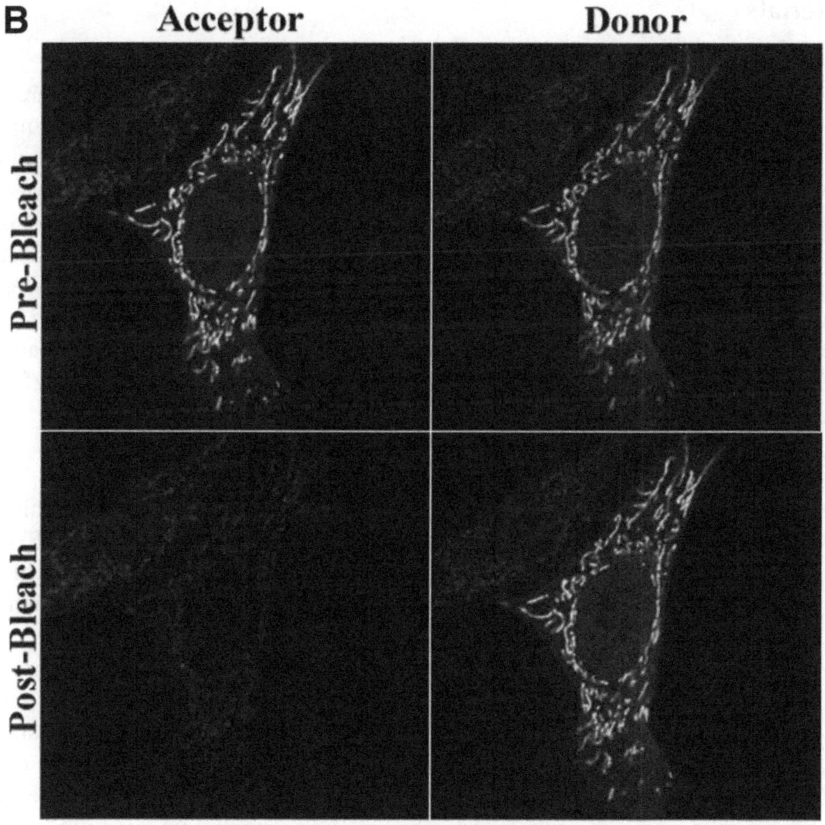
B
Acceptor
Donor
Pre-Bleach
Post-Bleach

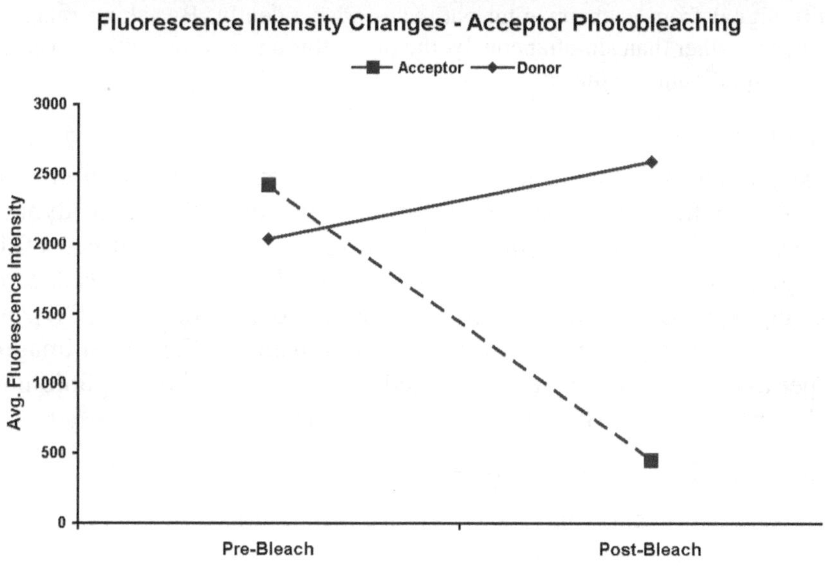
Fluorescence Intensity Changes - Acceptor Photobleaching
Acceptor
Donor
Avg. Fluorescence Intensity
3000
2500
2000
1500
1000
500
0
Pre-Bleach
Post-Bleach

2. Materials

2.1. Optics and Illumination

1. *Wide-field system*: A research grade microscope, with high NA optics that are corrected for spherical and chromatic aberrations. The microscope must be equipped with an epi-fluorescence attachment and a high-intensity light source (e.g. Olympus IX-70 microscope equipped with 60X PlanApo lens and mercury arc source).
2. *Scanning confocal system*: Confocal laser scanning imaging system with lasers appropriate for excitation of the chosen donor/acceptor FRET pairs, e.g., a Zeiss 510 laser scanning confocal microscope equipped with an argon ion laser emitting lines of 458 nm, 488 nm, and 514 nm for excitation of CFP, GFP, and YFP, respectively, green HeNe emitting at 543 nm for excitation of DsRed, rhodamine, and Cy3, and a red HeNe emitting at 633 nm for excitation of Cy5 and allophycocyanine.

2.2. Filters

1. *Wide-field system*: Excitation/emission filters for appropriate excitation/emission wavelength selection for the donor/acceptor to be employed (**Table 1**). Excitation and emission filters may be placed in cubes along with the dichroic filters. Alternatively, the excitation and emission filters may be placed in filter wheels for automated, rapid switching (*see* **Note 7**).
2. *Scanning confocal system*: The confocal system should have beamsplitters and emission filters in place such that specific pathways can be configured for acquisition of the acceptor signal only, the donor signal only, and the FRET signal. Acquisition of the FRET signal should be done by illuminating with the donor excitation wavelength but collecting the acceptor emission (*see* **Fig. 3A** for typical configuration). It is recommended to employ sequential imaging mode to acquire signal from each imaging pathway independently. By taking images in sequence rather than simultaneously, the bleed-through or cross-talk between filters is significantly reduced.

2.3. Detectors

The key features for choosing the right digital camera are high quantum efficiency over the entire visible spectrum of wavelengths, large dynamic range, low read noise, and minimum dark current. (*see* **Note 8**) It is useful to have a digital camera system with sufficient bit depth to be able to resolve very slight changes in fluorescence intensities. Charge-coupled device (CCD) cameras such as Hamamatsu Orca cameras are recommended. For FLIM imaging, it is imperative to use fast cameras coupled to microchannel plate photomultiplier tubes with very fast rise times (approx 0.3 ns).

2.4. Data Acquisition and Analysis

Image acquisition and analysis in fluorescence microscopy are now largely integrated in a few commercially available software systems. The essential

Table 2
Representative Websites of the Manufacturers[a]

Optics and Illumination	1. Coherent Inc.: http://www.coherent.com 2. Spectra-Physics: http://www.spectraphysics.com 3. Perkin Elmer: http://optoelectronics.perkinelmer.com 4. Hamamatsu Photonics: http://www.hamamatsu.com 5. Nikon Instruments Inc: http://www.nikon.com 6. Olympus America: http:// www.olympus.com 7. Carl Zeiss Microimaging: http:// www.zeiss.com 8. Melles-Griot: http:// www.mellesgriot.com/ 9. NewPort Corp.:http:// www.newport.com
Filters	1. Omega Optical: http:// www.omegafilters.com 2. Chroma Technology Corp.: http://www.chroma.com
Detectors	1. Hamamatsu Photonics: http://www.hamamatsu.com 2. Jobin Yvon: http:// www.jyhoriba.com 3. Perkin Elmer: http://optoelectronics.perkinelmer.com
Data acquisition	1. Universal Imaging Corp.: http://www.universal-imaging.com 2. Q Imaging: http://www.qimaging.com 3. Intelligent Imaging Innovations: http://www.intelligent-imaging.com 4. BioRad Laboratories: http://www.micoroscopy.bio-rad.com
Fluorophores	1. BD Biosciences Clontech: http:// www.clontech.com 2. Molecular Probes Inc.: http:// www.probes.com

[a] A partial list of manufacturers dealing with equipment needed for implementing FRET approaches in microsocope based systems.

requirement from any analysis package is that it should be able to select defined regions of interest and to measure the same regions between different images, for example, between acceptor/donor/FRET images in the case of sensitized emission or between pre- and postbleach images in the case of acceptor photobleaching approach. The measured intensities can then be transferred to spreadsheets for calculation of FRET efficiency. In our laboratory, journals have been written in MetaMorph (Universal Imaging Corp.) for both acquiring images as well as for further analysis to calculate normalized FRET (FRETN). Similar software packages are also available from other sources *(18)*. **Table 2** gives a comprehensive list of manufacturers and useful websites pertinent to fluorescence imaging and FRET microscopy.

2.5. Fluorophores

Table 1 summarizes a list of commonly used FRET pairs for studying protein–protein interactions. The choice of FRET pair depends on the spectral characteristics of donor and acceptor fluorophores. Regardless of the approach employed to measure FRET, these conditions remain the same (*see* **Note 1**). A vital advancement in the armamentarium of fluorophores for microscopy came with the discovery of green fluorescent protein (GFP) and its various spectral mutants ***(19,20)***. As these fluorescent proteins can be fused to the protein of one's interest within a living cell, FRET applications can include intracellular imaging and dynamic tracking of interactions between these proteins.

2.6. Reagents

1. PHEM: 60 m*M* PIPES, 25 m*M* HEPES, 10 m*M* EGTA, 2 m*M* $MgCl_2$.
2. Plasmid Mini-Prep Kit (Qiagen, Chatworth, CA).
3. DNA transfection reagent: Fugene (Roche).
4. Formaldehyde solution: 4% paraformaldehyde, pH 7.2–7.4, in PHEM buffer.
5. $NaBH_4$ (1 mg/mL).
6. Mounting medium: Vectashield (Vector) or Mowiol (Polysciences, Inc.) containing 1 mg/mL para-phenylene diamine.

3. Methods

3.1. Construct Preparation

1. In our laboratory, we employ mitochondrially targeted fluorescent proteins such as mCFP, mYFP, and mCGY (fusion protein of CFP and YFP linked through a polyglycine linker).
2. DNA transformation is carried out in bacterial cell cultures.
3. DNA is isolated using a Plasmid Mini-Prep Kit. DNA sequence for each vector is verified by an independent DNA sequencing laboratory. Glycerol stocks are made of transformed cells for future experiments.

3.2. Specimen Preparation

3.2.1. Cell Culture/Transfection

1. No. 1.5 glass coverslips are sterilized by alcohol wash and flaming or by autoclaving. Cells are then cultured on these coverslips in the preferred medium and at the appropriate temperature/atmosphere.
2. At approx 30–50% confluence, purified DNA is introduced into baby hamster kidney (BHK) cells with Fugene according to the instructions provided by the manufacturer. Typically, 2 μg DNA is mixed with 6 μL Fugene reagent in 100 μL serum-free OptiMem medium and incubated at room temperature for 45 min before adding to the cells.
3. After the transfection period, cells are rinsed and fresh medium is added to the cultures. Fluorescent proteins may be observed in cells between 6 and 24 h post-

transfection. At this time live cells may be transferred to a viewing chamber to measure FRET in the living cells. If cells are imaged live, it is recommended that a HEPES-buffered medium without phenol red be used instead of the conventional medium. Temperature should be maintained at 37°C using a heater stage. Alternatively, the cells may be fixed and mounted (approx 16–24 h post-transfection) for future use.

3.2.2. Fixation and Mounting of Cells for FRET Measurements

1. Medium is rinsed from transfected cells using PHEM or other appropriate buffer (*see* **Note 9** for situations that do not involve transfection but antibody labeling).
2. Cells are fixed for 20 min at room temperature with freshly prepared 4% paraformaldehyde, pH 7.2–7.4, in PHEM buffer.
3. After fixation, cells are rinsed twice with buffer.
4. Cells are incubated with 1 mg/mL $NaBH_4$, three changes for 5 min each. This step suppresses cellular autofluorescence significantly.
5. Cells are rinsed with buffer and mounted in Vectashield (Vector) or Mowiol (Polysciences, Inc.) containing 1 mg/mL para-phenylene diamine.
6. Excess mounting medium is removed and the coverslips are sealed with epoxy or nail hardener (*see* **Note 10**).
7. For extended use of fixed specimens, it is important that the slides are kept in dark and at 4°C.

3.3. Data Acquisition

3.3.1. Sensitized Emission FRET1

1. For detection of sensitized emission FRET, it is recommended that a three-sample/three-filter set approach be employed as detailed in Gordon et al. *(21)*. As stated previously, most sensitized emission measurements with filter-based microscopes suffer from problems of filter cross-talk or bleed through. The three-filter set method is the most conservative and the most general approach to the calculation FRETN that corrects for this cross-talk, and can be implemented on any microscope or microfluorimeter. For each measurement of FRET, three samples must be prepared—one containing donor-only (d), one containing acceptor-only (a), and one that contains both donor and acceptor (f). Each of the samples is imaged with an acceptor filter set (A), a donor filter set (D), and a FRET filter set (F).
2. For the correction algorithms to work effectively, the FRET filter set must be configured to have an exciter and dichroic that is *matched* with the donor-only filter set and an emitter filter that is *matched* with the acceptor-only filter set. (*see* **Notes 11** and **12** and **Table 1**).
3. Acquisition settings for the filter sets are determined by imaging a representative area of the donor sample with the donor filter set, the acceptor sample with the acceptor filter set, and the FRET sample with the FRET filter set. Exposure time, light intensity, and camera settings are adjusted to acquire images with

good signal-to-noise ratio. It is reasonable to select settings that will utilize approx 60–80% of the dynamic range to allow for sample variation. Once these settings are determined, they must remain constant for the filter set or imaging pathway regardless of the sample being observed. For every specimen at least 10 fields are imaged with each of the three filters sets.

3.3.2. Acceptor Photobleaching FRET

In the case of the acceptor photobleaching FRET approach, first an image of the donor fluorescence is acquired. Then the acceptor molecule is selectively photobleached by extensive illumination of the specimen with wavelengths of light centered at the absorption maximum of the acceptor. Photobleaching may be done in a wide-field microscope by continuously illuminating the sample using an acceptor-only filter set or in a confocal microscope with the laser line required for acceptor excitation—until the acceptor signal is destroyed. This may be easily performed in a confocal microscope implementing a time-lapsed bleach.

3.3.3. FLIM–FRET

In the case of either of the FLIM methods, FRET measurements involve monitoring only the donor lifetime for calculation of transfer efficiency according to eq. 6. Lifetimes can be calculated for the whole cell or in some specific regions of the cell depending on the requirements. This is the only method that does not obviously depend on the acceptor concentration and light path length. However, it may be important to monitor donor lifetimes at various acceptor concentrations to optimize the sensitivity of the system to FRET.

3.4. Analysis and Quantification

1. Each of the different FRET approaches discussed above yields two-dimensional images and the extent of FRET signal is calculated from changes in intensities or lifetimes due to FRET either in the whole cell or in smaller regions inside the cell.
2. In the case of sensitized emission FRET approach, first, the regions of local background in each image of a stack are measured to determine the average fluorescence contribution from this source. Next, regions of interest are marked. A detailed analysis procedure of these regions of interest is given in Gordon et al. ***(21)***.
3. Implementation of the three-sample/three-filter set approach and correction of the resulting FRET signals to determine FRETN results in a highly corrected value for FRET. Because this is an arbitrary number, it is very important to compare the FRETN value from a given experiment to a FRETN calculated from a negative control sample, one for which there should be little or no FRET. Whenever possible, it is also recommended that a positive control sample also be analyzed.
4. In the case of acceptor photobleaching FRET approach, donor fluorescence in regions of interest are measured from the prebleach and postbleach images. The

energy transfer efficiency is calculated according to eq. 4. It is to be noted that incomplete photobleaching (≤ 95%) of the acceptors can cause a significant error in the calculated energy transfer efficiency.

5. As mentioned earlier, energy transfer efficiency is a good quantitative measure of the FRET phenomenon. It is therefore a reliable parameter that should be monitored when comparing the FRET results for a given specimen measured by different FRET approaches described above.

By virtue of being a quantitative addendum to the conventional fluorescence methods, FRET microscopy approach allows analysis of cell component interactions in a rigorous quantitative framework. The FRET equation (eq. 2) is a nonlinear function of donor–acceptor separation and care has to be taken in interpreting small changes in intensity where many noise contributions are inevitable and may be significant (*see* **Note 13**). It is therefore critical to take into account the applicable corrections (optics/electronics) in designing a FRET experiment in the context of protein–protein interactions. Because the phenomenon of FRET is a stochastic process, failure to detect FRET in laboratory experiments does not necessarily rule out interactions among the molecules, but only sets a lower limit of sensitivity in detecting those interactions. This chapter is an attempt to set some general guidelines and protocols to enable sensitive and reliable detection of protein–protein interactions. Most of the discussions in this chapter have been limited to single-photon excitation. However, multiphoton-FRET *(22,23)* and single-particle FRET experiments *(24,25)* are finding their own place in biomedical applications in understanding biologically relevant interactions among the proteins such as conformation changes etc., at the atomic levels. With continual improvements in light sources, discovery of novel fluorophores and new detection devices, it is likely that FRET microscopy will emerge as an important biophysical tool in the biomedical sciences.

4. Notes

1. Forster distance R_0, is defined as that separation between the donor and acceptor, for which the energy transfer efficiency is 50% and is defined by the following expression:

$$R_0^6 = \left[8.75 \times 10^{-25} \times n^{-4} \times Q \times \kappa^2 \times J(\lambda)\right] \tag{N.1}$$

where n is the refractive index of medium in the range of overlap, Q is the quantum yield of the donor in the absence of acceptor, $J(\lambda)$ is the spectral overlap as shown in **Fig. 1B**. κ^2 is called the *orientation* factor, which depends on the angular orientation of the dipoles with respect to the vector separating them as well as with respect to each other (**Fig. 1A**) and is defined by

$$\kappa^2 = \left(\cos\theta_T - 3\cos\theta_A \times \cos\theta_D\right) \tag{N.2}$$

In general, this orientation factor can vary from 0 to 4, but is usually assumed to be 2/3, a value corresponding to a random orientation of the donors and acceptors. Typically, R_0 varies between 1 and 10 nm for various pairs of fluorophores (**Table 1**). The necessary criteria for the FRET pair (donor/acceptor) are summarized as follows: (i) The donor should be necessarily fluorescent with high quantum yield; (ii) acceptor excitation spectra (A_A)should have a substantial overlap with the donor emission spectra (E_D) but very little overlap with the donor excitation spectra (A_D); (iii) for a given donor–acceptor pair, the Forster distance is a constant in space and time. In order to measure a reasonable FRET signal, the donor–acceptor separation should be comparable to this value of R_0. For example, at $R = R_0$, the transfer efficiency is 50%, while this drastically reduces to 1.5% when $R = 2R_0$.

2. Energy transfer measured by all the FRET methods depends critically on the orientation factor that effectively represents an average angular profile of the fluorophores ***(26)***. Because it is not directly measured, this can easily cause uncertainty in quantification of the FRET signals. On the other hand, it can be argued that a large hydrophobic FRET probe that is covalently bound to an amino acid chain of a protein is deprived of true rotational freedom. This would cause minimal uncertainty in orientation factor thereby making FRET quantification less ambiguous.
3. Equation 3 assumes that there is no acceptor fluorescence by direct excitation of the acceptor at the donor excitation wavelength. This is seldom realized in practice because of the spectral cross talk between the donor excitation and acceptor-excitation spectra. Regardless of these inherent problems, this method of observing FRET is the most common in wide-field fluorescence microscopes. One of the main drawbacks of this method is that both the donor and acceptor fluorescence can be quenched during excitation that will further make quantification difficult.
4. Analysis of the change in donor intensity before and after photobleaching of the acceptor is performed on a pixel-by-pixel basis to determine the FRET efficiency. An advantage of this method is that it requires only a single sample and that the energy transfer efficiency can be directly related to both the donor fluorescence and the acceptor fluorescence.
5. A practical way of calculating anisotropy in a typical fluorescence microscope experiment is by using a set of polarizers (for detecting light with parallel and perpendicular polarization) at the emission port of a microscope. Polarized excitation can be achieved either by having an excitation polarizer next to an arc lamp or using the polarized light output of a laser. A set of images is collected with two crossed polarizers (with respect to the excitation polarizer) at the emission port. The corresponding intensities for these two images (for the entire cell or a selected region) are $I\|$ and $I\perp$, respectively. The anisotropy can be now calculated as

$$r = \frac{I_{\|} - I_{\perp}}{I_{\|} + 2I_{\perp}} \tag{N.3}$$

An advantage of using this approach is that it is possible to observe FRET between two identical donor molecules (homo- or single-color FRET) and this eliminates the spectral cross-talk issues between the donor and acceptor. Fluorescein is one of the classical examples of homo-FRET with Forster distance approx 44 Å ***(10)***. Typical applications of this approach are in membrane organization issues, protein proximity measurements, and diffusion-limited kinetics. However, it is to be noted that not all fluorophores exhibit observable change in anisotropy.

6. Measuring FRET by FLIM methodology is a reliable way although it is usually commented that it lacks the sensitivity that is possible with any of the spectral methods. Temporal sensitivity of any FLIM system is decided primarily by the excitation pulse widths (in the case of time-domain methods) and the fast response of detectors. There are lasers available to provide ultrafast, femtosecond pulses (e.g., Titanium: Sapphire, Coherent Inc.) and fast CCD cameras (e.g., Argus HiSCA CCD cameras) are able to provide time resolutions as low as a few picoseconds. This combination can make the lifetime measurements equally sensitive for FRET measurements. Another relatively less explored methodology is spectral imaging microscopy, which combines the spectral resolution (around a few nanometers) of a spectrograph and spatial resolution (about a few hundred nanometers) of a microscope to give a wealth of information about nanoscale protein interactions in biological systems.
7. Filter selection is critical to the success of FRET imaging. For example a donor imaging pathway should be configured with an excitor filter that has bandwidth specific for the donor fluorophore and little or no excitation of the acceptor fluorophores. The dichroic and emission filters should be selected to eliminate as much of the acceptor emission as possible. The acceptor imaging pathway should be configured with an excitor filter of bandwidth that exclusively excites the acceptor fluorophore and with dichroic and emssion filters that block any emission from the donor molecule. The FRET imaging pathway must be configured with an excitor filter and dichroic filter that match those of the donor imaging pathway and an emission filter that matches that of the acceptor pathway.
8. CCD-cooled digital cameras are now available with high quantum efficiency over the spectrum from 200 to 1000 nm. Robust spectral response characteristics and special digital contrast enhancement circuits increase versatility for even the most difficult imaging conditions. Of late, electron-bombardment CCD cameras have come into use which employ an innovative high-gain sensor to obtain high-gain images in very low light.
9. If antibody labeling is to be accomplished, rinse the cells with appropriate buffer media (PBS or PHEM) to remove the debris and serum components from the cells. Choose the appropriate donor/acceptor antibody concentration (e.g., Cy3/Cy5-labeled antibodies or FITC/TRITC-labeled antibodies) for labeling. A preliminary binding affinity curve for each of the antibodies (donor/acceptor) has to be measured. For FRET experiments, usually a low, constant concentration of donor/acceptor (approx 50 μg/mL) antibody is employed and FRET efficiency is measured as the concentration of the acceptor is varied at a constant overall anti-

body (donor plus acceptor) concentration (typically donor/acceptor molar ratios in the range 1:1 to 1:5). It is necessary to ensure that the measured transfer efficiency is not biased due to concentration artifacts. Centrifuge donor/acceptor-labeled proteins/antibodies before use to remove large aggregates. Normal serum from the secondary antibody host is preferably used as a blocking reagent. Alternately, one can use 0.5% bovine serum albumin (BSA) in PBS for blocking nonspecific binding of the antibodies. Prepare labeling mixtures (antibody + blocking reagent) in a buffer medium at physiological pH (approx 7.2). If cytoplasmic labeling has to be accomplished, then the membrane has to be permeabilized to allow antibodies into the cell. In fixed cells, either precipitation by anhydrous 100% methanol (–20 to 10°C) or cross-linking/permeabilization by 4% paraformaldehyde (37°C) and 0.1% Triton X-100 (4°C) can be used. In living cells, microinjection, transient hypoosmotic shock, ATP permeabilization, or other methods can be utilized.

10. Sealant should not contain acetone or other denaturants that might denature the fluorescent protein, thus destroying its ability to fluoresce.
11. When using a confocal laser scanning microscope, the same principle applies, the imaging pathway for the FRET sample must be a hybrid consisting of excitation pathway matched to the donor excitation pathway and the emission pathway matched to that of the acceptor. For example, the three imaging pathways used in a Zeiss 510 laser scanning confocal microscope are as represented in **Fig. 3A**. The acceptor-only signal is excited with the 514 nm line and collected in PMT 1, the donor-only signal is excited with the 458 nm and collected in PMT 2, and the FRET signal is excited by 458 nm and collected in PMT 1. The signals are collected in sequence using the "multitracking" configuration to minimize the crosstalk. Optics with both chromatic and spherical corrections will likely not have the most efficient transmission characteristics and therefore signal will be reduced. However, these corrections are important to ensure that excitation and emission are occurring in the same focal plane. This is particularly important when using a laser scanning confocal for FRET acquisition.
12. The various filters in a FRET set are chosen to minimize any potential cross talk between signals, although even with optimal selection of filters, this still remains a problem. A 12-bit confocal image would be optimized to a maximum of approx 3275 intensity units. This is repeated for the acceptor sample and acceptor filter set, and the FRET sample with the FRET filter set. Once the acquisition settings are determined for a filter set, they must not be altered regardless of the sample being acquired. Computer macros or journals may be written to "automate" acquisition once the initial parameters are established. While the fluorescence filters and neutral-density filters can be changed manually, it is far more time efficient to have motorized filter wheels that can be driven by the acquisition software such as is possible with Metamorph (Universal Imaging Corp., Downington, PA). Alternatively, confocal data sets may be acquired using "multitracking" (Zeiss LSM 510) or sequential imaging (Olympus FV500) that are existing functions of the operating software.

13. It is important to have precise intensity analysis to rule out noise contributions from autofluorescence, spectral cross-talk, and so on. Furthermore the effects of quenching or irreversible photobleaching have to be taken into account. In addition, usually the orientation factor is assumed to be 2/3 that is the average of a randomized ensemble of donor/acceptor dipoles. A few seminal works on FRET quantification take into account all the above features so as to give a reliable estimate of FRET efficiency in practice ***(21,27,28)***.

Acknowledgments

We gratefully acknowledge Yingpei Zhang, Marisa Lopez-Cruzan, and Eva Biener for their useful suggestions in this manuscript and Atsushi Masuda for providing us with the BHK specimens.

References

1. Pawley, J. B. (1995) *Handbook of Biological Confocal Microscopy*. Plenum Press, New York, NY.
2. Dyba, M. and Hell, S. W (2002) Focal spots of size λ/23 open up far-field fluorescence microscopy beyond 33 nm axial resolution. *Phys. Rev. Lett.* **88,** 163,901–163,904.
3. Herman, B. (1989) Resonance energy transfer microscopy. *Meth. Cell Biol.* **30,** 219–243.
4. Wu, P. G. and Brand, L. (1994) Resonance energy transfer: methods and applications. *Anal. Biochem. 218,* ***1–13.***
5. Clegg, R. M. (1995) Fluorescence resonance energy transfer. *Curr. Opin. Biotechnol.* **6,** 103–110.
6. Selvin, P. R. (2000) The renaissance of fluorescence resonance energy transfer. *Nat. Struct. Biol.* **7,** 730–734.
7. Lakowicz, J. R. (1999) *Principles of Fluorescence Spectroscopy*, 2nd Ed. Plenum, New York, NY.
8. Stryer, L. and Haughland, R. P. (1967) Energy transfer: a spectroscopic ruler. *Proc. Natl. Acad. Sci. USA* **58,** 719–726.
9. Jovin, T. M. and Jovin, D. J. A. (1989) FRET microscopy: Digital imaging of fluorescence resonance energy transfer. Application in cell biology, in *Cell Structure and Function by Microspectrofluorometry* (Kohen, E. and Hirschberg, J. G., eds.). Academic, San Diego, CA, pp. 99–116.
10. Krishnan, R. V, Varma, R., and Mayor, S (2001) Fluorescence methods to probe nanometer scale organization of molecules in living cell membrane. *J. Fluorescence* **11,** 211–226.
11. Weber, G. (1954) Dependence of the polarization of the fluorescence on concentration. *Trans. Faraday Soc.* **50,** 552–555.
12. Runnels, L. W. and Scarlata, S. F. (1995) Theory and application of fluorescence homotransfer to melitin oligomerization. *Biophys. J.* **69,** 1569–1583
13. Varma, R. and Mayor, S. (1998), GPI-anchored proteins are organized in submicron domains at the cell surface. *Nature* **394,** 798–801.

14. Gautier, I., Tramier, M., Durieux, C., et al. (2001) Homo-FRET microscopy in living cells to measure monomer-dimer transition of GFP-tagged proteins. *Biophys. J.* **80,** 3000–3008.
15. Bastiaens, P. I. H. and Squire, A. (1999) Fluorescence lifetime imaging microscopy: spatial resolution of biochemical processes in the cell. *Trends. Cell. Biol.* **9,** 48–52
16. Pepperkok, R., Squire, A., Geley, S., and Bastiaens, P. I. H. (1999) Simultaneous detection of multiple green fluorescent proteins in live cells by fluorescence lifetime imaging microscopy. *Curr. Biol.* **9,** 269–272.
17. Wang, X. F., Periasamy, A., Wodnicki, P., Gordon, G. W., and Herman, B. (1996) Time-resolved fluorescence lifetime imaging microscopy: instrumentation and biomedical applications in *Fluorescence Imaging spectroscopy and Microscopy* (Wang, X. F. and Herman, B., eds.) JohnWiley, New York, NY, pp. 313–350.
18. www.universal-imaging.com; www.qimaging.com; www.intelligent-imaging.com.
19. Tsien, R. Y. (1998) The green fluorescent protein. *Annu. Rev. Biochem.* **67,** 509–544, and references therein.
20. Hicks, B. W. (ed.) (2002) *Green Fluorescent Protein.* Humana, Totowa, NJ.
21. Gordon, G. W., Berry, G., Liang, X. H., Levine, B., and Herman, B. (1998) Quantitative fluorescence resonance energy transfer measurements using fluorescence microscopy. *Biophys. J.* **74,** 2702–2713.
22. Konig, K. (2000) Multiphoton microscopy in life sciences. *J. Microsc.* **200,** 83–104
23. Majoul, I., Straub, M., Duden, R., Hell, S. W., and Soling, H. D. (2002) Fluorescence resonance energy transfer analysis of protein–protein interactions in single living cells by multifocal multiphoton microscopy. *Rev. Mol. Biotechnol.* **82,** 267–277.
24. Ha, T., Enderle, T. H., Ogletree, D. F., Chemla, D. S., Selvin, P. R., and Weiss, S. (1996) Probing the interaction between two single molecules: fluorescence resonance energy transfer between single donor and a single acceptor. *Proc. Natl. Acad. Sci. USA* **93,** 6264–6268.
25. Ha, T., Ting, A. Y., Liang, J., et al. (1999) Temporal fluctuations of fluorescence resonance energy transfer between two dyes conjugated to a single protein. *Chem. Phys.* **247,** 107–118.
26. Dale, R. E., Eisinger, J., and Blumberg, W. E. (1979) The orientational freedom of molecular probes. The orientation factor in intramolecular energy transfer. *Biophys. J.* **26,** 161–193.
27. Nagy, P., Vamosi, G., Bodnar, A., Lockett,S. J., and Szollosi, J. (1998) Intensity-based energy transfer measurements in digital imaging microscopy. *Eur. Biophys. J.* **27,** 377–389
28. Xia, Z. and Liu, Y. (2001) Reliable and global measurement of fluorescence resonance energy transfer using fluorescence microscopes. *Biophys. J.* **81,** 2395–2402.

25

Monitoring Molecular Interactions in Living Cells Using Flow Cytometric Analysis of Fluorescence Resonance Energy Transfer

Francis Ka-Ming Chan

Abstract

Variants of the green fluorescence protein (GFP) are useful tools in many biological research applications. Two of the GFP spectral variants, the cyan and yellow fluorescence proteins (CFP and YFP), have compatible excitation and emission properties as fluorescence donor and acceptor in fluorescence resonance energy transfer (FRET). A protocol is described here to study the molecular association between p80 TNFR-2 and TRAF2 fusion proteins containing CFP and YFP using flow cytometric FRET measurement. The utility of this application in the study of protein–protein interaction is discussed.

Key Words

FRET; CFP; YFP; flow cytometry; TRAF; TNFR.

1. Introduction

The various biological functions of the cell are coordinated by the interaction of distinct molecular machineries. Numerous methods have been developed over the years to detect molecular associations both in vivo and in vitro. Many of these methods, however, do not allow the examination of dynamic processes in the living cells. The phenomenon of fluorescence resonance energy transfer (FRET) describes the transfer of energy from a fluorophore (donor) in an excited state to a neighboring fluorophore (acceptor) through dipole–dipole interaction *(1,2)* (**Fig. 1**). FRET can be a sensitive means to measure molecular distances within 100 Å **Fig. 1**) *(3,4)*. Many methods utilizing the concept of FRET have been used successfully for the measurement of biological interactions *(5–8)* (also see Chapter 24). For example, subunit associa-

From: *Methods in Molecular Biology, vol. 261: Protein–Protein Interactions: Methods and Protocols*
Edited by: H. Fu © Humana Press Inc., Totowa, NJ

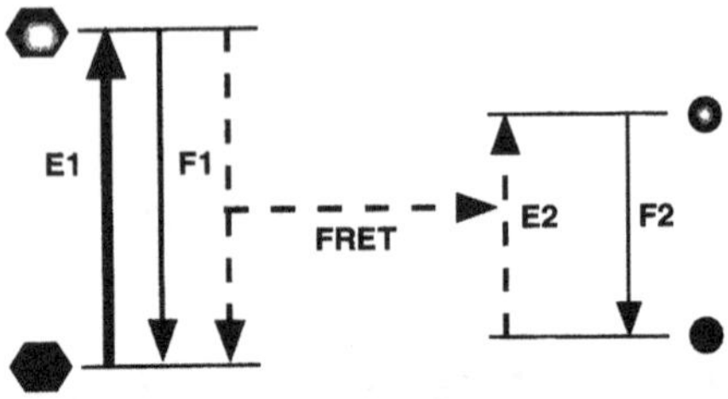

Fig. 1. Principles of fluorescence resonance energy transfer. The excitation of the first fluorophore (*E*1) results in the emission of energy (*F*1) at a lower wavelength as it returns to the ground state. Fluorescence emission from the first fluorophore (*F*1) can be "absorbed" by a nearby second fluorophore with an excitation wavelength (*E*2) that overlaps with the emission from the first fluorophore (*F*1) through dipole–dipole interactions. The net result is the transfer of energy in the form of fluorescence emission from the second fluorophore (*F*2 or FRET) and the "quenching" of fluorescence emission from the first fluorophore.

tions of the IL-2 receptor and EGF receptor have been detected with FRET-compatible, fluorescence-labeled antibodies using the flow cytometer ***(9,10)***. Furthermore, molecular interactions between the dimeric subunits of the transcription factor Pit-1 and the interaction between the apoptotic regulatory proteins Bcl-2 and Bax were determined using microscopic FRET measurement ***(11,12)***. These methods are superior to conventional biochemical methods in that they allow the investigators to monitor biological processes in living cells. However, their use was sometimes limited by the availability of the appropriate antibodies that allow FRET without disrupting the interaction under examination or the relatively small sample size of microscopic measurements.

Recently, several spectral variants of green fluorescence protein (GFP) ***(13)*** were developed that have compatible fluorescence excitation and emission properties for FRET analysis (**Table 1**). In particular, the excitation and emission spectra of the cyan (CFP) and yellow (YFP) versions of the GFP proteins allow them to be used in FRET analysis as donor and acceptor fluorophores, respectively. By creating fusion proteins to these GFP variants and introducing them into the appropriate cellular system, one can monitor the molecular associations of different biological machineries in living cells using flow cytometric methods. Combining FRET analysis with flow cytometry is advantageous as the investigator can screen a large number of cells within a short time. Unlike fluorescently labeled antibodies, the use of GFP fusion proteins permits the examination of extra- as well as intracellular associations. The versatility of flow cytometric FRET analysis is illustrated here using the interaction between p80 TNFR-2 and TRAF2 as an example. TRAF-2 is an important signal transduction component to many of the members of the TNF receptor superfamily.

Table 1
Spectral Properties of GFP Variants

Clone name	Mutations	Excitation peak (nm)	Emission peak (nm)
BFP (blue)	F64L, Y66H, Y145F	383	447
GFP (green)	F64L, S65T, H231L	488	507
CFP (cyan)	K26R, F64L, S65T, T66W, N146I, M153T, V163A, N164H, H231L	434 (452)	476 (505)
YFP (yellow)	S65G, S72A, T203Y, H231L	514	527

Upon activation with TNFα, TRAF2 is directly recruited to the preassembled TNFR-2 receptor (reviewed in ref. ***14*** and ***15***). This interaction will be monitored by the expression of fusion proteins of CFP and YFP to p80 TNFR-2 and TRAF2 in 293T cells (*see* **Note 1**).

2. Materials

1. Plasmids: pEF6-myc-HisB (*see* **Note 2**; Invitrogen), pECFP-N1, pEYFP-N1 (BD Clontech).
2. cDNAs for p80 TNFR-2, TRAF2, and SODD.
3. Oligonucleotide primers for polymerase chain reaction (PCR).
4. Restriction enzymes, T4 DNA ligase.
5. QIAEX gel purification kit (Qiagen).
6. Competent *Escherichia coli* cells for transformation (e.g., XL-1 blue from Stratagene).
7. HEK 293T cells (*see* **Note 3**).
8. Complete Dulbecco's modified Eagle's medium (DMEM) (Biowhitaker): DMEM (phenol red free), 10% fetal calf serum (FCS), 100 units of penicillin and streptomycin, 2 m*M* L-glutamine.
9. Fugene 6 (Roche).
10. Phosphate-buffered saline (PBS).
11. NP-40 lysis buffer: 10 m*M* Tris-HCl, pH 7.5, 150 m*M* NaCl, 1% NP-40.
12. Biorad protein assay reagent (Biorad).
13. 10% Bis-Tris NuPAGE protein gels (Invitrogen).
14. Rabbit polyclonal antiserum against TRAF2 (Santa Cruz).
15. 6 mL FACS tubes (Falcon).
16. Propidium iodide (PI) solution: 500 μg/mL.
17. FACS Vantage SE flow cytometer (Becton Dickenson).
18. FlowJo analytical software (Treestar Inc.) or other flow cytometry analytical softwares.

3. Methods

3.1. Design of the Constructs

The crystal structure of TRAF2 homotrimer bound to the TNFR-2 peptide reveals a "mushroom-like" conformation of TRAF2 with the carboxyl and amino termini resembling the cap and the stalk of a mushroom, respectively ***(16)***. The p80 TNFR-2 peptide makes contact with TRAF2 at the edge of the mushroom cap. To maximize the potential for interaction and therefore FRET, CFP was cloned at the carboxyl terminus of full-length p80 TNFR-2, and YFP was cloned at the amino terminus of the full-length TRAF2 protein. The silencer of death domain (SODD) protein ***(17)***, which binds to only TNFR-1 but not TNFR-2, was used as a noninteracting negative control.

3.1.1. Vectors and Generation of the CFP and YFP Vectors (see **Note 2**)

1. Design PCR primers for CFP and YFP cDNA inserts.
2. Perform PCR using pECFP-N1 or pEYFP-N1 as templates.
3. Resolve PCR products on 1% agarose gel. Excise and purify the correct fragments using QIAEX.
4. Digest PCR fragments with
   i. *BamHI*/*EcoR*V (YFP fragment).
   ii. *EcoR*V/*Xba*I (CFP and YFP fragments).
5. Digest vector pEF6-myc-His B with
   i. *BamHI*/*EcoR*V.
   ii. *EcoR*V/*Xba*I.
6. Ligate fragment from **step 4i** with vector from **step 5i**. Resulting plasmid is pEF6B-YFP-C. Ligate CFP or YFP fragment from **step 4ii** with vector from **step 5ii**. Resulting plasmids are pEF6B-CFP-N and pEF6B-YFP-N.
7. Transform into competent *E. coli* XL-1 blue cells.
8. Screen resulting clones for PCR inserts in minipreps.
9. Sequence miniprep DNAs to confirm sequence integrity.

3.1.2. Cloning of the cDNAs

1. PCR amplify cDNAs for p80 TNFR-2, TRAF2, and SODD.
2. Digest p80 PCR product with *BamHI*/*EcoR*V. Digest TRAF2 and SODD PCR products with *EcoR*V/*Xba*I.
3. Digest pEF6B-YFP-C with *EcoR*V/*Xba*I. Ligate TRAF2 and SODD cDNA inserts from **step 2** to create pEF6B-YFP-TRAF2 and pEF6B-YFP-SODD.
4. Digest pEF6B-CFP-N and pEF6B-YFP-N with *BamHI*/*EcoR*V. Ligate p80 fragment from **step 2** to create pEF6B-p80-CFP and pEF6B-p80-YFP.
5. Repeat **steps 7–9** in **Subheading 3.1.1.**

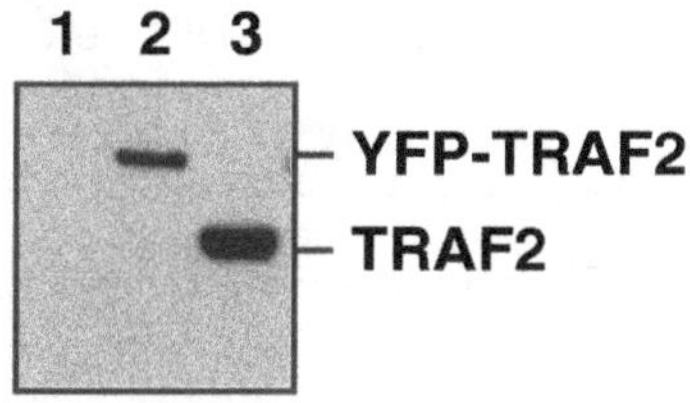

Fig. 2. Expression of YFP-TRAF2 in HEK 293T cells. HEK 293T cells were transfected with (1) pEF6-myc-HisB, (2) pEF6B-YFP-TRAF2, and (3) pcDNA3-TRAF2; 24 h later, cells were harvested for whole cell lysates. Equal amounts (50 μg) of lysates were loaded on a 10% Bis-Tris NuPAGE gel and resolved by electrophoresis. The expression of TRAF2 was examined in Western blotting using antibody against TRAF2. YFP-TRAF2 and TRAF2 were indicated.

3.2. Examination of the Expression of the Constructs

The next step in this process involves the confirmation of protein expression from the FRET donor and acceptor plasmids. To determine the expression of the plasmids, they were introduced into the human kidney epithelial cell line HEK 293T cells.

3.2.1. Transfection of 293T Cells (see **Note 3**)

1. 2.5×10^5 cells were seeded in each well of a 12-well plate in 1 mL of complete DMEM medium. Incubate at 37°C for 16–20 h.
2. 2 μg of each plasmid was transfected into HEK 293T cells using 6 μL of Fugene 6 to give a DNA/Fugene 6 ratio of 1:3.
3. Grow and incubate cells at 37°C for 24–48 h.
4. Harvest cells for subsequent analyses by Western blotting or flow cytometry.

3.2.2. Western Blotting Analysis

1. Aspirate medium from wells. Add 1 mL of PBS to each well. Gently resuspend cells in PBS and transfer cell suspension to microfuge tubes.
2. Spin cells for 5 min at 12,800*g*. Aspirate PBS.
3. Resuspend cell pellet in 100 μL of NP-40 lysis buffer and incubate for 15 min on ice.
4. Spin at 12,000*g* at 4°C in a microfuge for 10 min.
5. Transfer supernatant to a new microfuge tube. Measure lysate concentration using Biorad protein assay reagent.
6. Load 50 μg of cell lysates on a 10% Bis/Tris NuPAGE gel and transfer onto nitrocellulose membrane.
7. Probe membrane with antibody against TRAF2 and secondary HRP-conjugated antibody against rabbit IgG. **Figure 2** shows the expression of YFP-TRAF2 (lane 2) and untagged TRAF2 (lane 3) in HEK 293T cells (*see* **Note 4**).

3.2.3. Monitoring Fluorescence of the Expressed Proteins

As an alternative to Western blot analysis, expression of the transfected plasmids can be monitored by flow cytometry.

1. Wash cells twice in 1 mL PBS supplemented with 2% FCS by spinning 5 min at 4°C at 12,800*g*.
2. Resuspend in 1 mL PBS supplemented with 2% FCS.
3. Add 2 μL of PI solution (500 μg/mL) to cell suspension.
4. Acquire events on the flow cytometer.

3.3. Performing Flow Cytometric FRET Analysis

Once the expression of the FRET plasmids is confirmed, they can be tested in flow cytometric FRET analysis. Unlike the transfection procedure described in **Subheading 3.2.1.**, transfections with a combination of FRET donor and acceptor plasmids were performed here.

3.3.1. Transfections

The transfection procedure was similar to what was described in **Subheading 3.2.1.** using Fugene 6 (*see* **Note 5**):

Sample number	Sample description	
1	2 μg pEF6-myc-HisB	
2	0.5 μg pEF6B-p80-CFP	+ 1.5 μg pEF6-myc-HisB
3	1.5 μg pEF6B-p80-YFP	+ 0.5 μg pEF6-myc-HisB
4	1.5 μg pEF6B-YFP-TRAF2	+ 0.5 μg pEF6-myc-HisB
5	1.5 μg pEF6B-YFP-SODD	+ 0.5 μg pEF6-myc-HisB
6	0.5 μg pEF6B-p80-CFP	+ 1.5 μg pEF6B-p80-YFP
7	0.5 μg pEF6B-p80-CFP	+ 1.5 μg pEF6B-YFP-TRAF2
8	0.5 μg pEF6B-p80-CFP	+ 1.5 μg pEF6B-YFP-SODD

3.3.2. Cell Harvesting

1. After 24-48 h, harvest cells by washing and spinning twice in PBS supplemented with 2% FCS in FACS tubes.
2. Resuspend cells in PBS supplemented with 2% FCS.
3. Pass cells through a nylon filter prior to analysis.
4. Keep cells at 4°C until they are ready for analysis. Alternatively, if further manipulations of the cell samples are required (such as ligand stimulation), they can be kept at room temperature.

3.3.3. Acquisition of Events on the Flow Cytometer (see **Note 6**)

Cells were analyzed on a FACS Vantage SE flow cytometer. Two lasers are available on the FACS Vantage SE: an ILT air-cooled argon laser and a

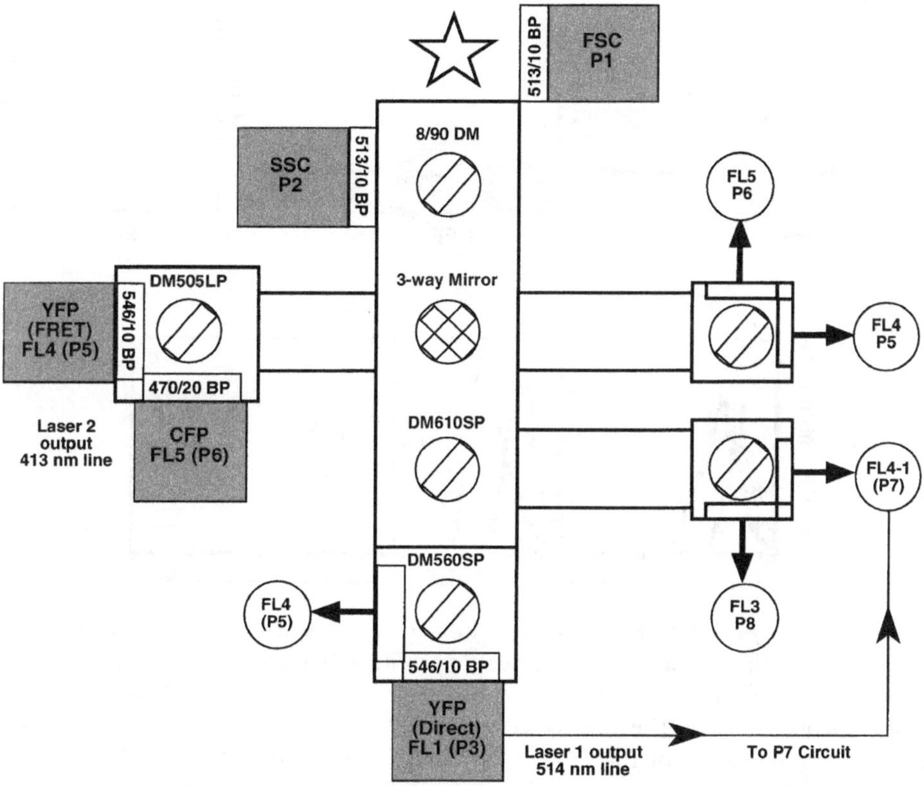

Fig. 3. Schematic setup of the FACS Vantage SE flow cytometer. Only the relevant filters and mirrors are labeled. The channels that were used for FRET analysis are colored gray. Other unused channels are shown for reference only.

krypton laser (Spectra Physics model 2060) that is equipped with violet optics (*see* **Note 7**). The argon laser was tuned to 514 nm for direct excitation of YFP (**Fig. 3**, laser 1). The krypton laser was tuned to 413 nm for excitation of CFP (**Fig. 3**, laser 2). Forward (FSC) and side scatter (SSC) filters were replaced with 513/10 nm bandpass (BP) filters. CFP fluorescence was detected in FL5 (P6) using a 470/20 nm BP filter. A 505LP dichroic mirror (DM) was used for separating CFP fluorescence and FRET emission from laser 2. FRET signal was detected in FL4 (P5) using a 546/10 nm BP filter. Direct YFP fluorescence was detected using a 546/10 BP filter in the FL1 (P3) channel but was directed to the P7 channel to allow P5-P7 interlaser compensation (**Fig. 3**).

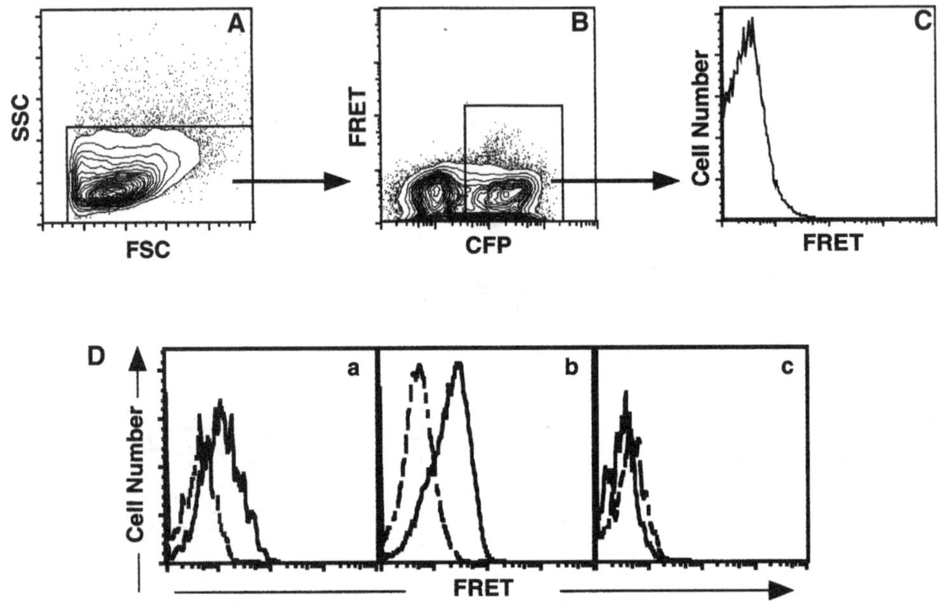

Fig. 4. Analysis of FRET data using Flowjo software. **(A)** Live cells (box) were determined by their forward and side-scatter profile. **(B)** The cells that were in the live gate in **(A)** were analyzed for their CFP fluorescence and FRET signal on two-dimensional dot plot analysis. The transfected cells (CFP-positive cells in the box) were gated for histogram analysis of FRET. **(C)** Histogram analysis of FRET on CFP-positive cells from **(B)**. **(D)** Histograms overlay showed that only cells co-expressing **(a)** p80-CFP and p80-YFP or **(b)** p80-CFP and YFP-TRAF2 exhibited significant level of FRET (solid lines). **(c)** Cells expressing p80-CFP and YFP-SODD did not exhibit any FRET (solid line). Dashed lines: baseline level of FRET in cells expressing only p80-CFP.

Samples were collected with fluidics pressure at 30 pound per square inch (psi). This was done to shorten the pulse timing between the argon laser and the krypton laser in order to allow interlaser compensation using the standard delay module with a maximum delay of 17.5 μs. One microgram per milliliter of propidium iodide was added prior to the acquisition of events; 50,000 live cells were collected.

3.3.4. Analyzing the Data Using Flowjo

1. Draw live cells gate by plotting SSC against FSC (**Fig. 4A**) or PI (detected in P8) against FSC.
2. Construct two-dimensional dot plot of CFP fluorescence (FL5/P6) versus FRET (FL4/P5) using cells in the live cells gate (**Fig. 4B**).

3. Use vector-transfected sample (sample no. 1 in **Subheading 3.3.1.**) as the negative control to set the cutoff between CFP-positive and CFP-negative populations. Draw the CFP positive gate accordingly.
4. Using cells in the CFP positive gate, perform histogram analysis for FRET intensity (*see* **Note 8**) (**Fig. 4C**).
5. Perform FRET histograms overlay using p80-CFP (sample no. 2; *see* **Subheading 3.3.1.**) as the negative control (**Fig. 4D**, dashed lines). Overlay histograms from sample nos. 6, 7, or 8 on histogram from sample no. 2. Expression of p80-CFP and p80-YFP (sample no. 6) resulted in a strong FRET signal (**Fig. 4D**, panel a). Co-expression of p80-CFP and YFP-TRAF2 (sample no. 7) also resulted in strong FRET signal (**Fig. 4D**, panel b). However, the noninteracting control YFP-SODD failed to generate any significant FRET signal with p80-CFP (sample no. 8) when co-transfected with p80-CFP (**Fig. 4D**, panel c).

4. Notes

1. The success of FRET is most dependent on two criteria: (a) the physical distance between the fluorophores and (b) the relative orientation of the fluorophores ***(18)***. In using CFP and YFP as FRET donor and acceptor molecules, the investigator should also pay particular attention to the spacer length between the CFP (or YFP) moiety and the protein polypeptide itself. Owing to the size of CFP and YFP (approx 30 kDa), it is not uncommon that their fusion to another protein may affect the protein function itself. In fact, fusion of CFP and YFP to the extracellular pre-assembly domain of TNFRs can severely hamper the ability of the receptor to bind ligand ***(19)***. However, this effect is ameliorated by increasing the spacer length between the YFP moiety and the receptor.
2. The vector pEF6-myc-HisB, which has an EF promotor, was chosen because of its ability to drive strong expression of the protein of interest in mammalian cells. The readers should note that the pECFP and pEYFP series of plasmids from BD Clontech can be used directly for the introduction of the cDNA insert of interest. This will save time in the cloning steps.
3. HEK 293T cells are ideal for high protein expression due to the presence of the SV40 large T antigen. However, other cell lines including Jurkat T cells have also been used successfully in FRET analysis. For 293T cells, Fugene 6 is the transfection reagent of choice because of its consistency in yielding high protein expression. Cell culture conditions should be carefully monitored as they can significantly affect the transfection efficiency. Typically, seeding 2.5×10^5 cells on 12-well plates will yield a 70–80% confluent culture after 16–20 h of incubation, which generally results in > 50% transfection efficiency.
4. To detect protein expression for the fusion proteins, monoclonal antibody against GFP can also be used. GFP-specific monoclonal antibody from Roche can cross-recognize both CFP and YFP in Western blotting analysis.
5. To optimize the FRET signal, it is necessary to have the acceptor molecule in slight excess relative to the donor molecule. To achieve that, a 1:3 ratio of CFP plasmids and YFP plasmids were used in the experiments described in this chap-

ter. However, the investigators should empirically determine the optimal DNA ratio for each pair of FRET plasmids. Because the phenol red present in most tissue culture media can sometimes affect the autofluorescence level of the cells, phenol red-free medium is recommended for culturing cells for FRET analysis.

6. The LSR model of flow cytometer from Becton Dickenson has an optional fourth laser that can be used for CFP excitation and therefore will be compatible for FRET analysis as well. Other models of flow cytometer may also work provided that they have the proper lasers installed. The investigators should consult the individual vendor for more information.
7. The more standard 488 nm laser found in most flow cytometers can sufficiently excite the YFP protein and therefore can replace the 514 nm argon laser. However, the 488 nm laser is not optimal for CFP excitation. Therefore, a separate laser tuned to ≤ 440 nm is needed (*see* **Subheading 3.3.**) for CFP excitation.
8. In the example given in this chapter, I used two-dimensional dot plot of CFP versus FRET to define the CFP positive gate for subsequent FRET signal analysis. However, other ways of defining the transfected populations such as plotting CFP fluorescence versus YFP fluorescence and gating on the double positive population are plausible alternatives for determining the transfected populations from which FRET is analyzed.

Acknowledgements

The author would like to thank Michael Lenardo, Richard Siegel, Kevin Holmes, and Ruth Swofford for discussions and technical help.

References

1. Forster, T. (1946) Energiewanderung und Fluoreszenz. *Naturwissenschaften* **6,** 166–175.
2. Forster, T. (1948) Zwischenmolekulare Energiewanderung und Fluoreszenz. *Ann. Phy. (Leip.)* **2,** 55–75.
3. dos Remedios, C. G. and Moens, P. D. (1995) Fluorescence resonance energy transfer spectroscopy is a reliable "ruler" for measuring structural changes in proteins. Dispelling the problem of the unknown orientation factor. *J. Struct. Biol.* **115(2),** 175–185.
4. Hillisch, A., Lorenz, M., and Diekmann, S. (2001) Recent advances in FRET: distance determination in protein-DNA complexes. *Curr. Opin. Struct. Biol.* **11(2),** 201–207.
5. Tron, L., Szoolosi, L., Damjanovich, S., Helliwell, S. H., Arndt-Jovin, D. J., and Jovin, T. M. (1984) Flow cytometric measurement of fluorescence resonance energy transfer on cell surfaces. Quantitative evaluation of the transfer efficiency on a cell-by-cell basis. *Biophys. J.* **45(5),** 939–946.
6. Tron, L., Szollosi, J., and Damjanovich, S. (1987) Proximity measurements of cell surface proteins by fluorescence energy transfer. *Immunol. Lett.* **16(1),** 1–9.
7. Szollosi, J., Matyus, L., Tron, L., et al. (1987) Flow cytometric measurements of fluorescence energy transfer using single laser excitation. *Cytometry* **8(2),** 120–128.

8. Szollosi, J., Damjanovich, S., and Matyus, L. (1998) Application of fluorescence resonance energy transfer in the clinical laboratory: routine and research. *Cytometry* **34(4),** 159–179.
9. Carraway, K. L. and Cerione, R. A. (1991) Comparison of epidermal growth factor (EGF) receptor-receptor interactions in intact A431 cells and isolated plasma membranes. Large scale receptor micro-aggregation is not detected during EGF-stimulated early events. *J. Biol. Chem.* **266(14),** 8899–8906.
10. Damjanovich, S., Bene, L., Matko, J., et al. (1997) Preassembly of interleukin 2 (IL-2) receptor subunits on resting Kit 225 K6 T cells and their modulation by IL-2, IL-7, and IL-15: a fluorescence resonance energy transfer study. *Proc. Natl. Acad. Sci. USA* **94(24),** 13134–13139.
11. Day, R. N. (1998) Visualization of Pit-1 transcription factor interactions in the living cell nucleus by fluorescence resonance energy transfer microscopy. *Mol. Endocrinol.* **12(9),** 1410–1419.
12. Mahajan, N. P., Linder, K., Berry, G., Gordon, G. W., Heim, R., and Herman, B. (1998) Bcl-2 and Bax interactions in mitochondria probed with green fluorescent protein and fluorescence resonance energy transfer. *Nat. Biotechnol.* **16(6),** 547–552.
13. Tsien, R. Y. (1998) The green fluorescence protein. *Annu. Rev. Biochem.* **67,** 509–544.
14. Wallach, D., Varfolomeev, E. E., Malinin, N. L., Goltsev, Y. V., Kovalenko, A. V., and Boldin, M. P. (1999) Tumor necrosis factor receptor and Fas signaling mechanisms. *Annu. Rev. Immunol.* **17,** 331–367.
15. Chan, F. K. M., Siegel, R. M., and Lenardo, M. J. (2000) Signaling by the TNF receptor superfamily and T cell homeostasis. *Immunity* **13,** 419–22.
16. Park, Y. C., Burkitt, V., Villa, A. R., Tong, L., and Wu, H. (1999) Structural basis for self-association and receptor recognition of human TRAF2. *Nature* **398,** 533–538.
17. Jiang, Y., Woronicz, J., Liu, W., and Goeddel, D. V. (1999) Prevention of constitutive TNF receptor 1 signaling by silencer of death domains. *Science* **283,** 543–546.
18. Miyawaki, A. and Tsien, R. Y. (2000) Monitoring protein conformations and interactions by fluorescence resonance energy transfer. *Methods Enzymol.* **327,** 472–500.
19. Chan, F. K. M., Siegel, R. M., Zacharias, D., et al. (2001) Fluorescence resonance energy transfer analysis of cell surface receptor interactions and signaling using spectral variants of the green fluorescent protein. *Cytometry* **44,** 361–368.

26

Fluorescence Correlation Spectroscopy

A New Tool for Quantification of Molecular Interactions

Keith M. Berland

Abstract

Fluorescence correlation spectroscopy (FCS) provides a powerful method to measure molecular dynamics and interactions in a wide variety of experimental systems and environments. In this article we focus on the use of FCS methods to quantify molecular interactions, including the use of diffusion analysis and molecular counting. Both autocorrelation and cross-correlation FCS measurements are discussed.

Key Words

Fluorescence correlation spectroscopy; molecular interactions; fluctuations.

1. Introduction

Fluorescence correlation spectroscopy (FCS) is rapidly growing in popularity as a research tool in biological and biophysical research owing to its versatile and highly sensitive capability to characterize the chemical, physical, and dynamic properties of biomolecules and biological systems ***(1–4)***. This chapter will focus on the capability to quantify protein–protein and other molecular interactions using FCS methods. We begin with a general introduction to fluorescence correlation spectroscopy and then discuss how to apply FCS to quantify molecular interactions, including the use of both autocorrelation and cross-correlation measurements.

1.1. General Concepts

Fluorescence correlation spectroscopy is a form of fluctuation spectroscopy in which the properties of an experimental system are revealed through statistical analysis of spontaneous equilibrium fluctuations in the fluorescence sig-

From: *Methods in Molecular Biology, vol. 261: Protein–Protein Interactions: Methods and Protocols*
Edited by: H. Fu © Humana Press Inc., Totowa, NJ

nal measured from within a minute sample volume. Fluctuations may arise due to molecules diffusing into or out of the volume, as well as from chemical or photophysical reactions. This chapter will focus on fluctuations due to diffusive motion only. FCS methods are essentially counting experiments, where the fluorescence signal from within a small optically defined volume serves as a measure of the number of molecules (N) within that volume. As molecules diffuse in and out of the observation volume, the fluorescence signal fluctuates. **Figure 1** shows a schematic of this process and the associated number fluctuations. There are two main sources of information in the fluctuation signal, each which can be used to quantify molecular interactions. First, the duration of the average fluctuation is related to the time individual molecules occupy the volume. This time, which can be measured with FCS, can be used to recover a diffusion coefficient (D) for the fluorescent species in the volume. Because the diffusion coefficient depends on the molecular mass, changes in diffusion rate following molecular association/dissociation can be quantified with FCS. Second, the fluctuation amplitudes obey Poisson statistics and the amplitude of the average fluctuation is proportional to $\sqrt{N}$, where N is the average number of independently diffusing molecules in the volume. Fluctuation amplitudes thus provide a direct measure of the average number of independently diffusing molecules occupying the volume (i.e., N). Because the number of independent diffusing molecules will change when molecules associate (e.g., there are half as many independent diffusers following a monomer–dimer association reaction), this count can directly quantify the extent of molecular interactions.

To harvest the information content of the fluctuation signal, temporal correlation functions (**Fig. 1C**) are calculated from the fluctuation data. A correlation function compares a signal with either itself (autocorrelation) or a related signal (cross-correlation). The amplitude of the correlation function is inversely related to the average number of molecules within the observation volume (approx $1/N$) (*see* **Note 1**), and the temporal decay of the correlation function specifies the average occupancy time of molecules diffusing through the volume. Detailed models have been derived to accurately recover this information through curve fitting, as discussed below. Correlation functions often contain additional information regarding chemical or photophysical dynamics present in addition to diffusive motion, but, as noted above, these features will not be addressed in this chapter.

1.2. Theoretical Models for FCS

For a detailed introduction to FCS theory we recommend ref. *5*. Here we focus on a few key features of fluorescence correlation functions. For a particular molecular species, the average fluorescence signal F_i is proportional to

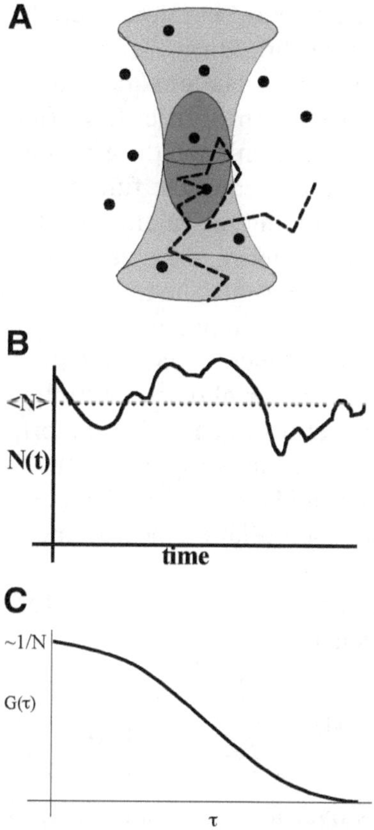

Fig. 1. **(A)** The FCS observation volume is optically defined. The light gray cone represents the excitation laser profile, and the dark gray ellipsoid is the observation volume. The volume is defined by the laser and fluorescence detection optics. Brownian motion of individual molecules will cause them to diffuse into and out of the volume. The dashed line represents the trajectory of a single particle. **(B)** As molecules diffuse in and out of the volume, the number in the volume will fluctuate as described in the text. **(C)** Correlation functions calculated from the fluctuation data are analyzed to determine diffusion coefficients and molecular concentrations.

the concentration of fluorescent molecules and the size of the volume, which can also be written as simply the number of molecules that occupy the volume:

$$\langle F_i \rangle = \psi_i \langle C_i \rangle \; V_{\text{eff}} = \psi_i \langle N_i \rangle \tag{1}$$

Here C_i represents molecular concentration, V_{eff} is the volume for fluorescence detection, and N_i represents the number of independent molecules within the volume. Angular brackets represent time-averaged quantities. The parameter ψ represents the average "brightness" of a molecular species (the average number of detected fluorescence photons/molecule/s), and is determined by the product of intrinsic molecular properties (e.g., fluorescence lifetime, quantum yield, absorption cross section), and the efficiency of the instrumentation for exciting, collecting, and detecting the fluorescence. ψ is one of the most critical parameters in determining the fidelity of FCS measurements ***(6)***. It is relatively straightforward to determine the value of ψ using FCS when only a single fluorescent species is present. The inverse amplitude of the correlation function [$G(0)\sim 1/N$] specifies the number of molecules in the volume, so the product of the average fluorescence signal and $G(0)$ determines ψ. (More precisely, this product yields the brightness multiplied by a constant factor γ introduced below.) It is useful to generalize the brightness parameter to the format $\psi_{i,n}$, where the double index represents the brightness of a particular molecular species i in a particular detection channel n (e.g., a "red" or "green" detection channel), as cross-correlation measurements require multiple detector channels.

Using the above notation, the theoretical form for fluorescence correlation functions, $G(\tau)$ can be generalized as:

$$G_{mn}(\tau) = \gamma \frac{\sum_i \psi_{i,m} \psi_{i,n} \langle N_i \rangle A_i(\tau)}{\left[\sum_i \psi_{i,m} \langle N_i \rangle\right]\left[\sum_i \psi_{i,n} \langle N_i \rangle\right]} \tag{2}$$

where the summation is over all molecular species i within the sample. For autocorrelation measurements, detectors m and n are the same, and for cross-correlation they are two different detection channels. The constant geometrical prefactor γ is determined by the shape of the observation volume and is typically determined experimentally. The temporal dependence of $G(\tau)$ is contained in the functions $A_i(\tau)$. For free diffusion with a "3D Gaussian" shaped volume (*see* **Note 2**), $A_i(\tau)$ can be written explicitly as

$$A_i(\tau) = \left(1 + a\varphi_i\right)^{-1} \left(1 + a\varphi_i \frac{\omega_0^2}{z_0^2}\right)^{-1/2} \tag{3}$$

where the dimensionless parameter and $\varphi_i = 4D_i\tau/\omega_0^2$, D_i, ω_0 and z_0 are the diffusion coefficient of species i and the radial and axial dimensions of the observation volume, respectively. In the above expression, the parameter a has the value 1 for one-photon excited fluorescence, and the value 2 for two-photon excited fluorescence. This model is applied to fit experimental data and recover information about dynamics and molecular interactions. It is useful to state explicitly that A_i functions all have unit value for short correlation times

($\tau << \omega_0^2/D$). Thus, the amplitude of the correlation functions $G(0)$ can be written explicitly as a sum over molecular species, weighted by the molecular brightness of each species. For example, the $G(0)$ for the autocorrelation function from two molecular species A and B can be written as:

$$G(0) = \gamma \frac{\psi_A^2 \langle N_A \rangle + \psi_B^2 \langle N_B \rangle}{\left[\psi_A \langle N_A \rangle + \psi_B \langle N_B \rangle\right]^2} \tag{4}$$

Note that the contribution of each molecular species to the $G(0)$ value is weighted by the square of its molecular brightness.

2. Materials

2.1. Instrumentation

The common strategy behind various implementations of FCS is to monitor time-dependent fluctuations in the fluorescence signal from minute observation volumes (approx 1 fL) characteristic of confocal or two-photon fluorescence microscopes *(2,7)*. For dilute sample concentrations (approximately nanomolar), the total number of molecules present in the observation volume at any one time is quite small (approx 1–1000), which ensures that fluctuations will be resolvable relative to the average fluorescence signal levels. Observation of such small molecular populations requires appropriately designed optical systems that have single-molecule sensitivity. A schematic of the two-photon FCS hardware used in the author's lab is shown in **Fig. 2**. High-performance FCS instrumentation is also commercially available. Instrumentation designed (or modified) specifically for FCS is generally required as commercial confocal microscopes and associated photodetectors are often not sufficiently sensitive for optimal FCS measurements.

2.2. Sample Preparation

There are many standard procedures for fluorescence labeling of biomolecules *(8,9)*. Green fluorescent proteins and other intrinsically fluorescent biomolecules can also work nicely in FCS measurements *(10)*. Specific details regarding labeling protocols are beyond the scope of this chapter, but there are several general considerations that are important in preparing samples for FCS measurements. First, clean samples are essential (*see* **Note 3**). Because only very small sample concentrations are used (nanomolar), impurities in the sample are more detrimental to FCS measurements than for many other techniques. Each fluorescent species contributes to the overall FCS measurement as the square of its molecular brightness; thus, bright fluorescent "junk" or aggregates in dirty samples can easily overwhelm the signal of inter-

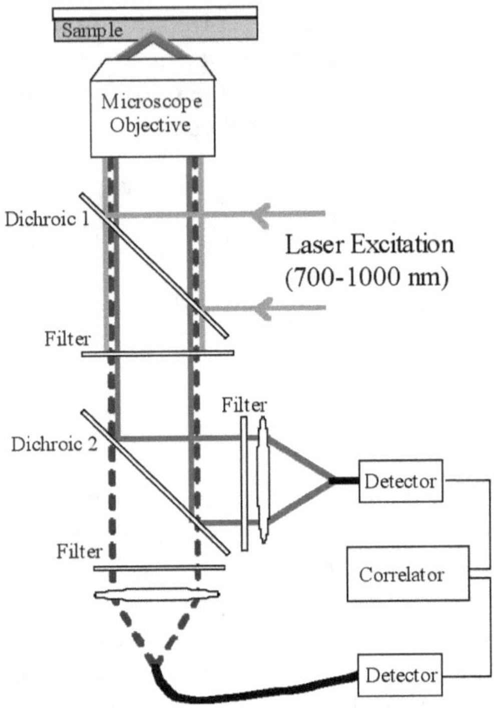

Fig. 2. Experimental setup for two-photon FCS used by the author. A mode-locked Ti:sapphire laser (100 fs pulses, 80 MHz pulse rate) serves as the two-photon excitation source. The laser is ported to a home-built inverted fluorescence microscope and focused onto the sample by the objective lens (Olympus 63X, 1.2 or Zeiss 63X, 1.2 water immersion lenses). Fluorescence emission is filtered by an E700SP filter to block scattered laser excitation, and then split into two detection channels with a second dichroic mirror. Each channel can be further filtered with bandpass filters as needed. Finally, the fluorescence signal is coupled into 100 μm optical fibers and directed to the photodetectors (EGG SPCM avalanche photodiodes or Hamamatsu H7421-40 photomultiplier tubes) The detector output is sent directly to an ALV-6000 correlator where auto- and cross-correlation functions are calculated.

est. Second, one should strive to have uniform fluorescent labeling of the proteins. The quality of an FCS signal will be far better with uniformly labeled molecular populations (e.g., a single fluorophore per protein monomer). Finally, if extrinsic dyes are used to label molecules, unconjugated free dyes should be removed to the fullest extent possible (*see* **Note 4**). Sample mounting procedures can also be very important. When working with the very low sample concentrations necessary for FCS measurements, sample loss to the container walls can become an important problem (*see* **Note 5**). It is thus

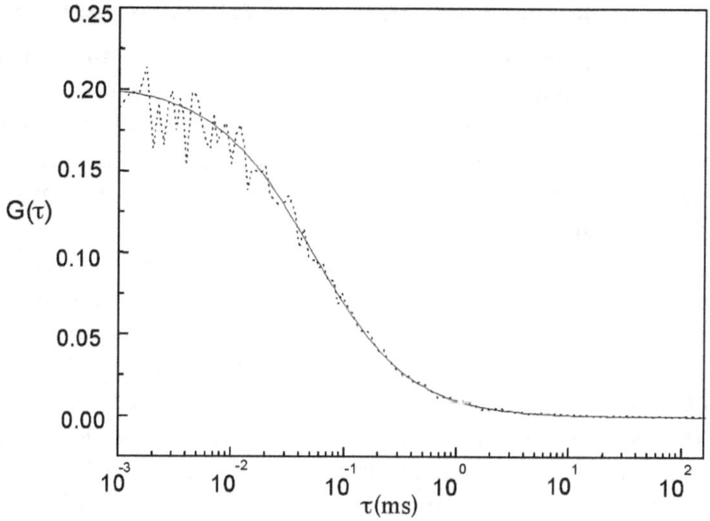

Fig. 3. Calibration of FCS instrumentation using rhodamine 6G, which has a known diffusion coefficient of 3×10^{-6} cm^2 s^{-1}. Rhodamine 6G was dissolved in ultrapure water and filtered using 0.01 μm syringe filters before being diluted to appropriate concentrations and mounted in a microbridge sample holder. The measured autocorrelation curve is shown here. Fitting these data to eqs. (2) and (3) yields a beam waist, ω_0, of 0.3 μm for 780 nm two-photon excitation. The ratio of the axial and radial dimensions was fixed at 10. The amplitude of the correlation curve is related to the sample concentration.

desirable to choose sample containers that have low surface area to volume ratios. In addition, container surfaces can be covered with various blocker proteins (e.g., BSA, casein) to minimize surface adsorption. Typical mounting strategies include placing samples in microwell plates with coverglass bottoms, hanging drop slides, or other chambered volumes. Fluid drops placed directly on coverslips can also work nicely for quick measurements, although sample evaporation will cause problems if measurements are taken over any substantial time period.

3. Methods

3.1. Instrument Calibration

As noted above, FCS measurements determine the average residency time for molecules in the observation volume. Because this time depends on both the diffusion rate and the size of the volume, the volume must be calibrated before diffusion coefficients can be recovered from FCS data. This is easily achieved by acquiring FCS data for molecules with known diffusion coefficients. Such calibration measurements are essential, and should be performed routinely. On such example is shown in **Fig. 3**, using rhodamine 6G in water.

3.2. Measuring Molecular Interactions With Diffusion Analysis

The capability to quantify protein–protein interactions by diffusion analysis is based on the molecular weight dependence of diffusion coefficients. This diffusion based approach resembles fluorescence polarization assays in that each method monitors a shift in diffusion coefficients following molecular association. However, fluorescence polarization (anisotropy) assays measure rotational diffusion and FCS methods monitor translational diffusion. Polarization measurements are not particularly effective for larger proteins because they rotate too slowly relative to the fluorescence lifetime of standard fluorophores. This gives FCS the important advantage that there is no restriction on the total molecular weight of the molecular complex being investigated, and FCS can thus be applied even for very large molecules. On the other hand, since diffusion is not a particularly sensitive measure of molecular weight (molecular size varies approximately as the cube root of molecular weight), diffusion-based analysis is best applied to quantify interaction between molecules that have substantially different molecular weights. In particular, changes in diffusion are easily resolved with FCS when a small fluorescent molecule interacts with a much larger molecule. On the other hand, it is difficult to accurately detect binding of a large fluorescent protein to a small protein or ligand using diffusion analysis. The estimated change in diffusion time ($\omega_0^2/4D$) required to resolve interactions is approximately a factor of 2 *(11)*.

The example of this type of measurement shown here quantifies the interaction of SV40 nuclear localization signal (NLS) peptides with the nuclear import receptor importin-α. These molecules play an important role in targeting specific proteins to the nucleus, and thus in cellular regulation and control. The SV40 NLS peptides were each tagged with a single fluorescein dye molecule, and the importin-α proteins were unlabeled. Importin-α has a molecular weight near 60 kDa, much larger than the 20-amino-acid NLS peptide. Thus, upon binding, the NLS–importin-α complex diffuses more slowly than the free NLS molecule. **Figure 4** shows FCS curves for two cases, where either all NLS molecules are bound to importin-α or none are bound. The NLS–importin-α complex diffuses more slowly than free NLS, and thus has a slower decay in the measured correlation function. It is clear that the two measurements are easily resolved. After using these two curves to determine the diffusion coefficients for the two species (bound and unbound), a series of FCS measurements was performed with fixed NLS concentration and serial dilutions of importin-α. The measurements were then fit to recover fraction of NLS molecules bound at various importin-α sample concentrations. The resulting binding curves are also shown in **Fig. 4**. This type of measurement provides very accurate quantitation of the protein–protein interactions, and has many useful applications (*see* **Notes 6** and **7**).

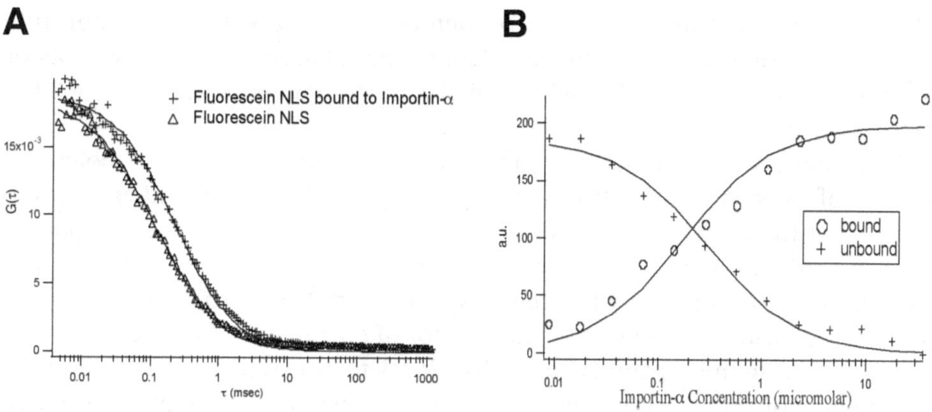

Fig. 4. Diffusion-based analysis of fluorescein NLS binding to importin-α. (**A**) FCS curves for free NLS and NLS fully bound to importin-α. The importin-α bound form diffuses more slowly and the FCS curve is thus shifted to the right (longer correlation times. (**B**) A series of similar measurements were performed for fixed concentration of NLS and serial dilutions of importin-α. The FCS curves were then fit, using the known diffusion times for free and bound NLS molecules, to determine the fraction bound and unbound at various importin-α concentrations. The full set of measurements was then fit to a binding curve, yielding a dissociation constant for the interaction of 180 n*M*. Fluorescein NLS peptides and purified importin-α protein were provided by Dr. Alec Hodel (Emory University, Department of Biochemistry). Binding curves were measured with fixed concentration of fluorescein NLS (50 n*M*) and successive dilutions of importin-α concentration.

3.3. Measuring Interactions By Molecular Counting

The concept behind the FCS "molecular counting" approach to detecting molecular associations is very simple, yet quite powerful. The amplitude of the measured correlation functions $G(0)$ is directly proportional to the inverse number of independently diffusing molecules in the volume ($1/N$). If molecules associate, this number changes; for example, by a factor of 2 in a complete monomer–dimer transition. The concentration of independent diffusers (measured by FCS) compared with the total molecular concentration (experimentally available from the average fluorescence intensity, which is proportional to the overall sample concentration independent of interactions) provides a direct measure of the extent of molecular interactions. For cases where the sample is not monodisperse, the concept remains unchanged; the amplitude of the correlation functions is still related to the average concentration of various molecular species (weighted by their molecular brightness).

Molecular associations can thus be detected by analyzing the amplitude of the correlation function—independent of the relative molecular weights of the interacting species. This is one of the more powerful strengths of FCS measurements.

In the molecular counting approach (unlike diffusion analysis) all molecular species of interest should be fluorescent. They can be labeled either with the same color fluorophore for autocorrelation, or spectrally distinct fluorophores for use in dual-color cross-correlation measurements. For cross-correlation measurements ($m \neq n$), only molecular species with "nonzero" values for both $\psi_{i,m}$ and $\psi_{i,n}$ will contribute to the numerator of eq. (2). It is this discrimination that gives rise to the capability for multicolor detection schemes to "tune-in" to particular molecular components of an heterogeneous sample population. In using dual-color cross-correlation, experiments should use carefully chosen fluorophores and filter sets to ensure cross-talk between detection channels does not degrade the signal quality ***(12)***.

To highlight the strength of molecular-counting-based applications of FCS, we here compare diffusion analysis with dual-color cross-correlation methods to quantify the hybridization of nucleic acids. In most dual-color FCS applications, interactions between two different molecular species labeled with spectrally distinct chromophores are studied. A slight variation on this scheme is shown here, where two spectrally distinct fluorescent molecules interact with the same nonfluorescent third molecular species. The altered experimental configuration does not change the details of the FCS measurement, but does highlight the possibility of monitoring the formation of molecular complexes even in the case where a particular component of the complex is not fluorescent. The experimental system, shown in **Fig. 5** consists of two 18mer single-stranded DNA molecules, denoted **A** and **B**, labeled with **R**ed (Cye 3.5) or **G**reen (Oregon Green) dyes, respectively. Each of these oligonucleotides is complementary to adjacent regions on a nonfluorescent 40mer DNA strand (**X**). The hybridization reaction observed is $A + B + X \leftrightarrow ABX$, i.e., all three species bind together to form a single complex.

Both autocorrelation and dual-color cross-correlation measurements were performed on samples with and without the target sequence X. **Figure 6** shows fits to the autocorrelation curves for each detection channel. The measured diffusion coefficient for the Oregon Green single-stranded molecule is $D = 1.2 \times 10^{-6}$ cm^2/s. This value decreases to $D = 8 \times 10^{-7}$ upon binding to its target sequence. A similar shift is observable in the red channel for the Cy3.5 oligo, although less pronounced. While there is a small change in diffusion upon binding to the target sequence, the similarity of the bound (with target) and unbound (no target) curves demonstrates the difficulty in recovering information about molecular interactions based on diffusion analysis when the

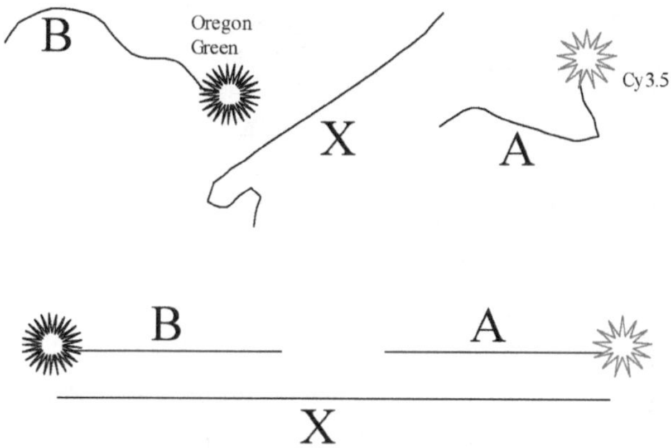

Fig. 5. The three molecular components of the oligonucleotide hybridization reaction; the top of the figure shows the individual strands, and the bottom the geometry of the trimeric complex. Species *A* is labeled with Cy3.5, *B* with Oregon Green 488, and *X* is unlabeled. The sequences, selected from the gamma tubulin gene, were chosen to minimize self-dimerization and hairpin formation. All samples were diluted to an appropriate concentration (typically 10–50 n*M*) and mounted in chambered cover glass slides or microbridges precoated with a casein buffer solution. FCS measurements were carried out at room temperature.

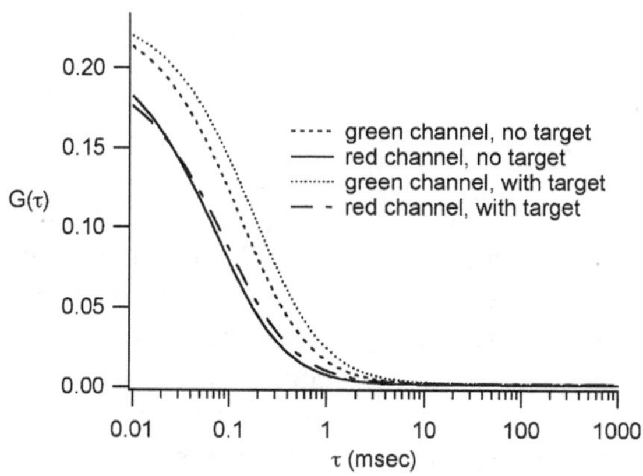

Fig. 6. Typical data fits to autocorrelation traces for mixtures of species *A* and *B*, both with and without the target sequence *X*. A shift in the diffusion coefficient from $D = 1.2 \times 10^{-6}$ cm^2/s to D = 8×10^{-7} cm^2/s is apparent for species *B* following the hybridization reaction. A similar shift, although less clearly resolved, occurs for species *A*. Clearly this interaction can barely be resolved using diffusion-based analysis. On the other hand, dual-color cross-correlation measurements easily resolve this interaction as shown in **Fig. 7**. The concentration of each molecular species in these measurements was 10 n*M*. The amplitude of the red channel autocorrelation functions is somewhat smaller due to the presence of free dye.

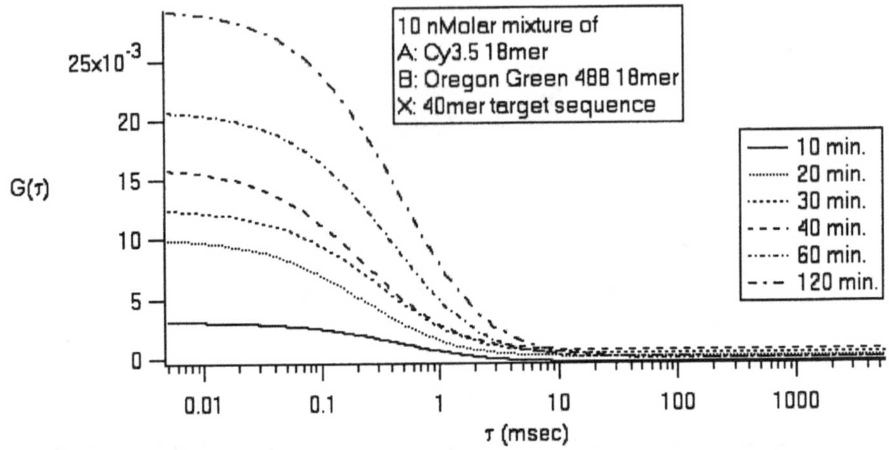

Fig. 7. Cross-correlation measurements for mixtures of all three oligonucleotides (*A*, *B*, *X*) that hybridize to form double-stranded DNA. Equal concentrations of the three single-stranded species (*A*, *B*, *X*) were mixed together prior to performing FCS measurements. The increasing population of double-stranded DNA, reflecting the progress of the hybridization reaction, is readily apparent from the growing amplitude of these dual-color cross-correlation curves. This clearly demonstrates the effectiveness of the dual-color FCS method for quantifying the extent of this molecular interaction. Note that the curves shown here are the data fits to measurements recorded at sequential times as noted in the figure legend.

interacting species have similar size. On the other hand, quantifying the interactions works extremely well using the amplitude of the correlation functions, as shown in **Fig. 7**. This result highlights the strength of dual-color cross-correlation methods for resolving molecular interactions, in this case the formation of the *ABX* (Cy3.5–Oregon-Green–Target) double-stranded complex. The time-dependent population of double-stranded DNA steadily increases as the hybridization reaction proceeds, and the concentration of the trimolecular DNA complex (*ABX*) can be monitored directly through the amplitude of the cross-correlation curves.

See **Notes 8–10** for additional considerations when performing the FCS analysis.

4. Notes

1. This is for a single molecular species. The expression for the amplitude with multiple species is shown in eq. (2).

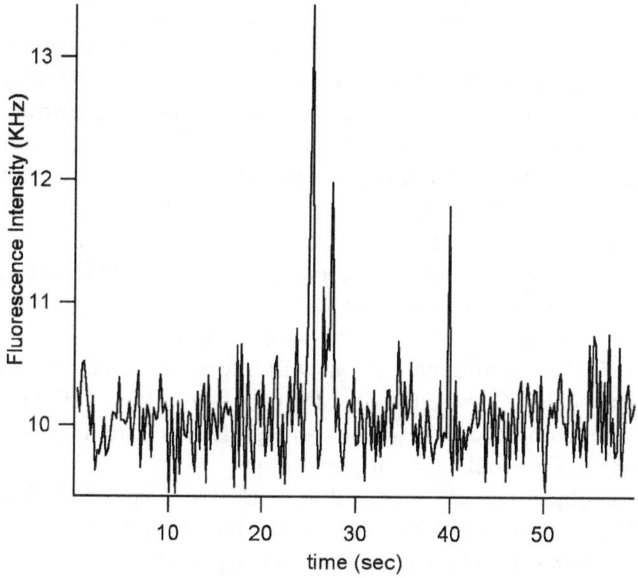

Fig. 8. Aggregates or other bright particles in the sample can lead to spikes in the fluorescence signal with amplitudes greatly exceeding the average fluctuation. These spikes will typically distort the measured correlation curves substantially, and sample preparation procedures that eliminate such spikes are necessary.

2. The 3D Gaussian is the most commonly used model for the shape of the observation volume in FCS measurements: The fluorescence excitation probability varies throughout the volume as

$$S_{3DG}(r,z) = \exp\left(-2r^2/\omega_0^2\right)\exp\left(-2z^2/z_0^2\right) \quad (5)$$

   Here r and z are the radial and axial cylindrical coordinates, and the parameters ω_0 and z_0 specify the dimensions of the volume.
3. Clean samples are critical for accurate FCS measurements. Use of clean solvents and repeated filtration steps can be very important. Large spikes in the average fluorescence signal, as shown in **Fig. 8**, or rapid changes in the amplitude or shape of correlation curves that appear during data acquisition most likely indicate the presence of aggregates or impurities that contaminate the data. While fluorescent impurities are highly detrimental, other molecules of interest that are not fluorescent need not be removed from the system.
4. Free dye molecules that remain after labeling proteins of interest can greatly complicate interpretation of FCS data. Care should be taken to purify out free dye to the fullest extent possible.

5. Sample loss to container walls can cause many problems when working with the low sample concentrations used for FCS measurements. Surface adsorption can be minimized by precoating container surfaces and coverslips with blocking buffers.
6. Binding curves, such as the one shown in **Fig. 4**, are best measured with serial dilutions in a single-sample container. One would typically start with the highest concentration of interest, remove one-half the sample volume, and replace the removed volume with the appropriate solvent or protein solution before each subsequent measurement, and continue this process until the lowest concentration of interest is reached. By using a single container the effects of sample loss to the container walls are minimized.
7. In monitoring molecular interactions with FCS, it is important to consider and measure how the specific interactions will affect the molecular brightness of the component species. Curve fitting results may be inaccurate if the effects of fluorescence quenching and/or resonance energy transfer are not taken into account (i.e., changes in the molecular brightness, ψ).
8. Microscope objectives are designed to work with a specific thickness of cover glass. Make sure to use the appropriate thickness for each lens (usually no. 1.5 coverslips). Both the glass thickness and depth of focus into the sample can affect the size and shape of the observation volume. If FCS data from samples mounted in separate containers are to be compared, it is often helpful to focus a fixed distance into each sample to minimize variations in sample volume.
9. It is important to limit the total number of fitting parameters in fitting any given FCS data set—typically not exceeding three or four free parameters. Usually the volume parameters are fixed following calibration experiments. In analyzing FCS data from heterogeneous sample populations, it is also useful to have independent knowledge of component diffusion coefficients and/or concentrations in order to eliminate some fitting parameters, perhaps obtained from separate FCS experiments. Use of global fitting routines can also help constrain fitting parameters, improving the accuracy of data fitting procedures.
10. It is very useful to calibrate the experimental system with the same calibration sample every time (i.e., use the same fluorophore, concentration, solvent, and so on). In performing calibration runs, one should pay attention not only to the volume parameters recovered (e.g., ω_0), but also to the amplitude of the correlation function and the average fluorescence signal. Any major changes in these parameters for a given data set can be very useful in identifying misalignments or other hardware problems.

Acknowledgment

This work was partially supported by the American Heart Association and the Emory University Research Fund.

References

1. Magde, D., Elson, E., and Webb, W. W. (1972) Thermodynamic fluctations in a reacting system. Measurement by fluorescence correlation spectroscopy. *Phys. Rev. Lett.* **29,** 705–708.

2. Eigen, M. and Rigler, R. (1994) Sorting single molecules: application to diagnostics and evolutionary biotechnology. *Proc. Natl. Acad. Sci. USA* **91,** 5740–5747.
3. Maiti, S., Haupts, U., and Webb, W. W. (1997) Fluorescence correlation spectroscopy: diagnostics for sparse molecules. *Proc. Natl. Acad. Sci. USA* **94,** 11,753–11,757.
4. Rigler, R. and Elson, E. L. (2001) *Fluorescence Correlation Spectroscopy: Theory and Applications*. Springer, New York, NY.
5. Thompson, N. L. (1991) Fluorescence correlation spectroscopy, in *Topics in Fluorescence Spectroscopy*, Vol. 1. (Lakowicz, J. R., ed.), Plenum, New York, NY, pp. 337–378.
6. Koppel, D. E. (1974) Statistical accuracy in fluorescence correlation spectroscopy. *Phys. Rev. A* **10,** 1938–1945.
7. Berland, K. M., So, P. T. C., and Gratton, E. (1995) Two-photon fluorescence correlation spectroscopy: method and application to the intracellular environment. *Biophys. J.* **68,** 694–701.
8. Hermanson, G. T. (1996) *Bioconjugate Techniques*. Academic Press, New York, NY.
9. Haugland, R. P. (2002) *Handbook of Fluorescent Probes and Research Products* (also at www.probes.com). Molecular Probes, Eugene, OR.
10. Tsien, R. Y. (1998) The green fluorescent protein. *Annu. Rev. Biochem.* **67,** 509–544.
11. Meseth, U., Wohland, T., Rigler, R., and Vogel, H. (1999) Resolution of fluorescence correlation measurements. *Biophys. J.* **76,** 1619–1631.
12. Schwille, P., Meyer-Almes, F. J., and Rigler, R. (1997) Dual-color fluorescence cross-correlation spectroscopy for multicomponent diffusional analysis in solution. *Biophys. J.* **72,** 1878–1886.

27

Confocal Microscopy for Intracellular Co-Localization of Proteins

Toshiyuki Miyashita

Abstract

Confocal laser scanning microscopy is the best method to visualize intracellular co-localization of proteins in intact cells. Because of the point scan/pinhole detection system, light contribution from the neighborhood of the scanning spot in the specimen can be eliminated, allowing high *Z*-axis resolution. Fluorescence detection by sensitive photomultiplier tubes allows the usage of filters with a narrow bandpath, resulting in minimal cross-talk (overlap) between two spectra. This is particularly important in demonstrating co-localization of proteins with multicolor labeling. Here I describe the methods outlining the detection of transiently expressed tagged proteins and the detection of endogenous proteins. Ideally, the intracellular co-localization of the two endogenous proteins should be demonstrated. However, when antibodies raised against the protein of interest are unavailable for immunofluorescence, or the available cell lines do not express the protein of interest sufficiently enough for immunofluorescence, an alternative method is to transfect cells with expression plasmids that encode tagged proteins and stain the cells with anti-tag antibodies after transient transfection. However, it should be noted that the tagging of proteins of interest or their overexpression could potentially alter the intracellular localization or the function of the target protein.

Key Words

Confocal microscopy; immunofluorescence; multicolor labeling; laser; fluorophore.

1. Introduction

Confocal laser scanning microscopy is the best method to visualize intracellular co-localization of proteins in intact cells. Confocal laser scanning microscopy offers significant advantages for viewing subcellular localization of proteins compared with conventional fluorescence microscopy.

First of all, the point scan/pinhole detection system eliminates light contribution from the neighborhood of the scanning spot in the specimen (**Fig. 1**).

From: *Methods in Molecular Biology, vol. 261: Protein–Protein Interactions: Methods and Protocols*
Edited by: H. Fu © Humana Press Inc., Totowa, NJ

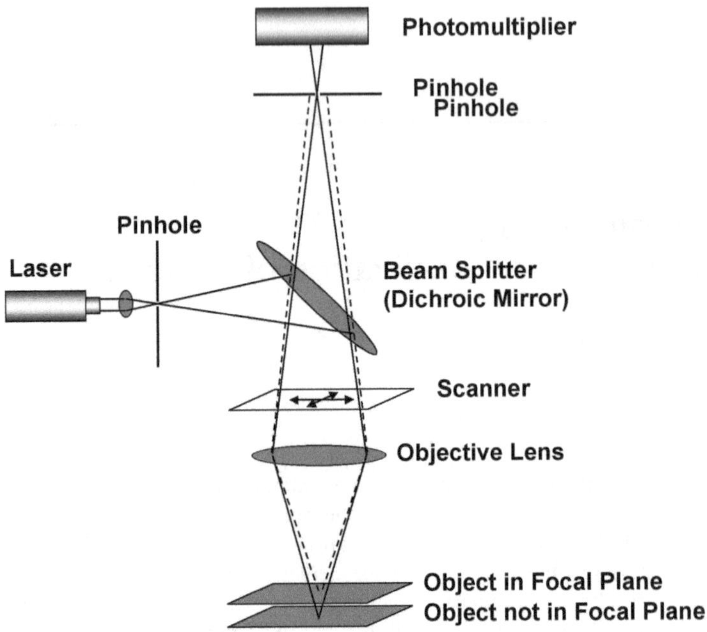

Fig. 1. Schematic diagram of the light path.

Therefore, high *Z*-axis resolution as well as *X-Y* image can be obtained. Fluorescence detection by sensitive photomultiplier tubes (PMT) allows the usage of filters with a narrow bandpath, resulting in minimal cross-talk (overlap) between two spectra. This is particularly important in demonstrating co-localization of proteins with multicolor labeling. In addition, digital images obtained with PMT are easy to record, modify, and transfer. Transmitted light images can be obtained by use of a transmitted light detector, but the images will not be confocal images. The transmitted light image can be used in combination with a fluorescence image to demonstrate fluorescent localization. It should kept in mind that co-localization of two proteins does not necessarily mean their physical association. Other methodologies, such as co-immunoprecipitation, should be combined to confirm the interaction.

2. Materials

1. Confocal laser scanning microscope (CLSM) (Olympus FLUOVIEW FV300 or equivalent).
2. Tissue culture equipment.
3. HeLa cells.
4. Plasmids encoding GFP-tagged or HA-tagged proteins of interest.

5. Lab-Tek II chamber slide with cover (Nalge Nunc International; Naperville, IL).
6. Effectene transfection reagent (Qiagen; Valencia, CA).
7. Phosphate-buffered saline (PBS).
8. 4% paraformaldehyde in PBS.
9. Permeabilization solution: 0.1% TritonX-100 in PBS.
10. Preblock solution: 10 m*M* Tris-HCl, pH 8.0, 150 m*M* NaCl, 0.1% Tween 20, 5% skim milk, 2% bovine serum albumin (should be filtered through a filter paper just before use).
11. Anti-HA mouse monoclonal antibody (HA.11, MMS-101R, CRP Inc.; Berkeley, CA) (*see* **Note 1**).
12. Secondary antibodies: TRITC (tetramethylrhodamine-isothiocyanate)-labeled rabbit anti-mouse immunoglobulin (R0270, Dako; Carpinteria, CA), Alexa Fluor 488-labeled goat anti-mouse immunoglobulin (A-11029, Molecular Probes; Eugene, OR), Alexa Fluor 546-labeled goat anti-rabbit immunoglobulin (A-11035, Molecular Probes).
13. Vectorshield® mounting medium (Vector Laboratories, Inc., Burlingame, CA).
14. Micro cover glass.
15. Rubber cement or nail polish.

3. Methods

The methods described below outline (1) the detection of transiently expressed tagged proteins and (2) the detection of endogenous proteins. To obtain multicolor images with optimum contrast and minimal cross-talk, appropriate combinations of fluorophores should be selected. The spectral properties of frequently used fluorophores and recommended lasers are listed in **Table 1**.

3. 1. Detection of Transiently Expressed GFP- and HA-Tagged Proteins

When antibodies raised against the protein of interest are unavailable for immunofluorescence, or the available cell lines do not express the protein of interest sufficiently enough for immunofluorescence, an alternative method is to transfect cells with expression plasmids that encode tagged proteins and stain the cells with anti-tag antibodies after transient transfection. However, it should be noted that the tagging of proteins of interest or their overexpression could potentially alter the intracellular localization or the function of the target protein. It is also mandatory to confirm the proper molecular weight of the expressed protein by Western blotting after verifying the plasmid by DNA sequencing *(1)*. To demonstrate that the protein of interest is localized in a particular organelle, a number of organelle-specific probes or vectors as well as organelle-specific monoclonal antibodies are available. To visualize mitochondria, for example, MitoTracker probes (Molecular Probes) and mitochondria localization vectors, such as pECFP-Mito (Clontech; Palo Alto, CA), can be used.

Table 1
Approximate Spectral Properties of Representative Fluorophores Used for Multicolor Labeling[a]

Fluorophore	Absorption (nm)	Emission (nm)	Laser	Notes
Group 1 (Green)				
FITC	490	512	Ar.488	Most widely used green fluorescent dye, easy to photobleach
EGFP	488	507	Ar.488	Brighter than wild-type GFP, for generating fusion proteins
Alexa Fluor 488	495	519	Ar.488	Brighter and more photostable than FITC
Group 2 (Red)				
TRITC	541	572	HeNe543 or Ar.Kr.568	Commonly used red fluorescent dye in combination with FITC
Texas Red	596	620	HeNe543 or Ar.Kr.568	Good spectral separation from FITC, slightly higher background staining *(2)*
Cy3	552	565	HeNe543 or Ar.Kr.568	Brighter than TRITC
Alexa Fluor 546	556	573	HeNe543 or Ar.Kr.568	Brighter than TRITC or Cy3
Propidium iodide	530	615	HeNe543 or Ar.Kr.568	For DNA/RNA staining
Group 3 (Blue)				
DAPI	345	425	UV.Ar.364	For nuclear staining
Hoechst33342	355	465	UV.Ar.364	For nuclear staining

[a] For multicolor labeling, choose one fluorophore from each group depending on the laser sources equipped in the CLSM, thus allowing up to three-color labeling.

3.1.1. Transfection

1. The day before transfection, seed HeLa cells at a density of 7×10^4 cells/well (when two-well chamber slides are used).
2. Incubate the cells at 37°C and 5% CO_2 in DMEM medium with 10% fetal bovine serum.
3. Prepare 0.5 μg of each plasmid that encodes GFP- or HA-tagged protein. The minimum DNA concentration should be 0.1 μg/μL.
4. Dilute the DNA in 60 μL of Buffer EC (Effectene transfection reagent kit). Add 6 μL of Enhancer and mix by vortexing for 1 s (*see* **Note 2**).
5. Incubate at room temperature for 2–5 min and spin down the mixture for a few seconds.
6. Add 10 μL of Effectene Transfection Reagent to the mixture. Mix by vortexing for 10 s.
7. Incubate the samples for 5–10 min at room temperature to allow complex formation.
8. During **step 7**, aspirate the medium from the slides and wash the cells once with PBS. Add 1 mL of fresh growth medium to each well.
9. Add the transfection complexes drop-wise onto the cells. Gently swirl the plates to ensure uniform distribution of the complexes.
10. Incubate the cells with the complexes at 37°C and 5% CO_2 for 24 h to allow for gene expression.

3.1.2. Immunofluorescence

1. Carefully aspirate the medium and add 0.5 mL/well of 4% paraformaldehyde in PBS drop-wise onto the cells (*see* **Note 3**).
2. To fix the cells, incubate the samples for 1 h at 4°C.
3. Carefully replace the solution with 0.5 mL/well of permeabilization solution.
4. Incubate the samples for 5 min at room temperature.
5. Carefully replace the solution with 0.5 mL/well of preblock solution and incubate the samples for 1 h at room temperature.
6. Carefully aspirate the preblock solution and add 0.5 mL/well of anti-HA mouse monoclonal antibody diluted with preblock solution at 1:200.
7. Incubate the samples for 1 h at room temperature (*see* **Note 4**).
8. Carefully aspirate the antibody solution and add 1 mL/well of PBS.
9. Incubate the samples for 10 min at room temperature. Gentle swirling facilitates washing efficiency.
10. Replace the solution with another 1 mL/well of fresh PBS and repeat **step 9**.
11. Carefully aspirate PBS and add 0.5 mL/well of TRITC-labeled rabbit anti-mouse immunoglobulin diluted with preblock solution. A 1:20 dilution is recommended.
12. Incubate the samples for 1 h at room temperature.
13. Carefully aspirate the antibody solution and wash the samples three times as described in **steps 8** and **9** (*see* **Note 5**).
14. Blot off excess PBS with a tissue, taking care not to allow the cells to dry out.

15. Place one drop of Vectorshield® over each well of cells and then place a micro cover glass over the entire well, taking care to minimize air bubbles.
16. Secure the cover glass with rubber cement or nail polish.

3.1.3. Confocal Laser Scanning Microscopy

Because the operation procedure of a CLSM depends on manufacturers and systems, it is not possible to generalize this procedure. This chapter is focused on the Olympus CLSM. The manufacturer's manual should be consulted for details.

1. Choose the correct combination of laser, barrier filters, and excitation dichroic mirrors for the dyes in use according to the manufacturer's recommendation. (For GFP/TRITC, the following combination is appropriate: laser combination: Ar488 nm + HeNe543 nm; barrier filters: BA510IF + BA530RIF for channel 1, BA565IF + BA590 for channel 2; dichroic mirror: DM488/543.)
2. Set the power switch of each unit to ON. To stabilize the laser beam output, allow the system to warm-up for at least 10 min after turning the laser power ON.
3. Start the FLUOVIEW software by double-clicking the FLUOVIEW icon on the desktop.
4. Place the specimen in an inverted position (in combination with an inverted microscope) on the microscope stage and turn the light path selector to "Binocular" section.
5. Engage the optimum cube for specimen dye by operating the cube turret and focus on the specimen initially using transmitted light and then quickly observe the cells with fluorescence.
6. When the area to be observed with confocal microscopy is determined, turn the light path selector to "Side port" position and rotate the cube turret so that no cube is engaged.
7. Choose the highest scan speed and set the zoom ratio to "X1." Set the channel to be acquired (brightness should be adjusted one channel at a time).
8. Click the "Focus" button to acquire repeated images at a high speed.
9. Focus on the cells of interest and adjust the image brightness. Parameters to be considered at this step are "PMT voltage," "Offset," and "Gain," each of which can be adjusted independently. The ND filters can also be selected using the LASER INTENSITY turret. Select the optimum ND filters according to the brightness of the specimen. A strong laser intensity, however, makes the fluorescence fade quickly.
10. When it is necessary to observe the detail of a specific area, the image of a limited area can be acquired by using the "Zoom" scale and the "Pan" buttons.
11. Once the appropriate parameters have been determined, stop repeated scanning, set a lower scan speed, and acquire the image by selecting the "Once" button (*see* **Notes 6** and **7**).
12. After the desired image has been acquired, save it to disk. It is possible to save images acquired with more than one channel at a time. The user can select the file

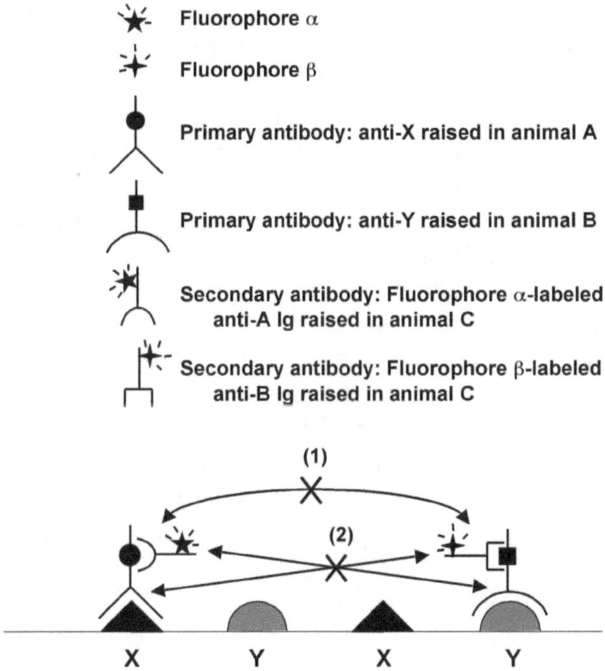

Fig. 2. Schematic diagram of multicolor detection of endogenous proteins.

types used for saving an image in a file. Fluoview Multi Tiff format is used for image analysis and processing on FLUOVIEW. Single TIF is a format for maximum portability between Macintosh and IBM PCs and programs such as Adobe Photoshop. To show the co-localization of two proteins, the merged image of the two proteins in a single view should be demonstrated as well as displaying images of the two channels side by side. For example, when the expression of a GFP-fused protein is displayed in green and that of an HA-tagged protein is shown in red, then co-localization of the two proteins is visualized in yellow in the merged image (**Figs. 3A–C**).

3. 2. Detection of Endogenous Proteins Using Labeled Secondary Antibodies

Demonstrating the intracellular co-localization of the two endogenous proteins is more convincing than showing subcellular localizations of epitope- or GFP-tagged proteins. However, the results should be obtained with highly specific primary antibodies. For example, antibodies that detect major cross-reactive bands in Western blotting are not suitable for immunofluorescence. Antibodies should also be affinity-purified prior to use to reduce background.

In addition, the cell lines used for this analysis should express the proteins of interest sufficiently enough for immunofluorescence. Two primary antibodies used in this experiment should be raised in two different species. Secondary antibodies used for simultaneous detection of more than one antigen should also be selected carefully according to the following criteria. (1) Secondary antibodies are derived from the same host species so that they do not recognize one another [**Fig. 2(1)**]. (2) They do not cross-react with other primary antibodies used in the assay system [**Fig. 2(2)**]. (3) They do not cross-react with endogenous proteins present in the cell lines under investigation. Secondary antibodies absorbed against the sera of a number of species to minimize cross-reactivity are commercially available. They are usually marked as "highly cross-absorbed" or "for multiple labeling." When cells growing in suspension are to be stained, they should be cytocentrifuged prior to the fixation using a Cytospin (Shandon; Pittsburgh, PA) or equivalent that allows low-speed centrifugal force to separate the cells on slides while maintaining cellular integrity.

1. Carefully aspirate the medium and add 0.5 mL/well of 4% paraformaldehyde in PBS dropwise onto the cells (*see* **Note 3**).
2. To fix the cells, incubate the samples for 1 h at 4°C.
3. Carefully replace the solution with 0.5 mL/well of permeabilization solution.
4. Incubate the samples for 5 min at room temperature.
5. Carefully replace the solution with 0.5 mL/well of preblock solution and incubate the samples for 1 h at room temperature.
6. Carefully aspirate the preblock solution and add 0.5 mL/well of anti-protein X mouse monoclonal antibody diluted with preblock solution. Because staining protocols vary with application, the appropriate dilution of antibody should be determined empirically.
7. Incubate the samples for 1 h at room temperature.
8. Carefully aspirate the antibody solution and add 1 mL/well of PBS.
9. Incubate the samples for 10 min at room temperature. Gentle swirling facilitates washing efficiency.
10. Replace the solution with another 1 mL/well of fresh PBS and repeat **step 9**.
11. Carefully aspirate PBS and add 0.5 mL/well of Alexa Fluor 488-labeled goat anti-mouse immunoglobulin diluted with preblock solution (1:200 or higher dilution).
12. Repeat **steps 7–10**.
13. Carefully replace the solution with 0.5 mL/well of preblock solution and incubate the samples for 1 h at room temperature. This step can be omitted depending on the antibody.
14. Carefully aspirate the preblock solution and add 0.5 mL/well of anti-protein Y rabbit polyclonal antibody diluted appropriately with preblock solution.
15. Repeat **steps 7–10**.
16. Carefully aspirate PBS and add 0.5 mL/well of Alexa Fluor 546-labeled goat anti-rabbit immunoglobulin diluted with preblock solution (1:200 or higher dilution).
17. Incubate the samples for 1 h at room temperature.

18. Carefully aspirate the antibody solution and wash the samples three times as described at **steps 8** and **9** (*see* **Note 5**).
19. Blot off excess PBS with a tissue, taking care not to allow the cells to dry out.
20. Place one drop of Vectorshield® over each well of cells and then place a micro cover glass over the entire well, taking care to minimize air bubbles.
21. Secure the cover glass with rubber cement or nail polish.

Confocal laser scanning microscopy can be performed using the same combinations of lasers, barrier filters, and dichroic mirror as described in **Subheading 3.1.3.** (channel 1 for Alexa Fluor 488, channel 2 for Alexa Fluor 546). Examples of multicolor labeling of HeLa cells are demonstrated in **Fig. 3D–I**.

4. Notes

1. Other representative anti-tag antibodies are described below. Recommended dilutions to start with are indicated in parentheses.
   a. Anti-FLAG (M2) mouse monoclonal antibody (1:100; F3165, Sigma; St. Louis, MO).
   b. Anti-His (His-probe, H-15) rabbit polyclonal antibody (1:200; sc-803, Santa Cruz; Santa Cruz, CA).
   c. Anti-c-Myc (9E10) mouse monoclonal antibody (1:100; sc-40, Santa Cruz).
2. To achieve optimal transfection efficiency for every new cell line/plasmid DNA combination, it is recommended to optimize the amounts of Effectene Reagent, DNA, and the cell number prior to transfection according to manufacturer's protocol. Other transfection reagents such as FuGene 6 (Roche) can also be used.
3. The alternative method to fix and permeabilize cells is to use methanol or acetone. The method should be selected empirically depending on the cell lines and antibodies.
4. Fluorophore-conjugated anti-tag antibodies are also available. In this case, go to **step 13**.
5. If the confocal laser scanning microscope is equipped with a UV laser, it is recommended to add 1.25 μ*M* of Hoechst33342 (**Table 1**) to the second wash for the nuclear counterstaining. This allows observation of the location of the nucleus.
6. To improve the image quality by reduction of noise, the Kalman accumulation is recommended. It allows acquiring of images for the specified number of times while averaging the images.
7. For a double-labeled specimen, there are two alternatives to obtain the image, sequential acquisition and simultaneous acquisition. With the former image-capturing method, the image slice of each laser excitation is obtained using one PMT at a time. This scan is of advantage in acquiring an image with a lower cross-talk between two wavelengths. In contrast, with a simultaneous scan, images without a time lag between two wavelengths are obtained.

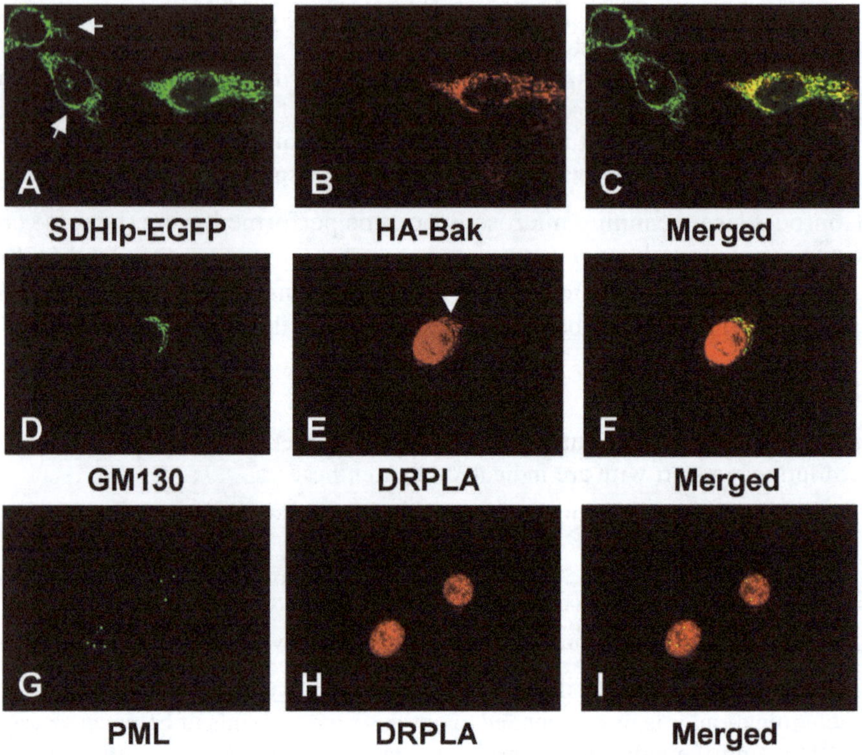

Fig. 3. Multicolor labeling of HeLa cells. **(A–C)** HeLa cells were transfected with pSDHIp-EGFP and pHA-Bak that encode the EGFP-tagged iron sulfur subunit of human succinate dehydrogenase *(3)* and HA-tagged proapoptotic Bcl-2 family member, Bak *(4,5)*, respectively. At 24 h after transfection, cells were stained with anti-HA antibody followed by TRITC-labeled secondary antibody. Panels **A** and **B** show the subcellular localizations of SDHIp and Bak, respectively. Merged pictures are shown in panel **C**. SDHIp is a mitochondrial protein and Bak is shown to co-localize with SDHIp. Arrows indicate cells transfected with only pSDHIp-EGFP. **(D–F)** HeLa cells were incubated with mouse anti-GM130 monoclonal antibody (BD transduction Laboratories, San Diego, CA) and rabbit anti-DRPLA antibody *(6)*. The primary antibodies were detected by secondary antibodies conjugated to Alexa Fluor 488 or 546. DRPLA protein, a product of the gene responsible for a genetic neurodegenerative disease, dentatorubral-pallidoluysian atrophy (DRPLA) *(7)*, is localized in the nucleus *(8)*. However, a closer observation reveals that it is also localized in the juxtanuclear region (arrowhead) where it co-localizes with GM130 **(F)**, which is a *cis*-Golgi network marker *(9)*. **(G–I)** HeLa cells were treated similarly as in **D–F** except that anti-PML mouse monoclonal antibody (Santa Cruz) was used as one of the primary antibodies. The promyelocytic leukemia (PML) protein is a major component of nuclear dot-like structures known as PML nuclear bodies (NBs) or PML oncogenic domain (POD) *(10,11)* **(G)**. These panels show that the nuclear DRPLA protein is not recruited to NBs or POD.

Acknowledgments

I thank Yuko Ohtsuka and Mami U for their technical assistance. I am also grateful to Drs. Yoshiaki Shikama and Yuko Okamura-Oho for their valuable discussions.

References

1. Sambrook, J. and Russell, D. W. (2001) *Molecular Cloning, A Laboratory Manual*, 3rd Ed. Cold Spring Harbor Laboratory Press, Cold Spring Harbor, New York, NY.
2. Wessendorf, M. W. and Brelje, T. C. (1992) Which fluorophore is brightest? A comparison of the staining obtained using fluorescein, tetramethylrhodamine, lissamine rhodamine, Texas red, and cyanine 3.18. *Histochemistry* **98,** 81–85.
3. Kita, K., Oya, H., Gennis, R. B., Ackrell, B. C., and Kasahara, M. (1990) Human complex II (succinate-ubiquinone oxidoreductase): cDNA cloning of iron sulfur (Ip) subunit of liver mitochondria. *Biochem. Biophys. Res. Commun.* **166,** 101–108.
4. Chittenden, T., Harrington, E. A., O'Connor, R., et al. (1995) Induction of apoptosis by the Bcl-2 homologue Bak. *Nature* **374,** 733–736.
5. Kiefer, M. C., Brauer, M. J., Powers, V. C., et al. (1995) Modulation of apoptosis by the widely distributed Bcl-2 homologue Bak. *Nature* **374,** 736–739.
6. Miyashita, T., Okamura-Oho, Y., Mito, Y., Nagafuchi, S., and Yamada, M. (1997) Dentatorubral pallidoluysian atrophy (DRPLA) protein is cleaved by caspase-3 during apoptosis. *J. Biol. Chem.* **272,** 29,238–29,242.
7. Nagafuchi, S., Yanagisawa, H., Ohsaki, E., et al. (1994) Structure and expression of the gene responsible for the triplet repeat disorder, dentatorubral and pallidoluysian atrophy (DRPLA). *Nat. Genet.* **8,** 177–182.
8. Miyashita, T., Nagao, K., Ohmi, K., Yanagisawa, H., Okamura-Oho, Y., and Yamada, M. (1998) Intracellular aggregate formation of dantatorubral-pallidoluysian atrophy (DRPLA) protein with the extended polyglutamine. *Biochem. Biophys. Res. Commun.* **249,** 96–102.
9. Nakamura, N., Rabouille, C., Watson, R., et al. (1995) Characterization of a cis-Golgi matrix protein, GM130. *J. Cell Biol.* **131,** 1715–1726.
10. Dyck, J. A., Maul, G. G., Miller, W. H. J., Chen, J. D., Kakizuka, A., and Evans, R. M. (1994) A novel macromolecular structure is a target of the promyelocyte-retinoic acid receptor oncoprotein. *Cell* **76,** 333–343.
11. Weis, K., Rambaud, S., Lavau, C., et al. (1994) Retinoic acid regulates aberrant nuclear localization of PML-RAR alpha in acute promyelocytic leukemia cells. *Cell* **76,** 345–356.

28

Mapping Biochemical Networks With Protein-Fragment Complementation Assays

Ingrid Remy and Stephen W. Michnick

Abstract

Cellular biochemical machineries, what we call pathways, consist of dynamically assembling and disassembling macromolecular complexes. Although our models for the organization of biochemical machines are derived largely from in vitro experiments, do they reflect their organization in intact, living cells? We have developed a general experimental strategy that addresses this question by allowing the quantitative probing of molecular interactions in intact, living cells. The experimental strategy is based on protein-fragment complementation assays (PCA), a method whereby protein interactions are coupled to refolding of enzymes from cognate fragments where reconstitution of enzyme activity acts as the detector of a protein interaction. A biochemical machine or pathway is defined by grouping interacting proteins into those that are perturbed in the same way by common factors (hormones, metabolites, enzyme inhibitors, and so on). In this chapter we review some of the essential principles of PCA and provide details and protocols for applications of PCA, particularly in mammalian cells, based on three PCA reporters, dihydrofolate reductase, green fluorescent protein, and β-lactamase.

Key Words

Protein fragment complementation assays; dihydrofolate reductase; green fluorescent protein; TEM β-lactamase; two-hybrid; protein–protein interactions; methotrexate; CCF2/AM; nitrocefin; fluorescein; flow cytometry; CHO; COS; HEK 293 cells.

1. Introduction

A first step in defining the function of a novel gene is to determine its interactions with other gene products in an appropriate context; that is, because proteins make specific interactions with other proteins as part of functional assemblies, an appropriate way to examine the function of the product of a novel gene is to determine its physical relationships with the products of other genes. This is the basis of the highly successful yeast two-hybrid system ***(1–6)***.

From: *Methods in Molecular Biology, vol. 261: Protein–Protein Interactions: Methods and Protocols*
Edited by: H. Fu © Humana Press Inc., Totowa, NJ

The central problem with two-hybrid screening is that detection of protein–protein interactions occurs in a fixed context, the nucleus of *Saccharomyces cerevisiae*, and the results of a screening must be validated as biologically relevant using other assays in appropriate cell, tissue, or organism models. Although this would be true for any screening strategy, it would be advantageous if one could combine library screening with tests for biological relevance into a single strategy, thus tentatively validating a detected protein as biologically relevant and eliminating false-positive interactions immediately. It was with these challenges in mind that our laboratory developed protein-fragment complementation assays (PCA). In this strategy, the gene for an enzyme is rationally dissected into two fragments. Fusion proteins are constructed with two proteins that are thought to bind to each other, fused to either of the two probe fragments. Folding of the probe protein from its fragments is catalyzed by the binding of the test proteins to each other, and is detected as reconstitution of enzyme activity. We have already demonstrated that the PCA strategy has the following capabilities: (i) Allows for the detection of protein–protein interactions in vivo and in vitro in any cell type; (ii) allows for the detection of protein–protein interactions in appropriate subcellular compartments or organelles; (iii) allows for the detection of induced versus constitutive protein–protein interactions that occur in developmental, nutritional, environmental, or hormone-induced signals; (iv) allows for the detection of the kinetic and equilibrium aspects of protein assembly in these cells; (v) allows for screening of cDNA libraries for protein–protein interactions in any cell type.

In addition to the specific capabilities of PCA described above, there are special features of this approach that make it appropriate for screening of molecular interactions, including the following: (i) PCAs are not a single assay but a series of assays; an assay can be chosen because it works in a specific cell type appropriate for studying interactions of some class of proteins. (ii) PCAs are inexpensive, requiring no specialized reagents beyond that necessary for a particular assay and off-the-shelf materials and technology. (iii) PCAs can be automated and high-throughput screening could be done. (iv) PCAs are designed at the level of the atomic structure of the enzymes used; because of this, there is additional flexibility in designing the probe fragments to control the sensitivity and stringencies of the assays. (v) PCAs can be based on enzymes for which the detection of protein–protein interactions can be determined differently including by dominant selection or production of a fluorescent or colored product.

The selection of enzymes and design of PCAs have been discussed in detail *(7)* and here we will review only the most basic ideas. Polypeptides have evolved to code for all of the chemical information necessary to spontaneously fold into a stable, unique three-dimensional structure ***(8–10)***. It logically follows that the folding reaction can be driven by the interaction of two peptides that together contain the entire sequence, and in the correct order of a single peptide that will fold. This was demonstrated in the classic experiments of Richards ***(11)*** and Taniuchi and Anfinsen ***(12)***. In practice this does not easily work because while the major driving force for protein folding is the hydrophobic effect, so also is nonspecific aggregation of unfolded peptides. However, if one adds soluble interacting proteins to the fragments that by interacting increase the effective concentration of the fragments, correct folding could be favored over any other nonproductive process ***(13–15)***. If the protein that folds from its constitutive fragments is an enzyme, whose activity could be detected in vivo, then the reconstitution of its activity can be used as a measure of interaction of the interacting proteins (**Fig. 1A**). Furthermore, this binary, all-or-none folding event, provides for a very specific measure of protein interactions dependent on not mere proximity, but on the absolute requirement that the peptides must be organized precisely in space to allow for folding of the enzyme from the cognate fragments. We select proteins to dissect into fragments that are not capable of spontaneously folding from their complementary fragment into a functional and complete protein. These facts distinguish the PCA strategy from complementation of naturally occurring and weakly associating subunits of enzymes ***(16)***, in which some spontaneous assembly occurs, as illustrated in **Fig. 1B**.

We have already developed five PCAs based on dominant-selection, colorimetric, or fluorescent outputs ***(7)***. Specific features and applications of PCAs are given in **Table 1**. Here, we will discuss the most well-developed PCA, based on the enzymes murine dihydrofolate reductase (mDHFR) and TEM β-lactamase, but also provide protocols for a new PCA based on green fluorescent protein (GFP).

The DHFR PCA can be used in a variety of applications to perform both simple survival-selection as readout and simultaneously, a fluorescent assay allowing quantitative detection and the cellular localization of protein interactions can be performed ***(17–19)***. The β-lactamase assay can be used as a very sensitive in vivo or in vitro quantitative detector of protein interactions as, unlike DHFR, one measures the continuous conversion of substrate to colored or fluorescent product ***(20)***. However it should be noted that generation of a product by an enzyme does not guarantee that signal to background would be superior to that of fixed fluorophore reporters like GFP and fluorescein-conjugated methotrexate (fMTX) bound to DHFR. Observable signal to background depends, for example, on the quantum yield of the fluorophore, retention of fluorophore by a cell, the optical properties of the cells used, and the extent to which fluorophores are retained in individual cellular compartments. For instance, in spite of no enzymatic amplification,

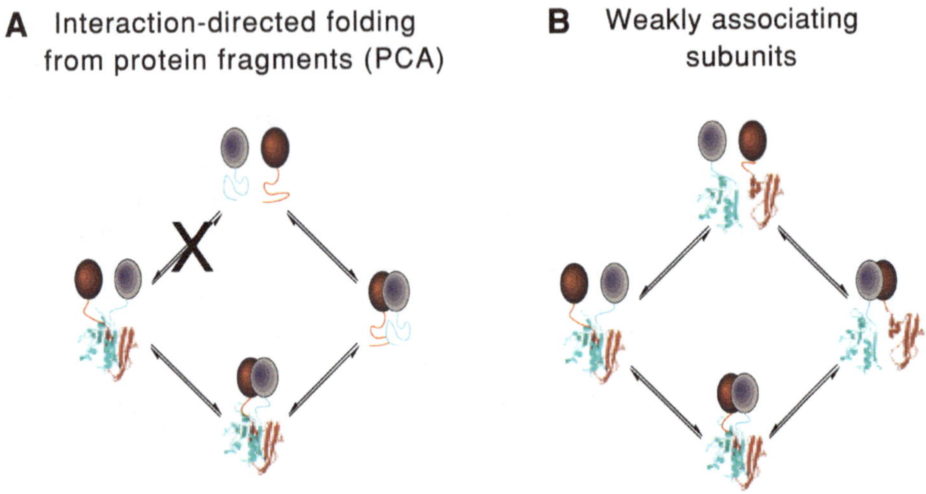

Fig. 1. Two alternative strategies to achieve complementation. **(A)** The PCA strategy requires that unnatural peptide fragments be chosen that are unfolded prior to association of fused interacting proteins. This prevents spontaneous association of the fragments (pathway X) that can lead to a false signal. **(B)** Naturally occurring subunits that are already capable of folding can be mutated to interact with lower affinity. However, to some extent, this will always occur, requiring the selection of cells that express protein partner fusions at low enough levels that background is not detected.

the DHFR fluorescence assay requires between only 1000 to 3000 molecules of reconstituted DHFR to clearly distinguish a positive response from background.

Reconstitution of DHFR activity can be monitored in vivo by cell survival in DHFR-negative cells (CHO-DUKX-B11, for example) grown in the absence of nucleotides. The principle of the DHFR PCA survival assay is that cells simultaneously expressing complementary fragments of DHFR fused to interacting proteins or peptides will survive in media depleted of nucleotides. This is an extraordinarily sensitive assay. In mammalian cells survival is dependent only on the number of molecules of DHFR reassembled, and we have determined that this number is approx 25 molecules of DHFR per cell *(17)*. The second approach is a fluorescence assay based on the detection of fMTX binding to reconstituted DHFR. The basis of the DHFR PCA fluorescence assay is that complementary fragments of DHFR, when expressed and reassembled in cells, will bind with high affinity (K_d = 540 pM) to fMTX in a 1:1 complex.

Table 1
Existing Protein-Fragment Complementation Assays (PCA) and Their Applications

Enzyme	Molecular weight	Function	Assays	Demonstrated Applications	Organism Restrictions
DHFR	21 kDa/monomeric	Reduces dihydrofolate to tetrahydrofolate	Fluorescence Survival selection	Survival selection Localization of protein interactions in living cells Quantitation of induced associations in vivo Translocation	None (Universal)
β-Lactamase	27 kDa/monomeric	Hydrolyzes β-lactam antibiotics (e.g., cephalosporin, ampicillin)	Survival selection in bacteria Fluorescence Colorimetric	In vivo fluorescence (e.g. with CCf2/AM) In vitro colorimetric assay (nitrocefin)	None (Universal)
GFP	28 kDa/monomeric	Spontaneously fluorescent protein	Intrinsic fluorescence	Localization of protein interactions in living cells Quantitation of induced associations in vivo Translocation	None (Universal)
Aminoglycoside phosphotransferase	35 kDa/monomeric	Phosphorylation of aminoglycosides (e.g., neomycin/G418) antibiotic	Survival selection in many cell types (bacteria, yeast mammalian, etc.)	Survival selection	None (Universal)
Hygromycin B phosphotransferase	35 kDa/monomeric	Phosphorylation of hygromycin B	Survival selection in many cell types (bacteria,yeast mammalian, etc.)	Survival selection	None (Universal)

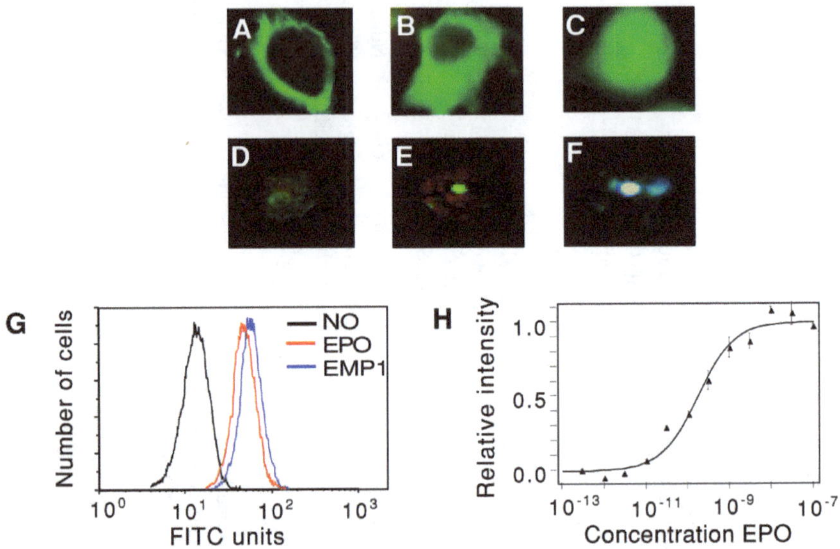

Fig. 2. Applications of the DHFR PCA to detecting the localization of protein complexes and quantitating protein interactions. **(A–C)** different protein pairs showing **A**, plasma membrane; **B**, cytosol; and **C**, whole cell localization in transiently transfected COS cells. **(D–F)** Localization of a protein complex in plant protoplasts. Potato protoplasts expressing two proteins implicated in response to salicylic acid (SA) fused to DHFR fragements. These are *NPR1/NIM1-DHFR (F[1,2])* and *TGA2-DHFR (F[3])* examined by fluorescent microscopy in the presence of fMTX and DAPI (4,6-diamidino-2-phenylindole). **(D)** A protoplast that has not been treated with SA or **(E)** treated with SA shows that that complex is induced to relocalize from cytosol to nucleus by SA. **(F)** Nuclear counterstaining with DAPI in the same protoplast. **(G)** FACS results of DHFR PCA. CHO cells expressing the erythropoietin receptor fused to complementary DHFR fragments. Receptor activation (conformation change) induced by erythropoietin (EPO) or a peptide agonist (EMP1) lead to an increase in fluorescence. **(H)** Dose–response curve for Epo-induced fluorescence as detected by FACS results in **G**.

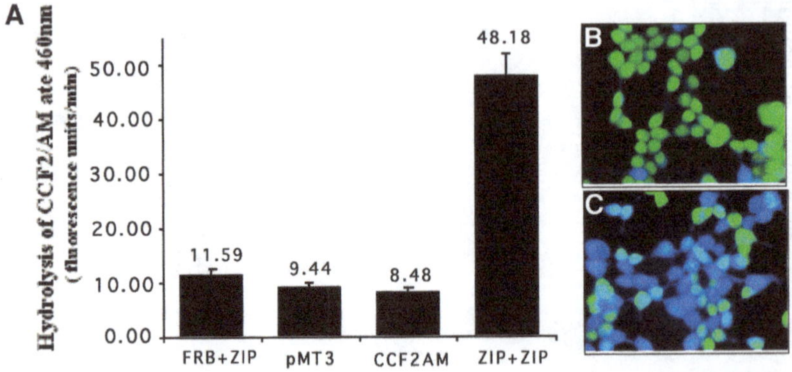

fMTX is retained in cells by this complex, while the unbound fMTX is actively and rapidly transported out of the cells ***(17,21,22)***. In addition, binding of fMTX to DHFR results in a 4.5-fold increase in quantum yield. Bound fMTX, and by inference reconstituted DHFR, can then be monitored by fluorescence microscopy, fluorescence-activated cell sorting (FACS), or spectroscopy ***(17–19)***. It is important to note that, although fMTX binds to DHFR with high affinity, it does not induce DHFR folding from the fragments in the PCA. This is because the folding of DHFR from its fragments is obligatory; if binding of the oligomerization domains does not induce folding, no binding sites for fMTX are created. Therefore, the number of complexes observed as measured by number of fMTX molecules retained in the cell is a direct measure of the equilibrium number of oligomerization domain complexes formed, independent of binding of fMTX ***(17)***. The other obvious application of the DHFR PCA fluorescence assay is in determining the location in the cell of interactions as illustrated in a number of cell types (**Fig. 2**). The GFP assay can be used for this purpose as well, but has the distinct advantage that no additional fluorophore is necessary to do this assay. However, readers should be cautioned that this assay is only appropriate for high-affinity, very stable complexes and applications to studying transient assembly and disassembly of protein complexes is very limited, owing to the slow folding of the protein and maturation of the fluorophore.

β-Lactamase is strictly a bacterial enzyme and has been genetically deleted from many standard *Escherichia coli* strains. It is not present at all in eukaryotes. Thus, the β-lactamase PCA can be used universally in eukaryotic cells and many prokaryotes, without any intrinsic background. Also, assays are based on catalytic turnover of substrates with rapid accumulation of product. This enzymatic amplification should allow for relatively weak molecular interactions to be observed. The assay can be performed simultaneously or serially in a number of modes, such as the in vitro colorimetric assay or the in vivo fluorescence assay (**Fig. 3**) or the survival assay in bacteria. Assays can be performed independent of the measurement platform and can easily be adapted to high-throughput formats requiring only one pipeting step.

Fig. 3. *(opposite page)* β-Lactamase PCA using the fluorescent substrate CCF2/AM. (**A**) ZIP (GCN4 leucine zipper-forming sequences) are tested in HEK 293 cells as described in the text. FRB (rapamycin-FKBP binding domain of FRAP) is used as a negative control. pMT3 is the expression vector alone and ZIP + ZIP is the positive control. Data recorded in white microtitre plates on a Perkin Elmer HTS 7000 plate reader. (**B,C**) Fluorescent micrographs of cells expressing β-lactamase PCA showing negative (B, FRB + ZIP) or positive (C, ZIP +ZIP) response.

2. Materials

2.1. DHFR PCA Survival Assay

1. Twelve-well plates, tissue culture treated (Corning Costar, cat. no. 3513); six-well plates, tissue culture treated (Corning Costar, cat. no. 3516).
2. Minimum essential medium: alpha medium without ribonucleosides and deoxyribonucleosides (α-MEM) (Life Technologies, cat. no. 12000022).
3. Dialyzed fetal bovine serum (Hyclone, cat. no. SH30079-03).
4. Adenosine (Sigma, cat. no. A-4036); desoxyadenosine (Sigma, cat. no. D-8668); thymidine (Sigma, cat. no. T-1895).
5. Lipofectamine Plus reagent (Life Technologies, cat. no. 10964013).
6. Trypsin-EDTA (Life Technologies, cat. no. 253100062).
7. Cloning cylinders (Scienceware, cat. no. 37847-0000).

2.2. DHFR PCA Fluorescence Assay

1. Twelve-well plates, tissue culture treated (Corning Costar, cat. no. 3513).
2. Dulbecco's modified Eagle's medium (DMEM) (Life Technologies, cat. no. 12100046); minimum essential medium: alpha medium without ribonucleosides and deoxyribonucleosides (α-MEM) (Life Technologies, cat. no. 12000022).
3. Cosmic calf serum (Hyclone, cat. no: SH3008703); dialyzed fetal bovine serum (Hyclone, cat. no. SH30079-03).
4. Lipofectamine Plus reagent (Life Technologies, cat. no: 10964013).
5. Fluorescein-conjugated methotrexate (fMTX) (Molecular Probes, cat. no. M-1198).
6. Dulbecco's phosphate-buffered saline (PBS) (Life Technologies, cat. no. 21600069).
7. Geltol aqueous mounting medium (Immunon, cat. no. 484950).
8. Trypsin-EDTA (Life Technologies, cat. no. 253100062).
9. Micro cover glasses, 18 mm circles, No. 2 (VWR Scientific, cat. no. 48382041).
10. Microscope slides, glass, 25 × 75 × 1.0 mm (any supplier).
11. 96-well white microtiter plates (Dynex no. 7905, VWR Scientific, cat. no. 62402-980).
12. Bio-Rad protein assay (Bio-Rad, cat. no. 500-0112).

2.3. GFP PCA Fluorescence Assay

1. Twelve-well plates, tissue culture treated (Corning Costar, cat. no. 3513).
2. Dulbecco's modified Eagle's medium (DMEM) (Life Technologies, cat. no. 12100046).
3. Cosmic calf serum (Hyclone, cat. no. SH3008703).
4. Lipofectamine Plus reagent (Life Technologies, cat. no. 10964013).
5. Dulbecco's phosphate-buffered saline (PBS) (Life Technologies, cat. no. 21600069).
6. Geltol aqueous mounting medium (Immunon, cat. no. 484950).
7. Trypsin-EDTA (Life Technologies, cat. no. 253100062).
8. Micro cover glasses, 18 mm circles, No. 2 (VWR Scientific, cat. no. 48382041).

9. Microscope slides, glass, 25 × 75 × 1.0 mm (any supplier).
10. 96-well black microtiter plates (Dynex no. 7805, VWR Scientific, cat. no. 62402-983).
11. Bio-Rad protein assay (Bio-Rad, cat. no. 500-0112).

2.4. β-Lactamase PCA Colorimetric Assay

1. Twelve-well plates, tissue culture treated (Corning Costar, cat. no. 3513).
2. Dulbecco's modified Eagle's medium (DMEM) (Life Technologies, cat. no. 12100046).
3. Cosmic calf serum (Hyclone, cat. no. SH3008703).
4. Fugene 6 transfection reagent (Roche Diagnostics, cat. no. 1814443).
5. Trypsin-EDTA (Life Technologies, cat. no. 253100062).
6. Dulbecco's phosphate-buffered saline (PBS) (Life Technologies, cat. no. 21600069).
7. Nitrocefin (Becton Dickinson Microbiology Systems, cat. no. 89-7065-0).
8. Dulbecco's phosphate-buffered saline (PBS) (Life Technologies, cat. no. 21600069)
9. 96-wells plates (Corning Costar, cat. no. 3595).

2.5. β-Lactamase PCA Fluorometric Assay

1. Twelve-well plates, tissue culture treated (Corning Costar, cat. no. 3513).
2. Dulbecco's modified Eagle's medium (DMEM) (Life Technologies, cat. no. 12100046).
3. Cosmic calf serum (Hyclone, cat. no. SH3008703).
4. Fugene 6 transfection reagent (Roche Diagnostics, cat. no. 1814443).
5. Trypsin-EDTA (Life Technologies, cat. no. 253100062).
6. Dulbecco's phosphate-buffered saline (PBS) (Life Technologies, cat. no. 21600069).
7. CCF2-AM (kindly provided by Roger Tsien).
8. 96-well white microtiter plates (Dynex no. 7905, VWR Scientific, cat. no. 62402-980).
9. Normal saline: 140 m*M* NaCl, 5 m*M* KCl, 2 m*M* $CaCl_2$, 10 m*M* HEPES, 6 m*M* sucrose, 10 m*M* glucose, pH 7.35.
10. Physiological saline solution: 10 m*M* HEPES, 6 m*M* sucrose, 10 m*M* glucose, 140 m*M* NaCl, 5 m*M* KCl, 2 m*M* $MgCl_2$, 2 m*M* $CaCl_2$, pH 7.35.
11. 15 mm glass coverslip (Ted Pella, Inc., cat. no. 26021).

3. Methods

3.1. DHFR PCA Survival Assay

1. Split CHO DUKX-B11 (DHFR-negative; could also be done in other cells lines, *see* **Note 1**) cells 24 h before transfection at 1×10^5 in 12-well plates in α-MEM medium enriched with 10% dialyzed fetal bovine serum and supplemented with 10 μg/mL of adenosine, desoxyadenosine, and thymidine.

2. Co-transfect cells with the PCA fusion partners (*see* **Note 2**) using Lipofectamine Plus reagent according to the manufacturer's instructions.
3. Forty-eight hours after the beginning of the transfection, split cells at approx 5×10^4 in six-well plates in selective medium consisting of α-MEM enriched with dialyzed FBS but without addition of nucleotides (*see* **Notes 3** and **4**).
4. Change medium every 3 d. The appearance of distinct colonies usually occurs after 4–10 d of incubation in selective medium. Colonies are observed only for clones that simultaneously express both interacting proteins fused to one or the other complementary DHFR fragment. Only interacting proteins will be able to achieve normal cell division and colony formation.

For further analysis of the interacting protein pair:

5. Isolate three to five colonies per interacting partners by trypsinization (trypsin-EDTA) using cloning cylinders and grow them separately.
6. Select the best expressing clone by immunoblot (Western blot) or using the DHFR PCA fluorescence assay (*see* **Subheading 3.2.**). Amplification of the expressed gene using methotrexate resistance can be done afterward if desired, to obtain clones with increased expression ***(23)***.
7. Carry out functional analysis of the clone stably expressing your interacting protein pair fused to the complementary DHFR fragments by using the DHFR PCA fluorescence assay.

3.2. DHFR PCA Fluorescence Assay

3.2.1. Fluorescence Microscopy

1. Split COS cells (this assay can be used with any other cell line, *see* **Note 5**) 24 h before transfection at 1×10^5 on 18-mm-circle glass coverslips in 12-well plates in DMEM medium enriched with 10% Cosmic calf serum.
2. Transiently co-transfect cells with the PCA fusion partners (*see* **Note 2**) using Lipofectamine Plus reagent according to the manufacturer's instructions.
3. The next day, change medium and add fMTX to the cells at a final concentration of 10 μ*M* (*see* **Note 6**).

   For CHO DUKX-B11 cells (or other cell line) stably expressing PCA fusion partners, seed cells to approx 2×10^5 on 18-mm glass coverslips in 12-well plates in α-MEM medium enriched with 10% dialyzed FBS. The next day, fMTX is added to the cells at a final concentration of 10 μ*M*.

4. After an incubation with fMTX of 22 h at 37°C, remove the medium and wash the cells with 1X PBS and re-incubate for 15–20 min at 37°C in the culture medium to allow for efflux of unbound fMTX (*see* **Note 7**). Remove medium and wash the cells four times with cold 1X PBS on ice and finally mount the coverslips on microscope glass slides with an aqueous mounting medium.
5. Fluorescence microscopy is performed on live cells (*see* **Note 8**). These experiments must be performed within 30 min of the wash procedure. If the negative control (untransfected cells treated with fMTX) is too fluorescent, the wash procedure must be modified (*see* **Note 9**).

3.2.2. Flow Cytometry Analysis

Preparation of cells for fluorescence-activated cell sorting (FACS) analysis is the same as described for fluorescence microscopy, except that following the 1X PBS wash (just twice in this case), cells are gently trypsinized (trypsin-EDTA), suspended in 500 µL of cold 1X PBS, and kept on ice prior to flow cytometric analysis within 30 min. Data are collected on a FACS analyzer with stimulation with an argon laser tuned to 488 nm with emission recorded through a 525 nm bandwidth filter.

3.2.3. Fluorometric Analysis

Preparation of cells for fluorometric analysis is the same as described for fluorescence microscopy, except that following the 1X PBS wash (just twice in this case), cells are gently trypsinized (trypsin-EDTA). Plates are put on ice and 100 µL of cold PBS is added to the cells. The total cell suspensions are transferred to 96-well white microtiter plates (Dynex) and kept on ice prior to fluorometric analysis. The assay can be performed on any microtiter plate reader; we use a Perkin-Elmer HTS 7000 Series Bio Assay Reader in the fluorescence mode. The excitation and emission wavelengths for the fMTX are 497 nm and 516 nm, respectively. Afterward, the data are normalized to total protein concentration in cell lysates (Bio-Rad protein assay).

3.3. GFP PCA Fluorescence Assay

All procedures describe for the DHFR PCA fluorescence assays are the same for GFP PCA, except that there is no use of fMTX and no washing steps. The wash procedure is obviously irrelevant in the case of the GFP PCA, where the folded/reassembled protein is a fluorophore itself.

3.4. In Vitro β-Lactamase PCA Colorimetric Assay

1. Split COS or HEK 293 T cells (this assay can be use with any other cell line) 24 h before transfection at 1×10^5 in 12-well plates in DMEM medium enriched with 10% Cosmic calf serum.
2. Transiently co-transfect cells with the PCA fusion partners (*see* **Note 10**) using Fugene 6 Transfection reagent according to the manufacturer's instructions.
3. Forty-eight hours after transfection, cells are washed three times with cold PBS and resuspended in 300 µL of cold PBS and kept on ice. Cells are then centrifuged at 4°C for 30 s, the supernatant discarded and cells resuspended in 100 µL of 100 m*M* cold phosphate buffer pH 7.4 (β-lactamase reaction buffer).
4. Freezing in dry ice/ethanol for 10 min and thawing in a waterbath at 37°C for 10 min, then lyse cells with three cycles of freeze and thaw. Cell membrane and debris are removed by centrifugation at 4°C for 5 min (10,000*g*). The supernatant whole cell lysate is then collected and stored at –20°C until assays are performed.

5. Assays are performed in 96-well microtiter plates. For testing β-lactamase activity, 100 μL of phosphate buffer (100 m*M*, pH 7.4) is allocated into each well. To this add 78 μL of H_2O and 2 μL of 10 m*M* nitrocefin (final concentration of 100 μ*M*). Finally, add 20 μL of unfrozen cell lysate (final buffer concentration of 60 μ*M*).
6. The assays can be performed on any microtiter plate reader; we use a Perkin-Elmer HTS 7000 Series Bio Assay Reader in the absorption mode with a 492 nm measurement filter.

3.5. In Vivo Enzymatic Assay and Fluorescent Microscopy With CCF2/AM

1. Split COS or HEK293 T cells 24 h before transfection at 1×10^5 in 12-well plates in DMEM medium enriched with 10 % Cosmic calf serum.
2. Transiently co-transfect cells with the PCA fusion partners (*see* **Note 10**) using Fugene 6 Transfection Reagent according to the manufacturer's instructions.
3. Twenty-four hours after transfection, cells are split again to ensure 50% confluency the following day (1.5×10^5) (*see* **Note 11**). The cells are split either onto 12-well plates for suspension enzymatic assay or onto 15-mm glass coverslips for fluorescent microscopy.
4. Forty-eight hours after transfection, cells are washed three times with PBS to remove all traces of serum (*see* **Note 12**).
5. Cells are then loaded with the following: 1 μ*M* of CCF2/AM diluted into a physiologic saline solution for 1 h.

For in vivo enzymatic assay:

6. Cells are then washed twice with the physiologic saline and resuspended into the same solution; 1×10^6 cells are aliquoted into a 96-well fluorescence white plate and are read for blue fluorescence with a Perkin Elmer HTS 7000 Series Bio Assay Reader in the fluorescence Top reading mode with a 409 nm excitation filter and a 465 nm emission filter.

For fluorescence microscopy:

7. Cells are washed twice with the physiologic saline as in **step 5**, prior to examination under the microscope (*see* **Note 13**).

We used two substrates to study the β-lactamase PCA. The first one is the cephalosporin called nitrocefin. This substrate is used in the in vitro colorimetric assay. β-Lactamase has a k_{cat}/k_m of 1.7×10^4 mM^{-1} s^{-1}. Substrate conversion can easily be observed by eye; the substrate is yellow in solution whereas the product is a distinct ruby red color. The rate of hydrolysis can be monitored quantitatively with any spectrophotometer by measuring the appearance of red at 492 nm. Signal to background, depending on the mode of measurement, can be greater than 30 to 1.

We have also developed an in vivo fluorometric assay using the substrate CCF2/AM ***(25,26)***. Although not as good a substrate as nitrocefin (k_{cat}/k_m of

1260 mM^{-1} s^{-1}), CCF2/AM has unique features that make it a useful reagent for in vivo PCA. First, CCF2/AM contains butyryl, acetyl, and acetoxymethyl esters, allowing diffusion across the plasma membrane where cytoplasmic esterases catalyze the hydrolysis of its ester functionality releasing the polyanionic (four anions) β-lactamase substrate CCF2. Because of the negative charge of CCF2, the substrate becomes trapped in the cell. In the intact substrate, fluorescence resonance energy transfer (FRET) can occur between a coumarin-donor and fluorescein-acceptor pair covalently linked to the cephalosporin core. The coumarin donor can be excited at 409 nm with emission at 447 nm, which is within the excitation envelope of the fluorescence acceptor (maximum around 485 nm), leading to remission of green fluorescence at 535 nm. When β-lactamase catalyzes hydrolysis of the substrate the fluorescein moiety is eliminated as a free thiol. Excitation of the coumarin donor at 409 nm then emits blue fluorescence at 447 nm, whereas the acceptor (fluorescein) is quenched by the free thiol.

4. Notes

1. Alternatively, recessive selection can be achieved in eukaryotic cells by using DHFR fragments containing one or more of several mutations (for example, F31S mutation, *see* below) that reduce the affinity of refolded DHFR to the antifolate drug methotrexate and growing cells in the absence of nucleotides with selection for methotrexate resistance.
2. The best orientations of the fusions for the DHFR PCA are protein A-DHFR[1,2] + protein B-DHFR[3] or DHFR[1,2]-protein A + protein B-DHFR[3], where proteins A and B are the proteins to test out for interaction. We typically insert a 10-amino-acid flexible polypeptide linker consisting of $(Gly.Gly.Gly.Gly.Ser)_2$ between the protein of interest and the DHFR fragment (for both fusions). DHFR[1,2] corresponds to amino acids 1–105, and DHFR[3] corresponds to amino acids 106–186 of murine DHFR. The DHFR[1,2] fragment that we use also contains a phenylalanine to serine mutation at position 31 (F31S), rendering the reconstituted DHFR resistant to methotrexate (MTX) treatment.
3. It is crucial that cell density is kept to a minimum and cells are well separated when split. To avoid cells "harvesting" nutrients from adjacent cells on dense plates or colonies might appear to be forming from clumps of cells that were not sufficiently separated during the splitting procedure.
4. The choice of dialyzed FBS manufacturer is crucial. Cells need very little nucleotide in the medium to propagate, and this will result in false-positives. The Hyclone-dialyzed FBS has proven a particularly reliable source.
5. The fluorescence DHFR PCA assay is universal and in theory can be used in any cell type or organism. This assay has already been shown to work in several mammalian cell lines as well as in plant and insect cells.
6. A stock solution of 1 m*M* fMTX should be prepared as follows: Dissolve 1 mg of fMTX in 1 mL of dimethyl formamide (DMF). To facilitate the dissolution, incu-

bate 15 min at 37°C and mix by vortexing every 5 min. Protect the tube from light. Keep at –20°C.

7. Complementary fragments of DHFR fused to interacting protein partners, when expressed and reassembled in cells, will bind with high affinity (K_d = 540 p*M*) to fMTX in a 1:1 complex. fMTX is retained in cells by this complex, while the unbound fMTX is actively and rapidly transported out of the cells.
8. All of the work reported to date has been performed in live cells. Although cells can be fixed, there is a significant reduction in observable fluorescence.
9. Particular attention must be given to optimizing the fMTX load and "wash" procedures. Important variables include the time of loading, temperatures at which each wash step is performed, the number and length of wash steps, and the time between washing and visualization. Too little washing will mean that background cannot be distinguished from a positive result. One should scrutinize the relevant parameters in the same sense as one would for, say, a Western blot. Results may also vary with the way the cells are plated and the types of cells used. Generally, as in other fluorescent microscopy procedures, the shape of cells and the localization of the fluorophore will result in better or worse results. For stable cell lines, the intensity of fluorescence will also depend on the levels of expression of the fusion proteins. The loading times and concentrations of fMTX (22 h, 10 μ*M*) used may result in a nonspecific and punctate fluorescence that is observed with any filter set. We do not know the source of this background, but it should not be mistaken for the real fluorescence signal produced by the PCA, which should be observed strictly with a filter that is optimal for observation of fluorescein. We have observed that loading fMTX for between 2 and 5 h at lower (5 μ*M*) concentrations prevents this nonspecific signal, although fewer cells are labeled. Loading times and concentrations must be optimized for specific cell types.
10. The best orientations of the fusions for the β-lactamase PCA are protein A-BLF[1] + protein B-BLF[2] or BLF[1]-protein A + protein B-BLF[2], where proteins A and B are the proteins to test out for interaction. We typically insert a 15-amino-acid flexible polypeptide linker consisting of $(Gly.Gly.Gly.Gly.Ser)_3$ between the protein of interest and the β-lactamase fragment (for both fusions). BLF[1] corresponds to amino acids 26–196 (Ambler numbering), and BLF[2] corresponds to amino acids 198–290 of TEM-1 β-lactamase.
11. The maximum loading efficiency of CCF2-AM is observed at 50% confluence.
12. Serum may contain esterases that can destroy the substrate.
13. We perform fluorescence microscopy on live HEK 293 or COS cells with an inverse Nikon Eclipse TE-200 (objective plan fluor 40× dry, numerically open at 0.75). Images were taken with a digital CCD cooled (–50°C) camera, model Orca-II (Hamamatsu Photonics) (exposure for 1 s, binning of 2 × 2, and digitalization 14 bits at 1.25 MHz). Source of light is a xenon lamp Model DG4 (Sutter Instruments). Emission filters are changed by an emission filter switcher (model Quantoscope) (Stranford Photonics). Images are visualized with ISee software (Inovision Corporation) on an O2 Silicon Graphics computer. The following

selected filters are used: filter set no. 31016 (Chroma Technologies); excitation filter: 405 nm (passing band of 20 nm); dichroic mirror: 425 nm DCLP; emission filter no. 1: 460 nm (passing band of 50 nm); emission filter no. 2: 515 nm (passing band of 20 nm).

References

1. Drees, B. L. (1999) Progress and variations in two-hybrid and three-hybrid technologies. *Curr. Opin. Chem. Biol.* **3,** 64–70.
2. Evangelista, C., Lockshon, D., and Fields, S. (1996) The yeast two-hybrid system—prospects for protein linkage maps. *Trends Cell Biol.* **6,** 196–199.
3. Fields, S. and Song, O. (1989) A novel genetic system to detect protein-protein interactions. *Nature* **340,** 245–246.
4. Vidal, M. and Legrain, P. (1999) Yeast forward and reverse ħ'-hybrid systems. *Nucleic Acids Res.* **27,** 919–929.
5. Walhout, A.J., Sordella, R., Lu, X., et al. (2000) Protein interaction mapping in *C. elegans* using proteins involved in vulval development. *Science* **287,** 116–122.
6. Uetz, P., Giot, L., Cagney, G., et al. (2000) A comprehensive analysis of protein-protein interactions in Saccharomyces cerevisiae. *Nature* **403,** 623–627.
7. Michnick, S.W., Remy, I., Campbell-Valois, F.-X., V.-Belisle, A., and Pelletier, J. N. (2000) Detection of protein-protein interactions by protein fragment complementation strategies, in *Methods in Enzymology*, Vol. 328, (Abelson, J. N., Emr, S. D., and Thorner, J., eds.), Academic Press, New York, NY, pp. 208–230.
8. Anfinsen, C. B., Haber, E., Sela, M., and White Jr., F. H. (1961) The kinetics of formation of native ribonuclease during oxidation of the reduced polypeptide chain. *Proc. Natl. Acad. Sci. USA* **47,** 1309–1314.
9. Anfinsen, C. B. (1973). Principles that govern the folding of protein chains. *Science* **181,** 223–230.
10. Gutte, B. and Merrifield, R. B. (1971) The synthesis of ribonuclease A. *J. Biol. Chem.* **246,** 1922–1941.
11. Richards, F. M. (1958). On the enzymatic activity of subtilisin-modified ribonuclease. *Proc. Natl. Acad. Sci. USA* **44,** 162–166.
12. Taniuchi, H. and Anfinsen, C. B. (1971) Simultanious formation of two alternative enzymically active structures by complementation of two overlapping fragments of staphylococcal nuclease. *J. Biol. Chem.* **216,** 2291–2301.
13. Pelletier, J. N., Campbell-Valois, F., and Michnick, S. W. (1998). Oligomerization domain-directed reassembly of active dihydrofolate reductase from rationally designed fragments. *Proc. Natl. Acad. Sci. USA* **95,** 12,141–12,146.
14. Pelletier, J. N. and Michnick, S. W. (1997) A protein complementation assay for detection of protein–protein interactions *in vivo. Protein Eng.* **10,** 89.
15. Johnsson, N. and Varshavsky, A. (1994) Split ubiquitin as a sensor of protein interactions in vivo. *Proc. Natl. Acad. Sci. USA* **91,** 10,340–10,344.

16. Rossi, F., Charlton, C. A., and Blau, H. M. (1997) Monitoring protein-protein interactions in intact eukaryotic cells by beta-galactosidase complementation. *Proc. Natl. Acad. Sci. USA* **94,** 8405–8410.
17. Remy, I. and Michnick, S. W. (1999) Clonal selection and in vivo quantitation of protein interactions with protein fragment complementation assays. *Proc. Natl. Acad. Sci. USA* **96,** 5394–5399.
18. Remy, I., Wilson, I. A., and Michnick, S. W. (1999) Erythropoietin receptor activation by a ligand-induced conformation change. *Science* **283,** 990–993.
19. Remy, I. and Michnick, S. W. (2001) Visualization of biochemical networks in living cells. *Proc. Natl. Acad. Sci. USA* **98,** 7678–7683.
20. Galarneau, A., Primeau, M., Trudeau, L. E., and Michnick, S. W. (2002) beta-Lactamase protein fragment complementation assays as in vivo and in vitro. *Nat. Biotechnol.* **20,** 619–622.
21. Israel, D. I. and Kaufman, R. J. (1993) Dexamethasone negatively regulates the activity of a chimeric dihydrofolate reductase/glucocorticoid receptor protein. *Proc. Natl. Acad. Sci. USA* **90,** 4290–4294.
22. Kaufman, R. J., Bertino, J. R., and Schimke, R. T. (1978) Quantitation of dihydrofolate reductase in individual parental and methotrexate-resistant murine cells. Use of a fluorescence activated cell sorter. *J. Biol. Chem.* **253,** 5852–5860.
23. Kaufman, R. J. (1990) Selection and coamplification of heterologous genes in mammalian cells. *Methods Enzymol.* **185,** 537–566.
24. O'Callaghan, C. and Morris, A. (1972) Inhibition of beta-lactamases by beta-lactam antibiotics. *Antimicrob. Agents Chemother.* **2,** 442–448.
25. Zlokarnik, G., Negulescu, P. A., Knapp, T.E., et al. (1998). Quantitation of transcription and clonal selection of single living cells with beta-lactamase as reporter. *Science* **279,** 84–88.
26. Zlokarnik G. (2000) Fusions to beta-lactamase as a reporter for gene expression in live. *Methods Enzymol.* **326,** 221–244.

29

In Vivo Protein Cross-Linking

Fabrice Agou, Fei Ye, and Michel Véron

Abstract

In the cell, homo- and heteroassociations of polypeptide chains evolve and take place within subcellular compartments that are crowded with many other cellular macromolecules. In vivo chemical cross-linking of proteins is a powerful method to examine changes in protein oligomerization and protein–protein interactions upon cellular events such as signal transduction. This chapter is intended to provide a guide to the selection of the cell-membrane-permeable cross-linkers, the optimization of in vivo cross-linking conditions, and the identification of specific cross-links in a cellular context where the frequency of random collisions is high. By combining the chemoselectivity of the homo-bifunctional cross-linker and the length of its spacer arm with knowledge on the protein structure, we show that selective cross-links can be introduced specifically on either the dimer or the hexamer form of the same polypeptide in vitro as well as in vivo, using the human type B nucleoside diphosphate kinase as a protein model.

Key Words

In vivo cross-linking; oligomerization; NDPK-B; transcription factors; oncogenenesis.

1. Introduction

In vitro chemical cross-linking is a powerful technique that has been widely used for protein–protein interaction studies *(1–4)*. The literature abounds with examples of cross-linking experiments performed with pure or enriched proteins to determine their quaternary structure, to stabilize their function, to connect useful chemical entities, or to favor a conformational state of protein associated with desirable properties. In contrast, much less can be found on in vivo cross-linking performed with membrane-permeable cross-linkers on intact cells. This is mainly due to the difficulty of targeting intramolecular cross-links in specific proteins in a cellular environment like that of *Escherichia coli*, in which the total concentration of protein and RNA is 300–400 g/L *(5)*. Such a situation increases the frequency of random collisions that may lead to nonspe-

From: *Methods in Molecular Biology, vol. 261: Protein–Protein Interactions: Methods and Protocols*
Edited by: H. Fu © Humana Press Inc., Totowa, NJ

cific intermolecular cross-linking. For example, tetrameric hemoglobin was shown to oligomer up to octamers when intact erythrocyte cells were treated with a membrane-permeable cross-linker ***(6)***. Using structural information provided by X-ray crystallography or nuclear magnetic resonance (NMR), it is possible to overcome some of the problems related to cross-linking in living cells. Indeed, more selective cross-links can be introduced in a protein taking into account the spatial arrangement of its nucleophilic residues.

This chapter will present a brief general survey of in vivo cross-linking approaches based on structural design to probe the quaternary structure of the human type B nucleoside diphosphate kinase, an enzyme that is involved in the maintenance of the cellular nucleoside triphosphate (NTPs) pool ***(7,8)***. This enzyme has also been identified as a transcription factor that is involved in the regulation of the *c-myc* gene ***(9)***. Only the hexameric enzyme is active in the NTPs synthesis, whereas the dimeric protein is not ***(10,11)***. The dimer binds to DNA with a higher affinity than the hexamer and may then act as a more efficient transcription factor than the hexamer ***(12)***. As this dual activity depends on hexameric and dimeric structures, different oligomeric states of the protein were investigated in living cells. In the first attempt, S100 extracts from different tumor cells were analyzed by gel filtration method. However, the resolution of the gel was too low to separate the different oligomeric forms owing to the interconversion of different species during the protein elution. Thus, we developed an in vivo cross-linking method to monitor the quaternary structure of the NDPK-B.

We first showed by in vitro cross-linking that specific cross-links on cysteine residues could be introduced selectively on the NDPK dimer or hexamer, depending on the length of the spacer arm of the homobifunctional cross-linker. We used this difference in the protein reactivity with long and short cross-linkers to establish an in vitro direct correlation between the oligomeric state of protein and the amount of cross-linked dimeric subunits. Differential cross-linking was then performed with long and short membrane-permeable cross-linkers in HeLa cells after overexpressing either the dimer or the hexamer. A positive correlation could be established between the amount of an over-expressed dimeric mutant and the amount of cross-links quantified by immunoblotting. This shows that our in vivo protein cross-linking method based on structural design is highly selective and can be used to probe the quaternary structure of protein in intact cells.

2. Materials

2.1. Purified Recombinant Proteins

The recombinant wild-type hexamer was purified as described in ref. ***13***. The recombinant dimeric mutant P96S-A146 stop was constructed by overlap extension method ***(14)***. The point mutation P96S and the removal of the seven C-terminal amino acids in the crystallographic hexamer interface have been shown to greatly affect hexamerization of the NDPK ***(11)***. The double mutant was purified according to a procedure similar to the one used for the wild-type protein except that all purification buffers contained 0.05 m*M* of dodecyl-β-D-maltoside (DDM) detergent. Proteins were stored at –20°C in 20 m*M* potassium phosphate pH 7.0 buffer containing 1 m*M* dithioerythritol (DTE) and 50% (v/v) glycerol.

2.2. SDS-PAGE

1. Laemmli SDS-PAGE equipment and buffers (see Chapter 12).
2. Coomassie R250 dye and Coomassie blue staining buffers.

2.3. Western Blotting

1. Western blotting equipment: nitrocellulose membrane, transfer chamber, filter paper, X-ray film (X-Omat AR; Kodak).
2. Western blotting buffers: Tris-glycine transfer buffer, pH 8.3, blocking and washing buffers.
3. Specific antibodies against the protein of interest and anti-IgG-peroxidase (secondary antibody). In our experiments we routinely use polyclonal anti-NDPK-B antibodies at a final concentration of 10 μg/mL.
4. Enhanced chemiluminescence Plus reagents (ECL Plus, Amersham Biosciences) with a laser scanner system such as Storm (Amersham Biosciences) for detection and quantification.

2.4. Protein Cross-Linking via Cysteine Residues

1. Membrane permeable sulfydryl-reactive homobifunctional cross-linkers:
   a. Bis-maleimidoethane (BMOE; Pierce).
   b. Bis-maleimidohexane (BMH; Pierce) freshly dissolved in DMSO.
2. Membrane-impermeable and water-soluble sulfhydryl-reactive homobifunctional cross-linker: 1,8-bis-maleimidotriethyleneglycol (BM[PEO_3], Pierce) freshly dissolved in water according to instructions from Pierce.
3. Quenching and reaction buffers for in vitro cross-linking:
   a. 2X quenching buffer: 2X Laemmli loading buffer containing 100 m*M* DTE.
   b. 2X reaction buffer: 100 m*M* potassium phosphate, pH 7.2, and 300 m*M* potassium chloride.
4. 6X quenching buffer for in vivo cross-linking: 180 m*M* DTE in water.

2.5. Cell Culture

1. Cell culture equipment: plastic culture vessels, laminar flow hood, CO_2 incubator, water bath.

2. Complete culture medium: Dulbecco's modified Eagle's medium (DMEM), fetal calf serum (FCS).
3. HeLa cells (American Type Culture Collection, ATCC CCL-2) in serum (10% FCS) supplemented DMEM.
4. Ca^{2+}- and Mg^{2+}-free phosphate-buffered saline (CMF-PBS).
5. Trypsin-EDTA solution in CMF-PBS.

2.6. DNA Transfections

1. Mammalian expression vectors bearing the cytomegalovirus promoter (pcDNA3, Invitrogen).
2. Cationic liposome reagent (FuGene 6, Roche).

2.7. Making Whole Cell Extracts

1. Urea lysis buffer (SBLU): 62.5 m*M* Tris-HCl, pH 7.8, 2 % SDS, 6 *M* urea, 100 m*M* DTE, and 0.05 % bromophenol blue.

3. Methods

3.1. General Considerations: In Vivo Selectivity in Cross-Linking Based on Structural Design

Cross-links can be more efficiently introduced in a protein when the protein's three-dimensional structure is known. The criteria to be considered for optimal selectivity in in vivo protein cross-linking (hetero- or homobifunctional reagent; type of reactive group; length and rigidity of the linker structure) are almost the same as those previously described in vitro ***(1,4)***, except for two:

1. The cross-linker must diffuse across the membrane, and therefore its linker structure should be mostly hydrophobic.
2. Individual amino acids likely to react with the cross-linker should be limited in the 3D space to minimize nonspecific intermolecular bridges in vivo (*see* **Note 1**). The use of a low cells:cross-linker ratio is also advised to limit the extent of protein modifications, thereby enhancing the relative differences in chemical reactivity between the amino acids.

We use a noncleavable, membrane-permeable, homobifunctional reagent containing at both ends maleimide groups for introducing selective cysteine cross-links in vivo. In addition to the advantages of cysteine-based cross-linking, the extent of protein modification through cysteine residues is usually reduced owing to the low content of cysteine in proteins, minimizing nonspecific intermolecular cross-linking in vivo ***(15,16)***.

We chose human NDPK-B as a model protein because (i) its crystal structure has already been determined ***(17)*** and (ii) the protein only contains two cysteine residues (C145 and C109), which are differentially exposed to solvent in dimeric and hexameric states (*see* **Fig. 1** for details).

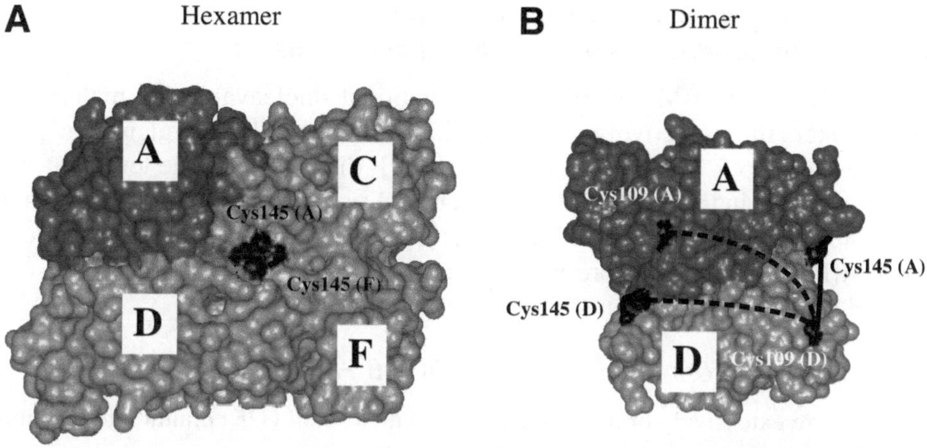

Fig. 1. Protein cross-linking based on structural design to probe the quaternary structure. **(A)** Side view of the hexameric structure of the NDPK-B. Out of the six identical subunits which compose the NDPK-B hexamer, only four subunits are clearly visible in this Connelly surface representation and are referred as *A*, *D*, *C*, and *F* subunits. **(B)** Connelly surface representation of the dimer rotated by 180° around the threefold axis (hexameric interface view). Each subunit contains two cysteines C109 and C145 (colored in black), which are differently exposed to solvent in dimeric and hexameric states. In the hexamer, only the C145 cysteines are solvent-accessible and close in the 3D space (4.5 Åhexameric contact between A chain and F chain), whereas in the dimer, C109 (*A*) of one monomer and C145 (*D*) of the second monomer are solvent-accessible and are distant from each other by 14.6 Åsolid line). All other cysteine pairs in the dimer are separated by more than 31 Ånd are indicated by dotted lines. Given the different spatial arrangement of cysteine pairs in the dimeric and hexameric states, selective cross-linking can be introduced on the dimer and on the hexamer using a homobifunctional reagent varying only in length and flexibility of the spacer arm that connects both specific cysteine reactive groups.

Whatever the protein studied, those who wish to use commercially available reagents are advised to read the Pierce reagent catalog, which includes a complete technical section and a discussion on cross-linking. This catalog, which is also accessible via the internet (http://www.piercenet.com; *Cross-linkers Selection Guide*) is a practical starting point for those who have no idea which reagents and protocols to cross-link their protein specifically in vivo, taking also into account all the criteria we have listed above.

3.2. In Vitro Protein Cross-Linking of NDPK-B via Cysteine Residues Using Variable Spacer Arms

BM[PEO]$_3$ and BMOE are homobifunctional uncleavable bis-maleimide cross-linkers that selectively react with thiols to form stable covalent thioether linkages at pH 6.5–7.5. BMOE and BM[PEO]$_3$ differ only in the length of the spacer arm and in their solubility in water. The two maleimide groups are 14.7 and 8 Å apart in BM[PEO]$_3$ and BMOE, respectively. In the following section, the BM[PEO]$_3$ will therefore be referred to as the long cross-linker and the BMOE as the short cross-linker.

3.2.1. In Vitro Cross-Linking of the NDPK-B Hexamer (see **Fig. 2**)

1. Dialyze extensively or desalt the protein with a small G25 column into reaction buffer to remove any trace of reducing agents.
2. Dissolve the long cross-linker in water according to Pierce instructions and the short cross-linker in DMSO, both at 1 m*M*.
3. Prepare several Eppendorf tubes each containing 25 µL of 2X reaction buffer, 1 or 10 µg of the protein of interest, and water to a final volume of 50 µL. Add also 10 µL of 2X quenching buffer in one extra tube to control the quenching efficiency.
4. Start the reaction by adding 5 µL of the cross-linker solution (1/10 volume) and mix well by vortexing.
5. Stop the reaction at different times by adding 10 µL of 2X quenching buffer and stir vigorously by vortexing.
6. Heat samples 3–5 min at 100°C.
7. Load samples on a Laemmli gel with a desired percentage of acrylamide. Generally, use 5% gels for SDS-denatured proteins of 60–200 kDa, 10% gels for 16–70 kDa, and 15% gels for 12–45 kDa. Visualize different cross-linked species with Coomassie blue staining according to the standard protocol (*see* **Note 2**).

3.2.2. In Vitro Cross-Linking of the NDPK-B Dimeric Mutant (P96S-A146stop) (see **Fig. 3**)

Gel filtration analyses show that the purified P96S-A146stop mutant forms a stable dimer in the protein concentration range of 0.5–50 µ*M* (subunit concentration). Experimental procedures to cross-link the NDPK-B dimeric mutant with the long and short cross-linkers are similar to those used for the wild type hexamer except that the reaction buffer contains 0.05 m*M* of DDM detergent, to stabilize the dimer.

Obviously the in vitro characterization of protein cross-linking with purified components as illustrated in **Fig. 4** is not a prerequisite step to carry out in vivo cross-linking. This is shown here for a direct comparison of the selectivity of protein cross-linking in vitro and in a cellular context.

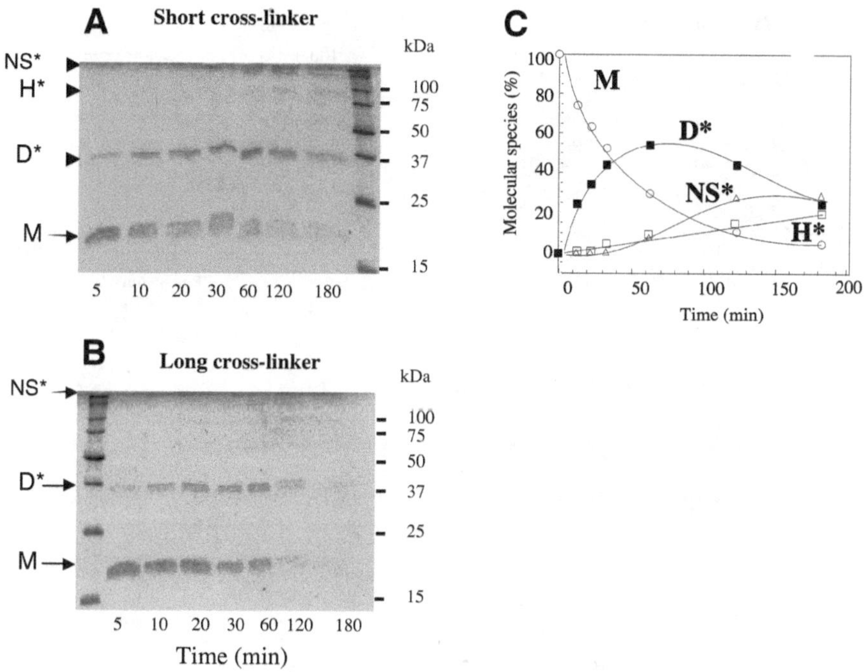

Fig. 2. In Vitro cross-linking of NDPK-B hexamer with short and long cysteine specific cross-linkers. NDPK-B, 0.02 mg/mL (1.2 μ*M*, subunit concentration), was reacted with either **(A)** the short (BMOE) or **(B)** the long cross-linker (BM[PEO]$_3$) using a concentration of 0.1 m*M* following the experimental procedure described in **Subheading 3.2.1.** At the time indicated, the reaction was quenched by the SDS-PAGE loading buffer and samples were subjected to 15% SDS-PAGE stained by Coomassie staining. M, D*, H*, and NS* refer to the positions of monomers, cross-linked dimers, cross-linked hexamers, and nonspecific higher-order cross-linked species of NDPK-B subunits, respectively. **(C)** Scans of the SDS-PAGE gel shown in **A** in order to quantify cross-linked species. The hexameric cross-linked species was very low as compared to the cross-linked dimer even though the protein at 0.02 mg/mL was a hexamer as judged by the equilibrium sedimentation ***(13)***. This reflects the very low efficiency of these short and long cross-linkers to cross-bridge all six subunits together. Patterns were similar using a higher concentration of cross-linker or using different protein concentrations (0.002 or 0.2 mg/mL), confirming the low cross-linking efficiency in the formation of the hexameric cross-linked species.

3.3. Cross-Linking of NDPK-B in Living Cells (see **Note 3**)

Because the long cross-linker BM[PEO]$_3$ is membrane-impermeable, an analogous reagent such as BMH that crosses the cell membrane is used for in vivo protein cross-linking. Like BM[PEO]$_3$, BMH is a thiol-specific, non-

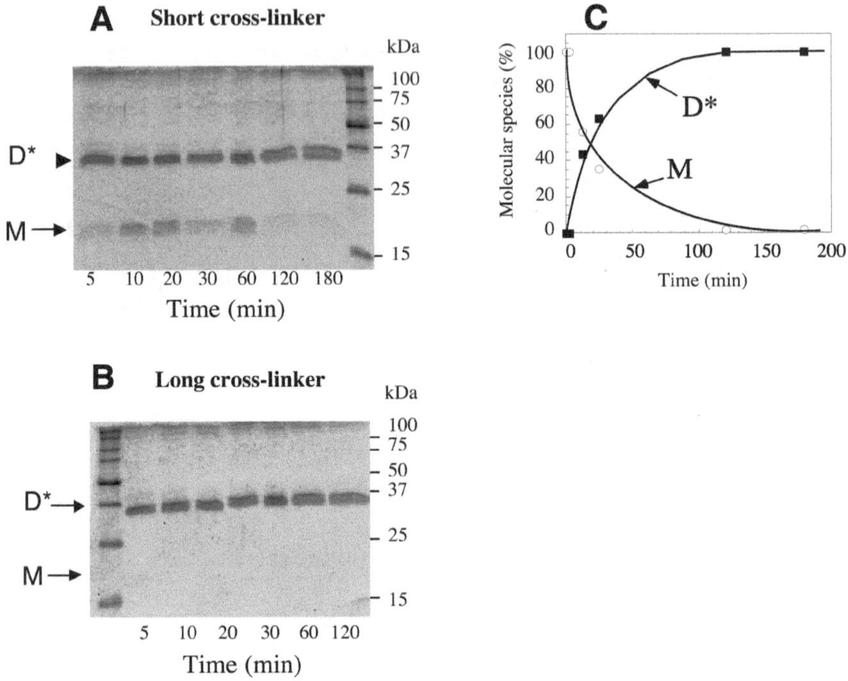

Fig. 3. In vitro cross-linking of the dimeric mutant of NDPK-B with short and long cysteine specific cross-linkers Purified dimeric mutant, 0.02 mg/mL (1.2 μ*M*, subunit concentration) was treated with either the short (BMOE) (**A**), or the long cross-linker (BM[PEO]$_3$) (**B**), according to the instructions described in **Subheading 3.2.2.** and using 0.1 m*M* of the cross-linker. The reaction was quenched by the SDS-PAGE loading buffer at the time indicated. Cross-linked species were then analyzed after SDS-PAGE and Coomassie blue staining. M and D* denote the monomer and the cross-linked dimer. (**C**) Scans of the SDS-PAGE gel shown in **A**. The graph represents the percentage of each species with time and each curve was fitted with a monoexponential (solid line) (*see* **Note 3**). Scans of the SDS-PAGE gel shown in **B** is not represented under these experimental conditions because the reaction was too fast.

cleavable, homobifunctional cross-linker that connects both maleimide groups via a spacer arm whose length is close to that of BM[PEO]$_3$ (16.1 Å In the following protocol of in vivo protein cross-linking, the long and short cross-linkers are referred to BMH and BMOE, respectively. To check whether overexpressing the stable dimeric mutant P96S-C145stop in HeLa cells induces a shift toward an hexameric state, GFP fusion proteins with the hexamer wild type or the dimeric mutant are made. After transient transfections of HeLa cells, analysis by fluorescence imaging shows different protein localization patterns.

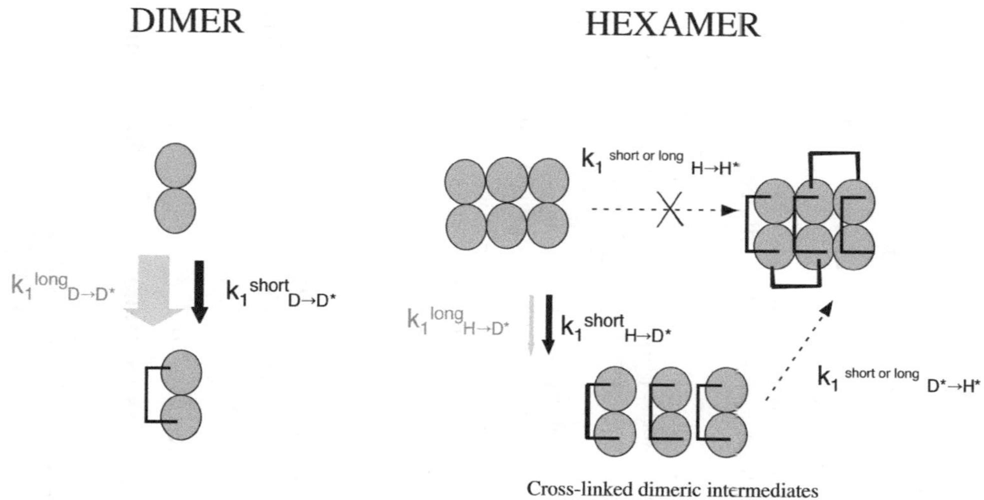

Fig. 4. Diagram of in vitro cross-linking efficiencies of NDPK-B with short and long cysteine-specific cross-linkers as a function of the oligomeric states. Bimolecular rate constants relative to short (BMOE) and long ($BM[PEO]_3$) cross-linkers with the dimer and the hexamer were determined using three independent cross-linker concentrations as described in **Subheading 3.2.** (*see* **Note 3**). The arrow surfaces relative to cross-linking reactions with short (in black) and long cross-linkers (in gray) are proportional to the magnitude of each rate constant. The reaction $D \rightarrow D^*$ is the formation of the dimeric mutant into the cross-linked dimer; $H \rightarrow D^*$ is the formation of the hexamer into two cross-linked subunits; $D^* \rightarrow H^*$ is the formation of the cross-linked dimer into the cross-linked hexamer; and $H \rightarrow H^*$ is the direct conversion of the wild-type into the cross-linked hexamer. The rate constants $k_1^{long}\ D \rightarrow D^*$ and $k_1^{short}\ D \rightarrow D^*$ were $900 \pm 50\ M^{-1}\ s^{-1}$ and $9.6 \pm 0.5\ M^{-1}\ s^{-1}$, respectively, meaning that the cross-linking efficiency on the dimer is enhanced about 100-fold with the long cross-linker compared to the short cross-linker. The cross-linking efficiency corresponding to the formation of two cross-linked subunits within the hexamer is lower with both reagents. Bimolecular rate constants, $k_1^{long}\ H \rightarrow D^*$ and $k_1^{short}\ H \rightarrow D^*$, were $0.9\ M^{-1}\ s^{-1}$ and $3\ M^{-1}\ s^{-1}$, respectively, indicating that the short cross-linker is slightly more effective as compared to the long cross-linker in the hexamer cross-linking. Nonmeasurable rates of the cross-linked hexamer or the intermediate cross-linked species (trimer, tetramer, or pentamer) are shown as dotted arrows.

Whereas the wild type hexamer-GFP fusion protein is mainly located in the cytosol, the dimeric mutant-GFP fusion protein is present in both compartments (nucleus and cytosol, data not shown). These results suggest that the dimeric mutant is not totally converted into a hexamer when highly expressed in HeLa cells.

3.3.1. Transient Transfections of HeLa Cells to Overexpress the Dimeric Mutant and the Wild-Type Hexamer

1. The day before the transfections, seed HeLa cells at the appropriate plating density in a complete medium so that they are 50–60% confluent on the day of the transfection. For cationic lipid-mediated transfections, a six-well plate is used, with each well containing 0.2×10^6 HeLa cells in a total complete media volume of 2 mL (DMEM, 10% FCS).
2. On the day of the transfection, add the required volume of serum-free medium as diluent to a total volume of 100 µL in a sterile polystyrene or polypropylene tube. Add the cationic lipid transfection reagents (i.e., FuGene 6; Roche). Tap gently to mix.
3. Add the DNA solution to a final concentration of 0.1 µg/µL to the prediluted FuGene 6 reagent. In our experiments, a 3:1 (µL/µg) ratio of FuGene 6 reagent to DNA is used for optimal transfection efficiency into HeLa cells. Mix gently and let the DNA:lipid complex form for 45 min at room temperature.
4. While DNA:lipid complexes are forming, replace medium on cells with the appropriate volume of fresh complete medium.
5. Add DNA–lipid complexes directly to each well with the desired range of DNA. In our experiments 0 µL, 2 µL, 5 µL, 10 µL, and 20 µL of transfection medium containing 0.1 µg/mL of DNA are added directly onto the cells. Distribute the DNA–lipid mixture around each well and swirl the six-well plate to ensure even dispersal. Incubate for 24 h at 37°C in 5% CO_2.

3.3.2. In Vivo Protein Cross-Linking (see **Fig. 5**)

On the day of the treatment with cross-linkers, HeLa cells should be approx 80% confluent.

1. Dissolve the long and short cross-linkers in 20 m*M* DMSO.
2. Trypsinize each well with 100 µL of trypsin-EDTA solution and incubate for 5 min at 37°C. Stop the reaction by adding 400 µL of complete medium and count the cells.
3. Immediately transfer the cells still in suspension into a sterile polystyrene or polypropylene tube. Wash the cells twice with 500 µL of complete medium by centrifugation. Incubate cells for 30 min in ice.
4. Distribute HeLa cells in sterile polystyrene or polypropylene tubes at a density of approx 0.8×10^6 cells in 20 µL of complete medium.
5. Add the cross-linker directly to the cell suspension at a final concentration of 1 m*M* (1 µL of 20 m*M* stock in DMSO in a total medium volume of 20 µL).

Immediately mix gently. Place one-half of cells without cross-linker treatment and add the same volume of DMSO as a control. To check the quenching efficiency, add the cross-linker in a tube already containing 30 m*M* of DTE. Incubate samples for 1 h at 37°C in 5% CO_2. Swirl the suspension occasionally during the 37°C incubation (*see* **Notes 4** and **5**).
6. Stop the cross-linking by adding 4 µL of 6X quenching buffer with a final concentration of 30 m*M* DTE. Let cells incubate for 10 min at 37°C to allow the quencher to fully diffuse into the cells.
7. Visualize and compare cells, which are treated and mock treated, by light microscopy in order to check for cellular death.
8. Pellet cells by centrifugation at 4°C and wash twice with cold CMF-PBS buffer at 4°C.

3.3.3. Preparation of Whole Cell Extracts

1. Prepare protein extracts by directly adding 40 µL of SBLU 1X buffer, vortex well, and boiling cells until lysis is complete.
2. Let cell lysates return to room temperature before centrifuging at 16,000*g* for 10 min at 4°C. Transfer the soluble protein fraction to a new microcentrifuge tube without touching the pellet (*see* **Note 6**).
3. Samples are then analyzed by SDS-PAGE after appropriate dilution with SBLU buffer (in general 1/10) and cross-linked proteins are identified by Western blotting according to standard protocols.

3.3.4. Quantification of Cross-Linked Species on Western Blots

The technique we currently use for immunodetection of cross-linked species is based on a chemiluminescent substrate system for horseradish peroxidase (HRP). This system can be visualized on both film and blot imaging system (ECL Plus, Amersham Biosciences). Immunodetection is carried out according to the supplied instructions and quantification is performed using Storm (Amersham Biosciences), which is better than film detection.

3.4. Concluding Comments

In vivo protein cross-linking based on structural design can greatly facilitate the introduction of selective cross-links on proteins in order to study protein oligomerization or protein–protein interactions. The combination of the chemoselectivity and regioselectivity of the reagent with the protein structure can even allow examination of changes in protein oligomerization upon cellular events and specific protein–protein interactions upon ligand binding. By considering the specificity of the maleimide group for cysteine residues and the spatial arrangement of these residues, a selective cross-linking is introduced specifically on either the dimer or the hexamer in vitro as well as in intracellular environments that are crowded with many other cellular macro-

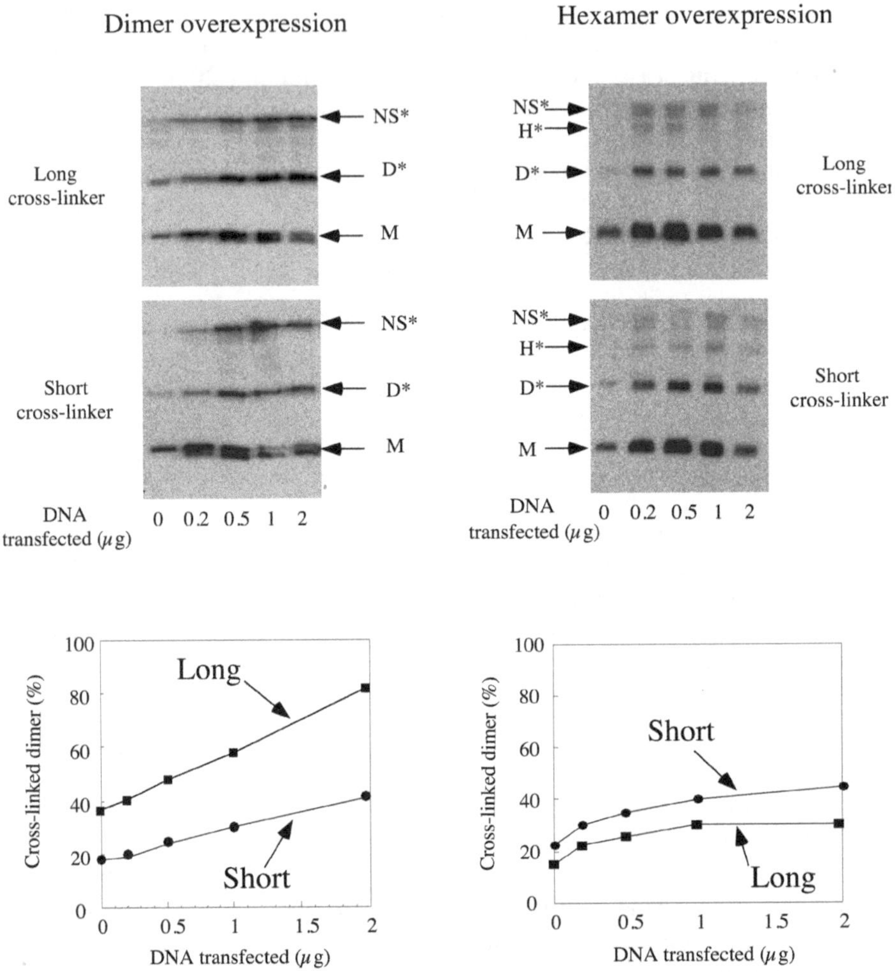

Fig. 5. In vivo cross-linking of NDPK-B in HeLa cells. HeLa cells were treated for 1 h with long (BMH, top gels) or short (BMOE, bottom gels) membrane permeable cross-linkers after transient transfections by a variable amount of DNA encoding either the dimeric mutant or the wild-type hexamer as indicated. After quenching the reaction with DTE, samples were taken up in SDS-urea gel-solubilizing extraction buffer (SBLU) and were analyzed by SDS-PAGE and Western blotting using a polyclonal antibody against NDPK-B as described in **Subheading 3.3.3.** M, D*, H*, and NS* refer to the electrophoretic migrations of monomers, cross-linked dimers, cross-linked hexamers, and higher-order nonspecific multimers of NDPK-B subunits, respectively. Dimeric cross-links were quantified using a Storm and percentages of cross-linked dimers related to either the dimer (left) or the hexamer (right) overexpressions were represented versus the amount of DNA transfected in HeLa cells.

(Continued on next page.)

molecules. The method described here allows investigations into the oligomerization of the NDPK-B protein in metastatic, tumor, and normal cells (unpublished results). This approach can be extended to other proteins, and we have used it successfully with the NEMO protein, an essential modulator involved in the NF-κB pathway ***(15)***. Given the high sensitivity of this method to detect unstable homo- and heteroassociations of proteins, the researcher will also have to address the question of whether the cross-links correspond to a mature protein association or to an assembly intermediate. This is particularly true when in vivo cross-linking results are compared within different cells at variable stages of proliferation.

4. Notes

1. Precautions should be taken when interpreting the in vivo chemical cross-linking results. If cross-links are observed in vivo, it may be specific (reflecting a relevant structure) or it may be nonspecific. Cross-linking on living cells increases the probability of detecting unstable oligomers, but it also increases the probability of cross-linking proteins together nonspecifically. When cross-links are observed, the assay should be repeated with reduced concentrations of cross-linker, at a reduced temperature, or for shorter periods of time to detect preferential cross-links. If identical results are obtained over a broad concentration range of cross-linkers or with several reagents, the in vivo cross-links are likely to be specific.
2. Often a monomer containing an internal cross-link forms a complex with SDS that has a smaller Stoke's radius than does the uncross-linked monomer–SDS complex. This cross-linked complex will then be observed on SDS-PAGE as a band migrating with lower apparent molecular weight than does the uncross-linked monomer. This slight change of the electrophoretic migration can also be observed for the cross-linked species of dimer, trimer, etc., on Laemmli SDS-PAGE.

Fig. 5. *(continued)* Note the double position of monomers especially when HeLa cells overexpressing the dimer were treated with the short cross-linker. This was due to unmodified and one-point modified monomers (*see* **Note 2**). Note also the considerable increase of dimeric cross-linked species with the dimer overexpression in HeLa cells treated by the long cross-linker. This increase is not as significant when the same transfected cells are treated with the short cross-linker. In cells that overexpress the hexamer, there are only small increases of dimer cross-links and the cross-linking is slightly more efficient with the short cross-linker. Taken together these results are very similar to those obtained in vitro (**Fig. 3**), indicating the high selectivity of our protein cross-linking in a cellular context.

3. General equation for in vivo protein cross-linking: The formation of a specific cross-linking with a homo- or heterocross-linker can be written as follows:

$$[\mathrm{P}] \xrightarrow{k_1 = k_0[\mathrm{CL}]} [\mathrm{P}^*] \xrightarrow{k_2} [\mathrm{P}^{**}] \tag{1}$$

where [P] is the unmodified protein, [CL] the cross-linker, $[\mathrm{P}^*]$ the protein cross-linked to one end of the bifunctionnal reagent, and $[\mathrm{P}^{**}]$ the protein cross-linked to both ends of the reagent. Usually for protein cross-linking $[\mathrm{CL}]_{\mathrm{Total}} >> [\mathrm{P}]_{\mathrm{Total}}$, we can then approximate [CL] by $[\mathrm{CL}]_{\mathrm{Total}}$ and we can consider the first reaction as pseudofirst-order. The general equation for the variation of $[\mathrm{P}^{**}]$ with time is

$$[\mathrm{P}^{**}] = \left\{ 1 + \frac{k_1 \exp(-k_2 t) - k_2 \exp(-k_1 t)}{k_2 - k_1} \right\} [P_0] \tag{2}$$

where t is the time, k_1 the pseudofirst-order rate constant, and P_0 the protein concentration at time 0. The first reaction is usually the rate-determining step of the reaction with a bimolecular reactant, meaning that whenever a P^* protein is formed, it is rapidly converted into P^{**} ($k_2 >> k_1$). The equation 1 then becomes a simpler expression:

$$[\mathrm{P}^{**}] = \left\{ 1 - \exp(-k_1 t) \right\} [P_0] \tag{3}$$

4. Because many proteins contained in the serum can interfere with in vivo cross-linking, it is recommended to carry out the reaction in PBS buffer, if cells can endure such a treatment up to 2 h. There is no general rule to calculate the effective concentration of cross-linker in serum-containing media. In our experiments with NDPK-B the effective concentration of both cross-linkers in PBS is 10-fold less compared to that in the media containing 10% serum.
5. It is possible to incubate the cross-linker for longer periods. We performed a range of experiments with different times of incubation, 10 min, 60 min, and 120 min. The cross-linking efficiency was optimal at 60 and 120 min, whereas the reaction was not complete at 10 min. This obviously depends on the chemoselectivity of the cross-linker as well as the types of functional group that are accessible on the protein. It also depends mostly on the subcellular localization of the protein. The in vivo cross-linking of the NEMO protein, a cytosolic signaling protein that is located near the membrane, was optimal after an incubation time of only 10 min ***(18)***.
6. It is important to note that the use of a high cross-linker concentration can reduce the total amount of extracted proteins. This is due to the in vivo formation of nonspecific intermolecular cross-links with a high molecular mass that are removed by centrifugation. It is then crucial for any in vivo protein cross-linking to determine the minimal effective concentration of the cross-linker in order to reduce the probability of random collisions and therefore to get a satisfactory yield of protein after extraction. To examine the extent of global protein cross-

linking, it is also advised to analyze by SDS-PAGE followed by Coomassie or silver staining the crude extracts from the mock-treated and the reagent-treated cells. If only a few proteins were entirely cross-linked into species of high molecular mass, the cross-linking is likely specific.

Acknowledgments

The authors are very grateful to Drs. D. Fourmy, P. England, and F. Traincard for their critical reading of the manuscript. This work was supported in part by a grant from the Association de la Recherche sur le Cancer (ARC n°5795).

References

1. Ji, T. H. (1983) Bifunctional reagents. *Methods Enzymol.* **91,** 580–609.
2. Staros, J. V. and Anjaneyulu, P. S. (1989) Membrane-impermeant cross-linking reagents. *Methods Enzymol.* **172,** 609–628.
3. Doms, R. W. (1990) Oligomerization and protein transport. *Methods Enzymol.* **191,** 841–854.
4. Kluger, R. (1997) Chemical cross-linking and protein function, in *Protein Function: A Practical Approach* (Creighton, T. E., ed.), IRL Press, Oxford, p.185.
5. Zimmerman, S. B. and Minton, A. P. (1993) Macromolecular crowding: biochemical, biophysical, and physiological consequences. *Annu. Rev. Biophy. Biomol. Struct.* **22,** 27–65.
6. Wang, K. and Richards, F. M. (1975) Reaction of dimethyl-3,3'-dithiobispropionimidate with intact human erythrocytes. Cross-linking of membrane proteins and hemoglobin. *J. Biol. Chem.* **250,** 6622–6626.
7. Parks, R. E. J. and Agarwal, R. P. (1973) Nucleoside diphosphokinases. *The Enzymes* **8,** 307–334.
8. Lacombe, M.-L., Wallet, V., Troll, H., and Veron, M. (1990) Functional cloning of a nucleoside diphosphate kinase from Dictyostelium discoideum. *J. Biol. Chem.* **265,** 10,012–10,018.
9. Postel, E. H., Berberich, S. J., Flint, S. J., and Ferrone, C. A. (1993) Human c-myc transcription factor PuF identified as nm23-H2 nucleoside diphosphate kinase, a candidate suppressor of tumor metastasis. *Science* **261,** 478–480.
10. Mesnildrey, S., Agou, F., Karlsson, A., Deville-Bonne, D., and Veron, M. (1998) Coupling between catalysis and oligomeric structure in NDP kinase. *J. Biol. Chem.* **273,** 4436–4442.
11. Karlsson, A., Mesnildrey, S., Xu, Y., Morera, S., Janin, J., and Veron, M. (1996) Nucleoside diphosphate kinase. Investigation of the intersubunit contacts by site-directed mutagenesis and crystallography. *J. Biol. Chem.* **271,** 19,928–19,934.
12. Mesnildrey, S., Agou, F., and Veron, M. (1997) The in vitro DNA-binding properties of NDP kinase are related to its oligomeric state. *FEBS Lett.* **418,** 53–57.
13. Agou, F., Raveh, S., Mesnildrey, S., and Veron, M. (1999) Single strand DNA specificity analysis of human nucleoside diphosphate kinase B. *J. Biol. Chem.* **274,** 19,630–19,638.

14. Pogulis, R. J., Vallejo, A. N., and Pease, L. R. (1996) In vitro recombination and mutagenesis by overlap extension PCR. *Meth. Mol. Biol.* **57,** 167–176.
15. Taggart, A. K. and Pugh, B. F. (1996) Dimerization of TFIID when not bound to DNA. *Science* **272,** 1331–1333.
16. Jackson-Fisher, A. J., Chitikila, C., Mitra, M., and Pugh, B. F. (1999) A role for TBP dimerization in preventing unregulated gene expression. *Molecular Cell* **3,** 717–727.
17. Morera, S., Lacombe, M.-L., Xu, Y., LeBras, G., and Janin, J. (1995) X-ray structure of nm23 human nucleoside diphophate kinase B complexed with GDP at 2 Å resolution. *Structure* **3,** 1307–1314.
18. Agou, F., Ye, F., Goffinont, S., et al. (2002) NEMO trimerizes through its coiled-coil C-terminal domain. *J. Biol. Chem.* **277,** 17,464–17,475.

V

Proteomics-Based Approaches

30

Computational Prediction of Protein–Protein Interactions

John C. Obenauer and Michael B. Yaffe

Abstract

Eukaryotic proteins typically contain one or more modular domains such as kinases, phosphatases, and phoshopeptide-binding domains, as well as characteristic sequence motifs that direct post-translational modifications such as phosphorylation, or mediate binding to specific modular domains. A computational approach to predict protein interactions on a proteome-wide basis would therefore consist of identifying modular domains and sequence motifs from protein primary sequence data, creating sequence specificity-based algorithms to connect a domain in one protein with a motif in another in "interaction space," and then graphically constructing possible interaction networks. Computational methods for predicting modular domains in proteins have been quite successful, but identifying the short sequence motifs these domains recognize has been more difficult. We are developing improved methods to identify these motifs by combining experimental and computational techniques with databases of sequences and binding information. *Scansite* is a web-accessible program that predicts interactions between proteins using experimental binding data from peptide library and phage display experiments. This program focuses on domains important in cell signaling, but it can, in principle, be used for other interactions if the domains and binding motifs are known. This chapter describes in detail how to use *Scansite* to predict the binding partners of an input protein, and how to find all proteins that contain a given sequence motif.

Key Words

Protein–protein interactions; phosphorylation sites; peptide library; phage display; sequence motifs; bioinformatics.

1. Introduction

Genomic sequence data for an ever-expanding collection of organisms continues to accumulate at a remarkably rapid pace. The emerging challenge is to develop techniques that elucidate the role of each genetically encoded protein

From: *Methods in Molecular Biology, vol. 261: Protein–Protein Interactions: Methods and Protocols*
Edited by: H. Fu © Humana Press Inc., Totowa, NJ

and to create a robust collection of tools that accurately identify protein–protein interactions important for cell function, survival, proliferation, and repair. The ultimate molecular analysis toolkit would combine experimental and computational techniques with systems analysis modules and database archiving schemes so that information flow down metabolic and signaling pathways can be visualized, annotated, and applied toward drug discovery and therapeutic intervention.

A number of experimental techniques have successfully been used for medium- or high-throughput identification of protein–protein interactions, including two hybrid methods (cf. ref. ***1***), phage display ***(2,3)***, and affinity tag purification combined with mass spectroscopy ***(4–6)*** (see related chapters in this volume). In addition, there currently exist several world wide web–accessible databases in which molecular interaction information emerging from these types of experiments, or from focused experiments on individual proteins, can be found, including the MINT, DIP, and BIND databases ***(7–9)*** (see Chapter 35). Similar progress has been made in collating known protein phosphorylation sites through the PhosphoBase database ***(10)***, although high-throughput screening efforts aimed at deducing all of the cellular targets for a particular protein kinase have, so far, remained elusive. Computational approaches to identify protein–protein interactions or protein phosphorylation sites *a priori*, based on protein sequence information alone, remain largely unexplored, although efforts are being increasingly made in this direction ***(11–14)***.

At the most basic level, protein sequences can be parsed into two fundamental components, modular domains capable of autonomous function (catalytic domains, DNA binding domains, protein-protein interaction domains) and short linear sequence motifs that bind to domains or become covalently modified by phosphorylation, methylation, acetylation, and so on, often in a reversible manner. Many proteins that function within cell signaling cascades contain one or more examples of each component. A number of growth-factor receptors, for example, contain tyrosine-kinase domains that undergo intramolecular autophosphorylation on tyrosine-containing motifs outside the catalytic domain. Other signaling proteins containing SH2 or PTB domains can then bind to these phosphotyrosine-containing motifs on the receptor to form large multimolecular signaling complexes. One of these receptor-associated proteins is the Src proto-oncogene product, which has a tyrosine-kinase domain, an SH2 domain, and an SH3 domain. When the Src SH2 domain binds to a specific sequence motif such as pY-E-E-I on the phosphorylated growth factor receptor, its SH3 domain can engage P-X-X-P sequences on other proteins, targeting them for Src-mediated phosphorylation.

If precise, quantitative knowledge of sequence motifs were available for every protein kinase and for every binding domain, it would theoretically be possible to predict all protein–protein interactions and modifications mediated through domain–motif pairs. Networks of interacting proteins could be inferred by connecting one protein's known domain types to its recognized sequence motif present in other proteins, and then continuing to connect domain–motif pairs as far as possible. These types of computational predictions could then be refined by comparison with known protein–protein interactions and modifications contained within the databases mentioned above, allowing delineation of signaling pathways within cells in a feed-forward manner.

How far along are we in this process? Bioinformatics techniques including sequence profiling for molecular signatures ***(15–18)***, sparse pattern recognition ***(19)***, and the use of hidden Markov models ***(20)*** have greatly improved our ability to identify modular signaling domains within protein sequence queries with a high degree of accuracy. There has been only limited success to date, however, in computationally identifying the short linear sequence motifs that these domains bind to within individual proteins or within protein sequence databases ***(11–14)***. One way to identify these sequence motifs involves construction of position-specific scoring matrix representations (PSSMs) of the linear amino acid sequences bound by particular modular signaling domains such as SH2, SH3, PDZ, or FHA domains, or phosphorylated by specific protein kinases such as Src, PKA, and AKT. These matrices, constructed from experimental oriented peptide library and phage display data, can then score query sequences for predicted interaction motifs. In this chapter we describe in detail how to use a web-based protein interaction motif prediction tool, *Scansite* (http://scansite.mit.edu), that predicts sequence motifs using this PSSM approach ***(11)***.

There are several caveats to the use of PSSMs for prediction of protein–protein interactions or protein modification by protein kinases. The fundamental assumption underlying this method is that short linear stretches of amino acid sequence are sufficient to determine whether a site is phosphorylated, or whether it constitutes a binding site for modular signaling domains. In many cases this appears to be true ***(10,21)***, although undoubtedly other factors including secondary and tertiary structure and surface accessibility also play a role ***(22)***. Another consideration is whether optimal phosphorylation- or binding-motif sequence data are effectively captured in PSSMs, depending on the type of experimental data upon which PSSMs are based. Both oriented peptide library and phage display screening for protein kinase phosphorylation motifs or modular domain binding motifs utilize large collections of degenerate peptides containing 10^7–10^{13} amino acid sequences with all amino acids (except Cys and sometimes Trp) in each degenerate position. In peptide library screening the entire ensemble of phosphorylated (for kinases) or bound (for binding domains) peptides is sequenced simultaneously to obtain preference values for each amino acid at each position within a motif ***(23,24)***. These preference values provide the raw data for PSSMs in which favorable or unfavorable contributions from individual amino acids are weighed quantitatively. However, because all selected peptides in a screening experiment are sequenced together, information on subtle second-order correlations between residues within a particular motif is lost. In contrast, in phage display experiments, the sequence of single peptides expressed on each phage is determined individually, capturing this type of information. However, because only a limited number of phage can realistically be sequenced, an inevitable sampling bias is introduced. Furthermore, some optimal sequences may not emerge from phage display experiments as a result of biological constraints that restrict which sequences are efficiently incorporated into the phage head or tail fibers. Phage display cannot be used when a post-translational modification such as phosphorylation is required for peptide-domain interaction to occur. Finally, because phage display only reveals positively interacting sequences, these types of experiments provide no information on residues that are selected against at particular positions within a motif. Oriented peptide library and phage display data are therefore complementary, and should be combined, when possible, in constructing PSSMs for sequence motif prediction. Additional improvements in computational motif prediction that retain position-correlation information within a motif using hidden Markov models, weigh evolutionary conservation of motifs in the scoring function, and consider the structure of the query protein and the motif surface accessibility are in development and should result in improved predictive accuracy. One final caveat of computational motif prediction is the assumption that motifs do not occur superfluously within protein sequences—that is, a short stretch of amino acids that matches a phosphorylation or domain-interaction motif, if present in a protein sequence, actually functions for that purpose. In the majority of cases, this appears to be true ***(11)***, although exceptions to this general principle are known to occur.

The *Scansite* program focuses on domains important in cell signaling, but can in principle be used for other interactions if the binding motif is known. Dozens of motifs are currently in *Scansite's* database, including versions of kinase domains, PDZ, SH2, SH3, 14-3-3, and PTB domains, with more being added as new data are generated. *Scansite* can be used to show all the potential sequence motifs in any input protein, or all proteins in the public databases that contain a given motif. Instructions for both uses will be provided in this chapter. Please note that some *Scansite* menu options or labels may change slightly over time. These directions match our website at the time of publication. Most changes will be additions of more matrices or new databases, which will not alter the steps outlined here. Assuming minor changes in subsequent years, these directions should still be a useful guide to using our programs.

2. Materials

Scansite is a web-accessible program, so any web-enabled computer (Unix, Macintosh, Windows, or other) will suffice. There is no requirement for Java or other plug-ins. Information about your protein or your own binding motif will be needed for some of the programs as listed below.

2.1. Motif Scan of a Protein from a Public Database

To scan your protein of interest using its entry in public databases (Swiss-Prot, TrEMBL, Genpept, Ensembl), you will need its accession number or ID in that database.

2.2. Motif Scan of an Input Protein Sequence

To scan your protein by copying its sequence directly into *Scansite*, you will need a text file containing this sequence. It can be in any text format. If numbers, spaces, or other invalid characters are included, *Scansite* will remove them.

2.3. Database Search Using a Scansite Motif

Both the database and the motifs are provided by *Scansite* in this case, so no other input materials are needed.

2.4. Database Search Using an Input Motif

To use your own binding motif in a database search, you will need to define it in a text file and import the text file into *Scansite*. The format is a matrix of typically 20 columns and 15 rows. The columns are scores for the 20 amino acids at each position, and the 15 rows are positions of each residue in the binding pocket. *Scansite* requires one residue to be invariant in the motif sequence, such as a tyrosine for motifs recognized by tyrosine kinases. (For motifs that recognize more than one residue in this position, such as kinases that phosphorylate both serine and threonine, multiple residues can be treated as equivalent in the fixed position.) The middle position, row 8, holds the fixed residue in the matrix. Rows 1 through 7 hold the scores for positions to the left (N-terminal side) of the fixed position, and rows 9 through 15 hold the scores for those to the right (C-terminal) side. Finally, at the top of this 20 by 15 matrix should be a row of column headings indicating the one-letter amino acid code.

The actual number of columns in the matrix can be higher or lower than 20. If your peptide library screen did not include some residues, such as cysteine or tryptophan, these columns can be left out of your matrix and *Scansite* will assign them default values of 1 everywhere except in the fixed position, where the default score is 0. Alternatively, if your motif targets the C terminus (as

PDZ domains do), you can include an extra column giving scores for the C terminus, labeled with the symbol "*" (asterisk). The N terminus can be given scores in a column labeled "$" (dollar sign). (When absent, *Scansite* assigns the N and C terminus columns values of zero by default.) Selenoprotein researchers can add a column of scores labeled "U" for selenocysteine (the U symbol is present in recent releases of the Genpept database). When this column is absent, *Scansite* scores selenocysteines as cysteines. Finally, columns can be included for the degenerate symbols "B" (aspartate/asparagine), "Z" (glutamate/glutamine), and "X" (any residue) to score sequences containing these symbols in the public databases, although there is probably no research reason to include these in an input matrix. As a result of all these options, an input matrix can contain as many as 26 columns, or even only one column naming the fixed residue (such a matrix would not be very informative, however). The order of the residue columns is not important ("A" does not have to be first, and so on).

An example will make this more clear. On the next page is a sample matrix. The full matrix might have 21 columns in this case (20 residues plus the C terminus), but only seven are shown here in order to fit well on a printed page. The first line in the text file contains the column headers (residue letters and the asterisk). These should be separated by tabs. In practice, it is easiest to do this by creating the whole matrix in a spreadsheet program and saving it as a tab-delimited text file. The next 15 lines are the 15 positions relative to the fixed center at row 8. This motif is specific for tyrosine (or phosphotyrosine) in the center position, so tyrosine is the fixed residue. It must be given the value 21 in row 8 for *Scansite* to recognize it as the fixed residue. All other values in row 8 must be zero. (For multiple residue possibilities at the fixed position, each of them must have the value 21 and the remaining residues must all be zero.) This motif also requires the C terminus to be four residues away from the center tyrosine (so the column "*" is given the value 21 at that position). In all the other rows, scores are given to represent the relative binding affinity for each residue type at each position in the sequence.

The scoring system ranges from 0 to roughly 21. Giving an individual amino acid a score of 1 at one position in the motif indicates that no preference exists, positive or negative, for that particular amino acid in that position. Giving all amino acids in one position of the motif a score of 1 (i.e., making all values in a single row of the matrix equal to 1, as in the first two rows in the sample matrix shown above) indicates no preference exists for *any* particular residue type at that position in the motif. A score of 15 means that residue has 15 chances out of 21 to be the selected residue, which is a very strong affinity. It is not necessary to normalize the scores at one position to add up to 21; *Scansite*

Sample Matrix

A	C	D	E	....W	Y	*
1	1	1	1	1	1	0
1	1	1	1	1	1	0
1.63	0.29	9.08	9.4	....2.91	3.67	0
8.88	0.83	0.28	3.92	....5.93	1.45	0
1.56	5.81	6.49	3.31	....5.35	7.07	0
0.11	5.57	15.0	0.03	....0.31	4.95	0
6.79	0.38	8.2	2.17	....3.37	0.59	0
0	0	0	0	0	21	0
5.12	3.61	5.53	1.91	....8.63	7.03	0
2.18	4.53	5.52	6.11	....8.94	0.71	0
0.63	4.07	5.34	2.51	....4.49	3.63	0
4.74	5.9	1.56	3.1	....2.39	6.82	0
0	0	0	0	0	0	21
1	1	1	1	1	1	0
1	1	1	1	1	1	0

will normalize them during program execution. Values higher than 21 are permitted if you wish to indicate very strong affinities. Values between 0 and 1 can be used for very low affinities, but negative numbers cannot. Beware that the scoring function uses natural logarithms, so values less than 1, particularly those less than 0.5, strongly penalize for that particular residue in a motif. In fact, the penalty of negative selection from a matrix value of 0.1 in the final score is equivalent (although opposite) to the positive selection obtained with a value of 10 for another residue in the motif. Consequently, choose values less than 1 (other than 0) carefully. This matrix favors the C terminus, but for motifs not showing any special affinity for the N or C terminus, scores in the "$" and "*" columns (if present) should be zero. No commas or other punctuation should separate matrix values. Only tabs are permitted between columns.

While the center position must have a fixed residue, it is also possible to fix other positions as well (as done in this example for the C terminus). As another example, suppose your motif requires aspartate in row 5 (three residues N-terminal to the fixed center). To do this, give that aspartate the value 21 in row 5, and make all other amino acid values equal to 0. Conversely, you should avoid giving residues the special score of 21 if they are not intended to be fixed positions; use values like 20.9 or 21.1 instead.

2.5. Database Search Using Quick Matrix Method

If you have insufficient information to make a full matrix of binding affinities, but you have a tentative consensus sequence describing your motif, you can still put this into *Scansite* and search the database for matches. To use this option, just have the consensus sequence available so you can enter it.

3. Methods

3.1. Motif Scan of a Protein from a Public Database

This program will scan one protein for all of the motifs in *Scansite* (or a subset of them, at your choosing).

1. In a web browser, go to the URL http://scansite.mit.edu. You will see the *Scansite* home page as shown in **Fig. 1**.
2. Under the "Motif Scan" heading, click "Scan a Protein by Accession Number or ID."
3. Enter the accession number or ID in the text field labeled "Protein ID or Accession Number" (*see* **Note 1**).
4. Select which public database you will be accessing (Swiss-Prot, TrEMBL, Genpept, or Ensembl) from the drop-down box.
5. Choose which motifs in *Scansite*'s database to scan. To scan for all motifs, click the checkbox labeled "Look for all motifs." To scan only for motifs you specify, click the checkbox labeled "Look only for motifs and groups selected below." Select one or more items in the "Individual Motifs" list, and/or one or more items in the "Motif Groups" list.
6. Choose the stringency level desired: high, medium, or low. This sets how high a sequence must score to be reported. These thresholds are set based on the scores of all subsequences that match the motif within the entire vertebrate collection of Swiss-Prot proteins. High stringency indicates that the motif identified in the query sequence is within the top 0.2% of all matching sequences contained in vertebrate Swiss-Prot proteins. Medium and low stringency scores correspond to the top 1% and 5% of sequence matches, respectively. Sites identified under high-stringency scoring are likely to be correct, although there is a possibility that real sites will fail to be identified (i.e., a nonzero false-negative selection rate). In contrast, medium and low stringency scoring has a much lower rate of false-negative predictions, but tends to over-call motif sites, resulting in increasing numbers of false-positive hits (*see* **Note 2**).
7. To show domains recognized in your sequence, check the box labeled "Show predicted domains in sequence." Otherwise, uncheck it.
8. Click "Submit Request" (*see* **Note 3**).

Figure 2 shows an example of the output. Your protein is drawn as a thin rectangle. If any sites were found, they are marked above the rectangle with a short-hand name of the domain type (such as "Y_Kin," "SH2," or "PDZ"). If you requested the domains in your sequence to be shown, they will be marked

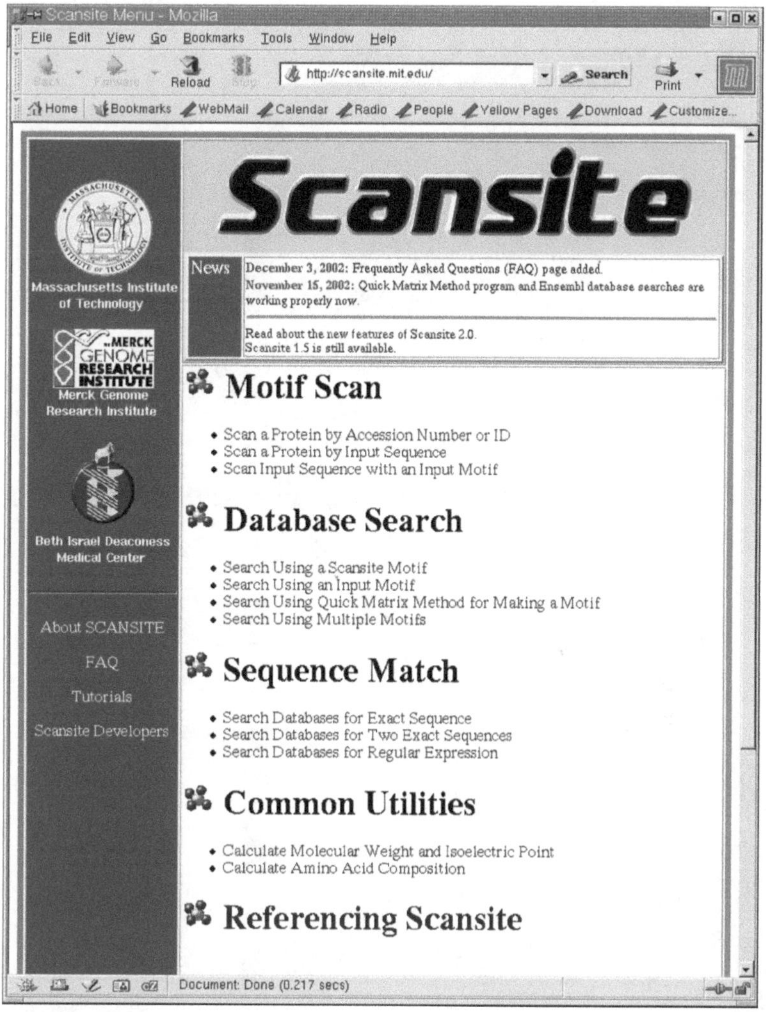

Fig. 1. The *Scansite* home page (http://scansite.mit.edu). The main programs are organized under the headings labeled "Motif Scan" and "Database Search." This tutorial covers examples of both.

as colored boxes with their names and residue ranges annotated below the rectangle. If phosphorylation and domain-binding sites already known from the literature are present, these will be marked below the domain names in a row labeled "Mapped sites" (none are present in this example). Further down, a plot of the surface accessibility indicates residues that are likely to be near the protein surface and thus able to interact with other proteins. At the bottom, a simple ruler indicates every hundredth position in the input protein sequence.

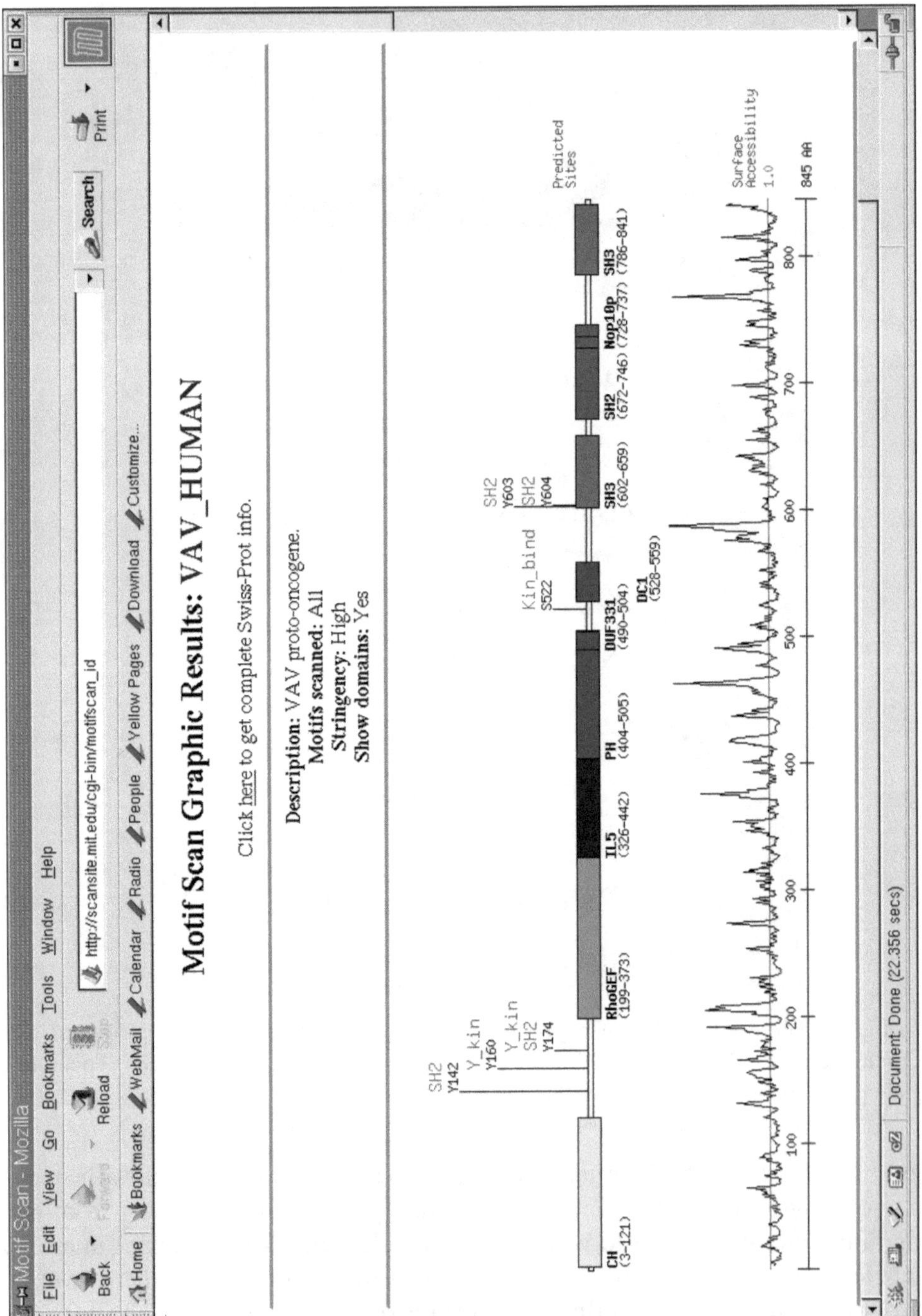
Motif Scan - Mozilla
File Edit View Go Bookmarks Tools Window Help
http://scansite.mit.edu/cgi-bin/motifscan_id
Motif Scan Graphic Results: VAV_HUMAN
Click here to get complete Swiss-Prot info.
Description: VAV proto-oncogene.
Motifs scanned: All
Stringency: High
Show domains: Yes
SH2 Y142
Y_kin Y160
Y_kin SH2 Y174
Kin_bind S522
SH2 Y603
SH2 Y604
Predicted Sites
CH (3-121)
RhoGEF (199-373)
IL5 (326-442)
PH (404-505)
DUF331 (490-504)
DC1 (528-559)
SH3 (602-659)
SH2 (672-746)
Nop10p (728-737)
SH3 (786-841)
Surface Accessibility 1.0
845 AA
Document: Done (22.356 secs)

Below the protein image is a table listing the details of the sites found (*see* **Fig. 3**). Similar motifs are grouped together (for example, all tyrosine-kinase domains). The table indicates the motif name and Gene Card (if one exists) for each site found. The next line lists each site found for that motif, with its score, the percentile that protein's score falls into compared with all vertebrate proteins in Swiss-Prot, the sequence surrounding that site, and the solvent accessibility at that position. Clicking on the Gene Card takes you to that entry on the Gene Card site *(25)*. Clicking on the near-site sequence displays the full protein sequence with the site location highlighted. Clicking on the score displays a histogram showing where this score falls in the distribution of all vertebrate Swiss-Prot proteins that have been scored for this motif (*see* **Note 4**).

3.2. Motif Scan of an Input Protein Sequence

If your protein is not in the public databases (or at least not in the ones *Scansite* uses), you can use this program to enter your sequence directly. Although it differs from the previous program in the input method, they are otherwise identical.

1. In a web browser, go to the URL http://scansite.mit.edu. You will see the *Scansite* home page as shown in **Fig. 1**.
2. Under the "Motif Scan" heading, click "Scan a Protein by Input Sequence."
3. Enter the protein name in the text box labeled "Protein Name."
4. Open the text file containing your protein sequence and copy it into the text box labeled "Sequence."
5. Choose which motifs in Scansite's database to scan. To scan for all motifs, click the checkbox labeled "Look for all motifs." To scan only for motifs you specify, click the checkbox labeled "Look only for motifs and groups selected below." Select one or more items in the "Individual Motifs" list, and/or one or more items in the "Motif Groups" list.
6. Choose the stringency level desired: high, medium, or low. This sets how high a sequence must score to be reported.
7. To show domains recognized in your sequence, check the box labeled "Show predicted domains in sequence." Otherwise, uncheck it.
8. Click "Submit Request."

The output will resemble that in **Figs. 2** and **3** as before. Some differences are that a description of the protein is no longer shown, and no mapped sites will

---

Fig. 2. *(opposite page)* Graphical output from Motif Scan. The protein is displayed schematically as a linear sequence, with modular domains colored and labeled based on Pfam information. Above the protein, sites that were found are marked with the type of motif that binds there. Below the protein, a surface accessibility plot indicates which sites are likely to be accessible for binding.

Site	Score	Percentile	Sequence	SA
Tyrosine kinase group (Y_kin)				
Src Kinase			Gene Card SRC	
Y174	0.1746	0.024 %	EAEGDEI**Y**EDLMRSE	1.040
Src Kinase			Gene Card SRC	
Y160	0.2289	0.155 %	VEEDEDL**Y**DCVENEE	0.329
Lck Kinase			Gene Card LCK	
Y174	0.2460	0.011 %	EAEGDEI**Y**EDLMRSE	1.040
Src homology 2 group (SH2)				
Lck SH2			Gene Card LCK	
Y174	0.0976	0.059 %	EAEGDEI**Y**EDLMRSE	1.040
Shc SH2			Gene Card SHC1	
Y174	0.1276	0.019 %	EAEGDEI**Y**EDLMRSE	1.040
Shc SH2			Gene Card SHC1	
Y142	0.1448	0.053 %	SVGDEDI**Y**SGLSDQI	0.460
Itk SH2			Gene Card ITK	
Y604	0.1457	0.168 %	MEVFQEY**Y**GLPPPPG	1.230
PLCg C-terminal SH2			Gene Card PLCG1	
Y604	0.1475	0.117 %	MEVFQEY**Y**GLPPPPG	1.230
Shc SH2			Gene Card SHC1	
Y603	0.1564	0.143 %	KMEVFQE**Y**YGLPPPP	1.377
Src SH2			Gene Card SRC	
Y174	0.1485	0.193 %	EAEGDEI**Y**EDLMRSE	1.040
Kinase binding site group (Kin_bind)				
PDK1 Binding			Gene Card PDK1	
S522	0.0624	0.019 %	GHDFQMF**S**FEETTSC	0.683

be displayed even if some sites have been mapped for your protein, because this information cannot easily be inferred from the input sequence alone.

3.3. Database Search Using a Scansite Motif

This program searches all proteins in a selected database for matches to a *Scansite* motif (*see* **Note 5**).

1. In a web browser, go to the URL http://scansite.mit.edu. You will see the *Scansite* home page as shown in **Fig. 1**.
2. Under the "Database Search" heading, click "Search Using a Scansite Motif." You will see a page of options as shown in **Fig. 4**.
3. In the list box labeled "Select Motif to Use," scroll through the list of available motifs and select the one you want to search with.
4. Select the database to search using this matrix (Swiss-Prot, TrEMBL, Genpept, or Ensembl).
5. If you are looking for proteins with specific characteristics, you can restrict the search to get a shorter list of more relevant results. **Steps 6** through **12** guide you through these choices. However, if you do not want to specify any of these, you can skip ahead to **Step 13**.
6. In the "Organism class" drop-down list, select which class to search. The choices are Mammals, Vertebrates, Invertebrates, Plants, Fungi, Bacteria, Viruses, or All. The default choice is Mammals. If you do not want to restrict your search by class, choose All. For a description of each class, click on the "Class descriptions" link.
7. To search only among proteins in a given species, write the genus and species in the "Single species" text box, such as "Homo sapiens," "Caenorhabditis elegans," and so on. Abbreviations and wild cards can be used here. Click the "Examples" link for more details (*see* **Note 5**). To avoid any restrictions on species, leave this text box blank.
8. For targeting proteins in a range of molecular weights, fill in weights in daltons in the two boxes labeled "Molecular weight range," such as "20,000" to "25,000." To avoid molecular weight restrictions, leave these fields blank (*see* **Note 6**).
9. If you are seeking only proteins with a characteristic isoelectric point, fill in these values in the two boxes labeled "Isoelectric point range," such as "5.5" to "6.0." To avoid isoelectric point restrictions, leave these fields blank (*see* **Note 6**).
10. For proteins likely to be phosphorylated at one or more sites, click on the number of phosphorylations you want to target under "Phosphorylated sites" (0 to 3 phosphorylations are allowed). This will affect the calculated molecular weights and isoelectric points if you are using either of those restrictions above . If you do not need to account for phosphorylations, leave this option set at 0 (*see* **Note 6**).

Fig. 3. *(opposite page)* Output table from Motif Scan. Details about the sites found are listed here, including the sequence position, score, percentile compared to vertebrate proteins in Swiss-Prot, the sequence receiving this score, and the surface accessibility value.

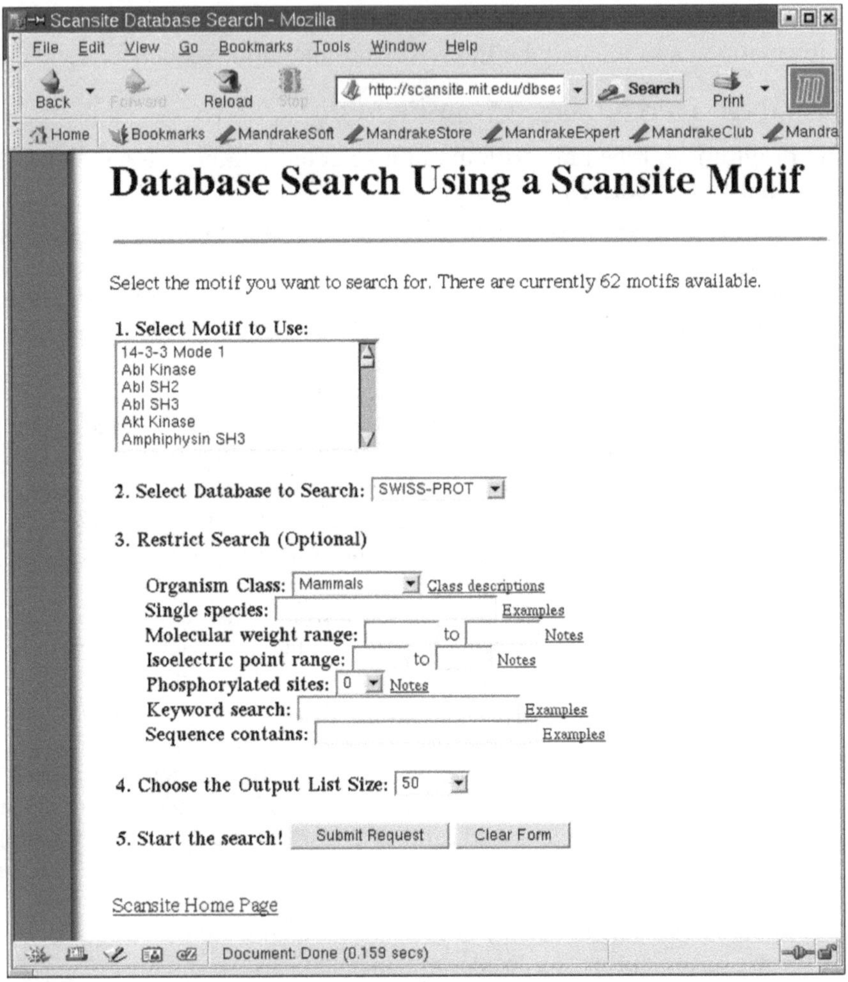

Fig. 4. Database Search options. The Database Search programs primarily ask which motif(s) to use and which database to search. Many useful options exist, however, for narrowing the output list to proteins you are most interested in. You can restrict your search to proteins from a given species, or within a range of molecular weights or isoelectric points, or containing a certain keyword in its description, or having specified sequence characteristics (that need not be related to the binding motif).

11. To look for proteins in a functional category, enter a search term in the "Keyword search" field, such as "oxidoreductase," "cytochrome," "kinase," "membrane," or perhaps "hypothetical" (to focus on novel genes). For other examples, click on the "Examples" link. Leaving this blank will skip the keyword search function (*see* **Note 7**).

12. To look for proteins that contain a consensus sequence or other sequence part, enter it in the "Sequence contains" text box. Wild cards are permitted here. For wild card details, click the "Examples" link. Leave this field blank to avoid restricting your search by sequence parts (*see* **Note 8**).
13. Now that all the restriction options have been specified (or no restrictions at all, if you so chose), select how large of an output list you want. The choices are 50, 100, 200, 300, 400, 500, 1000, or 2000 proteins.
14. Click on "Submit" to start the search.

This program can take several minutes to run if you select the larger databases (TrEMBL, Genpept). When it finishes, you will see output like that shown in **Fig. 5**. The proteins are sorted by score, but you can view them sorted by molecular weight or isoelectric point by clicking the "Sort by Molecular Weight" or "Sort by Isoelectric Point" links, which are near the top of the page. For each protein retrieved, its score, ID or accession number, description, site position, site sequence, molecular weight, and isoelectric point are shown. Clicking the score will show a histogram of how good this score is relative to all proteins that were scored in the selected database or database subset that you searched. Clicking the ID or accession number will take you to this protein's entry in the database that was searched. Finally, clicking the small "Submit" button on the left of any entry will submit it to *Scansite's* Motif Scan program described earlier.

3.4. Database Search Using an Input Motif

As in the last program, this too searches all proteins in a database for matches to a motif, but in this one you can use a motif you have made yourself (*see* **Note 9**; *see* **Subheading 2.** for information on how to make a motif).

1. In a web browser, go to the URL http://scansite.mit.edu. You will see the *Scansite* home page as shown in **Fig. 1**.
2. Under the "Database Search" heading, click "Search Using an Input Motif."
3. In the text box labeled "Motif Name," enter a name to identify your motif.
4. In the text box labeled "File of matrix values," type the location of your matrix file on your file system. Click the "Browse" button to select it from a directory listing. When finished, click the "SUBMIT" button (*see* **Subheading 2.** above for instructions on how to make the matrix file).
5. Your matrix will be displayed at this point. Verify that it is in the correct format (*see* **Subheading 2.**). If everything looks correct, click on "Yes, I would like to continue with this matrix." If some editing is required, click on "Yes, but I would like to edit this matrix." If it looks wrong, click on "No, I will upload the file again," and return to **Step 4**. If you have chosen to continue, you will see a page similar to **Fig. 4**, with the name of your motif displayed at the top.
6. Select the database to search using this matrix (Swiss-Prot, TrEMBL, Genpept, or Ensembl).

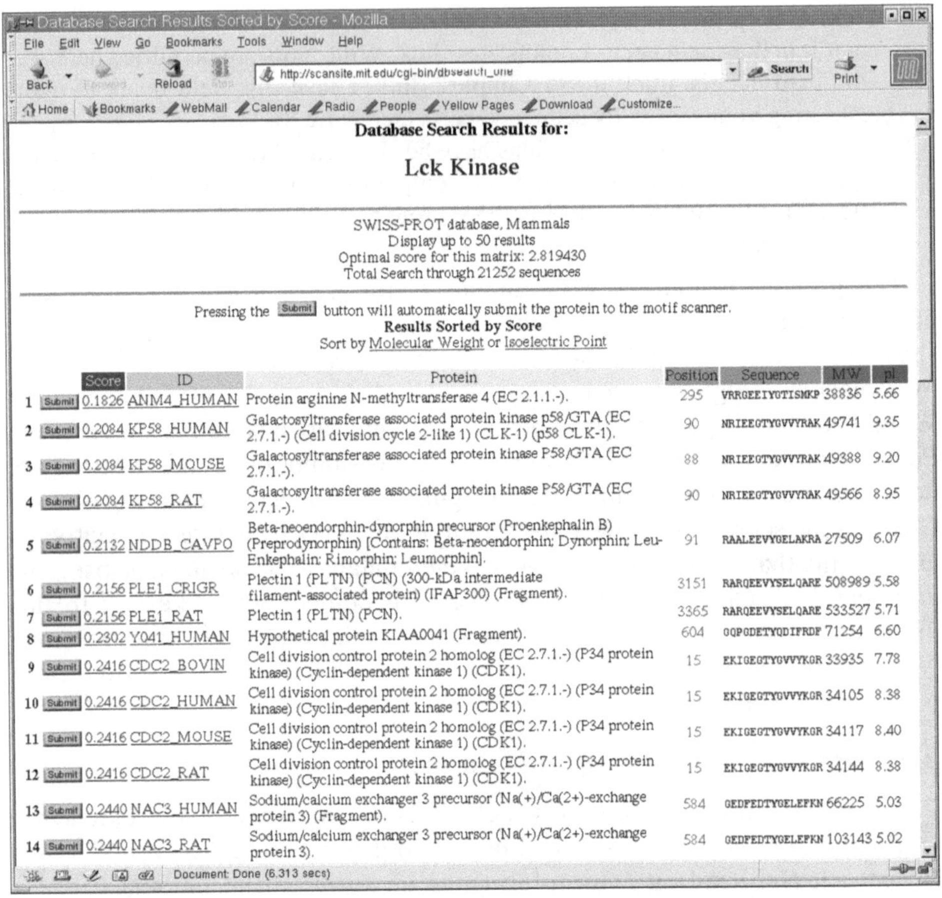

Fig. 5. List of Database Search results. The name of the selected motif and values of other options are shown at the top, and the proteins matching that motif are listed and ranked by score. Each line lists the protein's score, its accession number or ID, its description, the position and sequence that matches the motif, and the molecular weight and isoelectric point of this protein. To the left is a button labeled "Submit," that lets you quickly do a Motif Scan on any of these entries.

7. If you are looking for proteins with specific characteristics, you can restrict the search to get a shorter list of more relevant results. **Steps 8** through **14** guide you through these choices. These options are identical to those described previously in **Subheading 3.3.**, "Database Search Using a *Scansite* Motif." If you do not want to specify any restrictions, you can skip ahead to **Step 15**.
8. In the "Organism class" drop-down list, select which class to search. The choices are Mammals, Vertebrates, Invertebrates, Plants, Fungi, Bacteria, Viruses, or All.

The default choice is Mammals. If you do not want to restrict your search by class, choose All. For a description of each class, click on the "Class descriptions" link.

9. To search only among proteins in a given species, write the genus and species in the "Single species" text box, such as "Homo sapiens," "Caenorhabditis elegans," and so on. Abbreviations and wild cards can be used here. Click the "Examples" link for more details (*see* **Note 5**). To avoid any restrictions on species, leave this text box blank.
10. For targeting proteins in a range of molecular weights, fill in weights in daltons in the two boxes labeled "Molecular weight range," such as "20,000" to "25,000." To avoid molecular weight restrictions, leave these fields blank.
11. If you are seeking only proteins with a characteristic isoelectric point, fill in these values in the two boxes labeled "Isoelectric point range," such as "5.5" to "6.0." To avoid isoelectric point restrictions, leave these fields blank.
12. For proteins likely to be phosphorylated at one or more sites, click on the number of phosphorylations you want to target under "Phosphorylated sites" (0 to 3 phosphorylations are allowed). This will affect the calculated molecular weights and isoelectric points if you are using either of those restrictions above. If you do not need to account for phosphorylations, leave this option set at 0.
13. To look for proteins in a functional category, enter a search term in the "Keyword search" field, such as "oxidoreductase," "cytochrome," "kinase," "membrane," or perhaps "hypothetical" (to focus on novel genes). For other examples, click on the "Examples" link. Leaving this blank will skip the keyword search function.
14. To look for proteins that contain a consensus sequence or other sequence part, enter it in the "Sequence contains" text box. Wild cards are permitted here. For wild card details, click the "Examples" link. Leave this field blank to avoid restricting your search by sequence parts.
15. Now that all the restriction options have been specified (or no restrictions at all, if you so chose), select how large of an output list you want. The choices are 50, 100, 200, 300, 400, 500, 1000, or 2000 proteins.
16. Click on "Submit" to start the search.

The output will look like that from the previous Database Search program in **Fig. 5**.

3.5. Database Search Using the Quick Matrix Method

An alternative to making your own full matrix is to specify a consensus sequence as a binding motif. *Scansite* will then construct a rough matrix matching the characteristics of your consensus sequence, and this can be used to search the databases. The results in this case will be less quantitative. However, many users have found this program useful for quickly finding proteins with certain sequence characteristics.

1. In a web browser, go to the URL http://scansite.mit.edu. You will see the *Scansite* home page as shown in **Fig. 1**.

2. Under the "Database Search" heading, click "Search Using Quick Matrix Method for Making a Motif."
3. In the text box labeled "Motif Name," enter a name to identify the motif that will be created.
4. You will see two rows of small text boxes, labeled as positions –7 to +7, with 0 being the required fixed position. The top row is labeled "Primary Preference," and the second row is labeled "Secondary Preference." Start by entering your fixed residue in the only box at position 0. You can use the slash (/) to enter two fixed residues, such as "S/T."
5. In the "Primary Preference" row (top one), enter the residues of your consensus sequence in their position relative to your fixed residue. You can use the slash (/) to enter two residues, such as "D/E." Wild cards can be used, which are "$" for hydrophobic residues (G, A, V, I, L, M), "@" for aromatics (F, Y, W), "!" for neutral hydrophilics (S, T, W, Q), "#" for positive hydrophilics (H, K, R), and "&" for negative hydrophilics (D, E). *Scansite* will give residues in this top row a score of 9.0. For positions with no residue preference, leave it blank or use "X."
6. In the "Secondary Preference" row (bottom one), you can enter alternative residues at some positions if desired. These will be given a lower score of 4.5, and thus allows you to specify a weaker affinity for some residue types. The same wild cards can be used as in the last step. When you are finished, click the "Submit" button at the bottom of the page.
7. You will see a page similar to **Fig. 4**, with the name and schematic description of your consensus sequence displayed at the top.
8. Select the database to search using this matrix (Swiss-Prot, TrEMBL, Genpept, or Ensembl).
9. If you are looking for proteins with specific characteristics, you can restrict the search to get a shorter list of more relevant results. **Steps 10** through **16** guide you through these choices. They are identical to the options in the other Database Search programs, such as in **Subheading 3.3.** However, if you do not want to specify any of these, you can skip ahead to **Step 17**.
10. In the "Organism class" drop-down list, select which class to search. The choices are Mammals, Vertebrates, Invertebrates, Plants, Fungi, Bacteria, Viruses, or All. The default choice is Mammals. If you do not want to restrict your search by class, choose All. For a description of each class, click on the "Class descriptions" link.
11. To search only among proteins in a given species, write the genus and species in the "Single species" text box, such as "Homo sapiens," "Caenorhabditis elegans," and so on. Abbreviations and wild cards can be used here. Click the "Examples" link for more details (*see* **Note 5**). To avoid any restrictions on species, leave this text box blank.
12. For targeting proteins in a range of molecular weights, fill in weights in daltons in the two boxes labeled "Molecular weight range," such as "20,000" to "25,000." To avoid molecular weight restrictions, leave these fields blank.

13. If you are seeking only proteins with a characteristic isoelectric point, fill in these values in the two boxes labeled "Isoelectric point range," such as "5.5" to "6.0." To avoid isoelectric point restrictions, leave these fields blank.
14. For proteins likely to be phosphorylated at one or more sites, click on the number of phosphorylations you want to target under "Phosphorylated sites" (0 to 3 phosphorylations are allowed). This will affect the calculated molecular weights and isoelectric points if you are using either of those restrictions above. If you do not need to account for phosphorylations, leave this option set at 0.
15. To look for proteins in a functional category, enter a search term in the "Keyword search" field, such as "oxidoreductase," "cytochrome," "kinase," "membrane," or perhaps "hypothetical" (to focus on novel genes). For other examples, click on the "Examples" link. Leaving this blank will skip the keyword search function.
16. To look for proteins that contain a consensus sequence or other sequence part, enter it in the "Sequence contains" text box. Wild cards are permitted here. For wild card details, click the "Examples" link. Leave this field blank to avoid restricting your search by sequence parts.
17. Now that all the restriction options have been specified (or no restrictions at all, if you so chose), select how large of an output list you want. The choices are 50, 100, 200, 300, 400, 500, 1000, or 2000 proteins.
18. Click on "Submit" to start the search.

The output will again look like the Database Search results shown in **Fig. 5**.

4. Notes

1. This program accepts both accession numbers and protein IDs for databases that have both (Swiss-Prot and Genpept). TrEMBL and Ensembl assign only accession numbers to their entries. If you do not know the ID or accession number for your protein, there are links at the top of this page to find them, labeled "Search Swiss-Prot/TrEMBL for entry name" (and similar links for the Genpept and Ensembl databases). For fast program execution, local copies of all these databases are used and updated regularly. However, if an accession number or ID is not found in our databases because of an update delay, users are advised to enter the protein as a sequence instead (under the home page link titled "Scan a Protein by Input Sequence"). *See* **Subheading 3.2.** for details.
2. If motifs you expect are not found in your protein at *Scansite*'s default "High" threshold setting, try using the "Medium" or "Low" settings. The "Low" setting often overwhelms the graphical display unless you are scanning with a small number of selected motifs. However, even if the graphical display looks cluttered, the table of results is always easily readable (just longer).
3. Most of the program execution time for Motif Scan is spent retrieving the Pfam file for protein domains from the Pfam server in St. Louis ***(26)***. If you do not need domain information and have a lot of proteins to do, you can save time by turning off the "show domains" option. *Scansite* may switch to using a local copy of Pfam, which could speed up execution if network activity contributes significantly to the retrieval time.

4. *Scansite* scores are ranked on a 0 to ∞ scale, where 0 means a protein sequence perfectly matches the optimal binding pattern, and larger numbers indicate progressively poorer matches to the optimal consensus sequence ***(11)***. Lower scores in the output are thus better matches. In the matrices and during early parts of program execution, higher scores are better, so you should still use higher numbers in matrices to indicate high affinities.

   Clicking on the 15-residue sequence displayed in the results table shows the position of the site within the full protein sequence. Also, the page generated gives you a chance to submit this 15-mer peptide to BLAST. This can let you check whether this site's sequence is conserved in organisms expected to be physiologically similar to this hit.

   If the sites found by Motif Scan seem believable, you can use the motifs for those sites to search databases for other hits. In favorable cases, this can allow you to piece together parts of a pathway, if the interacting parts of different proteins can be connected.
5. The search options in this and the other two database programs are intended to address common problems with database searches. For many searches, you may only be interested in matches from humans or a model organism. Restricting your search to proteins from a single species is done by entering that species name in the "Single species" text box. *Scansite* uses a MySQL database, and the regular expressions syntax supported by MySQL allows certain helpful wild cards. For example, if you're tired of writing out "Caenorhabditis elegans," you can use "C.* elegans" instead. In a regular expression, the period (.) matches any single character, and the asterisk extends that match to multiple characters (or even zero characters). You could also do genus-wide searches, by entering "Rattus" for example. If you try doing that with "Mus," you will accidentally match "Ther<u>mus</u> aquaticus" as well, but you can avoid that by entering "^Mus." The caret symbol (^) requires the text to match at the beginning of the species name. Another pitfall to avoid is specifying an invertebrate species like "Drosophila melanogaster" when your Organism Class setting is "Mammals." You will get no hits, because no entry in Genpept has a source organism that is both a mammal and a fruit fly. (At least, the Genpept curators hope no entry like that is present.)
6. The molecular weight, isoelectric point, and phosphorylation options are intended for use in conjunction with two-dimensional gel electrophoresis experiments. When you find a few spots appearing reproducibly on a 2D gel under a particular test condition and not under the control, you could use *Scansite* to find what proteins are expected to be in that region of the 2D gel by putting in a molecular weight range and isoelectric point. You could simultaneously constrain the species to match the cell line you used in the experiment. If it is an experiment involving possible phosphorylation events, you can see how much a putative phosphorylation would move it on the gel.
7. The keyword search is primarily only useful for searching Swiss-Prot, because of its detailed annotations. It might be useful in Genpept if you are searching for novel proteins, in which case you could search for phrases like "hypothetical."

8. The "Sequence contains" text field is a quick way to restrict your search to proteins containing a consensus sequence. Unlike the protocol herein titled "Database Search Using a Consensus Sequence," in this case the desired consensus sequence does not need to be part of the motif being searched for. Regular expressions can be used here too. For example, the sequence "PXXP" is represented as "P..P" in regular expression syntax. More details about regular expression options are given in the "Examples" link to the right of the "Sequence contains" field.
9. *Scansite's* database searches can be made much more relevant to your own research by creating your own motifs. A motif can be read into *Scansite* as a text file. See **Subheading 2.** for instructions on making a motif. To avoid the common pitfalls, make sure your matrix passes these tests:
   1. Fixed residues should have the value 21, and non-fixed residues should avoid this value.
   2. The center position, row 8, should have a fixed residue.
   3. The matrix should contain 15 rows of values.
   4. For positions that you have no affinity information for, or that you know have a negligible role in selection, use 1's for all the residues. If you have included the N and C terminus columns ("$" and "*"), use 0's at this position.

   Making an effective matrix can be a challenging task. You will often need to change some values to keep the resulting output reasonable. If your affinity values are from an experimental source, we recommend that your changes preserve the rank ordering of the raw values, so that your motif is as strongly grounded in experiment as possible.

References

1. Uetz, P. and Hughes, R. E. (2000) Systematic and large-scale two-hybrid screens. *Curr. Opin. Microbiol.* **3,** 303–308.
2. Zucconi, A., Panni, S., Paoluzi, S., Castagnoli, L., Dente, L., and Cesareni, G. (2000) Domain repertoires as a tool to derive protein recognition rules. *FEBS Lett.* **480,** 49–54.
3. Kay, B. K., Kasanov, J., and Yamabhai, M. (2001) Screening phage-displayed combinatorial peptide libraries. *Methods* **24,** 240–246.
4. Ho, Y., Gruhler, A., Heilbut, A., et al. (2002) Systematic identification of protein complexes in *Saccharomyces cerevisiae* by mass spectrometry. *Nature* **415,** 180–183.
5. Gavin, A. C., Bosche, M., Krause, R., et al. (2002) Functional organization of the yeast proteome by systematic analysis of protein complexes. *Nature* **415,** 141–147.
6. Link, A. J., Eng, J., Schieltz, D. M., et al. (1999) Direct analysis of protein complexes using mass spectrometry. *Nat. Biotechnol.* **17,** 676–682.
7. Zanzoni, A., Montecchi-Palazzi, L., Quondam, M., Ausiello, G., Helmer-Citterich, M., and Cesareni, G. (2002) MINT: a Molecular INTeraction database. *FEBS Lett.* **513,** 135–140.

8. Xenarios, I., Salwinski, L., Duan, X. J., Higney, P., Kim, S. M., and Eisenberg, D. (2002) DIP, the Database of Interacting Proteins: a research tool for studying cellular networks of protein interactions. *Nucleic Acids Res.* **30,** 303–305.
9. Bader, G. D. and Hogue, C. W. (2000) BIND—a data specification for storing and describing biomolecular interactions, molecular complexes and pathways. *Bioinformatics* **16,** 465–477.
10. Kreegipuu, A., Blom, N., and Brunak, S. (1999) PhosphoBase, a database of phosphorylation sites: release 2.0. *Nucleic Acids Res.* **27,** 237–239.
11. Yaffe, M. B., Leparc, G. G., Lai, J., Obata, T., Volinia, S., and Cantley, L. C. (2001) A motif-based profile scanning approach for genome-wide prediction of signaling pathways. *Nat. Biotechnol.* **19,** 348–353.
12. Brannetti, B., Via, A., Cestra, G., Cesareni, G., and Helmer-Citterich, M. (2000) SH3-SPOT: an algorithm to predict preferred ligands to different members of the SH3 gene family. *J. Mol. Biol.* **298,** 313–328.
13. Blom, N., Gammeltoft, S., and Brunak, S. (1999) Sequence and structure-based prediction of eukaryotic protein phosphorylation sites. *J. Mol. Biol.* **294,** 1351–1362.
14. Tong, A. H., Drees, B., Nardelli, G., et al. (2002) A combined experimental and computational strategy to define protein interaction networks for peptide recognition modules. *Science* **295,** 321–324.
15. Luthy, R., Xenarios, I., and Bucher, P. (1994) Improving the sensitivity of the sequence profile method. *Protein Sci.* **3,** 139–146.
16. Henikoff, S., Henikoff, J. G., and Pietrokovski, S. (1999) Blocks+: a nonredundant database of protein alignment blocks derived from multiple compilations. *Bioinformatics* **15,** 471–479.
17. Altschul, S. F., Madden, T. L., Schaffer, A. A., et al. (1997) Gapped BLAST and PSI-BLAST: a new generation of protein database search programs. *Nucleic Acids Res.* **25,** 3389–3402.
18. Marchler-Bauer, A., Panchenko, A. R., Shoemaker, B. A., Thiessen, P. A., Geer, L. Y., and Bryant, S. H. (2002) CDD: a database of conserved domain alignments with links to domain three-dimensional structure. *Nucleic Acids Res.* **30,** 281–283.
19. Hart, R. K., Royyuru, A. K., Stolovitzky, G., and Califano, A. (2000) Systematic and fully automated identification of protein sequence patterns. *J. Comput. Biol.* **7,** 585–600.
20. Ponting, C. P., Schultz, J., Milpetz, F., and Bork, P. (1999) SMART: identification and annotation of domains from signalling and extracellular protein sequences. *Nucleic Acids Res.* **27,** 229–232.
21. Kemp, B. E. and Pearson, R. B. (1990) Protein kinase recognition sequence motifs. *Trends Biochem. Sci.* **15,** 342–346.
22. Pinna, L. A. and Ruzzene, M. (1996) How do protein kinases recognize their substrates? *Biochim. Biophys. Acta* **1314,** 191–225.
23. Songyang, Z. and Cantley, L. C. (1998) The use of peptide library for the determination of kinase peptide substrates. *Methods Mol. Biol.* **87,** 87–98.

24. Yaffe, M. B. and Cantley, L. C. (2000) Mapping specificity determinants for protein-protein association using protein fusions and random peptide libraries. *Methods Enzymol.* **328,** 157–170.
25. Rebhan, M., Chalifa-Caspi, V., Prilusky, J., and Lancet, D. (1998) GeneCards: a novel functional genomics compendium with automated data mining and query reformulation support. *Bioinformatics* **14,** 656–664.
26. Bateman, A., Birney, E., Durbin, R., Eddy, S. R., Finn, R. D., and Sonnhammer, E. L. (1999) Pfam 3.1: 1313 multiple alignments and profile HMMs match the majority of proteins. *Nucleic Acids Res.* **27,** 260–262.

31

Affinity Methods for Phosphorylation-Dependent Interactions

Greg Moorhead and Carol MacKintosh

Abstract

14-3-3s are a highly conserved protein family that exert many regulatory roles in eukaryotic cells by binding to phosphopeptide motifs in diverse target proteins. Here, we describe 14-3-3 affinity binding procedures that can be used to purify and identify 14-3-3–binding phosphoproteins; monitor how their phosphorylation and 14-3-3 binding is regulated by extracellular stimuli; define the functional effects of 14-3-3s on individual targets; and identify relevant protein phosphatases and kinases. In principle, these methods could be adapted to characterize other types of protein–protein interaction that depend on covalent modification of target sites.

Key Words

Protein–protein interactions; affinity chromatography; phosphorylation; phosphopeptide.

1. Introduction

A growing number of signaling modules and proteins, including FHA and WD40 domains and 14-3-3 proteins, have been discovered to exert regulatory effects on cellular processes by binding onto phosphorylated target sites on proteins. Moreover, other post-translational modifications, such as acetylation or methylation of lysine residues, are also being found to mediate protein–protein interactions (http://www.mshri.on.ca/pawson/domains.html). Defining the specificity and cellular regulation of these diverse interactions is critical to understanding the dynamics of cellular control. Here, we focus on the 14-3-3 proteins, and describe methods that were developed to identify and characterize the cellular regulation of their phosphoprotein binding partners. In principle, these methods could be adapted to characterize other protein–protein interactions that depend on covalent modification of target sites.

From: *Methods in Molecular Biology, vol. 261: Protein–Protein Interactions: Methods and Protocols*
Edited by: H. Fu © Humana Press Inc., Totowa, NJ

14-3-3 proteins are a family of highly conserved eukaryotic proteins comprising two isoforms in *Saccharomyces cerevisiae*, *Drosophila*, and *Caenorhabditis elegans*, seven in mammals, and at least 10 in *Arabidopsis thaliana* ***(1,2)***. 14-3-3s have many regulatory roles, which they exert by binding to phosphorylated serine and threonine residues in over 100 diverse target proteins ***(1,3–8)***. By way of example, we purified 14-3-3 isoforms for their ability to inhibit phosphorylated plant nitrate reductase (NR) ***(9)***, a mechanism that explains the inhibition of NR in leaves in the dark. In mammalian cells, 14-3-3s are best known for promoting cell survival through their interactions with phosphorylated signaling proteins ***(1,10)***, including the kinase Raf-1, Bcr, BAD ***(11,12)*** and many more.

14-3-3s function as homo- and heterodimers that are shaped like saddles with a broad central groove. Crystal structures, and the effects of synthetic phosphopeptides, site-directed mutagenesis, and protein (de)phosphorylation show that phosphorylated target proteins all dock into conserved sites located one at each side of the central target binding groove ***(12–14)***. Using oriented phosphopeptide libraries, two distinct 14-3-3 binding motifs have been defined RSXpSXP (mode 1) and RXXXpSXP (mode 2) ***(14,15)***, though other binding motifs are being discovered (for example, refs. ***16–19***).

Here we use 14-3-3 as a test case for an affinity purification strategy that was developed to isolate novel proteins that exhibit phosphorylation-dependent binding to the protein of interest ***(20)***. 14-3-3 binding proteins can be purified by 14-3-3 affinity chomatography in sufficient quantity for structural analysis and identification. Extracts are prepared in the presence of protein phosphatase inhibitors to preserve phosphorylation of proteins, and mixed with recombinant 14-3-3 proteins bound to Sepharose, so that the 14-3-3–Sepharose binds to target proteins in competition with endogenous 14-3-3s. After extensive washing, proteins that have bound specifically to the 14-3-3–Sepharose column are eluted by competition with a 14-3-3–binding phosphopeptide.

In addition, we explain how 14-3-3 overlays (Far-Westerns; *also see* **Chapter 12**) can be used to identify proteins that bind directly to 14-3-3s ***(20,21)***. The overlays are sufficiently sensitive that they can be adapted to monitor how the phosphorylation and 14-3-3–binding status of targets change in response to physiological stimuli. The 14-3-3–affinity-binding reagents described here can also assist in identifying the functional effects of 14-3-3 binding to individual target proteins.

2. Materials

2.1. Buffers

1. Buffer A: 20 m*M* Tris-HCl pH 7.5, 500 m*M* NaCl.
2. Buffer B: 50 m*M* HEPES–NaOH pH 7.5, 50 m*M* NaF, 5 m*M* sodium pyrophosphate, 1 m*M* dithiothreitol (DTT), 1 m*M* phenylmethylsulphonyl fluoride (PMSF), 1 m*M* benzamidine, and 1% (w/v) insoluble polyvinylpolypyrollidone.

3. Buffer C: 50 m*M* HEPES, pH 7.5 and 1 m*M* DTT.
4. Buffer D: 25 m*M* Tris-HCl, pH 7.5 and 1 m*M* DTT.

2.2. Preparation of Recombinant 14-3-3 Proteins

6xHis-tagged BMH1 and BMH2 14-3-3 isoforms from *S. cerevisiae* can be expressed in *Escherichia coli* DH5α (*see* **Note 1**). Twenty milligrams of each protein is typically purified from 2 L of culture using nickel-agarose (Qiagen) *(9)* and following the manufacturer's instructions. Purified proteins are dialyzed into phosphate-buffered saline (PBS).

2.3. Preparation of 14-3-3–Sepharose Affinity Matrix

BMH1 and BMH2 (5 mg of each) are reacted separately for 1 h at room temperature with 2.5 mL of swollen activated CH-Sepharose 4B (Amersham Pharmacia Biotech). Because coupling is through free amino groups on the protein, it is essential that the protein be dialyzed into an amino-free buffer such as PBS or $NaHCO_3$ before use.

1. Swell 2 g of activated CH-Sepharose in water for 15 min, then wash with ice-cold 1 m*M* HCl on a scintered glass funnel. Wash with 100 m*M* $NaHCO_3$, pH 8.2.
2. Add 14-3-3 proteins (2 mg 14-3-3/mL matrix) to activated CH-Sepharose, and rotate end-over-end for 1.5 h at 25°C.
3. Wash away excess ligand with 100 m*M* $NaHCO_3$, pH 8.2, block unreacted groups with 0.1 *M* Tris-HCl, pH 8.2 for 1 h, then wash alternately with 50 m*M* Tris-HCl, pH 8.0, 0.5 *M* NaCl and with 50 m*M* acetate, pH 4.0, 0.5 *M* NaCl. Finally wash with and store in 20 m*M* Tris-HCl, pH 7.5 plus 0.02% (w/v) sodium azide.
4. Check the coupling efficiency (typically >90%) by performing a Bradford protein assay on the supernatant after the 1.5 h reaction.

2.4. Preparation of DIG-Labeled 14-3-3 Proteins

The 6xHis-tagged BMH1 and BMH2 are labeled with the ester of digoxygenin-3-*O*-methylcarbonyl-ε-aminocaproic-acid-*N*-hydroxysuccinimide (DIG) (Roche). Coupling of DIG to proteins involves reaction with free amino groups and it is therefore essential that proteins be dialyzed into PBS and away from buffers with amino groups.

1. Mix 25 μg each of BMH1 and BMH2 in PBS (200 μL), pH 8.5, for 10 min to allow for potential heterodimer formation, then add 8 μg of DIG from a freshly prepared stock solution of 4 mg/mL in DMSO.
2. Mix and incubate for 3 h at room temperature, then add Tris-HCl, pH 7.2, to 20 m*M* to react any remaining DIG. After 1 h, dialyze versus PBS, pH 7.2, overnight with one change of buffer.
3. Dilute to 1 μg/mL in Buffer A plus 2 mg/ mL bovine serum albumin (fraction V, Sigma) for blotting (*see* **Note 2**).

3. Methods

3.1. BMH1– and BMH2–Sepharose Chromatographies of Phosphorylated Extracts

1. Homogenize the tissue of interest in ice-cold buffer at an appropriate pH and in the presence of inhibitors that block proteases from that tissue/cell type. The protocol described below is for plant tissue. Plant cells are homogenized in Buffer B and clarified by centrifugation at 12,000*g* for 20 min (*see* **Note 3**).
2. After filtration through glass wool and two layers of miracloth, mix the clarified extract end-over-end for 1 h at 4°C with 5 mL of BMH1/BMH2–Sepharose equilibrated in Buffer C plus 150 m*M* NaCl. Pour the mixture into a column and wash with Buffer C plus 500 m*M* NaCl (at least 500 mL, or until the eluate contains less than 5 µg/mL protein).
3. Perform an optional control mock elution by incubating the column in 10 mL of irrelevant phosphopeptide that does not bind 14-3-3 proteins in Buffer C plus 500 m*M* NaCl, for example, 1 m*M* GM peptide SPQPSRRGpSGSSEDAPA (single letter code, where pS is phosphorylated serine) *(20)*.
4. Selectively elute proteins that are bound to the phosphopeptide binding sites on the 14-3-3 column by incubating the column for 30 min in 1 m*M* of the synthetic 14-3-3–binding phosphopeptide ARAApSAPA dissolved in Buffer C plus 500 m*M* NaCl (*see* **Note 4**). Concentrate the eluted protein by centrifugation in Centricon-10 concentrators (Amicon) or equivalent (*see* **Note 5**). A typical preparation beginning with 4000 mg of soluble cauliflower proteins yields approx 600 µg of mixed 14-3-3 column eluate *(20)* (*see* **Note 3**).
5. If wished, the 14-3-3–binding proteins can be further fractionated by anion chromatography on a SMART Mono Q column (5 × 0.16 cm), or by 2D gel electrophoresis. Once fractionated, those purified proteins that bind directly to 14-3-3s can be visualised by the 14-3-3 overlay method (*see* **Notes 6** and **7**). In parallel, individual protein bands of spots can be processed for protein identification via Edman sequencing or MALDI-TOF mass spectrometry. Readers are referred to other papers for details on these protein identification procedures *(20–22)*.

3.2. Dephosphorylation of 14-3-3–Binding Proteins

1. Prior to electrophoresis, dephosphorylate samples by treatment with protein phosphatase 2A (PP2A) (Upstate Biotechnology). Incubate 25 µg of the concentrated sample from the 14-3-3 affinity column with PP2A (50 mU/mL) for 30 min at 30°C.
2. Stop the reaction with 1 µ*M* of the PP2A inhibitor microcystin-LR (Calbiochem). Perform control incubations in which the microcystin-LR is mixed with the PP2A before adding to the sample.
3. Dephosphorylated samples should not bind to the DIG-14-3-3 probe (**Fig. 1**; *see* **Note 8**).

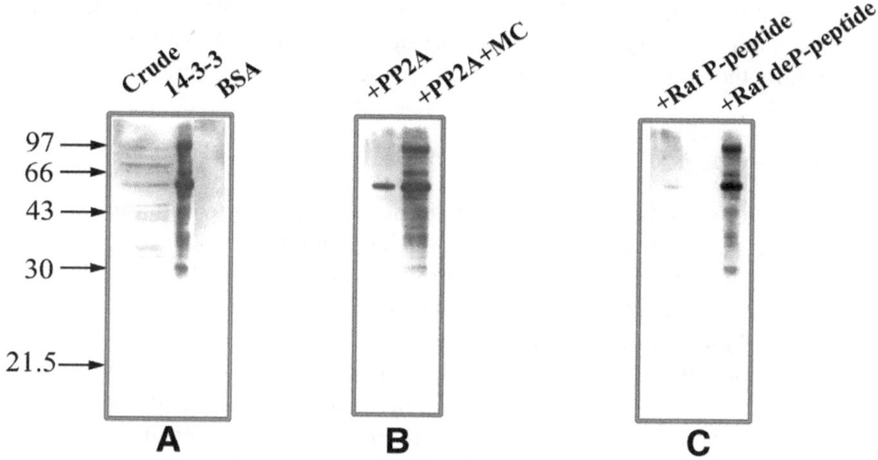

Fig. 1. 14-3-3 binding is blocked by target protein dephosphorylation or the presence of the Raf-1 phosphopeptide, but not the unphosphorylated peptide. Samples were electrophoresed on a 12% polyacrylamide SDS gel, transferred to nitrocellulose, and probed with DIG–14-3-3s as described in **Subheading 3.3.** (**A**) Samples of cauliflower crude extract (40 μg), the 14-3-3–Sepharose column eluate (1 μL or 3 μg), and an eluate from a control BSA–Sepharose column (1 μL; *see* **Note 6**) were probed with DIG–14-3-3. (**B**) 3 μg of 14-3-3 column eluate were treated with active PP2A or inhibited PP2A (PP2A +1 μ*M* microcystin-LR) prior to electrophoresis, and the blot probed with DIG-14-3-3. (**C**) The Raf-1 phosphopeptide (*see* **Note 4**) was used to block 14-3-3 binding by prior incubation with the probe (P-peptide). The requirement for phosphorylation is shown by the lack of effect of incubating the probe with dephosphorylated peptide (**deP-peptide**). Molecular mass standards are shown in kDa.

3.3. Identification of 14-3-3–Binding Proteins Using DIG-Labeled BMH1 and BMH2 as Probes

1. Run affinity purified, concentrated samples (*see* **Note 9**) on SDS-PAGE, transfer for 200 V h to nitrocellulose membranes and block nonspecific sites by incubating the blot for 16 h at ambient temperature with 5% (w/v) skimmed milk powder in Buffer A.
2. Incubate the nitrocellulose membranes for 2 h at ambient temperature with DIG-BMH1 and DIG-BMH2 diluted to 1 μg/mL in Buffer A containing 2 mg/mL BSA, then wash several times with Buffer A over a 2 h period to remove unbound probe.
3. Incubate blots for 45 min with anti-DIG antibodies (Roche) conjugated with peroxidase (diluted 5000X) in buffer A plus 2 mg/mL bovine serum albumin, then wash for 2 h with Buffer A and visualize with the enhanced chemiluminesence system (Amersham-Pharmacia Biotech).

4. The specificity of the Far Western interaction can be explored by probing samples that are dephosphorylated prior to electrophoresis (*see* **Subheading 3.2.**), and by blocking the interaction with the 14-3-3–binding phosphopeptide (*see* **Fig. 1**; *see* **Note 10**).

4. Notes

1. 14-3-3 isoforms display different affinities for binding to different target proteins, which is likely to be important in fine regulation in vivo. We have found, however, that the 14-3-3 isoforms from *S. cerevisiae* (BMH1 and BMH2) are useful general reagents for binding to target proteins from a variety of plant, yeast, and mammalian cells and tissues.
2. Preparations of DIG-14-3-3 proteins sometimes contain denatured protein that binds nonspecifically to blots, giving a high background. The denatured protein can be removed and the signal-to-noise ratio improved by preincubating the DIG–14-3-3 proteins with a washed, milk blocked membrane prior to use.
3. Because 14-3-3 proteins bind their targets through a phosphorylated motif, all extracts should be prepared in the presence of protein phosphatase inhibitors (NaF, NaPPi, and microcystin-LR) to prevent dephosphorylation. Some potential partners may not be phosphorylated when the cells/tissues are harvested, which is why, when this method was first developed, we incubated extracts with MgATP to promote target phosphorylation *(20)*. However, we now prefer to omit the MgATP step to avoid the risk of promoting phosphorylations in the extract that may not be physiologically relevant. The final yield of each target protein will depend on the initial amount of that target that is phosphorylated and capable of binding to 14-3-3 proteins, how well the 14-3-3 column competes with endogenous 14-3-3 proteins in the extract for binding to the protein, and competition between peptide and target proteins at the elution step.
4. While the phosphopeptide (ARAApSAPA) binds with much higher affinity than the dephosphorylated peptide (ARAASAPA), this specificity may be lost when using these peptides at higher concentrations. For example, we use a relatively high concentration of phosphopeptide (1 m*M*) to ensure that all binding proteins are eluted from the 14-3-3 affinity column. However, at such a high concentration, unphosphorylated peptide would also start to bind 14-3-3 proteins (for example, Fig. 5c in ref. ***20***). We have never tested whether lower concentrations of phosphopeptide would be equally effective for protein elution. Note that previously we used the Raf-1 phosphopeptide (LSQRQRSTpSTPNVHMV) for binding to 14-3-3 proteins (***20*** and *see* **Fig. 1**).
5. Proteins that have bound nonspecifically to the immobilized 14-3-3 proteins will remain bound to the 14-3-3 column, which should be discarded.
6. While the phosphopeptide elution and 14-3-3 overlay analysis should be sufficient to ensure the binding specificity of the purified proteins, additional controls could be used. For example, it would be expected that the purified proteins should not bind to columns of bovine serum albumin (fraction V, Sigma), any other irrelevant protein, or 14-3-3 proteins that have their binding sites mutated so that

they do not bind target proteins *(14)*. In addition, the chromatographic behavior of *bone fide* 14-3-3–binding proteins that are known components of the extracts could be tested by Western blotting.

7. The proteins that are purified by 14-3-3–affinity chromatography will include those that bind directly to the phosphopeptide binding site on the 14-3-3 proteins, and proteins that bind indirectly to 14-3-3s as part of multiprotein complexes. The 14-3-3 overlay method provides one means of distinguishing whether any individual protein can bind directly to DIG-labeled 14-3-3 proteins. Proteins have to be clearly separated from each other, so that proteins identified by mass spectrometry or Western blotting can be aligned clearly with proteins that do or do not give a DIG–14-3-3 binding signal. Two-dimensional gels would be another way to separate the binding proteins, although our experience is that only the more abundant proteins in the mixture can be visualized by this technique.
8. While the majority of reported target proteins have to be phosphorylated to bind 14-3-3 proteins, there are a few exceptions (for example, ref. *23*) .
9. The 14-3-3 overlay method is also sensitive enough to detect phosphorylated target proteins in crude extracts of plant or human cells, and to track how the phosphorylation and 14-3-3–binding status of cellular proteins change in response to nutrients, hormones, stresses, and drugs (for example, *see* ref. *21*). This approach is analogous to the popular method of using phospho-specific antibodies to track the phosphorylation status of a single site. The strength of using 14-3-3 overlays as a screen is that any 14-3-3–binding proteins whose phosphorylation is observed to be regulated by extracellular stimuli can be purified by 14-3-3 affinity chromatography and identified. Overlays are useful for other protein-protein interactions (for example, ref. *22*).
10. At least for *Arabidopsis* plant cell extracts, all the proteins that can be detected by the 14-3-3 overlay method are also co-immunoprecipitated from the cell extracts with anti-14-3-3 antibodies *(22)*. This confirms that the overlay is detecting proteins that really do bind to endogenous 14-3-3 proteins in the cell extract.

Acknowledgments

We thank our colleagues whose work has contributed to refining and expanding the use of the methods outlines here, namely, Valérie Cotelle, Jean Harthill, Fiona C. Milne, Mercedes Pozuelo Rubio, and Barry Wong in the MRC Protein Phosphorylation Unit, Dundee.

References

1. Tzivion, G. and Avruch, J. (2002) 14-3-3 proteins: active cofactors in cellular regulation by serine/threonine phosphorylation. *J. Biol. Chem.* **277,** 3061–3064.
2. Rosenquist, M, Alsterfjord, M., Larsson, C., and Sommarin, M. (2001) Data mining the *Arabidopsis* genome reveals fifteen 14-3-3 genes. Expression is demonstrated for two out of five novel genes. *Plant Physiol.* **127,** 142–149.
3. Aitken, A. (1995) 14-3-3 proteins on the MAP. *Trends Biochem. Sci.* **20,** 95–97.

4. van Hemert, M. J., Steensma, H.Y., and van Heusden, G. P. (2001) 14-3-3 proteins: key regulators of cell division, signalling and apoptosis. *Bioessays* **23**, 936–946.
5. van Hemert, M. J., van Heusden, G. P., and Steensma, H. Y. (2001) Yeast 14-3-3 proteins. *Yeast* **18,** 889–895.
6. Aitken, A. (1996) 14-3-3 and its possible role in coordinating multiple signalling pathways. *Trends Cell Biol.* **6,** 341–347.
7. Chung, H. J., Sehnke, P. C., and Ferl, R. J. (1999) The 14-3-3 proteins: cellular regulators of plant metabolism. *Trends Plant Sci.* **4,** 367–371.
8. Roberts, M. R. (2000) Regulatory 14-3-3 protein-protein interactions in plant cells. *Curr. Opin. Plant. Biol.* **3,** 400–405.
9. Moorhead, G., Douglas, P., Morrice, N., Scarabel, M., Aitken, A., and MacKintosh, C. (1996) Phosphorylated nitrate reductase from spinach leaves is inhibited by 14-3-3 proteins and activated by fusicoccin. *Curr. Biol.* **6,** 1104–1113.
10. Tzivion, G., Shen, Y. H., and Zhu, J. (2001) 14-3-3 proteins; bringing new definitions to scaffolding. *Oncogene* **20,** 6331–6338.
11. Michaud, N. R., Fabian, J. R., Mathes, K. D., and Morrison, D. K. (1995) 14-3-3 is not essential for Raf-1 function: identification of Raf-1 proteins that are biologically activated in a 14-3-3- and Ras-independent manner. *Mol. Cell Biol.* **15,** 3390–3397.
12. Zha, J., Harada, H., Yang, E., Jockel, J., and Korsmeyer, S. J. (1996) Serine phosphorylation of death agonist BAD in response to survival factor results in binding to 14-3-3 not BCL-X(L). *Cell* **87,** 619–628.
13. Muslin, A. J., Tanner, J. W., Allen, P. M., and Shaw, A. S. (1996) Interaction of 14-3-3 with signaling proteins is mediated by the recognition of phosphoserine. *Cell* **84,** 889–897.
14. Yaffe, M. B., Rittinger, K., Volinia, S., et al. (1997) The structural basis for 14-3-3:phosphopeptide binding specificity. *Cell* **91,** 961–971.
15. Rittinger, K., Budman, J., Xu, J., et al. (1999) Structural analysis of 14-3-3 phosphopeptide complexes identifies a dual role for the nuclear export signal of 14-3-3 in ligand binding. *Mol. Cell* **4,** 153–166.
16. Jelich-Ottmann, C., Weiler, E. W., and Oecking, C. (2001) Binding of regulatory 14-3-3 proteins to the C terminus of the plant plasma membrane H^+-ATPase involves part of its autoinhibitory region. *J. Biol. Chem.* **276,** 39,852–39,857.
17. Wang, H., Zhang, L., Liddington, R., and Fu, H. (1998) Mutations in the hydrophobic surface of an amphipathic groove of 14-3-3zeta disrupt its interaction with Raf-1 kinase. *J. Biol. Chem.* **273,** 16,297–16,304.
18. Zhang, L., Wang, H., Liu, D., Liddington, R., and Fu, H. (1997) Raf-1 kinase and exoenzyme S interact with 14-3-3zeta through a common site involving Lysine 49. *J. Biol. Chem.* **272,** 13,717–13,724.
19. Yaffe, M. B., Rittinger, K., Volinia, S., et al. (1997) The structural basis for 14-3-3:phosphopeptide binding specificity. *Cell* **91,** 961–971.
20. Moorhead, G., Douglas, P., Cotelle, V., et al. (1999) Phosphorylation-dependent interactions between enzymes of plant metabolism and 14-3-3 proteins. *Plant J.* **18,** 1–12.

21. Cotelle,V., Meek, S. E., Provan, F., Milne, F. C., Morrice, N., and MacKintosh, C. (2000) 14-3-3s regulate global cleavage of their diverse binding partners in sugar-starved *Arabidopsis* cells. *EMBO J.* **19,** 2869–2876.
22. Stubbs, M. D., Tran, H., Atwell, A. J., Smith, C. S., Olson, D., and Moorhead, G. B. G. (2001) Purification and properties of *Arabidopsis thaliana* type 1 protein phosphatase (PP1). *Biochim. Biophys. Acta* **1550,** 52–63.
23. Masters, S. C., Pederson, K. J., Zhang, L., Barbieri, J. T., and Fu, H. (1999) Interaction of 14-3-3 with a nonphosphorylated protein ligand, exoenzyme S of *Pseudomonas aeruginosa. Biochemistry* **38,** 5216–5221.

32

Two-Dimensional Gel Electrophoresis for Analysis of Protein Complexes

Karin Barnouin

Abstract

Advances in genomic and proteomic technologies combined with molecular and cell biology have together enabled the identification of numerous genes and their products. Two-dimensional gel electrophoresis (2DE) is especially useful in the study of protein–protein interactions as it permits an improved separation of proteins as well as the detection of specific interacting protein isoform(s) of a protein resulting from post-translational modification. The investigation of interacting proteins using 2DE can be complemented by identification of the proteins by mass spectrometry. Here, I describe how protein complexes, isolated by methods such as immunoprecipitation, can be analyzed by 2DE using either isoelectric focusing (tube gels or immobilized pH gradient strips) or nonequilibrium pH gradient electrophoresis (NEPHGE) in the first dimension, SDS-PAGE in the second dimension, and gel staining (silver and Coomassie) or Western blotting for the final detection of the interacting proteins.

Key Words

Two-dimensional gel electrophoresis; proteomics; protein complexes; isoelectric focusing; nonequilibrium pH gradient electrophoresis (NEPHGE); immunoprecipitation; post-translational modifications.

1. Introduction

Two-dimensional gel electrophoresis (2DE) was developed more than 20 years ago *(1)* and remains one of the most powerful techniques in the analysis of protein complexes. 2DE is especially useful in the study of protein–protein interaction as it enables the detection of the specific interacting isoform(s) of a protein resulting from post-translational modification.

Protein complexes can first be isolated by co-immunoprecipitation or pull downs using GST or hexahistidine fusion proteins *(2–6)* (*see* also **Chapters 13**

From: *Methods in Molecular Biology, vol. 261: Protein–Protein Interactions: Methods and Protocols*
Edited by: H. Fu © Humana Press Inc., Totowa, NJ

and **23**). The complexes can then be separated on 2DE and analyzed using various detection methods. If cells were previously labeled with [^{35}S]methionine or [^{32}P]- or [^{33}P]orthophosphate, the separated complexes can be detected using a PhosphoImager or by autoradiography. If cells are unlabeled, nonradioactive detection methods such as gel staining can be used to visualize most of the components of the complex or Western blot analysis for the analysis of specific proteins. Finally, the identity of proteins and their isoforms can be determined by mass spectrometry.

Combining isoelectric focusing and SDS-PAGE allows for a greater separation of proteins than in one-dimensional (1D) SDS-PAGE gels. More than one protein may co-migrate in a single band in 1D gels thus interfering with the identification of a particular protein. This problem is reduced in large-format 2DE.

2DE separates proteins according to charge in the first dimension by isoelectric focusing and according to molecular weight in the second dimension by SDS-PAGE. During isoelectric focusing, proteins separate along a pH gradient according to their charge and will continue to migrate until they reach a point where their net charge is zero i.e., their isoelectric points (pI). pH gradients are generated with amphoteric molecules called carrier ampholytes. These molecules contain both acidic and basic groups and can exist at different charge states. Commercial ampholyte mixtures are now commonly used for isoelectric focusing and are available at many different pH ranges. The pH range in both methods described below is determined by varying the mixture of basic and acidic ampholytes.

Two isoelectric focusing techniques are currently used. One method utilizes tube gels, where the pH gradient is generated by an ampholyte mixture during electrophoresis. The other method uses immobilized pH gradient (IPG) strips, where a pH gradient is formed with acrylamide-bound ampholytes on a plastic support.

Tube gels are faster and cheaper than IPG strips and rely on the user to cast their own gels (one day is gained and the samples are immediately processed; *see* **Fig. 1**). Tube gels are preferable for hydrophobic proteins such as membrane proteins because these proteins do not enter IPG gels very efficiently. Tube gels used together with the nonequilibrium pH gradient electrophoresis (NEPHGE) method allows a better separation of very basic proteins (pH 9–12) *(7)*. The main drawback of tube gels is that they are fragile and the position of the proteins may vary slightly from gel to gel resulting in lower reproducibility.

The IPG strip was developed by Görg et al. *(8)* and has helped make 2D gels more accessible. The first dimensional gel is more robust and the final 2DE results are more reproducible. It is possible to hand-cast IPG strips *(9)* but the commercially available ones are very convenient to use.

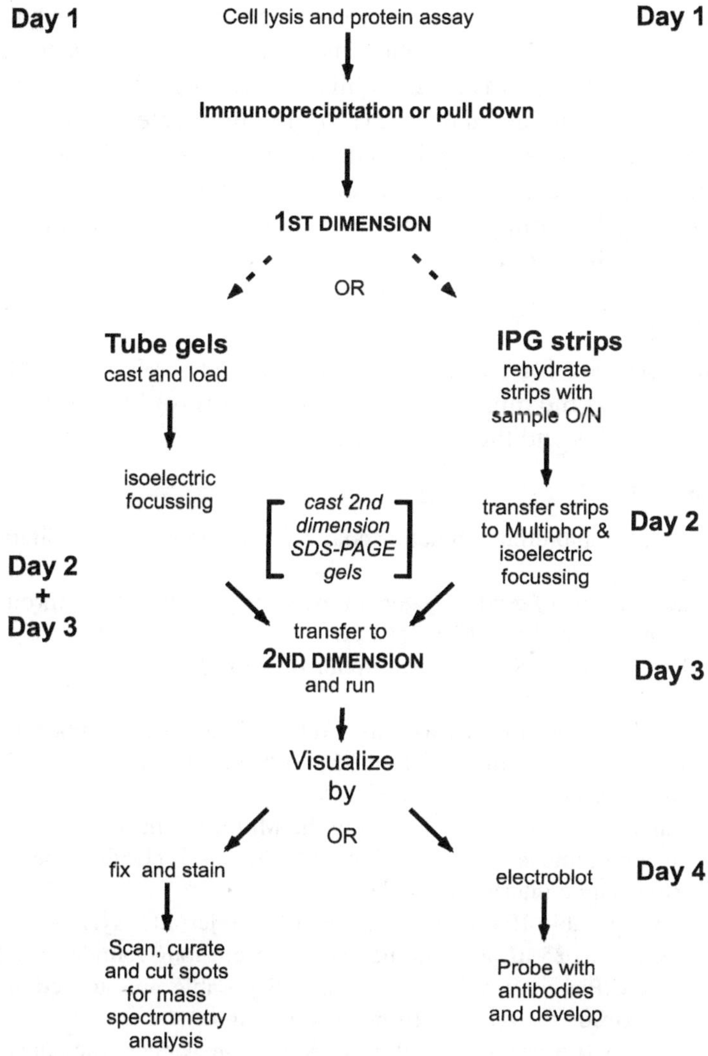

Fig. 1. Flow chart outlining the procedure of protein complex analysis using 2D-gel electrophoresis.

The recent revival in the use of 2DE has in part been due to the development of more sensitive mass spectrometers. Furthermore, advances in molecular and cell biology, the generation of ESTs, full-length gene and protein sequence databases, and analysis software that are freely available on the internet have enabled the isolation of genes and their products and facilitated their identification. In parallel to these advancements in technology, several companies, especially Amersham and Biorad, have developed instrumentation and reagents that have standardized 2D electrophoresis, making it a more accessible technique. Amersham has, for instance, published a 2D gel electrophoresis handbook that is freely available on the web ***(10)***. In this chapter, I will explain how protein complexes isolated by co-immunoprecipitation should be prepared for 2DE, and I will describe the subsequent detection of their components by Western blotting and gel staining and give examples of applications in the analysis of protein–protein interactions.

2. Materials

Materials listed are for 20 × 20 cm format 2D gels. We recommend doing large 2D gels versus the mini format (e.g., 7 × 6 cm) for better separation and resolution of proteins and their isoforms.

2.1. Equipment

1. Rotating wheel and cold cabinet or cold room for immunoprecipitation or pull downs.
2. Powerpack capable of gradually generating voltage up to 3500 V through a linear voltage ramp (Amersham: EPS 3501 XL).
3. For IPG strips or DryStrips, a Multiphor or IPG-phor (Amersham) is required (*see* **Note 1**).
4. For tube gels, glass tubes as well as a tube gel running chamber are required (Protean® II cell Glass tubes, 1.5 × 7.5 × 180 mm length; Biorad: Model 175 Tube Cell, Biorad).
5. Two circulating water baths for cooling the Multiphor during isoelectric focusing and second dimension gel running tanks (Grant: LTD6G). The temperature of IPGphors can be internally regulated.
6. Syringes: Disposable 10 mL syringes; Hamilton microliter syringes are required for volumes up to 25 μL and 100 μL with a 25-cm-long rounded needle to load the gels (SGC/Pierce & Warriner, UK); a disposable pressure-control syringe (Sigma) to extrude the tube gel from the glass tube.
7. Reswelling tray (for use with Multiphor, Amersham) or ceramic boats (IPGphor, Amersham).
8. Immobiline DryStrip kit that includes anode and cathode electrodes; tray and electrode holder; DryStrip plastic aligners and IEF electrode strips (Multiphor; Amersham).
9. Gel casting chamber (for two gels use Protean II XL system, for more than two gels use Biorad's PROTEAN II xi multi-gel casting chamber that enables casting of up to 10 single or 12 double 1.5-mm-thick gels).
10. Gel casting chamber that can hold 20 × 22 plates (Biorad). It is important to select the plates, spacers, and plate holders that are compatible with the chosen second dimension running tanks. Glass plates (20 × 20 cm and 20 × 22 cm) for SDS-PAGE (Web Scientific Ltd.). Each gel requires plates of two sizes to facili-

tate the transfer of first dimension tube gels or IPG strips to the second SDS-PAGE gels.

11. ProteanII XL spacers: 22.3 × 8 × 1.5 mm (Biorad).
12. Gradient mixer for gradient gels (Model 395 Gradient former: Biorad).
13. Fine forceps and palate knife with a rounded edge (BDH).
14. Biorad or Amersham gel running tanks compatible with plates and holders in 10.
15. For blotting: transfer tank large enough to accommodate 20 × 20 cm size gels: Electro-blot unit x3 (Consort, Web Scientific Ltd.)
16. Scanner to record gel images (Biorad GS700 scanner).

2.2. Reagents

All chemicals should be close to 99% pure where possible.

1. Phosphatase inhibitors: β-glycerophosphate (Sigma), sodium fluoride (BDH), and sodium orthovanadate (Sigma). 1 *M* stocks can be prepared for each of these compounds except sodium orthovanadate, which will be prepared at 100 m*M*, pH 10.0. Stock solutions can all be stored at 4°C.
2. Protease inhibitors: AEBSF, aprotinin, and leupeptin (Sigma), EDTA (BDH). Alternatively, complete protease inhibitor cocktail tablets (Roche) can be used. Store according to manufacturer's instructions.
3. Triton X-100 lysis buffer: 0.5 % Triton X-100, 150 m*M* NaCl, 2 m*M* DTT, 20 m*M* EDTA, 10 m*M* glycerophosphate, 1 m*M* sodium fluoride, 1 m*M* orthovanadate, 2 μg/mL aprotinin, 1 m*M* AEBSF, 2.5 μg/mL leupeptin (*see* **Note 2**).
4. DTT (Merck): Store at 4°C; use fresh or store 1 *M* stocks at –20°C.
5. Ampholytes (Amersham) or Resolytes (Biorad).
6. HPLC grade water (or double-distilled water).
7. Protein G or A Sepharose (Amersham).
8. 2D lysis buffer: 8 *M* urea (Merck ARISTAR), 2 *M* thiourea (Merck AnalR), 4% CHAPS, 65 m*M* DTT, 2% ampholytes or resolytes pH 3–10 (*see* **Note 3**).
9. Tube gel equilibration buffer: 62.5 m*M* Tris-HCl, pH 6.8, 3% SDS, 10% glycerol, 100 m*M* DTT, and 0.001% bromophenol blue.
10. IPG strip equilibration buffer: 6 *M* urea, 2% SDS (w/v), 1% DTT (w/v), 30% glycerol, and 50 m*M* Tris-HCl, pH 6.8 (*see* **Note 3**).
11. Mineral oil: DryStrip mineral oil (Amersham).
12. Ampholine mix (pH 3–10): 10% pH 2.5–5 (67 μL), 10% pH 3.5–5 (67 μL), 10% pH 4–6 (67 μL), 10% pH 7–9 (67 μL), 20% pH 5–7 (134 μL), and 40% pH 3–10 (168 μL).
13. CHAPS-NP40 stock solution: 0.3 g CHAPS, 0.1 mL NP40, and 0.9 g HPLC grade water.
14. Tube gel overlay solution: 12.012 g urea (8 *M*), 0.25 mL pH 6.5–9.0 Ampholine (1%), 5% NP40 (v/v), 0.386 g DTT (100 m*M*), and HPLC grade water to 25 mL. Aliquots can be stored at –20°C for future use.
15. Tris-HCl: pH 6.8, pH 7.4, pH 8.8 (in solution from Severn Biotech; dessicate from BDH, pH with HCl).

16. Acrylamide: 30% acrylamide and 0.8% bis for casting tube gels (Biorad); for casting second dimension gels, 40% solution PAGE, Plus-One (Amersham). Store at +4°C (*see* **Note 4**).
17. Immobiline Drystrips NL 3-10 (Amersham) or Biorad IPG strips (*see* **Note 5**). Store at –20°C.
18. Ammonium persulfate (Sigma): make fresh or store in 10% aliquots at –20°C.
19. TEMED (Sigma).
20. SDS-PAGE running buffer: 25 m*M* Tris base, 190 m*M* glycine, 0.1% SDS.
21. Second dimension overlay agarose: dissolve 0.5% molecular biology grade agarose in SDS-PAGE running buffer containing 0.001% bromophenol blue by microwaving at low heat until ready. Leave this solution to cool to about 50°C before applying to the gels.
22. For blotting: nitrocellulose or PVDF (Biorad), Whatman paper 3MM (Merck).
23. Transfer buffer: 20 m*M* Tris, 192 m*M* glycine, 0.1% SDS, 10% methanol.
24. Blocking solution: use 5% low fat milk or 5% BSA dissolved in TBST.
25. TBST: 20 m*M* Tris-HCl, pH 7.4, 138 m*M* NaCl, 0.05% Tween-20.
26. ECL Plus (NEN).
27. Silver staining solutions: prepare fresh solutions for all reagents

    Fixation solution: 50% ethanol or methanol, 10% acetic acid.

    50% ethanol or methanol.

    Sensitizer: 0.02% sodium thiosulfate pentahydrate.

    Silver: 0.1% silver nitrate.

    Developer: 0.05% formalin prepared from 37% formaldehyde, 2% sodium carbonate, 0.0004% sodium thiosulfate.

3. Methods

An overview of the procedure is shown in **Fig. 1**.

3.1. Sample Preparation

In this section, sample preparation for analyzing immunoprecipitated proteins, their subsequent separation by 2DE, and their detection by silver staining and Western blotting are described. If other methods such as GST or His-pull downs are used (*see* **Note 6**), prepare the samples for isoelectric focusing as described in previous chapters in this book and refer to **Subheading 3.1.2., steps 3–4** in this chapter for the final preparation. It is important to wear clean gloves at each major step of the procedure to ensure that the samples are free of external contaminants.

3.1.1. Cell Lysis

Cells are grown and treated according to the experimenter's needs.

1. Wash cells three times with PBS, scrape adherent cells, and transfer to a microcentrifuge tube (*see* **Note 7**).

2. Microcentrifuge cells for 3 s at 18,000*g* and remove the excess PBS.
3. Add lysis buffer (approx 100 μL per 1 × 10^6 cells and *see* **Note 8**), mix well by pipeting up and down, and incubate on ice for 15 min.
4. Microcentrifuge cells at 18,000*g*, 4°C for 15 min.
5. Separate the supernatant from the genomic DNA pellet by transferring it to a fresh microcentrifuge tube.
6. The supernatant is now ready to be assayed for protein content by a protein assay such as Bradford.

3.1.2. Isolation of Interacting Proteins: Immunoprecipitation With Antibodies Covalently Cross-Linked to Sepharose Beads

It is particularly useful for antibodies to be immobilized to Protein A or Protein G Sepharose beads when considering large-scale isolation of protein complexes. Overloading of immunoglobulins affects isoelectric focusing of proteins in the first dimension and also their separation in the second dimension. The immobilization of antibodies is able to overcome these problems (*see* **Note 9**).

1. Preclear the lysate: Wash Protein G or A Sepharose beads three times with the cell lysis buffer (20 μL of beads per sample). The Sepharose beads are usually delivered as a 50% slurry in 20% ethanol (*see* **Note 10**). Add 0.5–2 mg of protein lysate to 20 μL of Protein G or A Sepharose beads and rotate for 1 h at 4°C. This step removes any nonspecific binding of proteins to the beads.
2. Centrifuge the beads and transfer the precleared lysate to a microcentrifuge tube containing 10–20 μL of antibody cross-linked beads. Rotate for 2 h to overnight at 4°C.
3. Wash the beads three times with lysis buffer and once with a low-salt and detergent-free buffer (10 m*M* NaCl, 50 m*M* Tris-HCl, pH 7.4, 5 m*M* EDTA). Remove all excess buffer (*see* **Note 11**).
4. Add 2D-compatible lysis buffer containing 0.001% bromophenol blue and 2% pH 3–10 ampholytes at room temperature. For 18-cm-long IPG strips, the final volume should be 350 μL or follow manufacturer's instructions. If tube gels are used, the final volume does not have to be adjusted to 350 μL. The final loading volume depends on the space between the gel and the top of the glass tube. Use 2D lysis buffer to adjust the volumes. Mix the sample, incubate for a few minutes at room temperature, and centrifuge. Load the supernatant onto first dimension gels.

3.2. Isoelectric Focusing: the First Dimension

In this section, protocols for tube gels, pH range 3–10 (*see* **Subheading 3.2.1.**), NEPHGE tube gels, pH range 7–12 (*see* **Subheading 3.2.2.**), and IPG strips (*see* **Subheading 3.2.3.**), are given (*see* **Note 12**).

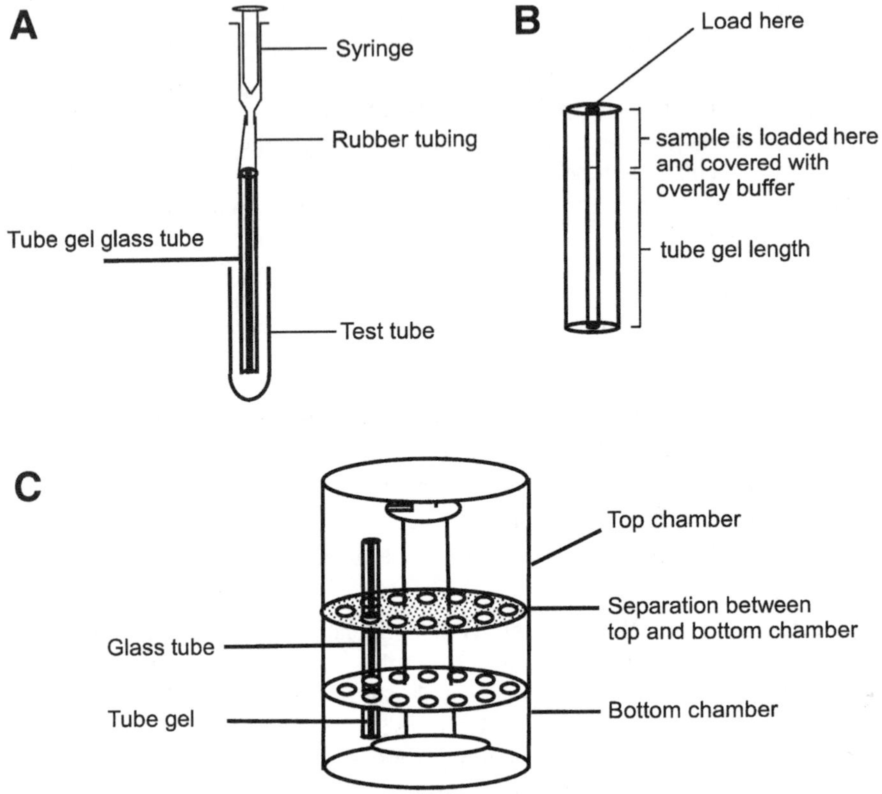

Fig. 2. Schematic of first dimension tube gel set up using Biorad's Tube Cell apparatus. **(A)** Tube gel casting setup. **(B)** Glass tube containing a tube gel. **(C)** Tube Cell apparatus containing one glass tube with gel.

3.2.1. Tube Gels, pH Range 3–10

1. For 10 tube gels, mix 6.68 g of urea (9.6 *M*), 1.688 mL of acrylamide stock (30% acrylamide, 0.8% bis-acrylamide), 688 µL of CHAPS-NP40 stock solution, and the following ampholytes: 135 µL pH 5–7, 135 µL pH 7–9, and 400 µL pH 3.5–10, and 4.666 mL of HPLC grade water in a small beaker stirring gently with a magnetic stirrer at room temperature (*see* **Note 3**). When dissolved, add 10 µL TEMED and 17 µL of 10% ammonium persulfate per 6 mL solution. Add 600 µL of the above solution to the test tube and place the glass tube in that test tube. Put some rubber tubing (about 3 cm long) on the end of an empty disposable 10 mL syringe and place the other end of the tubing onto the glass tube (*see* **Figs. 2A,B**). Slowly pull up the solution until it is 3 cm from the top of the glass tube. Leave the syringe on the tube for about 2 h or until the gel is set.

2. To load the samples, place the tube gels in the Tube Cell isoelectric focusing instrument and load the sample with a Hamilton syringe (*see* **Figs. 2B**, **2C**). Cover the sample with overlay solution until the top of the tube is reached. Rinse the syringe twice with 50% ethanol and twice with HPLC water between samples.
3. Fill the bottom chamber with 6 m*M* H_3PO_4 and the top chamber with 20 m*M* NaOH. Immerse the bottom and the top of the tubes with the corresponding buffer. The electrodes of the top chamber should come into contact with the top buffer (*see* **Fig. 2C**).
4. Electrophoresis conditions are as follows: 200 V for 2 h, 500 V for 5 h, 800 V for 4 h, 1200 V for 6 h, and 2000 V for 3 h. Set the power pack to no current check at start (*see* **Note 13**). Go to **Subheading 3.3.** for running the second dimension.

3.2.2. NEPHGE Tube Gels, pH Range 7–12

1. For six to eight gels mix: 4.12 g of urea, 0.975 mL of 30% acrylamide 0.8 % bis solution, 1.5 mL of HPLC water and the following ampholines: 50 µL pH 6.5–9, 140 µL pH 8–10.5, 50 µL 3.5–10, and 140 µL pH 5–8. Dissolve by gentle mixing at room temperature (*see* **Note 3**). Polymerize the gel with the addition of 10 µL of TEMED and 15 µL of 10% ammonium persulfate. Cast and load the tube gels as described in **Subheading 3.2.1., steps 1** and **2**.
2. Fill the bottom chamber with 20 m*M* NaOH and the top chamber with 10 m*M* H_3PO_4 (*see* **Fig. 2C**). See **Subheading 3.2.1., step 3** for more detail. Note that the buffers are inverted for NEPHGE. Electrodes are also inverted such that positive (red) goes to negative (black) and negative to positive. Electrophoresis is performed at 400 V for 12 h and 600 V for 8 h (*see* **Note 14**). It is necessary to set the power pack to no current check at start (*see* **Note 13**).
3. Go to **Subheading 3.3.** for running the second dimension.

3.2.3. IPG Strips

3.2.3.1. Loading and Rehydration of IPG Strips for Isoelectric Focusing With the Multiphor

1. IPG strips stored are –20°C should be allowed to thaw at room temperature for at least 15 min before loading the sample.
2. Ensure the IPG reswelling tray is level and orient it such that the pointed end is facing left. Spread the sample in 2D buffer (beads supernatant) along the whole length of a groove in the IPG reswelling tray. Remove plastic backing of an IPG strip and lay the strip gel-face down on the sample with the cationic end pointing left. The end is pointed if Amersham DryStrips or marked positive if Biorad IPG strips are used. No air bubbles should be trapped between the sample and the gel.
3. Cover the strip with 2 mL of DryStrip mineral oil to prevent sample evaporation and rehydrate for 10 h to overnight at room temperature (20–22°C).

3.2.3.2. Isoelectric Focusing With Multiphor (*see* **Note 15** and **Fig. 3**)

1. Transfer the strip to a plastic DryStrip aligner with the gel facing upward and the cationic end pointing left.

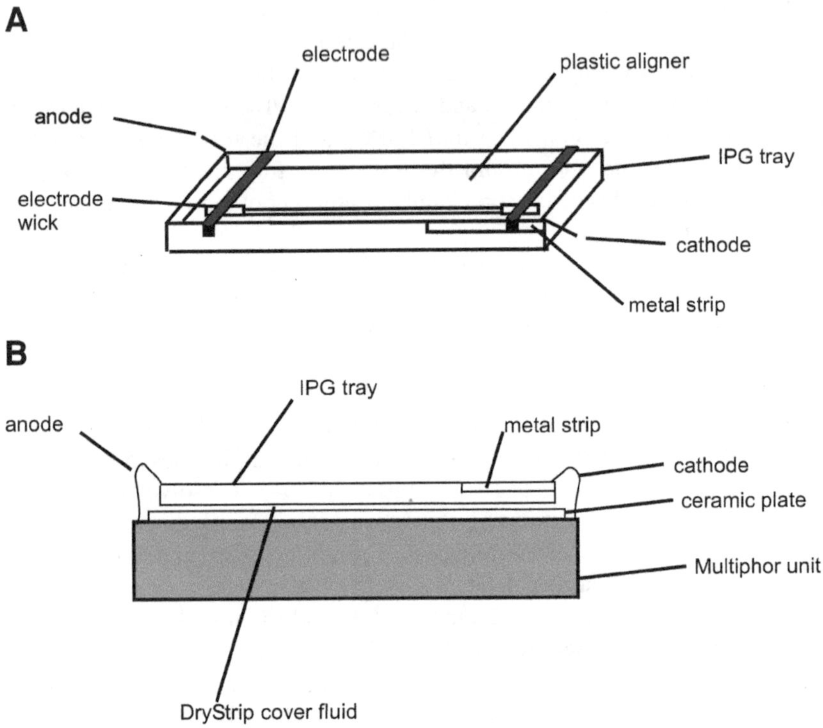

Fig. 3. Schematic of first dimension Multiphor setup. **(A)** IPG tray containing one IPG strip. **(B)** Side view of the Multiphor showing the position of the IPG kit components used for isoelectric focusing.

2. Cut electrode strips into 2-cm-long pieces, moisten them with 100 µL of HPLC water, and place a piece using forceps on each side of the strip, partially covering the gel (*see* **Fig. 3A**).
3. Pour DryStrip mineral oil onto the ceramic cooling plate on the Multiphor.
4. Place the Multiphor isoelectric focusing tray on the ceramic plate on top of the DryStrip mineral oil such that no air bubbles are trapped between the plate and the tray (*see* **Note 16** and **Fig. 3C**).
5. Place the DryStrip aligner containing the IPG strips loaded with sample on the tray. Fit the electrodes such that the red positive side is placed over the acidic side of the strips on the wicks (left) and the black negative over the wicks on the basic side (right). The electrodes should be in contact with the wicks (*see* **Fig. 3**).
6. Connect the electrode leads on the tray to the Multiphor.
7. Cover the strips completely with DryStrip mineral oil and close the Multiphor cover.
8. Set the water bath to 18°C.

Table 1
SDS-PAGE Gel Casting Protocol for One Gel (Approx 60 mL)[a]

	Gradient gel		Isocratic	
	8 % front	16% back	8% gel	12% gel
1.5 *M* Tris-HCl pH 8.8	8.50	8.50	15.0	15.0
40% Acrylamide	4.25	13.60	12.0	18.0
HPLC grade water	21.25	11.90	33.0	27.0
10% APS	0.12	0.12	0.24	0.24
TEMED	0.012	0.012	0.024	0.024

[a] All volumes are given in milliliters.

9. Set power pack to no current check at start (*see* **Note 13**). The following is a general focusing program for 18 cm IPG strips that can be varied as required: 300 V gradient for 30 min, 3500 V gradient for 6 h, and 3500 V step for 20 h. The current and power are maintained, respectively, at 5 mA and 10 W for each step.

3.3. Second Dimension: SDS-PAGE

3.3.1. Gel Casting

Cast the gels during the time the samples are undergoing isoelectric focusing. Depending on the proteins that are being analyzed, cast gradient or single percentage gels. The advantage of gradient gels is that proteins of a wide molecular weight range can be detected on one gel. For instance, 8–16% gels separate proteins in the range of 10 and 200 kDa. Single percentage gels, on the other hand, give a better purification of a particular mass range (12% gels separate 15–80 kDa proteins well) and, furthermore, the preparation of these gels is more reproducible. For 20 × 22 cm plates, about 60 mL of acrylamide mix is required. In **Table 1**, the components for either one 6–19% gradient gel or one 8% and one 12% gel are given. Pour the gels until the gel solution reaches 2–3 cm below the top of the shorter glass plate. Once cast, the gels are overlayed with 2 mL of HPLC water each and left to set overnight (*see* **Note 17**). Rinse the plates with water prior to transferring the first dimension gels to these second dimension ones.

3.3.2. Extrusion of Tube Gels From the Glass Tube and Transfer to the Second Dimension

Handle the tube gels carefully since they break easily.

1. Cut the top of a 200 μL tip and insert it onto the end of the syringe. Remove some of the overlay solution from the top of the tube gel, draw some water into the

syringe, and place the pipet tip into the orifice of the top of the tube gel as far as it will go and apply repetitive pressure on the syringe plunger to push the gel out of the glass tube (*see* **Note 18**). Deposit the gel into a small Petri dish containing tube gel equilibration buffer (*see* **Note 19**). Leave for a few minutes before transferring to the second dimension gels (*see* **Note 20**).
2. Pour out the water overlaying the second dimension gel.
3. Lay the tube gel onto the surface of the SDS-PAGE gel (if the tube gels break, they can be pieced together on the second dimension gel) (*see* **Note 21**).
4. Cover the tube gel with second dimension overlay agarose solution and let the agarose set before transferring the gels to the second dimension running tanks.

3.3.3. IPG Strip Transfer to the Second Dimension

1. Submerge the strips in IPG strip equilibration buffer for 15 min. Rock the strips gently avoiding their movement. (*see* **Notes 22** and **23**).
2. Pour out the overlaying water from the second dimension gel.
3. To keep the IPG strips in place during the second dimension run and to monitor the distance the proteins have run, pour as much as possible of the melted 2nd dimension overlay agarose solution over the top of the SDS-PAGE gel (*see* **Note 24**).
4. Transfer the strips with the acidic end pointing left before the agarose sets. Use forceps and a palate knife to help transfer the strips.
5. Let the agarose set before transferring the gels to the second dimension running tanks.

3.3.4. Running Second Dimension SDS-PAGE Gels

The SDS-PAGE tank is filled with SDS-PAGE running buffer until the bottom of the gel is covered and the top chamber is filled. Cool the running tanks to 15°C with a circulating water bath. Cooling ensures that the plates do not crack and the proteins separate well. At high temperatures protein streaking can occur. Electrophoresis at 30–40 mA per gel typically takes approx 3–4 h. At 5–7 mA it takes between 12 and 16 h. If gels are run at a high current setting, add more running buffer to the tank to ensure a better cooling of the gels.

3.4. Protein Detection: Gel Staining and Electroblotting for Immunodetection

At this point, the proteins in the SDS-PAGE gel can either be transferred to nitrocellulose or PVDF, or be fixed and stained. Transfer of proteins onto to nitrocellulose or PVDF permits blotting and analysis of specific interactions of a protein of interest using immunodetection. Fixing and staining of gels as well as staining blots permits the analysis of the number of interacting proteins. In both cases, protein spots can be cut out for analysis by mass spectrometry.

3.4.1. Protein Transfer to Nitrocellulose and Immunodetection (for PVDF, see **Note 25**)

1. Cut Whatman paper and nitrocellulose to 20 × 20 cm and soak in transfer buffer.
2. Open the transfer cassettes and submerge one side in a large basin containing transfer buffer. Place one sponge on the cassette and a piece of wet Whatman paper on the sponge.
3. Separate the glass plates, discard the strip and agarose, and cut the top right corner of the gel (basic end) to help orient the gel.
4. Transfer the gel onto the Whatman paper on the cassette so that the cut corner is on the top left, wet the nitrocelluose in transfer buffer, cut one corner, and place it on the gel. This way the same view on the blot is maintained as during the run. Roll out the air bubbles between the nitrocellulose and the gel, cover the gel with wet Whatman paper followed by the second moist sponge, close the cassette, and place it into the transfer tank such that the gel is facing the negative electrode and the nitrocellulose membrane toward the positive one.
5. Perform electrophoretic transfer at 20 V for 4 h to overnight.
6. Rinse the blot with HPLC grade water, remove any excess acrylamide, and incubate in blocking solution 1 h.
7. Rinse the membrane twice for 5 min each with TBST.
8. Add enough primary antibody to cover the blot and leave gently shaking for 1–2 h at room temperature or overnight at 4°C (*see* **Note 26**).
9. Wash the blot three times for 10 min each with TBST.
10. Add secondary antibody to the blot and leave to incubate with gentle shaking for 1 h.
11. Wash four times for 10 min each with TBST.
12. Perform ECL by mixing solutions 1 and 2 at a 1:1 ratio, i.e., mixing 20 mL of solution 1 with 20 mL of solution 2. Add the ECL mix to the blot and incubate for 1 min with gentle agitation. The same solution can be reused for other blots.
13. Place the blot on Saran wrap, absorb the excess ECL with paper towel, cover the blot with Saran wrap, place in a film cassette, and expose to film (*see* **Note 27** and **Fig. 4A**).

3.4.2. Silver Staining Gels

Not all silver staining methods are compatible with mass spectrometry analysis and those that are tend to be less sensitive. Several companies (e.g., Invitrogen) sell silver staining kits and many silver staining methods have been published. The following is a protocol modified from Shevchenko et al. ***(11)*** (*see* **Notes 28**, **29**, and **30**).

1. Fix the proteins in the SDS-PAGE gel in 10% acetic acid and 50% ethanol for 20 min at room temperature or at 4°C overnight.
2. Wash the gel for 10 min in 50% ethanol.
3. Wash the gel three times for 5 min each in HPLC grade water to remove the acid.
4. Incubate the gel for 1 min in sensitizer.

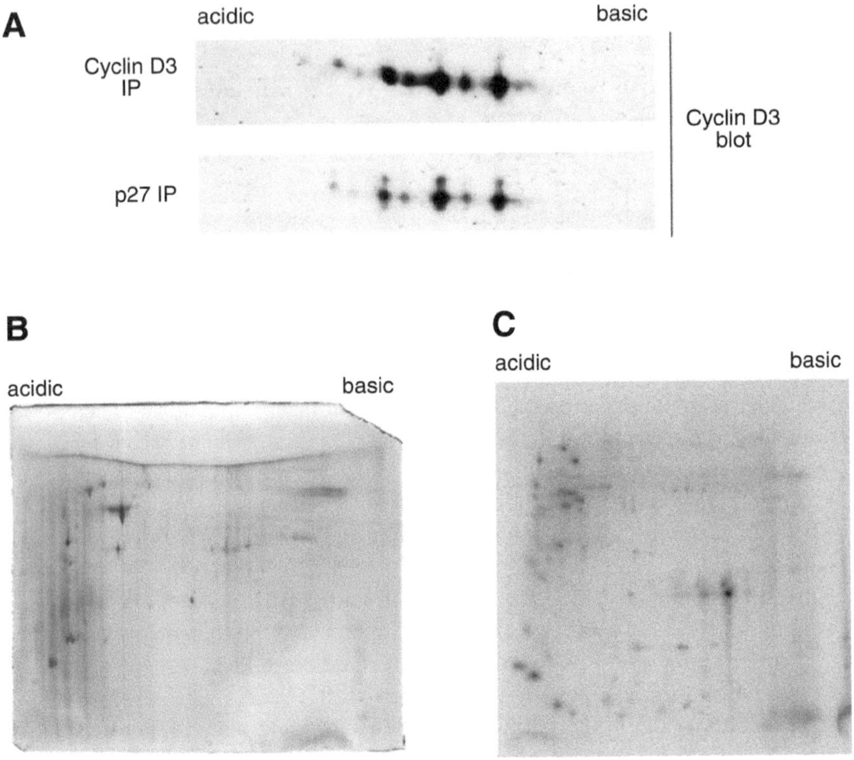

Fig. 4. Examples showing analysis of protein complexes by 2DE using pH 3.5–10 NL IPG strips. **(A)** Protein isoform specific interactions between $p27^{kip1}$ and cyclin D3. The top panel shows a 2D immunoblot of cyclin D3 immunoprecipitated from BL40 cell lysate using Mab DCS22 (Biosource) covalently cross-linked to Protein G beads and probed for cyclin D3. In the bottom panel, the same BL40 lysate was immunoprecipitated with a polyclonal $p27^{kip1}$ antibody (C19, Santa Cruz) immobilized to Protein G and the 2D blot probed for cyclin D3. **(B)** Silver stained gel of CAL51 lysate immunoprecipitated with polyclonal $p27^{kip1}$ antibody (C19, Santa Cruz) immobilized to Protein G. **(C)** ^{32}P detection of phosphorylated proteins binding to in vitro phosphorylated GST-$p27^{kip1}$ using BL40 lysate as kinase. GST was cleaved from $p27^{kip1}$ prior to loading an IPG strip. The gel was transferred to a blot and the blot exposed to a PhosphoImager screen.

5. Rinse the gel twice for exactly 1 min in HPLC grade water.
6. Submerge the gel in silver nitrate ($AgNO_3$) for 20 min.
7. Discard the $AgNO_3$ and rinse twice in HPLC water for exactly 1 min.
8. Incubate in developer solution. Replace the developer before it turns yellow with fresh solution. When the desired staining intensity is reached, discard the developer and soak the gel in 5% acetic acid.

9. Scan the gels for record keeping.
10. Store the gel in 1% acetic acid at 4°C until used, e.g., spot picking for mass spectrometric analysis (*see* **Fig. 4B**).

4. Notes

1. Mutiphor vs IPGphor: Both instruments are produced by Amersham. The Multiphor is more versatile and also has other applications besides isoelectric focusing. The Multiphor can also be used, for example, to separate ^{32}P-labeled phosphopeptides on thin layer chromatography plates ***(10)***.
2. Aliquots of the lysis buffer can be stored at –20°C because most protease inhibitors and reducing agents are unstable. Alternatively, a lysis buffer can be freshly prepared with the inhibitors and DTT added prior to use.
3. When preparing buffers containing urea, add water progressively because the urea will take up most of the volume. Do not heat the urea but let it dissolve slowly at room temperature, otherwise protein carbamylation occurs. Depending on the volume of the solutions they may take more than 1 h to prepare.
4. Other acrylamide stocks that are commonly used for 1D SDS-PAGE, such as 30% acrylamide/bis (37.5:1), can also be used.
5. The advantage of using NL pH 3–10 versus linear pH 3–10 IPG strips is that the focusing is continuous over the whole pH range, especially between pH 5 and 7. One typically sees a gap in the first dimension when linear pH 3–10 IPG strips are used.
6. Care must be taken if His-tagged proteins are used to pull down protein complexes. Both low pH or high imidazole concentrations required to elute the proteins off the IMAC beads are not compatible with isoelectric focusing. Imidazole can be used but should be removed by dialysis, by buffer exchange, or with Centricons (Amicon) prior to addition of lysis buffer.
7. If the post-translational modifications are unstable or if the proteins degrade rapidly, it may be necessary to lyse adherent cells directly on the tissue culture dish. In this case, wash cells three times with ice cold PBS, remove as much PBS as possible, add 200 μL of lysis buffer per 10 cm dish or enough to cover the cells, scrape the cells on ice and transfer the lysate to a microcentrifuge tube. Leave to incubate on ice for 15 min, microcentrifuge for 15 min at 18,000*g* at 4°C. Determine protein concentration of the supernatant.
8. Most standard lysis buffers used for SDS-PAGE can be used for lysing cells, e.g., NP40-based lysis buffers and RIPA. It must be noted that 2D lysis buffer is not compatible with immunoprecipitation because it prevents the binding of the proteins to the beads. Indeed, 2D lysis buffer is used to remove the complex from the antibody-bound beads (*see* **Subheading 3.1.2., step 4**).
9. Covalent cross-linking of antibodies to Sepharose A or G beads with dimethly pimelidate (DMP). With DMP, Protein A or G Sepharose beads can be used. It is important to check which subtype the antibody is and use the appropriate sepharose beads (*see* ref. ***12***) for more details). The following procedure is modified from ref. ***12***). With Protein G or A Sepharose beads, purified antibodies are

bound to Sepharose beads for 2 h by rotating at 4°C using about 500 μg of antibody per 500 μL of beads. The antibody is diluted to a volume that allows the antibody and the beads to mix well. The antibody-bound beads are washed three times with 0.1 *M* boric acid, pH 9.0, and resuspended in 10 mL of the wash solution in a 50 mL Falcon tube. In a separate tube, 0.5 g of DMP is dissolved in 40 mL of 0.1 *M* sodium tetraborate (borax), pH 9.0, and added to the antibody. It is important to note that DMP is to be added to the antibody within 5 min of dissolving it because it is rapidly inactivated. Furthermore, the ratio of boric acid to sodium tetraborate 1:4 (v/v) must be maintained for the cross-linking of antibody to beads to work. Owing to the unstable nature of DMP, a 1 g aliquot (Sigma) is dissolved in 1–2 mL of sodium tetraborate immediately before use and transferred equally into two 50 mL tubes. The DMP in each tube is diluted to 40 mL. Each tube is used for one set of beads in 10 mL of boric acid solution. The antibody-bound beads and dimethyl pimelidate cross-linking solution is left 1 h to rotate at room temperature. The beads are then resuspended in 0.2 *M* ethanolamine, pH 8.0, and left to rotate for 1 h at room temperature to wash off any antibody that did not become covalently bound to the beads. Finally, the beads are washed three times in PBS and stored in PBS containing 0.01% thimerasol (also known as merthiolate) or 0.005% sodium azide at 4°C. These beads can be used directly for immunoprecipitations.

10. This step consists of resuspending the beads in lysis buffer, centrifuging them at 18,000*g* for a few seconds, and aspirating the supernatant, carefully avoiding removal of the beads. This is repeated twice. Use of gel loading tips or a 23G needle helps reduce the loss of beads.
11. The final wash step is important because high detergent and salt concentrations are not compatible with isoelectric focusing.
12. It is useful to determine the pI of the proteins of interest. Commercial pI standards are available (e.g., Biorad) and can be run on a gel in parallel with the others. I tend to start my investigation by analyzing protein separation on a wide pH range i.e., between pH 3 and 10. The pH range can be narrowed down to pH ranges separated by one to two pH units. Narrowing down the pH range leads to a better purification of a specific protein of interest and reduces the number of proteins in a given spot. If only a few proteins are investigated, their theoretical pI can be determined using pI/MW determination on the EXPASY website (http://ca.expasy.org/tools/pi_tool.html) given that the net charge is zero. Post-translational modifications such as phosphorylation adds a negative charge to a protein and renders it more acidic.
13. At the start of isoelectric focusing, the current is initially undetectable. It is therefore necessary to have a power pack that will not stop when no load is detected.
14. With NEPHGE it is important to adhere to the given focusing time because a steady state is never attained.
15. Isoelectric focusing with the IPGphor is similar, but there is one step less than with the Multiphor. Loading and focusing are performed in ceramic tubs directly on the instrument. Loading can be performed at low voltage (30 V) to increase

sample uptake during rehydration. After 10–12 h, the isoelectric focusing program can be started. The same focusing program and temperature setting at 18°C as for the Multiphor is used.

16. The DryStrip mineral oil—on the ceramic plate and under the isoelectric focusing tray—helps ensure an even thermal transfer between the ceramic plate and the tray. Air bubbles prevent even thermal transfer.
17. Cover the gels with a moist paper towel and Saran wrap to prevent the water overlaying the gels to evaporate.
18. NEPHGE gels are more difficult to extrude from the tube because a bulge forms near the bottom during isoelectric focusing.
19. Gel equilibration ensures the efficient transfer of the proteins from the first dimension to the second dimension. It is important not to prolong the equilibration time as the proteins in the tube gel may diffuse out.
20. At this point, it is possible to store the tube gels in equilibration buffer at –80°C.
21. To facilitate this procedure, it is best at this point to carefully transfer the tube gels onto a clean glass plate, and slide the tube gel with the help of some equilibration buffer onto the second dimension gel by gently orienting the tube gel with a spatula that has rounded edges.
22. After isoelectric focusing, and prior to equilibration, it is possible to freeze the strips by covering them with aluminium foil and storing them at –80°C. When ready for IPG strip transfer to the second dimension, let them thaw and equilibrate as described in **Subheading 3.3.3.**
23. It is advisable to alkylate free cysteines. This will prevent vertical streaking of the proteins ***(13)*** and the formation of acrylamide adducts that complicate mass spectra. For alkylation, remove the equilibration buffer containing DTT and replace it with equilibration buffer containing 245 m*M* iodoaceta-mide (IAA) for 15 min. Equilibration buffer that does not contain DTT or IAA can be prepared in advance and stored at room temperature.
24. If many gels are run, keep the agarose heating and stirring at 50°C. Let the agarose set before transferring the gels to running tanks.
25. Protein transfer to PVDF membranes is similar to nitrocellulose except that the membrane first has to be activated by soaking in methanol prior to the transfer buffer.
26. If antibody probing is done in a box, at least 40 mL is needed to cover a 20 × 20 cm size blot. This can also be done in sealed plastic bags where only 5 mL of antibody solution is needed to cover the blot, provided that air bubbles are rolled out. If several blots are probed with the same antibody, they can be placed back to back so that 5 or 40 mL are sufficient for two blots.
27. Depending on the antibody and the quantity of blots analyzed, it may be useful to expose the blots to three or four films (i.e., placing the films on top of each other) simultaneously for 4 h to overnight. This way multiple exposures are obtained without testing for appropriate exposure times.
28. It is important to handle the gels as little as possible during silver staining to avoid fingerprint marks on the gel.

29. For high abundant proteins such as cytoskeletal proteins, they may be detected using Coomassie blue stain. Many inexpensive staining protocols are available, but GelCode (Pierce) reagent is preferable because it is rapid, gives a low background without a destaining step, and is compatible with mass spectrometry analysis. If protein spots stained with GelCode are to be cut out in the same day, it is not necessary to fix the gel. This may yield higher peptide concentrations for mass spectrometry. To fix the proteins in the SDS-PAGE gel, incubate the gel in 10% acetic acid and 50% ethanol for at least 15 min at room temperature or 4°C overnight. For staining with GelCode blue stain, wash the fixed gels four times for 5 min each with HPLC grade water to remove the SDS in the gel. Immerse the gel in GelCode blue stain for 1 h at room temperature to overnight (at 4°C) or until the required staining is achieved. Discard the GelCode solution and replace it with HPLC grade water for 30 min to 1 h. This last step enhances the staining.
30. For less abundant proteins, it is advisable to use silver stain, Sypro Rubie stain (requires a laser scanner to detect the proteins), or zinc reverse stain. Proteins from ^{35}S-methionine or ^{32}P/^{33}P-labeled cells can be detected with a PhosphoImager. A good zinc staining protocol is given in ref. ***14***. *See* **Fig 4C** for an application of a pull-down of ^{32}P-labeled phosphoproteins.

Acknowledgments

Research is funded by Kay Kendall Leukemia Fund and by the Ludwig Institute for Cancer Research. I would like to thank B. Geering, Drs A. Eddaoudi, V. Kanagasundaram, and G. Warnasuriya for their careful reading of the manuscript.

References

1. O'Farrell, P. H. (1975) High resolution two-dimensional electrophoresis of proteins. *J. Biol. Chem.* **250,** 4007–4021.
2. Gavin, A. C., Bosche, M., Krause, R., et al. (2002) Functional organization of the yeast proteome by systematic analysis of protein complexes. *Nature* **415,** 141–147.
3. Ho, Y., Gruhler, A., Heilbut, A., et al. (2002) Systematic identification of protein complexes in *Saccharomyces cerevisiae* by mass spectrometry. *Nature* **415,** 180–183.
4. Puig, O., Caspary, F., Rigaut, G., et al. (2001) The tandem affinity purification (TAP) method: a general procedure of protein complex purification. *Methods* **24,** 218–229.
5. Rigaut, G., Shevchenko, A., Rutz, B., Wilm, M., Mann, M., & Seraphin, B. (1999) A generic protein purification method for protein complex characterization and proteome exploration. *Nat. Biotechnol.* **17,** 1030–1032.
6. Gygi, S. P., Rist, B., Gerber, S. A., Turecek, F., Gelb, M. H., and Aebersold, R. (1999) Quantitative analysis of complex protein mixtures using isotope-coded affinity tags. *Nat. Biotechnol.* **17,** 994–999.
7. O'Farrell, P. Z., Goodman, H. M., and O'Farrell, P. H. (1977) High resolution two-dimensional electrophoresis of basic as well as acidic proteins. *Cell* **12,** 1133–1141.

8. Görg, A., Boguth, G., Obermaier, C., Posch, A., and Weiss, W. (1995) Two-dimensional polyacrylamide gel electrophoresis with immobilized pH gradients in the first dimension (IPG-Dalt): the state of the art and the controversy of vertical versus horizontal systems. *Electrophoresis* **16,** 1079–1086.
9. Gianazza, E. (1999) Casting immobilized pH gradients (IPGs). *Methods Mol. Biol.* **112,** 175–188.
10. Berkelman, T, and Stenstedt, T. (1998) 2D Electrophoresis Using Immoblized pH Gradients—Principles and Methods. http://www4.amershambiosciences.com/aptrix/upp00919.nsf/(FileDownload)?OpenAgent&docid=DA084A2108EB6FAFC1256A5E0053256A&file=80-6429-60AB_Version_May_2002.pdf.
11. Shevchenko, A., Wilm, M., Vorm, O., and Mann, M. (1996). Mass spectrometric sequencing of proteins silver-stained polyacrylamide gels. *Anal. Chem.* **68,** 850–858.
12. Harlow, E. and Lane, D., eds. (1988) *Antibodies: A Laboratory Manual.* Cold Spring Harbor Laboratory, Cold Spring Harbor, NY, pp. xiii, 726.
13. Görg, A., Postel, W., and Gunther, S. (1988). The current state of two-dimensional electrophoresis with immobilized pH gradients. *Electrophoresis* **9,** 531–546.
14. Matsui, N. M., Smith-Beckerman, D. M., and Epstein, L. B. (1999). Staining of preparative 2-D gels. Coomassie blue and imidazole-zinc negative staining. *Methods Mol. Biol.* **112,** 307–311.

33

Sample Preparation of Gel Electrophoretically Separated Protein Binding Partners for Analysis by Mass Spectrometry

Rainer Cramer, Malcolm Saxton, and Karin Barnouin

Abstract

The identification and characterization of binding partners from protein complexes is increasingly undertaken by mass spectrometry because of its high sensitivity and expedient elucidation of protein structure by accurate mass measurement. A variety of affinity purification methods including immunoprecipitation and glutathione-*S*-transferase (GST) pull-downs are commonly employed for the isolation of protein complexes and coupled to gel electrophoresis for further separation and basic information with regard to their constituents. For the successful analysis of gel-separated proteins by mass spectrometry, additional sample preparation steps involving sample clean-up, proteolysis, and peptide recovery are essential. This chapter describes the important procedure of in-gel digestion with particular emphasis on maximum peptide recovery and compatibility for subsequent mass spectrometric analysis.

Key Words

Mass spectrometry; in-gel digestion; trypsin; proteomics; protein binding partners.

1. Introduction

Mass spectrometry (MS) has become a very powerful analytical technique in the identification of proteins in complexes. Considerable advances have recently been made in the development of mass spectrometric analysis of nondissociated noncovalent protein complexes *(1)*. MS has been used extensively to study the interaction between small-molecular-weight ligands and a protein. However, it still remains difficult to ionize and therefore to analyze large protein–protein complexes *(1,2)*. Although recent technological advances have enabled more direct MS studies of intact protein complexes, identifica-

From: *Methods in Molecular Biology, vol. 261: Protein–Protein Interactions: Methods and Protocols*
Edited by: H. Fu © Humana Press Inc., Totowa, NJ

tion and characterization of protein binding partners is comparatively better achieved by affinity purification strategies and the so-called "bottom-up" approach involving enzymatic protein digestion and peptide analysis by MS *(3)*. Affinity purifications can be of simple or more complex nature involving secondary antibodies, specific cleavage of fusion proteins, and/or extra purification steps.

In basic immunoaffinity-based methods, proteins of interest are immunoprecipitated using well-defined antibodies for the co-purification and subsequent analysis of their associated binding partners (*see* **Chapter 23** on co-immunoprecipitation). Alternatively, fusion proteins may be used by adding a tag that can either be immunoprecipitated (e.g., FLAG-, VSV-, and Myc-tag; *see* **Chapter 23**) or pulled down with the appropriate affinity resin (GST- or HIS-tag) (*see* **Chapter 13** on GST-fusion protein). FLAG-tag and tandem affinity purification (TAP) have also been used successfully in the identification of novel protein binding partners as recently demonstrated in yeast using mass spectrometry *(4,5)*. The TAP method, being more elaborated, uses two affinity tags that are intercepted by a protease cleavage site. The purification method involves a two-step process prior to analysis by MS. TAP has been mainly used in yeast where relatively large amounts of protein can be obtained.

After the affinity purification step, protein binding partners can be either further purified/separated as intact proteins using protein separation techniques such as gel electrophoresis or digested with subsequent purification/separation of their peptides using, for instance, liquid chromatography. In multidimensional liquid chromatography, peptide digest mixtures from isolated protein complexes are often separated by ion exchange chromatography followed by reverse phase chromatography on-line coupled to electrospray ionization (ESI)-MS(/MS) *(6,7)*. One advantage of multidimensional liquid chromatography is that this method is amenable to automation *(8)*.

Gel electrophoretic systems, on the other hand, separate intact protein binding partners or, in the case of blue native gels, even the intact isolated protein complexes *(9,10)*. Proteins in the excised gel bands or spots are then digested before mass spectrometric analysis takes place. There are several advantages associated with the use of gel electrophoresis in the analysis of protein complexes. One-dimensional gel electrophoretic separation (e.g., SDS-PAGE) is relatively easy to perform and therefore frequently used in protein analysis. Generally, gel electrophoretic separation and its imaging is an excellent tool to visualize the occurrence of protein isoforms, modifications, degradation, or other types of protein changes. In particular, two-dimensional gel electrophoresis with its additional dimension of isoelectric focusing and therefore high separation and resolution power can provide extremely valuable information on protein changes (*see* **Chapter 32** on 2D gel electrophoresis). These and the obvious advantages of an additional purification/separation are probably the main reasons that gel electrophoresis is and will be an essential tool in the analysis of protein complexes for some time.

In proteomics, mass spectrometric analysis of proteins relies mainly on the generation of small peptides by proteolysis using endoproteases that specifically cleave at certain amino acid residues. Peptides in the mass range of 1–3 kDa are particularly amenable to both MS and tandem MS (MS/MS) analysis. Furthermore, the specificity of enzymatic peptide bond cleavage and the high mass accuracy in this mass range can efficiently restrict the data volume in protein database searches and, more important, increase the success rate and confidence in protein identification and characterization by both peptide mass mapping and analysis of fragmentation data, which have been obtained by tandem MS (for reviews *see* refs. ***11*** and ***12***).

However, peptide mixtures from in-gel digestion of proteins are contaminated with salts, gel-derived products, and other reagents and have to be purified by liquid chromatography before one of the two main "soft" ionization techniques in biological MS, ESI, can be successfully employed. The other main "soft" ionization technique, matrix-assisted laser desorption/ionization (MALDI), can be used alternatively because it is more tolerant of these contaminations. However, ESI is usually preferred for tandem MS because it produces multiply charged peptide precursor ions, from which simpler and for sequencing easier interpretable fragmentation spectra can be recorded. MALDI MS, although more tolerant of contaminations, would obviously also benefit from purified samples.

In summary, in-gel digestion is an important link between the gel electrophoretic purification/separation of binding partners from protein complexes and their successful analysis by mass spectrometry. Protocols for in-gel digestion should provide small enzyme-specific proteolytic peptides, quantitative reactions, a minimal level of sample contamination, and obviously little or no loss of sample.

2. Materials

2.1. Chemicals and Consumables

1. Ammonium bicarbonate (Sigma).
2. Dithiothreitol (DTT) (Perbio Science, Cheshire, UK).
3. Iodoacetamide (Sigma).
4. Acetonitrile (Rathburn).
5. Trifluoroacetic acid (TFA) (Fischer Scientific).
6. Formic acid (Fischer Scientific).
7. HPLC grade water (Rathburn).
8. Hydrogen peroxide (H_2O_2) (Sigma).

9. Modified porcine trypsin (Promega).
10. Natural siliconized tubes, 0.5 mL (Bioquote).
11. Multiflex round tips, 200 μL (Gel Loader Tips) (Bioquote).

2.2. Solutions Needed for Digest

1. Solution A: 5 m*M* ammonium bicarbonate, pH 8.0.
2. Solution B: 1% H_2O_2.
3. Solution C: 50% acetonitrile.
4. Solution D: 10 m*M* DTT made up in solution A.
5. Solution E: 50 m*M* iodoacetamide made up in solution A.
6. Solution F: Solution A containing 50 ng trypsin; the volume of solution F to be used depends on the size of the gel pieces (*see* **Subheading 3**).
7. Solution G: 50% acetonitrile/5% TFA.
8. Solution H: 0.1% formic acid.

3. Methods

3.1. General Considerations

Depending on the sample and its predigest treatment, various aspects have to be considered for an optimized in-gel digestion protocol for mass spectrometry. For proteins with few lysine residues, for instance, the use of endoprotease Lys-C for proteolysis could result in peptides with masses of more than 4 kDa. If some of these proteins, however, have a suitable distribution of aspartic acids, the use of endoprotease Asp-N could provide a peptide mixture with peptide masses in the optimal 1–3 kDa mass range. Gel samples, which have been stored for a long time, are often partially oxidized at their methionine residues. This results in ion signal splitting and therefore reduced analytical sensitivity. If this mass split is not to be used for its structural information or other reasons, the sample should be fully oxidized to optimize the detection of methionine-containing peptides and to reduce the complexity of the peptide mixture. These two examples show that prior knowledge of the status quo of the sample and the downstream analysis can help to define optimal protocols for further sample preparation after gel electrophoretic separation. Additionally, samples can be prepared and run on gels in a way that is more compatible with its subsequent treatment of in-gel digestion and mass spectrometric analysis. The following sections will discuss a few general considerations and give a protocol for a successful characterization of gel-separated proteins by MS using in-gel digestion.

3.1.1. Importance of Sample Purity

Under ideal conditions, MS can detect biomolecular ions at the attomole level. To achieve this level of analytical sensitivity, it is important that any sample preparation protocol for MS analysis ultimately creates sample conditions resembling these ideal conditions. In this context, sample purity is pivotal. Samples should be prepared in a clean environment, preferably in a clean room that has filtered ventilation. Furthermore, clean lab coats, (powder free) gloves, face masks, and hair covers should be worn to minimize sample contamination. Leaning over gels should be avoided as much as possible. It is essential that all materials used in the preparation of samples for MS starting with harvesting cells down to cutting gel plugs are clean and, where possible, even sterile. Cleanliness is important because dust particles, keratins from hair and skin, or other contaminants may not only produce additional spots and background noise on the gel image but also mask or suppress the mass spectrometric signal of the peptide ions of interest.

3.1.2. Isolation of Binding Partners From Protein Complexes

In the analysis of protein complexes by the combination of gel electrophoresis and mass spectrometry, it is advisable to avoid simple immunoaffinity-based methods where possible, especially if proteins of interest migrate at 25–30 and 50–60 kDa because antibody light and heavy chains co-migrate at these molecular weights. Furthermore, after in-gel digestion the presence of immunoglobulin peptides in the sample may mask peptide ion signals of interest resulting in low-quality data for the identification or characterization of the protein. For simple immunoaffinity purification methods, it is therefore advisable to covalently link antibodies to beads to reduce this problem (*see* **Chapter 32** on 2D gel analysis for the analysis of protein complexes). Other methods used in the isolation of protein complexes that are compatible with MS include the use of fusion proteins as outlined in **Subheading 1.** However, problems associated with the occurrence of antibody fragments or other contaminants can still occur. Therefore, the choice of the protein tag and elution conditions is obviously important. Unfortunately, no general guideline can be given because the right choice is highly dependent on the protein of interest and can often only be found through trial and error. Furthermore, in the case of fusion proteins, proteins that bind to the fusion protein may interfere with the accurate assembly of the protein complex leading to the loss of native binding partners or the addition of protein binding artifacts. To avoid these potential problems, it may be necessary to separate the protein from its tag by cleavage combined with an additional purification step as is done in the more complex TAP method ***(13)***.

3.1.3. Gel Electrophoresis

Polyacrylamide gels, on which samples are run for subsequent mass spectrometric analysis, should be left to polymerize fully, i.e., overnight. This reduces the amount of free acrylamide that can react with cysteine residues, and therefore the complexity of the peptide mixture. Ideally, samples should be reduced and alkylated prior to any electrophoretic separation ***(14)***. Besides decreasing the complexity of the peptide mixture and eliminating the reduction and alkylation step in in-gel digestion protocols, this early reduction and alkylation step can also be beneficial for gel electrophoresis ***(14,15)***. However, if SDS is present in the sample buffer, alkylation is severely suppressed. Alkylation is also suppressed in urea with the alkylation reagent iodoacetamide as reported by Galvani et al. ***(16)***. At least for one-dimensional SDS-PAGE, these suppression effects might be the reason that many groups reduce and alkylate cysteines after gel electrophoretic separation.

3.1.4. Gel Imaging

Several gel staining methods that are compatible with MS have been described ***(17)***. One of these is the popular Coomassie staining, although the detection limit of image analysis is around 500 fmol for an approx 70 kDa protein such as bovine serum albumin (BSA) loaded on the gel. Silver staining is more sensitive but the silver staining methods that are compatible with MS are among those that are less sensitive although still superior to Coomassie staining. With the SilverQuest™ silver staining kit (Invitrogen), 125 fmol of BSA can be detected. Another MS compatible silver staining method, based on Shevchenko et al. ***(18)***, is described in the chapter on two-dimensional gel electrophoresis for the analysis of protein complexes. Zinc acetate-reverse staining is also among the more sensitive methods and can detect low-molecular-weight proteins such as casein down to 60 fmol, but the proteins are difficult to see and the staining is unstable. Another good MS compatible method that detects proteins at a similar sensitivity level as silver uses the fluorescent dye Sypro Ruby™ (Cambridge BioScience) that requires fluorescence imaging equipment.

3.1.5. Proteolysis

Enzymes are most commonly used for proteolysis in in-gel digestion protocols, although chemical compounds such as cyanogen bromide, hydrogen chloride, or formic acid can also cleave peptide bonds ***(19,20)***. Enzymes, however, show a much greater specificity in proteolysis and are therefore generally preferred. It must be noted that post-translational modifications or other steric conformations may hinder enzymatic digestion resulting in missed cleavages and hence longer peptides (*see* **Table 1**). Of all proteases trypsin is usually the enzyme of choice as it cleaves at the carboxyl side of lysine and arginine residues. Proteins digested with trypsin generally result in peptides that fall in the above mentioned optimal mass range of 1–3 kDa and have C-termini that sup-

Table 1
Commonly Used Enzymes for Protein Digestions[a]

Enzyme	Preferred cleavage site	Notes
Trypsin	C-terminally to K or R	No cleavage occurs when P is C-terminal of K or R
Glu-C		
(bicarbonate)	C-terminally to E	No cleavage occurs when P is C-terminal of D or E
(phosphate)	C-terminally to D or E	
Arg-C	C-terminally to R	No cleavage occurs when P is C-terminal of R
Asp N	N-terminally to D	
Lys-C	C-terminally to K	No cleavage occurs when P is C-term of K
Chymotrypsin	C-terminally to F, L, M,W, Y; or	There are multiple versions of chymotrypsin.
	C-terminally to F, L, K, R, W, Y	No cleavage occurs when P is C-terminal of all of these residues as well as when P is N-terminal to Y
Pepsin	C-terminally to F, L, Y	Pepsin is active at pH 1–4. No cleavage occurs when A, G, or V is C-terminal of all of these residues

[a] For further information *see* also http://prowl.rockefeller.edu/recipes/contents.htm.

port mass spectrometric analysis owing to the positive charge of lysines and arginines. After an initial analysis it can be beneficial to use a different protease or a combination of proteases that can lead to the detection of additional peptide ions and therefore an increase in protein sequence coverage. Other enzymes often used include chymotrypsin, Glu-C, Asp-N, and Lys-C (*see* **Table 1**).

3.2. In-Gel Digestion Protocol

This protocol describes the in-gel digestion of electrophoretically separated proteins from excised gel pieces of one- or two-dimensional gels for subsequent mass spectrometric analysis using the endoprotease trypsin. It has been primarily designed to minimize unwanted contaminations and loss of sample through an excess of reagents and protocol steps (*see* **Note 1**).

Once the gel pieces have been excised for in-gel digestion, there are still a number of practical points to consider before starting the digest. First, if the excised gel pieces are overly large, they should be cut into smaller pieces, ideally not larger than 2 × 2 mm^2 (*see* **Note 2**). Also every stage of the digest should be carried out in natural (noncolored) siliconized microcentrifuge tubes and organic and acidic solutions should not be made-up and kept in plastic containers (*see* **Note 3**). When removing liquid from tubes containing gel pieces, it is advisable to use gel loader tips and to take care not to remove the gel pieces with the pipet tip. This is particularly important for small gel pieces. All solutions should be made up freshly in HPLC grade water (*see* **Note 4**). To avoid formation of storage by-products such as oxidized methionines, do not store gels for too long before performing in-gel digestion and mass spectrometric analysis.

For non-silver stained gels start at **step 2**.

1. Destain silver stained gels in sufficient solution B so that the gel pieces are just covered. Leave gel pieces in solution B for 10 min or until the gel pieces are completely destained (*see* **Note 5**).
2. Wash gel pieces in sufficient solution C so that the gel pieces are just covered. Remove solution C and repeat this washing step three times. For Coomassie and Sypro Ruby stained gels, wash the gel pieces until the stain is completely removed or until no further removal can be obtained. After three washes in solution C, the gel pieces should have shrunk and should appear opaque.
3. Completely dry gel pieces in a speed vacuum drier (*see* **Note 6**).
4. Reduce cysteines and disulfide bridges by adding sufficient solution D so that the gel pieces are just covered. Incubate the gel pieces in solution D for 45 min at 50°C.
5. Spin down gel pieces and remove solution D.
6. Alkylate cysteines by adding sufficient solution E so that the gel pieces are just covered. Incubate the gel pieces in solution E for 1 h at room temperature in the dark (*see* **Note 7**).
7. Spin down gel pieces and remove solution E.
8. Wash gel pieces in sufficient solution C so that the gel pieces are just covered. Remove solution C and repeat this washing step until the gel pieces have shrunk and are opaque. This usually occurs after three washes.
9. Completely dry gel pieces in a speed vacuum drier (*see* **Note 8**).
10. Completely rehydrate gel pieces by adding sufficient solution F, but avoid larger volumes that cannot be absorbed by the gel pieces. Once solution F has been fully absorbed by the gel pieces, add sufficient solution A so that the gel pieces are just covered (*see* **Note 9**).
11. Incubate the sample overnight at 37°C.
12. Spin down the gel pieces ensuring that all liquid is collected at the bottom of the tube and add an equal volume of solution G to the supernatant. Wait 5 min and transfer supernatant to a fresh microcentrifuge tube. Repeat this extraction step

twice by adding sufficient solution G to the gel pieces so that they are just covered, waiting 5 min and transferring supernatant to the earlier transferred supernatant. After the final extraction the gel pieces should begin to turn opaque (*see* **Note 10**).

13. Dry down the collected supernatant in a speed vacuum drier until all liquid is removed and store at –20°C (*see* **Note 11**).
14. Resuspend in 5 μL solution H just before mass spectrometric analysis (*see* **Note 12**).

4. Notes

1. Further in-gel digestion protocols can be found at http://donatello.ucsf.edu/ingel.html, http://www.abrf.org/JBT/2000/june00/jun00speicher.html, http://www.ccic.ohio-state.edu/ms/proteomics/digestion%20protocol.html, http://www.bio.vu.nl/vakgroepen/mnb/proteomics/6.in-geldigestion.html, http://www.chem.arizona.edu/facilities/msf/in-gel/ingel_index.html or in refs. (***18***,***21***, and ***22***).
2. The reason for cutting larger gel pieces is to increase the surface-to-volume ratio for better protein accessibility for reagents and enzymes. However, macerating gel pieces too much could result in the loss of sample through aspirating and therefore losing the gel pieces at various steps in the protocol.
3. The use of noncoated microcentrifuge tubes will result in the loss of peptides due to adhesion to the tube walls. Colored tubes should not be used because some of these tubes seem to introduce sufficient amounts of synthetic polymers or plasticizers that hinder the crystallization of MALDI samples and introduce additional contamination leading to elevated baselines and contamination signals in the mass spectra. Initial washing of the microcentrifuge tubes with HPLC grade methanol also reduces the extent of contamination. To minimize contamination of the solutions from the solution containers, these containers should be made of material that cannot be solubilized by the solution. Glass containers are suitable for use with the presented protocol.
4. Hydrogen peroxide solutions, for instance, have a limited shelf life.
5. The gel pieces should be left for at least 10 min in solution B if methionines are to be fully oxidized for simplified database searching.
6. Alternatively, a freeze drier can be used.
7. We have found no differences in the alkylating efficiencies of iodoacetic acid or iodoacetamide. However, we use iodoacetamide as it allows easier identification of the first isotope ion signal of cysteine-containing peptides when either iodoacetic acid or iodoacetamide has been used prior to gel running, which might already result in partial alkylation. Alternatively, acrylamide can also be used for alkylation of cysteines. In this case earlier potential alkylation by free acrylamide from the gel will be less problematic. Because the alkylation efficiency using iodoacetamide is optimal under basic conditions, it is advisable to check the pH value and adjust it, if necessary.

8. It is very important that the gel pieces are fully dried at this point to ensure optimal absorption of the trypsin solution and analyte digestion efficiency while minimizing trypsin autolysis (*see also* **Note 6**).
9. The exact adjustment of the added volume of the trypsin solution for this rehydration step is supposed to ensure that all trypsin is available in the gel for proteolysis. The amount of liquid added obviously depends on the size of the gel pieces and can vary from 3 μL for spots from a two-dimensional gel up to 15 μL for large one-dimensional gel bands. It is also advisable to check the pH at this step. The activity of trypsin is optimal at approx pH 8.0.
10. Some in-gel digestion protocols use the peptide digest mixture from the supernatant directly after proteolysis without additional peptide extractions for mass spectrometric analysis. This can minimize the loss of sample due to adhesion to the tube walls during the drying step. However, we have not seen an improvement in MS sensitivity compared to pooling additional extractions and concentrating the peptide mixture by drying it down. In addition, the storage of dried sample, in contrast to liquid sample, is more desirable.
11. The sample is preferentially analysed right after the in-gel digest.
12. Solution H (and G in **step 12**) can also be prepared with 1–2 m*M* *n*-octyl glucoside, which is MALDI-MS compatible and increases peptide/protein solubility resulting in lower losses due to adhesion to the tube walls ***(23)***.

Acknowledgments

We would like to thank the Ludwig Institute for Cancer Research for funding this research and the Kay Kendall Leukaemia Fund for supporting K. B.

References

1. Hernandez, H. and Robinson, C. V. (2001) Dynamic protein complexes: insights from mass spectrometry. *J. Biol. Chem.* **276,** 46,685–46,688.
2. Robinson, C. V. (2002) Protein complexes take flight. *Nat. Struct. Biol.* **9,** 505–506.
3. Reid, G. E. and McLuckey, S. A. (2002) 'Top down' protein characterization via tandem mass spectrometry. *J. Mass Spectrom.* **37,** 663–675.
4. Ho, Y., Gruhler, A., Heilbut, A., et al. (2002) Systematic identification of protein complexes in *Saccharomyces cerevisiae* by mass spectrometry. *Nature* **415,** 180–183.
5. Gavin, A. C., Bosche, M., Krause, R., et al. (2002) Functional organization of the yeast proteome by systematic analysis of protein complexes. *Nature* **415,** 141–147.
6. Link, A. J., Eng, J., Schieltz, D. M., et al. (1999) Direct analysis of protein complexes using mass spectrometry. *Nat. Biotechnol.* **17,** 676–682.
7. MacCoss, M. J., McDonald, W. H., Saraf, A., et al. (2002) Shotgun identification of protein modifications from protein complexes and lens tissue. *Proc. Natl. Acad. Sci. USA* **99,** 7900–7905.
8. Link, A. J., Eng, J., Schieltz, D. M., et al. (1999) Direct analysis of protein complexes using mass spectrometry. *Nat. Biotechnol.* **17,** 676–682.

9. Brookes, P. S., Pinner, A., Ramachandran, A., et al. (2002). High throughput two-dimensional blue-native electrophoresis: A tool for functional proteomics of mitochondria and signaling complexes. *Proteomics* **2,** 969–977.
10. Nijtmans, L. G., Henderson, N. S., and Holt, I. J. (2002) Blue native electrophoresis to study mitochondrial and other protein complexes. *Methods* **26,** 327–334.
11. Mann, M., Hendrickson, R. C., and Pandey, A. (2001) Analysis of proteins and proteomes by mass spectrometry. *Annu. Rev. Biochem.* **70,** 437–473.
12. Naaby-Hansen, S., Waterfield, M. D., and Cramer, R. (2001) Proteomics—post-genomic cartography to understand gene function. *Trends Pharmacol. Sci.* **22,** 376–384.
13. Puig, O., Caspary, F., Rigaut, G., et al. (2001) The tandem affinity purification (TAP) method: a general procedure of protein complex purification. *Methods* **24,** 218–229.
14. Herbert, B., Galvani, M., Hamdan, M., et al. (2001) Reduction and alkylation of proteins in preparation of two-dimensional map analysis: why, when, and how? *Electrophoresis* **22,** 2046–2057.
15. Galvani, M., Hamdan, M., Herbert, B., and Righetti, P. G. (2001) Alkylation kinetics of proteins in preparation for two-dimensional maps: a matrix assisted laser desorption/ionization-mass spectrometry investigation. *Electrophoresis* **22,** 2058–2065.
16. Galvani, M., Rovatti, L., Hamdan, M., Herbert, B., and Righetti, P. G. (2001) Protein alkylation in the presence/absence of thiourea in proteome analysis: a matrix assisted laser desorption/ionization-time of flight-mass spectrometry investigation. *Electrophoresis* **22,** 2066–2074.
17. Rabilloud, T. (2000) Detecting proteins separated by 2-D gel electrophoresis. *Anal. Chem.* **72,** 48A–55A.
18. Shevchenko, A., Wilm, M., Vorm, O., and Mann, M. (1996) Mass spectrometric sequencing of proteins silver-stained polyacrylamide gels. *Anal. Chem.* **68,** 850–858.
19. Li, A., Sowder, R. C., Henderson, L. E., Moore, S. P., Garfinkel, D. J., and Fisher, R. J. (2001) Chemical cleavage at aspartyl residues for protein identification. *Anal. Chem.* **73,** 5395–5402.
20. Patterson, S. D. (1995) Matrix-assisted laser-desorption/ionization mass spectrometric approaches for the identification of gel-separated proteins in the 5-50 pmol range. *Electrophoresis* **16,** 1104–1114.
21. Hellman, U., Wernstedt, C., Gonez, J., and Heldin, C. H. (1995) Improvement of an "in-gel" digestion procedure for the micropreparation of internal protein fragments for amino acid sequencing. *Anal. Biochem.* **224,** 451–455.
22. Stensballe, A. and Jensen, O. N. (2001) Simplified sample preparation method for protein identification by matrix-assisted laser desorption/ionization mass spectrometry: in-gel digestion on the probe surface. *Proteomics* **1,** 955–966.
23. Kussmann, M., Nordhoff, E., Rahbek-Nielsen, H., et al. (1997) Matrix-assisted laser desorption/ionization mass spectrometry sample preparation techniques designed for various peptide and protein analytes. *J. Mass Spectrom.* **32,** 593–601.

34

Quantitative Protein Analysis by Solid Phase Isotope Tagging and Mass Spectrometry

Huilin Zhou, Rosemary Boyle, and Ruedi Aebersold

Abstract

Here we describe a method for stable isotope labeling and solid-phase capture of cysteinyl peptides from complex protein mixtures. Site-specific, quantitative labeling of cysteine residues with tags that differ in isotopic content enables quantification of relative peptide abundance between samples. Labeling on a solid phase provides for simultaneous simplification of a complex peptide mixture by isolating cysteinyl, and subsequently tagged, peptides. Peptides from proteolytic digests of protein samples are labeled in preparation for analysis by microcapillary liquid chromatography and tandem mass spectrometry (μLC-MS/MS) to determine their sequences and relative abundance between samples. This approach enables rapid identification and accurate quantification of relative abundance of individual proteins from different biological contexts.

Key Words

Protein quantification; protein–protein interaction; mass spectrometry; proteomics; solid phase chemistry; isotope tagging; protein identification; comparative protein analysis; photocleavable chemistry.

1. Introduction

Relative quantification of proteins from different samples by mass spectrometry is based on the stable isotope dilution approach *(1–4)*. Proteins or peptides are labeled with chemically identical tags that differ in mass, owing to stable isotope content. In a typical experiment, proteins (or peptides derived from proteolytic digestion of proteins) from one sample are labeled with an isotopically heavy mass tag, whereas the light isotope tag is used to label the sample to be compared. The isotopically labeled peptides are combined, purified, or separated into fractions, and analyzed by mass spectrometry, which measures the mass and ion abundance of peptides. Because isotopically heavy

From: *Methods in Molecular Biology, vol. 261: Protein–Protein Interactions: Methods and Protocols*
Edited by: H. Fu © Humana Press Inc., Totowa, NJ

and light forms of a peptide of the same amino acid sequence are chemically identical, they generate responses with identical sensitivity from the mass spectrometer and are readily distinguished based on their mass difference. Therefore, the measured ion abundance ratio between heavy- and light-labeled peptides is the actual abundance ratio of this peptide from two different samples. In this way, the relative abundance of peptides, and thus proteins, in two different samples can be accurately determined.

Here we describe a method for site-specific stable isotope labeling of cysteinyl peptides in complex peptide mixtures via a solid phase capture and release process, and the concomitant isolation of the labeled peptides *(4)*. The recovered, tagged peptides are analyzed by microcapillary liquid chromatography and tandem mass spectrometry (μLC-MS/MS) to determine their sequences and relative abundance.

2. Materials

1. Amino propyl glass beads, 200–400 mesh, pore size 170 Å (*see* **Note 1**; Sigma, St. Louis, MO, cat. no. G4518).
2. Organic solvents: dimethylformamide (DMF) and dichloromethane (Aldrich).
3. 1-Hydroxybenzotriazole (HOBt; Nova Biochem, Laufelfingen, Switzerland, cat. no. 01-62-0008).
4. Fmoc-protected amino acids: Fmoc-aminoethyl photolinker (Nova Biochem, cat. no. 01-60-0042), Fmoc-γ-amino butyric acid (Fmoc-GABA; Nova Biochem).
5. Diisopropyl carbodiimide (DIC) (Aldrich, cat. no. D12540-7).
6. Acetic anhydride (Aldrich).
7. Pyridine (Aldrich).
8. Piperidine (Aldrich).
9. D6-γ-aminobutyric acid (d6-GABA) (Isotec, Inc.).
10. Fmoc-*N*-hydroxysuccinimide (Fmoc-Osu; Nova Biochem, cat. no. 01-63-0001).
11. Diisopropyl ethyl amine (DIPEA; Aldrich, cat. no. D12580-6).
12. Iodoacetic anhydride (Aldrich).
13. Micro Bio-Spin columns (Bio-Rad Labs, Hercules, CA, cat. no. 732-6204).
14. Blak-Ray longwave UV lamp (100 W; VWR Scientific, Inc.).
15. Tris[2-carboxyethyl]phosphine hydrochloride (TCEP; Pierce).
16. Trypsin, sequencing grade (Promega).

3. Methods

The methods described below outline the procedures for synthesis of the reagents and their uses.

3.1. Synthesis of the Solid-Phase Isotope Labeling Beads

A schematic diagram of the chemical structure of the solid phase reagent is shown in **Fig. 1**. Synthesis of the solid phase reagents is based on a method

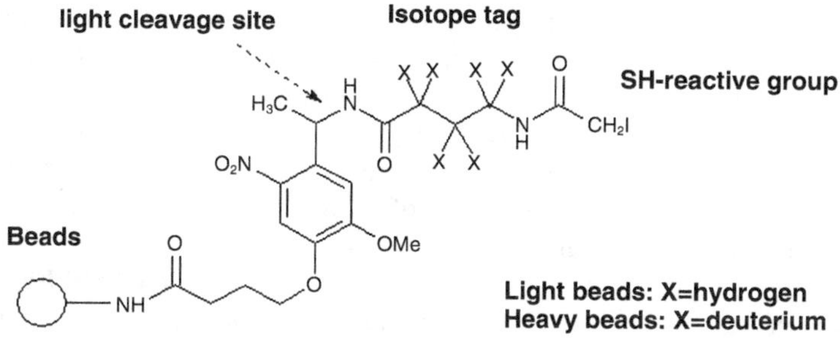

Fig. 1. A schematic diagram of the solid phase isotope tagging reagent, showing the chemical structure. The aminopropyl glass bead is first coated by a photo-cleavable linker molecule. Peripheral to the photo-cleavable linker, an amino acid, γ-aminobutyric acid, is used as an isotope-encoding mass tag that can be either nondeuterated (d0) or deuterated (d6). Following the isotope mass tag, an iodoacetyl group is used as a SH-reactive group to capture cysteinyl peptides. Following capture, photo-cleaving will lead to recovery of cysteinyl peptides with the isotope tags attached to their cysteine residues.

that has been published previously *(5)*. For synthesis of the beads with heavy isotope, Fmoc-d6-GABA was prepared from d6-GABA and Fmoc-OSu, as described in **Subheading 3.2.**, because the Fmoc-protected, deuterated amino acid is not commercially available. The methods described involve standard peptide chemistry, making it possible to use other amino acids with isotopically heavy or light forms.

1. Load 100 mg of aminopropyl glass beads in an empty Bio-Rad column or other column of suitable size. Wash beads once with one column volume of anhydrous DMF.
2. Form amino acid ester: Dissolve 120 μmol each of HoBt and Fmoc-aminoethyl photolinker in 0.8 mL of dry DMF completely. Add to this solution 120 μmol of DIC for 30 min. (Protect light-sensitive reagents from direct room light.)
3. Add the amino acid ester to the beads, mixing the beads by pipetting a few strokes. Incubate for 90 min.
4. Wash beads with three column volumes of DMF and two column volumes of dry dichloromethane. Always remove excess solvent between washes by applying a little pressure (squeeze a Pasteur pipetter bulb or apply house vacuum).
5. Block: Prepare a 1 mL mixture of 20% acetic anhydride, 30% pyridine, and 50% dichloromethane. Add this mixture to the beads for 30 min to block residual free amines on the beads.
6. Wash beads with two column volumes of dichloromethane and three column volumes of DMF. Remove excess DMF.
7. De-protect: Prepare 3 mL of 20% (v/v) piperidine/DMF solution. Add 1 mL to the beads and incubate for 30 min. Collect all of the 1 mL flow-through, containing

Fmoc released from the photo-linker. Calculate the capacity of the beads by measuring the absorbance of the released Fmoc. Use 20% piperidine/DMF solution as blank solution and measure absorbance at 290 nm (A_{290}) of the 1/100 dilution (by 20% piperdine/DMF) of the flow-through. The A_{290} should be between 0.6 and 0.8. Calculate capacity according to the formula: [A_{290} × dilution factor × flow-through volume (mL)]/[1.65 × weight of beads (mg)] = capacity (mmol/g).

8. Wash beads with five column volumes of dry DMF.
9. Repeat **steps 2–8** with Fmoc-d0-GABA or its heavy form (see **Subheading 3.2.**). The calculated capacity for GABA should be close to that for the photo-linker.
10. Attach iodoacetyl group to the beads ***(6)***: Dissolve 120 µmol of iodoacetic anhydride in 0.8 mL dry DMF and add to the beads. Immediately add 132 µmol of diisopropyl ethyl amine (DIPEA) to the beads and mix well by pipetting a few strokes. Let it incubate for 90 min.
11. Wash the beads with five column volumes of DMF and excess methanol; dry the beads in the Speed-Vac (covered in foil). The beads can be stored in the dark at room temperature or in the refrigerator indefinitely.

3.2. Synthesis of Fmoc-Protected Amino Acid

The following procedure permits custom synthesis of Fmoc-protected amino acid for attachment to solid phase as described above (*see* **Note 2**).

1. Dissolve 600 µmol d6-GABA in 3 mL of 9% sodium carbonate in H_2O, in a vial under stirring.
2. Dissolve 900 µmol Fmoc-OSu in 3 mL of DMF. Add to d6-GABA solution in one proportion. Continue to stir for 30 min at room temperature.
3. Divide the sample into six 1.5 mL Eppendorf tubes, and dry out DMF under reduced pressure in a Speed-Vac.
4. Add 1 mL of H_2O to dissolve the white powder as much as possible. Spin down and collect the supernatant. Repeat the water wash. Combine all of the supernatant, adding water until the final volume is 15 mL. Discard the insoluble material.
5. Add 600 µL of concentrated HCl to the supernatant very slowly. The solution should become cloudy immediately with foam due to carbon dioxide and precipitation of reagent. Check by pH paper that the final pH is approx 2.
6. Extract Fmoc-d6-GABA: Add 4 mL or more ethyl acetate to the acidified aqueous solution, wait for phase separation, and collect the ethyl acetate phase that contains Fmoc-d6-GABA. Repeat this extraction procedure three times and combine the extracts.
7. Wash the ethyl acetate extract once with 3 mL of 0.1% HCl in H_2O and once more with 3 mL of water. Dry the extract completely in a Speed-Vac. The dried sample can be used directly with the assumption of >90% yield.

3.3. Preparation of Protein Digest

For labeling with SH-specific solid phase reagents, it is advantageous to label peptides instead of proteins because proteins may possess tertiary structures that render some cysteine residues inaccessible to the solid phase reagent.

Protein digestion can be performed with any commercially available and suitable protease. Trypsin is most frequently used. For example, proteins in 100 μL of 0.2 *M* Tris-HCl, pH 8.0, can be digested by 1/50 (w/w) trypsin at 37°C overnight (*see* **Note 3**).

3.4. Capture and Release of Isotope-Labeled Peptides

Keep light-sensitive beads out of direct light as much as possible.

1. Reduce protein digest with 5 m*M* TCEP for 30 min at room temperature. Because TCEP is quite acidic, it is essential that there is sufficient buffering capacity; 200 m*M* Tris-HCl, pH 8.0, should be sufficient (*see* **Note 4**).
2. Weigh 5 mg each of isotopically light and heavy beads into tubes that are covered with foil to protect against light.
3. Add reduced protein digests to the beads and shake immediately for 15 min on a vortex mixer at lowest speed.
4. Quench the labeling reaction with 2 μL β-mercaptoethanol for 1–2 min (*see* **Note 5**).
5. Combine the beads by loading onto a foil-wrapped Bio-Spin column, rinsing with water and methanol to transfer all beads. Retain the flow-through of each labeling reaction separately for analysis of non-cysteine-containing peptides if so desired. Wash with:
   a. 2 × 1 mL 2 *M* NaCl
   b. 2 × 1 mL 0.1% TFA
   c. 2 × 1 mL 80% ACN/0.1% TFA
   d. 2 × 1 mL MeOH
   e. 2 × 1 mL 28% NH4OH:MeOH (1:9 v/v)
   f. 2 × 1 mL MeOH
   g. 2 × 1 mL water
6. Seal bottom of Bio-Spin column with a cap. Suspend beads in 200 μL of 20 m*M* Tris-HCl/1 mM EDTA, pH 8, and 4 μL β-mercaptoethanol with a magnetic stirrer on a stir plate.
7. Expose beads to UV light for 2 h (*see* **Note 6**), then collect the supernatant through the Bio-Spin column.
8. Wash the remaining beads with 5 × 100 μL 80% ACN/0.4% acetic acid, combining with the previous supernatant.
9. Reduce the sample volume in Speed-Vac to approximately 200 μL. Check that pH is acidic.

3.5. Sample Clean-Up and Mass Spectrometric Analysis (see **Notes 7–9**)

Although the labeled peptides appear to be highly pure, free from side-reactions of the peptides themselves, we have observed side products other than peptides following photo-cleaving of the beads. These residual products are likely to be impurities generated during synthesis of the solid phase reagent.

Because these products interfere with MS analysis of peptide samples, it is necessary to remove them prior to MS analysis. Additionally, we have observed that these side-products are not positively charged under acidic pH, whereas peptides are positively charged due to protonation of basic residues such as the N-terminus, histidine, lysine, and arginine. We therefore devised a strategy using cation-exchange chromatography to remove adducts from peptides. A disposable mixed cation-exchange (MCX) cartridge can be used.

1. Load sample onto MCX column (30 mg beads).
2. Wash with three column volumes of 0.1% TFA in water.
3. Wash with three column volumes of 80% ACN/0.1% TFA in water.
4. Wash by one column volume of water to prevent salt formation.
5. Elute in 500 μL of elution solvent consisting of one volume of ammonia solution (28% NH_4OH stock) and nine volumes of methanol.
6. Dry out ammonia and methanol in Speed-Vac and re-suspend the sample in 10 μL of water for MS analysis.

There are several advantages to this solid phase approach for isotopic labeling of peptides. First, isolation of cysteine-containing peptides and stable isotope incorporation are achieved in a single step. Therefore, the solid-phase method is rather simple. Second, the covalent attachment of peptides to a solid phase allows for the use of stringent wash conditions to remove noncovalently associated molecules. Third, this procedure is unaffected by the presence of proteolytic enzymes such as trypsin, or strong denaturants or detergents such as urea or SDS. There is no need for additional steps for their removal prior to peptide capture by the solid phase beads and it is easy to remove them by washing. Fourth, the standard solid phase peptide chemistry involved in the coupling process enables the use of a range of natural or unnatural amino acids in place of the d0/d6-GABA to function as the isotopic mass tag. This allows for synthesis of beads with a range of mass tags for analysis of multiple samples (i.e., more than two) in a single experiment if desired.

3.6. An Example of Protein Quantitation

We show an example of protein quantitation by this approach (*see* **Table 1**). Three proteins—glyceraldehyde 3-phosphate dehydrogenase from rabbit, bovine lactoalbumin, and ovalbumin from chicken—were prepared in different amounts and labeled by the solid phase isotope tagging reagents. Following light cleavage, the recovered peptides were analyzed by mass spectrometry, and the isotopically labeled peptides were identified and quantified as described ***(1,2,7)***. The agreement with expected values was generally within 20% and, for any given protein, consistent ratios were observed. Additional application of this method can be found elsewhere ***(4)***.

Table 1
Quantitation of Protein Mixture by Solid Phase Isotope Tagging and MS[a]

Gene name	Cys-containing peptides found	Observed ratio (light/heavy)	Expected ratio (light/heavy)
G3P_rabit	VPTPNVSVVDLTC*R	4.6	4.0
	IVSNASC*TTNC*LAPLAK	4.3	
LCA_bovin	DDQNPHSSNIC*NISC*DK	1.8	2.0
	FLDDDLTDDIMC*VK	1.9	
	LDQWLC*EK	2.1	
	ALC*SEK	2.0	
	C*EVFR	1.9	
Oval_chick	YPILPEYLQC*VK	1.0	1.0
	LPGFGDSIEAQC*GTSVNVHSSLR	0.9	
	ADHPFLFC*IK	1.1	

[a] Isotopically labeled cysteine residues are marked by asterisks.

4. Notes

1. The amine capacity of the aminopropyl glass beads should be measured in spite of the value quoted by the manufacturer. Other derivatized beads may be used in place of the glass beads if desired, provided that they have good swelling properties under aqueous condition.
2. During synthesis of the Fmoc-protected amino acid, it is important to acidify the sample very slowly and shake well. This should alleviate foaming due to the release of carbon dioxide. Also, one could use more ethyl acetate than prescribed in order to achieve better phase separation during extraction step.
3. For protein digestion by trypsin, proteins can be denatured by boiling for a few minutes if the protein concentration is not so high as to cause precipitation. In the current protocol, proteins are not reduced prior to digestion; however, it is possible that one could reduce proteins prior to digestion.
4. TCEP is quite acidic and the optimal pH for solid phase capture of cysteinyl peptides is 8.0. Therefore, it is essential that there is sufficient buffering capacity such as 200 m*M* Tris at pH 8.0. In this case, the pH of the solution would be not strongly affected by addition of 5 m*M* TCEP.
5. It is necessary to quench the capturing reaction after 15 min by mercaptoethanol or other excess SH-containing reagent because histidine side-chains or other nucleophilic functional groups could suffer potential side-reactions with the iodoacetyl group on the solid phase beads. The protocol for washing beads following the quenching reaction can be altered by individual investigators as we found the solid-phase-captured peptides are stable to a variety of washing conditions.

6. The UV light can be filtered through a copper sulfate solution that passes light of 300–400 nm. Although we used a long-wave UV lamp, the cleaving reaction can be accelerated by using a more powerful mercury arc lamp according to ref. **5**.
7. When small amount of peptides are expected, it is particularly important to remove labeling contaminants.
8. The use of β-mercaptoethanol in the photo-cleaving buffer prevents methionine oxidation.
9. Although very stringent washing steps were used to remove nonspecifically associated molecules from the solid phase after capturing, they may not be entirely necessary for all applications. The readers are encouraged to test different washing conditions for their own applications.

Acknowledgments

This work was supported in part by the NCI grant (CA84698), NIH Research Resource Center (RR11823), and NIH grant (GM 41109) to R.A.

References

1. Gygi, S. P., Rist, B., Gerber, S. A., Turecek, F., Gelb, M. H., and Aebersold, R. (1999) Quantitative analysis of complex protein mixtures using isotope-coded affinity tags. *Nat. Biotechnol.* **17,** 994–999.
2. Han D., Eng, J., Zhou, H., and Aebersold R. (2001) Quantitative profiling of differentiation induced membrane associated proteins using isotope coded affinity tags and mass spectrometry. *Nat. Biotechnol.* **19,** 946–951.
3. Smolka, M., Zhou, H., and Aebersold, R. (2002) Quantitative protein profiling using two-dimensional gel electrophoresis, isotope-coded affinity tag labeling, and mass spectrometry. *Mol. Cell Proteomics* **1,** 19–29.
4. Zhou, H., Ranish, J. A., Watts, J. D., and Aebersold, R. (2002) Quantitative proteome analysis by solid-phase isotope tagging and mass spectrometry. *Nat. Biotechnol.* **5,** 512–515.
5. Holmes, C. P. and Jones, D. G. (1995) Reagents for combinatorial organic synthesis: development of a new o-nitrobenzyl photolabile linker for solid phase synthesis. *J. Org. Chem.* **60,** 2318–2319.
6. Zhou, H., Watts, J. D., and Aebersold, R. (2001) A systematic approach to the analysis of protein phosphorylation. *Nat. Biotechnol.* **19,** 375–378.
7. Eng, J., McCormack, A. L., and Yates, J. R., 3rd. (1994) An approach to correlate tandem mass spectral data of peptides with amino acid sequences in a protein database. *J. Am. Soc. Mass Spectrom.* **5,** 976–989.

35

Internet Resources for Studying Protein–Protein Interactions

Shane C. Masters

Abstract

The Internet is now an essential tool for scientists. This chapter presents a beginner's guide to some of the valuable resources freely available on the Internet that are relevant to the study of protein–protein interactions. The format is designed to parallel a typical experimental project, indicating web sites to visit at each step. Although detailed instructions for using each site are not provided, the goal of each visit and what kind of information you can expect to obtain are indicated. A final section lists some directory sites that can point to additional web tools.

Key Words

Protein–protein interaction; Internet.

1. Getting Started

1.1. Start Simply

If you are not familiar with the protein you are going to be working with, you may want to start with a simplified overview of what is known about it. This can come from review articles, which you can get from PubMed (*see* below), but the Internet has some additional options.

- BioCarta (http://www.biocarta.com/genes/index.asp) maintains a collection of pathway diagrams for many well-studied proteins, with accompanying brief explanations.
- Bookshelf at NCBI (http://www.ncbi.nlm.nih.gov/entrez/query.fcgi?db=Books) is an expanding database of full-text, searchable reference books.

1.2. Learn More

Once you have a basic grasp of the subject you will be studying, getting more detailed information is the obvious next step. This will help you design

From: *Methods in Molecular Biology, vol. 261: Protein–Protein Interactions: Methods and Protocols*
Edited by: H. Fu © Humana Press Inc., Totowa, NJ

your experiment and evaluate its potential impact, indicate possible pitfalls, and avoid duplication of existing work.

- PubMed (http://www.ncbi.nlm.nih.gov/entrez/query.fcgi?db=PubMed) is one way to search the MEDLINE collection of journal articles. It will provide article abstracts and in many cases links to full-text journal articles.
- You might want to explore other ways of accessing the MEDLINE database (i.e., Ovid) or other bibliographic databases. Check with a librarian at your institution to find out what resources are available to you.

1.3. Gather Sequence Information

For many of the subsequent steps, you will need the sequences of your proteins. Try to gather sequence information from multiple species. If you later find that a particular feature is conserved between species, you may wish to place more weight on it. It is convenient to save the sequences on your local computer in the almost universally accepted FASTA format. Other sequence formats can be converted to FASTA using the Readseq tool, available from multiple sites, including http://iubio.bio.indiana.edu/soft/molbio/readseq/, which has the stand-alone program for your computer, and http://searchlauncher.bcm.tmc.edu/seq-util/readseq.html, which has a web-based interface. You may also want to make a table of sequence accession numbers, as some web sites can accept these numbers in place of the full sequence.

- When searched using the name of your protein, LocusLink (http://www.ncbi.nlm.nih.gov/LocusLink/) will provide links to NCBI reference sequences for the genomic DNA, mRNA, and protein. LocusLink is particularly valuable because it lists alternative names and mRNA forms for a gene, which can be confusing at the start of a project. However, it is limited in that it does not cover all major experimental organisms, notably yeast.
- As an alternative to LocusLink, you can try the Protein database at NCBI (http://www.ncbi.nlm.nih.gov/entrez/query.fcgi?db=Protein) or the Swiss-Prot database (http://us.expasy.org/sprot/). These databases are very large, and it is very likely that your sequence of interest can be found. However, it can be difficult to pick out the sequences you want when searching for a protein name due to the large number of hits. Consider searching by accession number if you found one in your literature search.

2. Database Mining

With the advent of high-throughput testing of protein–protein interactions, it is no longer sufficient to search a bibliographic database like MEDLINE to find out whether two proteins are known to interact. As more and more protein–protein interactions are identified, it is becoming increasingly difficult to access and analyze this information. Several databases are being developed to address this issue. At this time the content overlap between the major databases is not very extensive, and multiple sites should be searched.

A search for your protein of interest will return a list of proteins in the database that interact with the target and some properties of the interaction. In general, you can expect to find fields for alternate names for genes, links to the appropriate entries in sequence databases, some description of protein function, and subcellular localization. A list of methods used to establish the interaction and links to MEDLINE for articles in which the interaction was established are also standard. However, much of this information is incomplete at this time. Most sites include graphical tools that allow you to browse the database by following a chain of interactions and creating a "pathway." Searching these resources can produce several benefits. First, you may determine that the interaction you proposed to test has already been reported. In addition, knowledge of a protein's interactions can guide hypotheses of its function and may suggest interesting ligands.

- A Molecular Interactions Database (MINT; http://cbm.bio.uniroma2.it/mint/index.html) may be a good place to start because it is very easy to navigate, with relatively few options and an easy-to-use graphical interface. It also includes information on post-translational modifications and the regions of the proteins involved in an interaction when known.
- The Biomolecular Interaction Network Database (BIND, http://bind.ca/index.phtml) is not quite as easy to navigate as MINT. Like MINT, it lists post-translational modifications and interaction domains for some protein pairs. BIND is notable for its high level of detail for experimental conditions.
- General Repository for Interaction Datasets (GRID, http://biodata.mshri.on.ca/grid/servlet/Index) tends to focus on yeast protein–protein interactions. It can be navigated using an excellent graphical interface, which may also be used as a stand-alone program for generating pathway diagrams. It uses Gene Ontology categories to describe protein function and localization. (The Gene Ontology Consortium [http://www.geneontology.org/] aims to provide a controlled vocabulary for several aspects of the biosciences.)
- One of the notable features of the Database of Interacting Proteins (DIP, http://dip.doe-mbi.ucla.edu/) is the attempt to evaluate the strength of the evidence supporting each interaction. Entries in DIP also have an extensive list of links to other databases, including various protein domain listings. Consider installing the add-on JDIP graphical interface, available at http://dip.doe-mbi.ucla.edu/dip/jdip.cgi, to make browsing the database easier.

3. Domain Searches and Interaction Predictions

Domains are functional subunits that can confer specific properties on their host proteins. One such property is the ability to interact with other domains.

For example, SH2 domains can bind phosphotyrosine-containing domains. There are several well-established databases on the Internet that contain signature motifs for many different domains. Checking your protein against them can give insight into the possible functions and interaction partners of a protein.

- InterPro Scan (http://www.ebi.ac.uk/interpro/index.html) allows you to check your protein sequence against multiple motif databases at once, including Pfam and PROSITE, which are two of the largest and most commonly used collections of motifs. Each database differs in the way that motifs are defined and which motifs are included, so scanning multiple databases can improve the sensitivity of your search. The results are hyperlinked to pages that define each domain, summarize what is known about their functional significance, and allow you to search for other proteins with the same domains.
- iSpot (http://cbm.bio.uniroma2.it/ispot/) specifically addresses protein–protein interactions mediated by three common interaction domains. It uses results from published peptide binding studies to predict how well different PDZ, SH3, and WW domain containing proteins will bind to your target sequence.
- Interdom (http://interdom.lit.org.sg/) attempts to predict what protein domains will bind your protein. Given the sequence of your protein, Interdom will identify domains in your protein and report what other domains could interact with them. The program uses a variety of approaches to make these predictions, and will attempt to rank which interactions are most likely. Proteins containing these potential binding domains can be found by following the hyperlink to the Pfam database.
- ScanSite (http://scansite.mit.edu) searches protein sequences for domains and for potential sites of post-translational modification. Please refer to **Chapter 30** for a detailed description of this resource.

4. During the Experiment

After preliminary information has been gathered and experiments have been planned, the internet can still be an aid to the researcher. Several sites are available that can assist in locating reagents and protocols and in troubleshooting experiments. In addition, the advice of other researchers can be solicited through message boards and discussion lists.

- BioSupplyNet (http://www.biosupplynet.com) can aid you when you are looking for reagents or equipment. This site provides a searchable, categorized listing of scientific equipment manufacturers.
- Sites like Protocol Online (http://www.protocol-online.org/) can be a helpful resource when troubleshooting experiments. As the name suggests, at this site you can search for protocols. In addition, there is a discussion board where you can post questions and problems or answer the questions of others.

5. Springboards for Further Exploration of the Internet

The goal of this chapter is to describe a few of the more useful tools freely available on the Internet for researchers interested in protein–protein interactions. It cannot be more than a starting point because of the huge number of resources available and the rapid development of new tools. To provide additional avenues for exploration, a few high-yield directory sites are listed below.

- The Baylor College of Medicine Search Launcher (http://searchlauncher.bcm.tmc.edu/) links to many sequence analysis tools for both nucleic acids and proteins.
- The BioToolKit (http://www.biosupplynet.com/cfdocs/btk/btk.cfm) is made available at BioSupplyNet. It includes over 1000 annotated links to various protein and nucleic acid related web sites.
- ExPASy Molecular Biology Server (http://us.expasy.org/) is a service of the Swiss Institute of Bioinformatics that provides tools for protein sequence analysis, 2-D electrophoresis, and protein structure examination.
- The JCB Protein-Protein Interaction Website (http://www.imb-jena.de/jcb/ppi/) has annotated links to a broad spectrum of protein-protein interaction related resources.

Acknowledgments

I would like to thank Dr. Haian Fu for valuable discussions and editing and Dr. Tong Li-Masters for critical review of the manuscript.

Index

FSC
www.fsc.org
MIX
Papier | Fördert
gute Waldnutzung
FSC® C083411

Kolibri 360 GmbH
Ferdinand-Jühlke-Straße 7
99095 Erfurt, Deutschland
produktsicherheit@kolibri360.de